Epigenetics and Human Health

Volume 11

Series Editors

Rasime Kalkan, Faculty of Medicine, Department of Medical Genetics, European University of Lefke, Lefke Northern Cyprus, via Mersin, Turkey

Luis M. Vaschetto, Alta Gracia, Córdoba, Argentina

This books series concerns the quickly expanding field of epigenetics. It evokes novel research questions and how these are linked to human health. Each of its volumes covers a particular topic that is important to clinical practice and research, and to researchers in epigenetics who study the relevant medical issues of today. The individual books of the series complement each other and underline the importance of epigenetics in health-related disciplines such as toxicology, reproduction, psychiatry, and nutritional and environmental medicine.

Epigenetics and Human Health is indexed in SCOPUS, EMBASE, Reaxys, EMCare, EMBiology and SCImago.

Rasime Kalkan
Editor

Cancer Epigenetics

Editor
Rasime Kalkan
Faculty of Medicine, Department of
Medical Genetics
European University of Lefke
Lefke, Northern Cyprus, via Mersin, Turkey

ISSN 2191-2262   ISSN 2191-2270 (electronic)
Epigenetics and Human Health
ISBN 978-3-031-42364-2   ISBN 978-3-031-42365-9 (eBook)
https://doi.org/10.1007/978-3-031-42365-9

© The Editor(s) (if applicable) and The Author(s), under exclusive license to Springer Nature Switzerland AG 2023

This work is subject to copyright. All rights are solely and exclusively licensed by the Publisher, whether the whole or part of the material is concerned, specifically the rights of translation, reprinting, reuse of illustrations, recitation, broadcasting, reproduction on microfilms or in any other physical way, and transmission or information storage and retrieval, electronic adaptation, computer software, or by similar or dissimilar methodology now known or hereafter developed.

The use of general descriptive names, registered names, trademarks, service marks, etc. in this publication does not imply, even in the absence of a specific statement, that such names are exempt from the relevant protective laws and regulations and therefore free for general use.

The publisher, the authors, and the editors are safe to assume that the advice and information in this book are believed to be true and accurate at the date of publication. Neither the publisher nor the authors or the editors give a warranty, expressed or implied, with respect to the material contained herein or for any errors or omissions that may have been made. The publisher remains neutral with regard to jurisdictional claims in published maps and institutional affiliations.

This Springer imprint is published by the registered company Springer Nature Switzerland AG
The registered company address is: Gewerbestrasse 11, 6330 Cham, Switzerland

Paper in this product is recyclable.

This book is dedicated to my daughter, Ozra Ozkut.

Preface

Understanding the differences in nuclear organization, DNA methylation, and histone modification patterns between cancer cells and normal cells is made possible by the study of epigenetics. Epigenetic alterations are dynamic and reversible. Epigenetic modifications include DNA methylation, histone modifications, non-coding RNAs, and RNA modifications. Epigenetic modifications have numerous impacts on human health. Epigenetic modifications play a crucial role in cancer development and progression. Epigenetic modifications, however, have the potential to be therapeutic targets due to their reversible nature. Epigenetics is crucial in cancer research for detecting various cancer types at their early stages, monitoring treatment responses, and facilitating the development of new treatments.

This book has two major parts:

Part I provides a background and a general overview of epigenetic modifications, epigenetic enzymes and mutation of epigenetic enzymes, and basics of cancer epigenetics.

Part II contains cancer-specific alterations and epigenetic alterations in the diagnosis, prognosis, and therapy of cancer.

This book is intended for researchers who are studying epigenetics, cancer epigenetics, or any other research topic, as well as for physicians and other healthcare professionals from a variety of specialties. It is also intended to be a valuable resource for instructing graduate and undergraduate students.

Lefke, Northern Cyprus, via Mersin, Turkey Rasime Kalkan

Contents

Part I Background

1 **An Overview of Epigenetics Modifications in Normal and Cancer Cell** 3
   Satu Mäki-Nevala and Päivi Peltomäki

2 **Epigenetic Enzymes and Their Mutations in Cancer** 31
   Aysegul Dalmizrak and Ozlem Dalmizrak

3 **Introduction to Cancer Epigenetics** 77
   Ebru Erzurumluoğlu Gökalp, Sevgi Işık, and Sevilhan Artan

Part II Cancer Specific Epigenetic Alterations

4 **Epigenetics in the Diagnosis, Prognosis, and Therapy of Cancer** 137
   Leilei Fu and Bo Liu

5 **Clinical Studies and Epi-Drugs in Various Cancer Types** 165
   Taha Bahsi, Ezgi Cevik, Zeynep Ozdemir, and Haktan Bagis Erdem

6 **Epigenetic Regulation in Breast Cancer Tumor Microenvironment** 213
   Bhavjot Kaur, Priya Mondal, and Syed Musthapa Meeran

7 **The Epigenetics of Brain Tumors: Fundamental Aspects of Epigenetics in Glioma** 245
   Sevilhan Artan and Ali Arslantas

8 **Epigenetic Alterations in Pancreatic Cancer** 275
   Cincin Zeynep Bulbul, Bulbul Muhammed Volkan, and Sahin Soner

9	**Current Preclinical Applications of Pharmaco-Epigenetics in Cardiovascular Diseases**.. 295
	Chiara Papulino, Ugo Chianese, Lucia Scisciola, Ahmad Ali, Michelangela Barbieri, Giuseppe Paolisso, Lucia Altucci, and Rosaria Benedetti
10	**Epigenetic Alterations in Colorectal Cancer**.................... 331
	Brian Ko, Marina Hanna, Ming Yu, and William M. Grady
11	**Epigenetic Alterations in Hematologic Malignancies**............. 363
	Emine Ikbal Atli

About the Editor

Rasime Kalkan Rasime Kalkan is a Professor of Medical Genetics at the European University of Lefke. Professor Rasime Kalkan completed her PhD in the Department of Medical Genetics at the Eskisehir Osmangazi University in Turkey. She worked as a member of the Department of Medical Genetics at the Near East University between November 2011 and June 2022. She worked as the Chair of the Department of Molecular Biology and Genetics of the Arts and Sciences Faculty at Near East University (2020–2022). She was the founding Dean of the Faculty of Medicine at Cyprus Health and Social Sciences University in Cyprus between July 2022 and August 2023. She was Vice-Rector at the Cyprus Health and Social Sciences University (CHSSU) in Cyprus between July 2022 and August 2023. She has been a member of Turkish Medical Genetics Association since 2006 and a member of the board of directors of Turkish Cypriot Association of Human Genetics since 2022.

She worked on genetic and epigenetic alterations in brain tumors, especially in glioblastoma, during her PhD. She has published numerous articles in peer-reviewed journals and was the author of several chapters for various national and international books. She took part as project coordinator and assistant investigator in many research projects. Her research is focused on the identification of epigenetic and genetic alterations in cancer, obesity, and post-menopausal term.

List of Contributors

Ahmad Ali Department of Precision Medicine, University of Campania 'Luigi Vanvitelli', Naples, Italy

Lucia Altucci Department of Precision Medicine, University of Campania 'Luigi Vanvitelli', Naples, Italy
Biogem Institute of Molecular and Genetic Biology, Ariano Irpino, Italy
Institute of Endocrinology and Oncology 'Gaetano Salvatore' (IEOS), Naples, Italy

Ali Arslantas Department of Neurosurgery, Eskisehir Osmangazi University, Medical Faculty, Eskişehir, Turkey

Sevilhan Artan Department of Medical Genetics, Eskisehir Osmangazi University, Medical Faculty, Eskisehir, Turkey

Emine Ikbal Atli Faculty of Medicine, Department of Medical Genetics, Trakya University, Edirne, Turkey

Taha Bahsi Department of Medical Genetics, Ankara Etlik City Hospital, Ankara, Turkiye

Michelangela Barbieri Department of Advanced Medical and Surgical Sciences, University of Campania "Luigi Vanvitelli", Naples, Italy

Rosaria Benedetti Department of Precision Medicine, University of Campania 'Luigi Vanvitelli', Naples, Italy

Cincin Zeynep Bulbul Istanbul Nisantaşı University, Medical Faculty, Medical Biochemistry, Istanbul, Turkey
Istanbul Nisantaşı University, Health Ecosystem, Istanbul, Turkey

Ezgi Cevik Department of Medical Genetics, Ankara Etlik City Hospital, Ankara, Turkey

Ugo Chianese Department of Precision Medicine, University of Campania 'Luigi Vanvitelli', Naples, Italy

Aysegul Dalmizrak Department of Medical Biology, Faculty of Medicine, Balıkesir University, Balıkesir, Turkey

Ozlem Dalmizrak Department of Medical Biochemistry, Faculty of Medicine, Near East University, Nicosia, Cyprus

Haktan Bagis Erdem Department of Medical Genetics, Ankara Etlik City Hospital, Ankara, Turkey

Leilei Fu Sichuan Engineering Research Center for Biomimetic Synthesis of Natural Drugs, School of Life Science and Engineering, Southwest Jiaotong University, Chengdu, China

Ebru Erzurumluoglu Gokalp Department of Medical Genetics, Eskisehir Osmangazi University, Medical Faculty, Eskişehir, Turkey

William M. Grady Department of Medicine, University of Washington, School of Medicine, Seattle, WA, USA
Translational Science and Therapeutics Division, Seattle, WA, USA
Public Health Sciences Division, Fred Hutchinson Cancer Center, Seattle, WA, USA

Marina Hanna Department of Medicine, University of Washington, School of Medicine, Seattle, WA, USA

Sevgi Isik Department of Medical Genetics, Eskisehir Osmangazi University, Medical Faculty, Eskişehir, Turkey

Bhavjot Kaur Department of Biochemistry, CSIR-Central Food Technological Research Institute, Mysuru, Karnataka, India

Brian Ko Department of Medicine, University of Washington, School of Medicine, Seattle, WA, USA

Bo Liu State Key Laboratory of Biotherapy and Cancer Center, West China Hospital, Sichuan University, Collaborative Innovation Center of Biotherapy, Chengdu, China

Satu Mäki-Nevala Department of Clinical and Medical Genetics, Faculty of Medicine, University of Helsinki, Helsinki, Finland

Syed Musthapa Meeran Department of Biochemistry, CSIR-Central Food Technological Research Institute, Mysuru, Karnataka, India
Academy of Scientific and Innovative Research (AcSIR), Ghaziabad, India

Priya Mondal Department of Biochemistry, CSIR-Central Food Technological Research Institute, Mysuru, Karnataka, India
Academy of Scientific and Innovative Research (AcSIR), Ghaziabad, India

Zeynep Ozdemir Department of Medical Genetics, Ankara Etlik City Hospital, Ankara, Turkey

List of Contributors

Giuseppe Paolisso Department of Advanced Medical and Surgical Sciences, University of Campania "Luigi Vanvitelli", Naples, Italy
Mediterranea Cardiocentro, Naples, Italy

Chiara Papulino Department of Precision Medicine, University of Campania 'Luigi Vanvitelli', Naples, Italy

Päivi Peltomäki Department of Clinical and Medical Genetics, Faculty of Medicine, University of Helsinki, Helsinki, Finland

Lucia Scisciola Department of Advanced Medical and Surgical Sciences, University of Campania "Luigi Vanvitelli", Naples, Italy

Sahin Soner Istanbul Nisantaşı University, Health Ecosystem, Istanbul, Turkey
Department of Neurosurgery, Istanbul Nisantaşı University, Medical Faculty, Istanbul, Turkey

Bulbul Muhammed Volkan Istanbul Nisantaşı University, Health Ecosystem, Istanbul, Turkey
Istanbul Nisantaşı University, Medical Faculty, Histology and Embryology, Istanbul, Turkey

Ming Yu Translational Science and Therapeutics Division, Seattle, WA, USA
Public Health Sciences Division, Fred Hutchinson Cancer Center, Seattle, WA, USA

Abbreviations

2HG	2-hydroxyglutarate
2-OG	2-oxoglutarate
2-OGDD	2-oxoglutarate-dependent dioxygenases
5′-C	5′-residue of cytosine
5caC	5-carboxylcytosine
5fC	5-formylcytosine
5hmC	5-hydroxymethylcytosine
5-mC	5-methyl cytosine
5mC	5-methylcytosine
ADCC	Antibody-dependent cell-mediated cytotoxicity
ADGRE2	Adhesion G protein-coupled receptor E2
Ago-2	Argonaute-2
AKT	Protein kinase B
ALC1	Amplification in liver cancer 1
AMI	Acute myocardial infarction
AML	Acute myeloid leukemia
APAF1	Apoptotic peptidase activating factor 1
APCs	Antigen-presenting cells
APNG	Alkylpurine-DNA-N-glycosylase
AR	Androgen receptor
ARID2	AT-rich interactive domain 2
ASIR	Age-standardized incidence rate
ASMR	Age-standardized mortality rate
ATP	Adenosine triphosphate
ATPase	Adenosine triphosphatase
AUC	Area under the curve
BAF	BRG1/BRM-associated factor (nucleosome remodeling enzyme)
BAH1 and BAH2	Bromo-adjacent homology 1 and 2
BC	Bladder cancer

B-CLL	B-cell chronic lymphocytic leukemia
BER	Base excision repair
BET	Bromodomain and extra-terminal
BL	Burkitt lymphoma
BMI1	B-cell-specific Moloney murine leukemia virus insertion site1
BRD	Bromodomain
BRDs	Bromodomain-containing proteins
CAFs	Cancer-associated fibroblasts
CDC6	Cell-division cycle-6
CDK	Cyclin-dependent kinase
CDK9	Cyclin-dependent kinase-9
CDKN2A	Cyclin-dependent tumour inhibitor 2A
CHD	Chromodomain helicase DNA-binding
ChIP	Chromatin immunoprecipitation
CHK2	Checkpoint kinase 2
CIMP	CpG island methylator phenotype
CIN	Chromosomal instability
circRNAs	Circular RNA
CLL	Chronic lymphocytic leukemia
CMI	Cumulative methylation index
c-Myc	MYC proto-oncogene
CN-AML	Cytogenetically normal acute myeloid leukemia
CNS	Central nervous system
CR	Complete response
CRC	Colorectal cancer
CREB	cAMP-response-element-binding protein
CTCF	CCCTC-binding factor
CVD	Cardiovascular diseases
DCLK1	Doublecortin- like kinase 1
DCs	Dendritic cells
DGCR8	DiGeorge syndrome critical region gene 8
DIPGs	Diffuse intrinsic pontine gliomas
DLBCL	Diffuse large B-cell lymphoma
DMA	Direct miRNA analysis
DMAP1	DNA methyltransferase associated protein 1
DMRs	Differentially methylated regions
DNMT	DNA methyltransferase
DNMT1	DNA methyltransferase 1
DNMT3	DNA methyltransferase 3
DNMTi	DNA- methyltransferase inhibitors
DSB	Double-strand break
DSBH	Double-stranded β-helix
DUSP1	Dual specificity phosphatase 1 gene
E1	Ubiquitin-activating enzyme 1

Abbreviations

E2	Ubiquitin-conjugating enzyme 2
ECM	Extracellular matrix
EGFR	Epidermal growth factor receptor
EGR-1	Early growth response-1
EHMT1 / 2,	Euchromatin histone-lysine-n-methyltransferases I and II
EMT	Epithelial-mesenchymal transition
EOC	Epithelial ovarian cancer
ES	Embryonic stem cell
ESCC	Esophageal squamous cell carcinoma
EZH2	Enhancer of zeste homolog 2
FAD	Flavin adenine dinucleotide
FAL1	Focally amplified lncRNA on chromosome 1
FBW7	F-box and WD repeat domain containing 7
FIT	Fecal immunochemical test
FOXO1	Fork-headed Box Protein O1
GBM	Glioblastoma
GC	Guanine cytosine
G-CIMP	Glioma CpG island methylator phenotype
GDP	Guanin diphosphate
GENIE	Genomics Evidence Neoplasia Information Exchange
GEO	Gene Expression Omnibus
GNAT	GCN5-related n-acetyltransferase
GTP	Guanin triphosphate
H2Bub1	H2B monoubiquitination
H3	Histone 3
H4	Histone 4
HAT	Histone acetyltransferase
HBO1	Histone histone lysine acetyltransferase
HCC	Hepatocellular carcinoma
HDAC	Histone deacetylase
HDACi	HDAC inhibitors
HDLP	Histone deacetylase-like protein
HEP	The Human Epigenome Project
HIF1	Hypoxia-inducible factor 1
HIF-1α	Hypoxia-inducible factor-1α
HMG	High mobility group
HMTs	Histone methyltransferases
hnRNPL	Heterogeneous nuclear ribonucleoprotein L
HP1α	Heterochromatin protein 1α
hTERT	Telomerase reverse transcriptase
ICRs	Imprinting control regions
IDH	Isocitrate dehydrogenase
IDH1	Isocitrate dehydrogenase
ISWI	Imitation switch

JAK2	Janus kinase 2
JMJD	Jumonji domain containing protein
K	Lysine
KATs	Lysine acetyltransferases
KDM	Histone lysine demethylase
KDM1	Homolog lysine demethylase 1
KG	Ketoglutarate
KLF	Kruppel-like factor 1
KMT2C	Histone lysine methyltransferase 2C
KMTs	Lysine methyltransferases
KRAS	Kirsten rat sarcoma virüsvirus
Len	Lenalidomide
LINEs	Long interspaced nuclear elements
lncRNA	Long non-coding RNA
LOI	Loss of imprinting
LSD1/KDM1A	Lysine-specific demethylase 1A
m1A	N1-methyladenosine
m3C	N3-methylcytosine
m5C	5-methylcytosine
m6A	N6-methyl-adenosine
m7G	N7-methylguanosine
MAGE	Melanoma antigen gene
MBD	Methyl-CpG-binding domain
MBD2	Methyl-CpG-binding domain protein 2
MDS	Myelodysplastic syndrome
MDSCs	Myeloid-derived suppressor cells
MeCP	Methyl-CpG- binding domain protein
MeCP2	Methyl-CpG-binding protein 2
MET	Mesenchymal-to-epithelial transition
METTL3	Methyltransferase like 3
MGMT	O6-methylguanine methyltransferase
MHC-II	Major histocompatibility complex class II
MI	Myocardial infarction
miRISC	miRNA-induced silencing complex
miRNA	microRNA
MLL	Mixed lineage leukemia
MLL1	Mixed-lineage leukaemia 1
MORF	MOZ- related factor
MOZ	Monocytic leukemic zinc-finger protein
MPC1	Mitochondrial pyruvate carrier-I
MPN	Myeloproliferative neoplasms
MPTP	Mitochondrial permeability transition pores
mRNA	Messenger RNA
MS	Mass spectrometry

MSI	Microsatellite instability
MSP	Methylation-specific PCR
MTAP	Methylthioadenosine phosphorylase
MYC	Myelocytomatosis oncogene
NAD+	Nicotinamide adenine dinucleotide
NADP(+)	Nicotinamide adenine dinucleotide phosphate
ncRNA	Non-coding RNA
NEPC	Neuroendocrine prostate cancer
NLS	A nuclear localization sequences
NSCLC	Non-small cell lung cancer
NuRD	Nucleosome remodeling and deacetylase complex
ORC	Origin recognition complex
OS	Overall survival
PanIN	Pancreatic intraepithelial neoplasms
PARP1	Poly (ADP-ribose) polymerase 1
PBHD domain	Polybromo homology domain
PC	Pancreatic cancer
PCAF	p300/CBP-related factor
PCNA	Proliferating cell nuclear antigen
PDAC	Pancreatic ductal adenocarcinoma
PD-L1	Programmed death-ligand 1
piRNAs	PIWI-interacting RNAs
PLCD1	Phospholipase C delta1
PMDs	Partially methylated domains
PR C1	Polycomb repressive complex 1
PRC	Polycomb repressive complexes
PRMT	Protein arginine N-methyltransferase
PTM	Post-translational modification
PTPRR	ERK phosphatases protein tyrosine phosphatase receptor type R
qMSP	Quantitative methylation-specific PCR
R	Arginine
RAN	Ras-related nuclear protein
RFT domain	Replication foci targeting domain
RFXAP	Regulatory X-related protein
ROS	Reactive oxygen species
RRBS	Reduced-representation bisulfite sequencing
S	Serine
SAM	S-adenosyl methionine
SASP	Senescence-associated secretary phenotype
SATB	Special AT-rich sequence-binding protein
SCLC	Small-cell lung cancers
siRNA	Short interfering RNA
SIRT1	Sirtuin 1
SIRTs	Sirtuins
SMAD1	Small mothers against decapentaplegic homolog1

SMAD4	SMAD family member 4
SMYD	Set and MYND domain protein
SND1	Staphylococcal nuclease domain-containing protein 1
SNPs	Single nucleotide polymorphisms
snRNA	Small nuclear ribonucleic acid RNA
SRA	SET- and RING-associated
SS	Sezary syndrome
SSL	Serrated sessile lesions
SSP	Sessile serrated polyps
SUMO	Small ubiquitin-like modifier
SWI/SNF	Switching defective/sucrose non-fermenting complex
T	Threonine
T-ALL	T-cell acute lymphoblastic leukemia
TAM	Tumor-associated macrophage
TCGA	The Cancer Genome Atlas
TDG	Thymine DNA glycosylase
TEs	Transposable elements
TET proteins	Ten-eleven translocation methylcytosine dioxygenases
TET	Ten-eleven- translocation
TF	Transcription factor
TGF-β	Transforming growth factor β
TKI	Tyrosine- kinase inhibitor
TME	Tumor microenvironment
TMZ	Temozolomide
TNBC	Triple negative breast cancer
TNF-alpha	Tumor necrosis factor-alpha
TP53	Tumour protein 53
TRD	Target recognition domain
TS	Targeting sequence
TSA	Trichostatin A
TSG	Tumor suppressor gene
TSS	Transcription starting site
Ub	Ubiquitin
USPs	Ubiquitin-specific peptidases
VEGF	Vascular endothelial growth factor
WGBS	Whole-genome bisulfite sequencing
WNK2	WNK lysine deficient protein kinase 2
Y	Tyrosine
YBX1	Y-Box Binding Protein 1
ZEB1	Zinc finger E-box- binding homeobox 1
α-KG	α-ketoglutarate

Part-I
Background

Chapter 1
An Overview of Epigenetics Modifications in Normal and Cancer Cell

Satu Mäki-Nevala and Päivi Peltomäki

Abstract There is no need for all the approximate 20,000 genes of a human being to be active in every cell or all the time. Epigenetic regulation allows for selective expression with respect to, for example, type of cell, phase of development, or allelic origin. Epigenetic regulation involves covalent and non-covalent modifications of the DNA molecule and chromatin structure, without altering the actual base sequence of DNA. As normal cellular functions, including cell proliferation and interactions with adjacent cells, depend on proper epigenetic regulation, it is not surprising that cancer cells attempt to disturb this regulatory system in many ways to acquire and maintain neoplastic properties. This review provides an overview of the main epigenetic mechanisms—DNA methylation, DNA hydroxymethylation, histone modifications, nucleosome remodeling, and non-coding RNAs—in a normal and cancer cells. We summarize the essential mechanistic features of each epigenetic regulator and offer illustrative examples of their importance for normal and neoplastic states of human cells.

Keywords Epigenetics · DNA methylation · DNA hydroxymethylation · Histone modifications · Nucleosome remodeling · Non-coding RNAs · Cancer · Normal cell function

Abbreviations

5caC	5-Carboxylcytosine
5fC	5-Formylcytosine
5mC	5-Methylcytosine
5hmC	5-Hydroxymethylcytosine
BAF	BRG1/BRM-associated factor (nucleosome remodeling enzyme)

S. Mäki-Nevala (✉) · P. Peltomäki
Department of Clinical and Medical Genetics, Faculty of Medicine, University of Helsinki, Helsinki, Finland
e-mail: satu.maki-nevala@helsinki.fi; paivi.peltomaki@helsinki.fi

© The Author(s), under exclusive license to Springer Nature Switzerland AG 2023
R. Kalkan (ed.), *Cancer Epigenetics*, Epigenetics and Human Health 11, https://doi.org/10.1007/978-3-031-42365-9_1

CHD	Chromodomain helicase DNA-binding (nucleosome remodeling enzyme)
ChIP	Chromatin immunoprecipitation
CIMP	CpG island methylator phenotype
CIN	Chromosomal instability
CpG	A dinucleotide; 5′—C—phosphate—G—3′
CRC	Colorectal cancer
DNMT	DNA methyltransferase
EMT	Epithelial-to-mesenchymal transition
HDAC	Histone deacetylase
KDM	Histone lysine demethylase
lncRNA	Long non-coding RNA
mRNA	Messenger RNA
miRNA	Micro RNA
MS	Mass spectrometry
ncRNA	Non-coding RNA
NuRD	Nucleosome remodeling and deacetylase complex
piRNA	Piwi-interacting RNA
PTM	Post-translational modification
RRBS	Reduced-representation bisulfite sequencing
SASP	Senescence-associated secretary phenotype
siRNA	Short interfering RNA
SWI/SNF	Switching defective/sucrosenon-fermenting complex
TAM	Tumor-associated macrophages
TET	Ten-eleven translocation
TSG	Tumor suppressor gene
TSS	Transcription starting site

1.1 Introduction

Epigenetics refers to mechanisms that modify gene expression and determine the ultimate outcome of a genetic locus without altering the corresponding DNA sequence (Goldberg et al. 2007). Dr. Waddington was the first to introduce the term epigenetics as "epigenotype" in 1942 to describe those interactions that a straightforward connection between genotype and phenotype was unable to explain (Waddington 2012). Indeed, a straight translation of epigenetics is "above the genetics."

Epigenetic alterations may be covalent and non-covalent modifications and occur on the DNA molecule and on the chromatin structure. Chromatin is the crucial structure mediating through transcription factors, signaling pathways, and other signals to modify gene expression and further cellular functions. The most widely studied epigenetic alterations are DNA methylation and post-translational histone

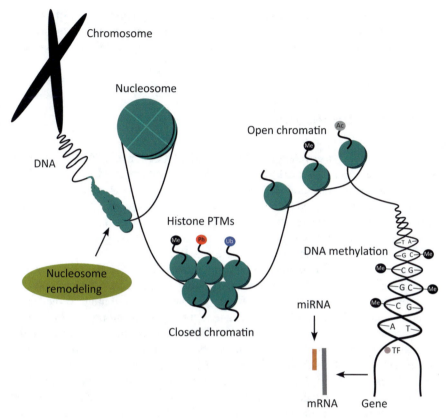

Fig. 1.1 Schematic overview of the epigenetic mechanisms at the DNA, chromatin, and nucleosome level

modifications. DNA hydroxymethylation, nucleosome remodeling, and non-coding RNAs also belong to the epigenetic contributors. It is possible to present these epigenetic modifiers separately, although their interplay in the epigenetic regulatory system is remarkable—as will be clear after reading this chapter.

Epigenetics is required for the normal development and function of a human cell (Goldberg et al. 2007; Zeng and Chen 2019). Although epigenetic modifications do not alter DNA sequence, they can be heritable. For the normal cell's efficient function, it is useful that epigenetic reprogramming is not repetitively required at every cell division. Besides the normal cellular functions, epigenetics plays a crucial role in neoplastic growth. Very recently, "nonmutational epigenetic reprogramming" was included in the universal cancer hallmarks as an enabling characteristic (Hanahan and Weinberg 2000, 2011; Hanahan 2022). Cancer hallmarks are functional capabilities that enable a normal cell to acquire neoplastic features that all cancer cells share.

In this chapter, we focus on human cells, although we present a few studies or examples on mouse models. We provide a mechanistic overview of the aforementioned epigenetic alterations (Fig. 1.1), including a brief review of methods to investigate those changes. Then we describe the importance of epigenetics in a normal and cancer cells by providing functional examples of each epigenetic regulator.

1.1.1 DNA Methylation

DNA methylation is a covalent chemical alteration of chromatin. DNA methylation is a reversible epigenetic mark that occurs in the fifth carbon of cytosines of the DNA. DNA methyltransferases (DNMTs) mediate the process in which a methyl group is covalently attached to the C-5 position of the cytosine ring (5mC). In human cells (and in other mammals), DNA methylation is mostly found at the CpG dinucleotides. After DNA replication, only symmetrical DNA methylation at the CpG sites remains (Greenberg and Bourc'his 2019).

Methylation in the promoter and distal regulatory regions of the genes controls gene expression. High DNA methylation levels are associated with suppressed transcription activity and there are several mechanisms contributing to regulatory events (Li and Tollefsbol 2021). For example, methylation can prevent binding of transcription factors and thus inhibit transcription initiation. In a recent study, approximately one-fifth of transcription factors (117/542) showed suppressed binding on their methylated sequence recognition target sites (Yin et al. 2017). Moreover, enzymes responsible for DNA methylation can interact with chromatin modifiers, and thus DNA methylation contributes to chromatin structure (Greenberg and Bourc'his 2019). In addition to promoter and distal regulatory elements, DNA methylation can also occur further from the transcription starting site (TSS), in the gene bodies and repetitive elements (a more detailed description is available in Sect. 1.2).

DNA methylation events include de novo methylation, maintenance methylation, and demethylation (Greenberg and Bourc'his 2019). DNMT3A and DNMT3B, which act after DNA replication, carry out de novo methylation. DNMT1 methylates the newly synthesized DNA strand during the replication and thus maintains methylation. A group of enzymes called ten-eleven translocation (TET) methylcytosine dioxygenases are responsible for active DNA demethylation by oxidizing 5meC to 5-hydroxymethylcytosine (5hmC) (Ito et al. 2010), 5-formylcytosine (5fC) and 5-carboxylcytosine (5caC) (Ito et al. 2011). These oxidized forms can be demethylated during the replication.

DNA methylation can indirectly alter the genome itself. Methylated CpG sites can undergo spontaneous mutations by deamination causing C to T transition. Thus, DNA methylation reduces CpG content of the genome (Greenberg and Bourc'his 2019). Rosic et al. discovered that DNMTs can cause toxic 3meC (alkylation damage) lesions in DNA, which can lead to replication stress (Rošić et al. 2018).

Various assays and platforms are available to study DNA methylation. The very first method to quantify DNA methylation was liquid chromatography (Wagner and

Capesius 1981). Since then, there has been a great development of technologies. Nowadays, when there is a specific locus or loci of interest, PCR-based methods are the most common, and electrophoresis-based methods, microarrays, and high-throughput parallel sequencing allow genome-scale investigation (Li and Tollefsbol 2021). Most applications require the pretreatment of DNA. The most common pretreatment is bisulfite conversion using sodium bisulfite (NaHSO$_3$) for denatured DNA. It converts unmethylated cytosines into uracils whereas methylated cytosines remain unaffected, and subsequent PCR amplifies uracils as thymines. Other pretreatments include methylation-sensitive restriction enzymes (endonucleases) and immunoprecipitation methods (Li and Tollefsbol 2021).

The gold standard of applications is bisulfite sequencing that involves treatment of DNA with sodium bisulfite, followed by PCR amplification and direct sequencing or sequencing after cloning. Bisulfite sequencing can utilize Sanger sequencing to study specific smaller genomic locations, or whole-genome sequencing for a global screen (Urich et al. 2015). Another sequencing method for genome-wide methylation profiling of CpG islands and repetitive sequences is reduced-representation bisulfite sequencing (RRBS). RRBS relies on both restriction enzymes and bisulfite conversion (Meissner et al. 2005). The advantage of RRBS is its eligibility for repetitive sequences, which are challenging for parallel sequencing data interpretation. Following the RRBS, parallel sequencing methods have emerged for genome-wide methylation studies, and different pretreatments stratify these into different technologies (Li and Tollefsbol 2021). Third-generation sequencing that can read long single-strand DNA molecules in real time belong to the newest promising methods (Simpson et al. 2017). These methods can directly detect methylated cytosine and other DNA modifications (such as 5hmC) by producing a specific signal compared to other nucleotides.

Microarray-based methods, such as Illumina BeadChip, are a good alternative to parallel sequencing-based methods, as microarray data does not require similar technical expertise for data analysis (Li and Tollefsbol 2021). The development of Illumina DNA methylation microarrays includes four assays: the GoldenGate (Bibikova and Fan 2009), the Infinium HumanMethylation27 (Bibikova et al. 2009), the Infinium HumanMethylation450 (Bibikova et al. 2011), the Infinium MethylationEPIC BeadChip (Pidsley et al. 2016), and as the most recent one the Infinium MethylationEPIC v2.0 (Noguera-Castells et al. 2023). The first one targeted 1536 CpG sites while the EPIC v2.0 targets over 935,000 CpGs in the human genome. Bisulfite-treated DNA is hybridized to the array which in the (newest) versions includes two different probes, one designed for methylated and one for unmethylated DNA. Fluorescently labeled dideoxynucleotides attach to the 3′CpG site of each probe. Scanning of the array allows the ratio of the fluorescent signals measured for each CpG site. The calculated beta value provides the proportion of DNA methylation for each locus.

1.1.2 DNA Hydroxymethylation

Another, less abundant form of DNA methylation is 5hmC, a DNA pyrimidine nitrogen base derived from cytosine. The origin of 5hmC is in the action of DNA demethylation by TET enzymes oxidizing 5mC to 5hmC (also more rarely to 5fC and 5caC) (Li and Tollefsbol 2021). Its function was described for the first time relatively recently in murine embryonic stem cells (Tahiliani et al. 2009) and neurons (Kriaucionis and Heintz 2009), although it was discovered in mammalian (rat) genomes several decades ago (Penn et al. 1972). The content of 5hmC varies between different tissues in humans, showing the highest level in brain (0.67%) followed by rectum (0.57%), liver (0.46%) and colon (0.45%), and lower levels in lung (0.14%), placenta (0.06%), heart (0.05%), and breast (0.05%) (Li and Liu 2011). 5hmC has been proven to be a stable modified form of cytosine and has its own functions as an epigenetic regulator (see Sect. 1.2.2) (Wu and Zhang 2017; Efimova et al. 2020; Li and Tollefsbol 2021).

Common methods to study DNA methylation, such as bisulfite sequencing and enzymatic assays, cannot differentiate between 5mC and 5hmC. Specific techniques allow investigation of 5hmC distribution at the whole-genome level (Wu and Zhang 2017; Li and Tollefsbol 2021). The first approaches utilized thin layer chromatography methods with radioactive labeling (Kriaucionis and Heintz 2009; Tahiliani et al. 2009), liquid chromatography-mass spectrometry (Münzel et al. 2010), and other liquid chromatography-based applications (Li and Tollefsbol 2021). Alternative methods may rely on affinity to antibodies, anti-5hmC and anti-5-methylenesulfonate (bisulfite-treated 5hmC) (Pastor et al. 2011), the latter one being more quantitative. Hydroxymethylated DNA immunoprecipitation (hMeDIP) combined with high-throughput sequencing (hMeDIP-seq) serves as a tool to determine genome-wide distribution of 5hmC (Tan et al. 2013). However, all these methods have limitations in their detection thresholds. Enzymatic beta-GT/azide-glucose/biotin-labeling is a more sensitive approach and is possible to combine with deep sequencing (Song et al. 2011) or TET-assisted bisulfite sequencing (Yu et al. 2012). Moreover, oxidative conversion using potassium perrhenate (equivalent to bisulfite conversion of cytosine) can chemically modify 5hmC, enabling subsequent combination with RRBS (Booth et al. 2012) or Infinium BeadChip assay (Stewart et al. 2015), for example. Most promising advances rely on new sequencing methods, some of which allow investigating 5hmC at the base level (Wu and Zhang 2017; Li and Tollefsbol 2021).

1.1.3 Histone Modifications

Histone proteins together with DNA molecules form a nucleosome, a key unit of the chromatin structure. DNA molecules wrapped around the histone octamer of two of H2A, H2B, H3, and H4 proteins each form the nucleosome. According to the current knowledge, chromatin structure is dynamic architecture rather than a stable package of DNA and proteins, and that is where histone modifications play a role. Histone modifications are covalent post-translational modifications (PTMs) altering genome function either in direct or indirect manner, of which the latter requires the involvement of other proteins (Bannister and Kouzarides 2011; Millán-Zambrano et al. 2022). On the other hand, a transcriptional process can underlie PTMs themselves, e.g., a polymerase can endorse the formation of histone PTM. Modifications can take place in both histone tails and cores. Histone core PTMs occur either in lateral surface of the octamer (contact with the DNA) or on the surface interacting with other histone proteins while forming the octamer structure. Those PTMs can directly affect the binding or affinity of DNA molecules and/or other histone proteins (Bannister and Kouzarides 2011; Millán-Zambrano et al. 2022).

PTMs are involved in transcription, recombination, replication DNA repair, and genomic organization. Genomic regions, e.g., genes, enhancers (transcriptionally active), euchromatin, and (transcriptionally inactive) heterochromatin show different histone PTMs (Millán-Zambrano et al. 2022). Table 1.1 describes different PTMs, of which the most common are methylation, phosphorylation, acetylation, and ubiquitin-like alterations. Various enzymes catalyze, maintain, and remove PTMs, acting as writers, readers, and erasers, respectively (Millán-Zambrano et al. 2022). Certain genomic regions are especially rich in these enzymes, and many of the enzymes require cofactors for their activity. The equilibrium of histone PTMs reflects that of these enzymes.

As an example of the nomenclature of the PTMs, H3K4me3 denotes tri-methylation of histone H3 at lysine 4. This well-known modification is a mark of transcriptionally active chromatin that occurs at the 5′ end of the genes located close to TSS and after the first exons of genes. We will give more examples in the further paragraphs describing the importance of histone PTMs in a normal and cancer cells (Sects. 1.2.3. and 1.3.3).

To study histone modifications, chromatin immunoprecipitation (ChIP)-based methods are nowadays the most common. ChIP utilizes the cross-linking of DNA-associated proteins, followed by digestion (e.g., sonication). Detection of proteins and modifications is by specific antibodies. After a purification step, quantitative PCR (qPCR), microarray analysis (chip), or deep sequencing (seq) follows (Li 2021). These methods allow both quantification and detection of the genomic location of the PTMs at single nucleotide resolution. However, antibody-based techniques have some disadvantages, such as cross-reaction with similar modifications, nearby PTMs can prevent an antibody to bind, former knowledge is required, and creation of antibodies can be difficult (Karch et al. 2013). Mass spectrometry can overcome these drawbacks and is especially suitable to study novel and concurrent PTMs (Karch et al. 2013).

Table 1.1 Types of histone PTMs

Modification type	Type of chemical alteration	Targeted amino acid
Acylations	Formylation	K
	Acetylation	K, S, T
	Propinylation	K
	Butyrylation	K
	Crotonylation	K
	Benzoylation	K
	2-Hydroxyisobutyrylation	K
	Hydroxybutyrylation	K
	Lactylation	K
	Malonylation	K
	Succinylation	K
	Glutarylation	K
Ubiquitin-like	Ubiquitylation	K
	Sumoylation	K
	Ufmylation	K
Others	Methylation	K, R
	Biotinylation	K
	ADP ribosylation	K, E
Non-lysine PTMs	Serotonylation	Q
	Dopaminylation	Q
	O-palmitoylation	S
	S-palmitoylation	C
	Isomerization	P
	Hydroxylation	Y
	O-GlcNAcylation	S, T
	Deamination	R
	Phosphorylation	S, T, Y, H
	N-terminal acetylation	S

Adapted from Millán-Zambrano et al. (2022)

1.1.4 Nucleosome Remodeling

As discussed above, nucleosome is the highly dynamic basic unit of chromatin where approximately 147 bp of DNA molecule is wrapped around the histone octamer. Nucleosome remodeling complexes with ATP-dependent enzymes modify the mobility, structure, shape, and organization of nucleosome, changing DNA accessibility and eventually altering gene expression (Allis and Jenuwein 2016; Clapier et al. 2017). They can interact directly with DNA and DNA-associated proteins. In addition, they can specifically recognize different histone variants (i.e., histone proteins differing with a small number of amino acids from their major counterparts) and move them into and out of chromatin (Mizuguchi et al. 2004). A common factor for all nucleosome remodeling complexes is an ATPase enzyme, but differences are between other protein subunits, catalytic domains, complex function, and recruited proteins. Nucleosome remodeling enzymes belong to the RNA/DNA helicase Superfamily 2 and include four major subfamilies: imitation switch (ISWI), Switch/sucrose non-fermentable (SWI/SNF) (also called BRG1/BRM-associated factor (BAF)), chromodomain helicase DNA-binding (CHD), and INO80, of which each has a different functionality (Clapier et al. 2017).

It is possible to study the structure of histone proteins by X-ray crystallography and nuclear magnetic resonance spectroscopy, which have defined details about protein interactions and nucleosome function (Karch et al. 2013). Micrococcal nuclease digestion of chromatin followed by sequencing (MNase-seq) reveals the regions protected by nucleosomes, providing information on nucleosome positioning (Gaffney et al. 2012).

1.1.5 Non-coding RNAs

Regulatory non-coding RNAs (ncRNAs) are small single-stranded RNA molecules that serve as a sequence-complementary mechanism to silence gene expression. RNA complementary to DNA sequence suppresses the activity of chromatin, and the state can last even after multiple cell divisions (Allis and Jenuwein 2016; Wei et al. 2017). Table 1.2 describes the different types of regulatory ncRNAs and their functions.

To study ncRNAs, there are several in silico, in vitro, in vivo, and wet lab experiments available depending on the study design and question. Expression studies can be based on sequencing technologies or microarrays, and validation can be performed by reverse transcriptome quantitative PCR, Northern blot analysis, or in situ hybridization. There are multiple databases available for non-coding RNAs with an increasing number of data, to guide in silico investigation. Recent reviews (Grillone et al. 2020; Sun and Chen 2020; Li 2021) are available for a detailed description of the different methods.

1.2 Epigenetics in Normal Cell

There are two fundamental points of human genetics where epigenetics plays a pivotal role. First, every cell in the human body includes the DNA molecule whose linear length is approximately 2 m. This molecule has to be packed into small nuclei of each cell while keeping it accessible. Second, all those cells have an identical genotype while they can still form a vast range of cell types and tissues in a human body. This is where epigenetics plays a major role, and cell differentiation is more about epigenetics than genetics. It is important to maintain the state and identity of the differentiated cells. According to the current knowledge, it is evident that differentiation is not irreversible, but plastic (Bitman-Lotan and Orian 2021).

Table 1.2 Characteristics and function of regulatory ncRNAs

ncRNA	Length (nt)	Function	Origin and associated complexes
miRNA Micro RNA	19–24	Regulates translation in a sequence-specific manner, 6–7 nucleotide complementarity at the 5′ end, by targeting mRNA leading to translation inhibition or mRNA degradation	Originates from miRNA genes and forms a self-folding hairpin structure; pre-miRNA is processed by dicer; binds to RISC complex
siRNA Short interfering RNA	21–25	Regulates translation in a complete sequence-specific manner by targeting mRNA leading to mRNA degradation	Originates from long dsRNA molecules (virus replication, transposon activity, and gene transcription); binds to RISC complex
piRNA Piwi-interacting RNA	26–31	Regulates chromatin regulation and transposon silencing through PIWI proteins	Originates from ssRNA precursor (e.g., transposons); binds to PIWI proteins
lncRNA long non-coding RNA	>200	Regulates gene expression through the interaction with DNA, mRNA, miRNAs, and proteins; many have specific targets; five different categories: sense, antisense, bidirectional, intronic, and intergenic	Originate from different sources, e.g., retrotransposition, chromosomal reorganization; no common mechanism

Adapted from Wei et al. (2017), Ferreira and Esteller (2018)
Abbreviations: *RISC* RNA interference silencing complex, *mRNA* messenger RNA, *ssRNA* single-stranded RNA, *dsRNA* double-stranded RNA

Here, two processes are going on: one allows expression from a set of genes required for the differentiated state, and another keeps irrelevant or counteracting genes silent. In essence, normal function of the human cell and the developmental process are dependent on the epigenetic modifiers. We cover these in the next paragraphs, divided according to the different epigenetic alterations.

1.2.1 DNA Methylation in Normal Cell

DNA methylation contributes to the formation of heterochromatin. Silent chromatin shows characteristic patterns of DNA methylation and its binding proteins, in combination with RNA and histone modifications. As discussed above, DNA methylation in the promoter regions and close to the TSS silences the transcriptional activity of a gene. However, outside the TSS regions, DNA methylation might contribute to transcription elongation and splicing, and control insulator and enhancer regions. The role of DNA methylation is also important in silencing repetitive sequences, such as *LINE1*, *Alu*, and retroviruses; moreover, methylation in the centromeric sequences prevents transposable elements to move in the genome

(Li and Tollefsbol 2021). This is a crucial contribution to chromosomal and genome stability.

Mammalian genomes show extensive DNA methylation: up to 70–80% of the genome is methylated. CpG islands have a particularly high content of CpG dinucleotides compared to bulk DNA. Human promoters are often rich in CpG islands, especially promoters of housekeeping genes and developmentally regulated genes, and they overlap with TSSs (Deaton and Bird 2011). DNA methylation is fundamental for normal development. Mouse models with DNMT deficiency have shown severely impaired development (Okano et al. 1999). There are two major phases of development where reprogramming of DNA methylation occurs: after fertilization and after germ cell specification (Zeng and Chen 2019). DNA methylation silences repetitive and centromeric sequences, represses transposons, takes part in X chromosome inactivation in females, and is responsible for genomic imprinting.

In each cell of females, one of the X chromosomes becomes inactive early in embryonic development to obtain a dosage compensation of X-linked genes compared to males (Pessia et al. 2012). Inactivated X chromosome is packed as a compact heterochromatin structure called the Barr body. Inactivation is coordinated through the regulatory locus called X inactivation centre (XIC). XIC encodes the long non-coding RNA X-inactive specific transcript (XIST), which is highly expressed by the future inactive X chromosome and spreads in cis along the inactive X chromosome (Engreitz et al. 2013). XIST recruits many factors required to form heterochromatin and colocalizes, especially in the domains characterized by a repressive histone modification, H3K27me3 (Nozawa et al. 2013). DNA methylation takes place relatively late in the process of highly organized events and serves as a final mechanism for already silenced genes. Taken together, inactive X chromosome takes the form of a repressive heterochromatin, with high levels of DNA methylation and H3K27me3 and H3K9me3 and low levels of H3K4me3 and histone acetylation (Chow et al. 2005).

1.2.2 DNA Hydroxymethylation in Normal Cell

Besides the fundamental role of 5mC as an epigenetic regulator, its oxidized form 5hmC, too, has recently been proven to play a role in gene regulation (Wu and Zhang 2017; Efimova et al. 2020; Li and Tollefsbol 2021). For example, 5hmC is enriched in the central nervous system and seems crucial for neuronal function and structure. Furthermore, it has a pivotal role in epigenetic reprogramming, cellular differentiation, gene expression regulation, and aging.

1.2.3 Histone Modifications in Normal Cell

Histone PTMs are important for the formation and reorganization of the chromatin structure and for nucleosome assembly. They are involved in packing DNA in the nuclei as well as in the regulation of transcriptional activity of the genes. As mentioned earlier, H3K4me3 is especially abundant close to the TSSs and is associated with transcriptional activity. However, studies show that the activation is context dependent (Cano-Rodriguez et al. 2016). H3K4me3 occurs in a mutually exclusive manner with DNA methylation in the CpG islands (Hughes et al. 2020), which might serve as a mechanism to prevent those regions from unwanted DNA methylation during development.

As mentioned above (Sect. 1.1), the function of DNMTs shows a connection to histone modifications. As a more detailed example of the H3K4me3 mark, ADD domain of DNMTs (DNMT3A and DNMT3B) normally binds to H3K4, but an increased number of methyl groups at the residue prevents its binding and does not allow DNA methylation to occur (Zhang et al. 2010). As an auto-inhibiting event, ADD domain binds to the MTase (highly conserved DNMT) domain of the DNMT3s when binding to H3K4 is not possible (Guo et al. 2015) (Table 1.3).

Similar to DNA methylation, histone PTMs seem heritable. In a mouse model, the embryo partially maintains paternal H3K4me3 modifications, and those marks affect transcription activity (Lismer et al. 2021). In addition to transcription, histone PTMs play a role in recombination events, such as in V(D)J recombination, homologous

Table 1.3 Some of the best-characterized histone PTMs in humans and their roles

PTM	Functionality	Genomic location
H3K4me1	Transcriptional activity	Enhancers and downstream from TSS
H3K4me2	Transcriptional activity	Cis-regulatory regions, enhancers, and promoters
H3K4me3	Transcriptional activity; "enhancer"	Promoters, around TSS
H3K9me3	Transcriptional silencing; a marker for constitutive heterochromatin	Broad distribution on inactive regions
H4K20me1	Transcriptional activity and elongation	Promoters
H3K27me3	Transcriptional silencing; facultative heterochromatin formation	Around inactive TSS
H3K36me3	Transcriptional activity; inhibits spurious intragenic transcription; regulates DNA repair	Gene bodies of actively transcribed genes
H3K79me2	Transcriptional activity	Transcribed regions of active genes
H3K9ac	Transcriptional activity	Promoters
H3K27ac	Transcriptional activity	Around active enhancers and TSS
H2Aub	Transcriptional silencing; facultative chromatin	Promoters of silenced genes

Adapted from Bannister and Kouzarides (2011), Talbert and Henikoff (2021), Millán-Zambrano et al. (2022)

recombination, and meiotic recombination (Bannister and Kouzarides 2011; Millán-Zambrano et al. 2022). Those events require large-scale rearrangement of DNA strands. Histone PTMs contribute to DNA repair as well, since DNA damage induces certain histone PTMs, leading to DNA damage response (Bannister and Kouzarides 2011; Millán-Zambrano et al. 2022).

1.2.4 Nucleosome Remodeling in Normal Cell

Nucleosome remodeling by ATP-dependent protein complexes is a key mechanism to control and maintain the dynamic nature of the chromatin. Complexes include different protein and DNA-binding sites allowing the interaction of DNA and different proteins. Gene families typically encode subunits of the complexes and thereby enable variation of the complexes. This variation ensures the emergence of complexes with specific functions. For example, BAF complexes play a role in cell differentiation, neural development, and embryonic development (Alfert et al. 2019). BAF complex exemplifies interactive network of epigenetic modifiers since it can bind stronger to histone modifications and is thus coordinated with regional modifications of the chromatin (Kadoch and Crabtree 2015). Another well-studied complex, a member of CHD family, nucleosome remodeling and deacetylase complex (NuRD) is involved in the regulation of genome integrity, cell cycle, and gene expression, also playing a role in embryonic development (Allen et al. 2013; Basta and Rauchman 2015). NuRD has dual enzymatic activity, as it contains histone deacetylase in addition to ATPase. As discussed above, nucleosome remodeling complexes can change the content of histone octamers by introducing histone variants. Multiple variants for H2A, H2B, and H3 histones exist, whereas H4 has only one recognized variant. Variants contribute to, e.g., DNA double-strand break repair, transcriptional regulation, chromosome segregation, and spermatogenesis (Kurumizaka et al. 2021).

1.2.5 Non-coding RNAs in Normal Cell

Most of the human genome (approximately 98%) is non-coding regions/genes and sites where many ncRNAs are produced (Lander 2011). ncRNAs, especially miRNAs, play a crucial role in the epigenetic regulation of protein-coding genes at both gene and chromosomal levels. The number of these molecules highlights their importance: to date, 2300 miRNAs (Alles et al. 2019) and more than 270,000 lncRNAs have been identified (Ma et al. 2019). ncRNAs regulate cellular events such as proliferation, apoptosis, and differentiation. In addition to being a crucial part of the epigenetic regulatory network, ncRNAs themselves are subject to epigenetic regulation, forming a sophisticated reciprocal regulation system. Here, we present a few examples of the role of ncRNAs in the normal cell—however, the regulation events are very diverse as the vast number of them can indicate. Table 1.2 describes the main characteristics of regulatory ncRNAs.

The *Xist*-derived lncRNA which plays an important role in X chromosome inactivation (please see Sect. 1.2.1) provides a well-known example of lncRNAs. Another example is H19, which is a lncRNA strongly associated with genomic imprinting, although the exact mechanism and function remain obscure (Wei et al. 2017). Besides the straightforward regulation of gene expression by targeting the mRNA molecules, ncRNAs, especially miRNAs, are able to contribute to chromatin remodeling by regulating remodeling enzymes, histone modification enzymes, and DNMTs (Wei et al. 2017; Arif et al. 2020). For example, the miR-29 family regulates DNMT3a and DNMT3b, which regulate genome-wide de novo DNA methylation. Both miRNA and siRNA can inhibit EZH2, a histone methyltransferase. Active EZH2 induces H3K27 methylation, which serves as a base for DNA methylation leading to silent chromatin (Wei et al. 2017; Arif et al. 2020).

1.3 Epigenetics in Cancer Cell

Cancer is a genetic disease, but more and more evidence has emerged during the last decades to suggest that cancer is also an epigenetic disease. Nowadays, the important interactive role of genetics and epigenetics in cancer development is indisputable, although the available details of genetic cancer-related alterations and their mechanisms are much more extensive and deeper. The latest "proof" of the importance of epigenetics is its inclusion as one of the cancer hallmarks (Hanahan 2022). Besides, some investigators suggest that only epigenome deregulation can plausibly induce all the classical cancer hallmarks (Flavahan et al. 2017). Epigenome deregulation can be caused indirectly by genetic mutations in epigenetic regulatory genes (chromatin modifiers), and directly by aberrant epigenetic modifications. Genomic stimuli or environmental factors, such as nutrition, aging, or cell microenvironment may induce epigenetic changes (Nebbioso et al. 2018). In this chapter, we describe the relation of epigenetic events and mechanisms to cancer hallmarks by giving functional examples (Fig. 1.2). Cancer hallmarks are indicated as bold text in the following chapters.

1.3.1 DNA Methylation in Cancer Cell

DNA methylation is one of the major epigenetic mechanisms in cancer and can be associated with all cancer hallmarks. Besides inactivating tumor suppressor genes (TSGs) by hypermethylation and activating proto-oncogenes by hypomethylation, and thereby leading to aberrant proliferation, i.e., sustaining proliferative signaling, DNA methylation affects other genes and pathways related to cancer hallmarks

1 An Overview of Epigenetics Modifications in Normal and Cancer Cell

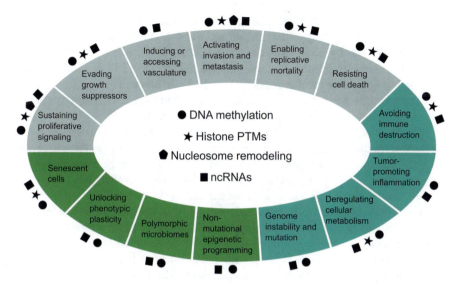

Fig. 1.2 Cancer hallmarks and epigenetic mechanisms. Gray backgrounds indicate basic hallmarks (Hanahan and Weinberg 2000), turquoise color stands for "next generation" hallmarks (Hanahan and Weinberg 2011) and green color for the most recent "new dimension" hallmarks (Hanahan 2022). Black symbols indicate the epigenetic mechanisms that can be associated with the hallmarks and are discussed or presented in this chapter

(Berdasco and Esteller 2010). In addition, enzymes carrying out DNA (de)-methylation can be altered and thus result in DNA methylation alterations (Berdasco and Esteller 2010; Rasmussen and Helin 2016).

Evading negative growth suppressors may occur by regulating Retinoblastoma (Rb) protein and p53 pathway which are some of the key players regulating cell division and cell cycle. DNA methylation of *RB1* often silences Rb expression in different cancers. *RB1* is one of the first TSGs found to be inactivated by DNA methylation (Berdasco and Esteller 2010). Methylation can also silence the *TP53* gene encoding p53 (Saldaña-Meyer and Recillas-Targa 2011), as well as its activating regulator p14-ARF (Esteller et al. 2001). The p53 protein is a key player regulating and activating cell death signaling due to DNA damage, and its silencing is therefore associated with resisting cell death, another cancer hallmark, as well.

Inducing the angiogenesis is a necessity for tumor growth. It is regulated, e.g., by a TSG, von Hippel-Lindau (VHL), the gene of which is found to be hypermethylated in many tumors (Berdasco and Esteller 2010). VHL downregulates many angiogenic molecules (Kaelin 2005). DNA methylation of *THBS1* is also a pro-angiogenetic mechanism in cancer (Berdasco and Esteller 2010).

Epithelial-to-mesenchymal transition (EMT) is one of the mechanisms associated with carcinogenesis, especially invasiveness and metastases of epithelial-derived tumors. Normally, EMT is a developmental event allowing cells to migrate, but some cancer cells go through the dedifferentiation program through altered gene expression (Nieto et al. 2016). DNA methylation plays a role in altering the expression of EMT-associated genes, such as silencing of *CHD1* encoding E-cadherin, a key regulator of EMT (Sun and Fang 2016). Cancer cell plasticity is also mainly driven by EMT (Nieto et al. 2016), and EMT is involved in immune escape, too (Terry et al. 2017; Benboubker et al. 2022). The EMT process is widely regulated by epigenetic mechanisms (Nieto et al. 2016; Sun and Fang 2016). Cancer cell plasticity also has other forms where the cell identity and functionality change due to transcriptional or epigenetic alterations (Yuan et al. 2019).

Protection of telomeric DNA enables replicative immortality in cancer cells. Epigenetic mechanisms regulate transcription of *hTERT* that encodes the telomerase reverse transcriptase enzyme responsible for maintaining telomere ends. Hypermethylation in *hTERT* promoter prevents its binding to repressors (WT1 and CTCF), resulting in hTERT activation (Lewis and Tollefsbol 2016).

In immune evasion, DNA methylation plays a crucial role by regulating the components of immune pathways. Hypomethylation of immune checkpoint genes and hypermethylation of co-stimulatory genes are frequent events in solid tumors (Berglund et al. 2020). For example, methylation of PD-L1, a ligand for immune checkpoint receptor PD-1, inversely correlates with PD-L1 expression (Cao and Yan 2020). In the presence of PD-L1, T cell exhaustion, i.e., functional impairment, takes place leading to immune evasion. Tumor-associated macrophages (TAMs) are very plastic immune cells regulated by different stimuli and epigenetic mechanisms (Larionova et al. 2020; Pan et al. 2020). TAMs are a key component of tumor microenvironment and are involved in tumorigenic events, such as angiogenesis, immune evasion, tumor-promoting inflammation, tumor growth and metastasis. DNA methylation is crucial for inducing inflammatory responses in TAMs (Larionova et al. 2020; Pan et al. 2020).

Tumor-promoting inflammation occurs when inflammation is chronic, and it activates multiple inflammatory pathways that promote tumorigenesis by controlling angiogenesis, proliferation, invasion, and apoptosis. Ulcerative colitis is an inflammatory bowel disease that increases the risk of colorectal cancer (CRC). Non-neoplastic colon tissue from ulcerative colitis patients shows aberrant promoter hypermethylation in, e.g., *CDKN2A* and *CDH1* (Hartnett and Egan 2012). In our recent study, we also observed some high methylation levels in *NTSR1* and CpG island methylator phenotype (CIMP)-associated genes already in normal mucosa of patients diagnosed with ulcerative colitis-associated CRC (Mäki-Nevala et al. 2021). Reactive oxygen species released by the inflammation process can contribute to both DNA hypermethylation of TSGs and global hypomethylation leading to genomic instability (Wu and Ni 2015). Recently, the CIMP phenotype was associated with different immunological subtypes, and other immunological features, such as lymphocyte infiltration and macrophage regulation (Yates and Boeva 2022).

Reprogrammed energy metabolism involving glucose metabolism, glutamine metabolism, and lipid biosynthesis is required for cancerous growth. In cancer cells, glucose uptake is increased, and its processing occurs mostly by anaerobic glycolysis instead of oxidative phosphorylation even if oxygen is available. This is known as the Warburg effect (Warburg et al. 1927). Glycolysis releases lactate into extracellular space resulting in activation of enzymes promoting cell migration and invasion. Elevated glucose uptake can be due to epigenetic silencing of certain genes. For example, *VHL* promoter hypermethylation can induce a constitutive expression of HIF1α in renal carcinoma, resulting in increased glycolysis, and DNA methylation of *DERL3* upregulates GLUT1, a glucose transporter (Llinàs-Arias and Esteller 2017). Hypoxia (due to impaired vascularization) can suppress the activity of TET enzymes and thus promote DNA hypermethylation at TSG promoters in cancer cells (Thienpont et al. 2016).

DNA methylation is an important mechanism for genome instability. In CRC, microsatellite instability by *MLH1* promoter methylation and chromosomal instability (CIN) by *LINE-1* hypomethylation are illustrative examples. *MLH1* is a DNA repair gene and its inactivation by promoter hypermethylation impairs DNA mismatch repair system and leads to a number of mutations in short repetitive sequences called microsatellites (Boland and Goel 2010). This increases genomic instability and leads to a hypermutable phenotype. *LINE-1* is the long interspersed nuclear element-1, a transposable repetitive sequence covering approximately 18% of human genome (Lander et al. 2001). Normally, it is heavily methylated but can be hypomethylated in cancer and is a marker for global hypomethylation. Global DNA hypomethylation facilitates a loss of genomic imprinting, CIN, and further increased mutation rates (Sahnane et al. 2015). As an indirect epigenetic mechanism, mutations in TET proteins (i.e., enzymes demethylating DNA) increase genomic instability and regulate DNA repair (Wu and Zhang 2017).

Microbiomes exist on tissue surfaces exposed to external environment, such as gastrointestinal tract and lungs. Microbiomes are highly polymorphic and differences between individuals and populations may affect tumorigenesis. Compounds released from microbial metabolism may influence epigenetic mechanisms by having an impact on compounds used for epigenetic modifications or for the activity of epigenetic regulatory enzymes. Zhao et al. review the roles of gut microbiota and its epigenetic regulation (DNA methylation, histone modifications, and ncRNAs) in CRC (Zhao et al. 2021). Most of the evidence derives from mouse studies.

Senescent cells undergo the biological process called senescence-associated secretory phenotype (SASP) resulting in cell cycle arrest. Senescent cells can arise due to the lack of nutrients or DNA repair, or impairments in cellular signaling processes. In general, senescence is a protective mechanism against cancer. However, recent studies indicate that senescent cells can contribute to tumorigenesis by SASP-induced molecules (Hanahan 2022). Epigenetic mechanisms, including DNA methylation and histone PTMs, regulate senescence and SASP (Crouch et al. 2022).

1.3.2 DNA Hydroxymethylation in Cancer Cell

There is a paucity of published studies on the impact of 5hmC on cancer hallmarks, which is why we do not discuss hydroxymethylation separately in this context. Overall, cancerous tissues show lower levels of 5hmC compared to normal tissues, most plausibly caused by reduced levels of TET enzymes (Skvortsova et al. 2019;

Xu and Gao 2020). In cancer, a lower level of 5hmC is associated with poorer survival (Skvortsova et al. 2019; Xu and Gao 2020).

1.3.3 Histone Modifications in Cancer Cell

Altered histone modifications regulate gene expression in cancer, and histone-modifying enzymes can be mutated or deregulated in cancer (Nebbioso et al. 2018). A very recent study shows that histone modifications affect the accumulation of somatic signatures (Otlu et al. 2023).

Repressive histone modifications may occur in TSGs, such as those regulating cell cycle and apoptosis being associated with resisting cell death and evading growth suppressors. Rb regulates the protein complexes needed in controlling of cell cycle and proliferation. For example, histone deacetylases (HDACs) belong to these repressor molecules. In some tumors, HDACs can be upregulated (Nebbioso et al. 2018). Sustaining proliferative signaling can be caused by histone modifications in proto-oncogenes, such as histone acetylation in fibroblast growth factor receptor 2 (FGFR2) in breast cancer (Zhu et al. 2009).

As discussed above, EMT contributes to invasiveness and metastases. Histone modifications regulate EMT-associated genes, such as *CDH1*, *SNA1*, and *TWIST1* (Sun and Fang 2016). Also, senescence-associated genes are regulated by histone PTMS (Crouch et al. 2022).

Active hTERT mainly regulates replicative immortality, as discussed above. Besides DNA methylation, histone PTMs regulate hTERT expression. For example, H3K4me3 mark is associated with active hTERT transcription and a knockdown of SMYD3 (H3K4-specific demethyltransferase) significantly decreased *hTERT* mRNA in cancer cells (Lewis and Tollefsbol 2016).

In immune evasion, an upregulated histone modification enzyme, EZH2, decreases immunogenicity by silencing antigen presentation complexes, and SETDB1 and KDM5B silence retroelements in melanoma (Benboubker et al. 2022). Moreover, histone modifications regulate TAMs causing immunosuppression (Larionova et al. 2020).

Similarly, histone modifications or deregulation of associated enzymes may contribute to reprogramming of metabolism. For example, several histone lysine demethylases (KDMs) show overexpression in various solid tumors promoting glycolysis (Miranda-Gonçalves et al. 2018). KDMs interact with promoters of glycolytic genes resulting in, e.g., H3K9me2 demethylation and transcriptional activation of the gene(s). In the case of HDACs, the role of Sirtuins in cellular metabolism is the best characterized. For example, SIRT6 inhibits the HIF1α and MYC-dependent glycolysis and glutaminolysis, and SIRT6 is deleted in various solid tumors, resulting in increased H3K9ac, i.e., transcriptional activation of glycolytic genes (Miranda-Gonçalves et al. 2018). Additionally, different metabolites alter the levels of epigenetic enzymes.

1.3.4 Nucleosome Remodeling in Cancer Cell

Nucleosome remodeling complexes are involved in carcinogenesis and their inactivation may promote sustaining the proliferative signaling and contribute toward the invasiveness and metastatic nature of the tumor, especially through enabling the EMT. They can also regulate gene expression of TSGs and proto-oncogenes and modulate cell fate. The genes encoding the subunits are often mutant. For example, subunits of the BAF complex, encoded by *ARID1A* and *SMARCA4,* act as tumor suppressors, and their mutations are frequent in many cancers (Kadoch and Crabtree 2015). As another example, the genes encoding subunits of NuRD complex often exhibit changes in cancer (Basta and Rauchman 2015). Alterations include, e.g., chromosomal deletions of *MBD3* in serous endometrial cancer, CNVs of *MTA3* in brain tumors, *CHD4* mutations and CNVs in serous endometrial cancer, and overexpression of *MTA1/2* (Basta and Rauchman 2015). The development of exome sequencing methods has contributed to these findings, and even 20% of human cancers may harbor mutations in SWI/SNF complex members (Kadoch and Crabtree 2015).

1.3.5 Non-coding RNAs in Cancer Cell

ncRNAs, especially miRNAs and lncRNAs, are often found to be deregulated in various cancers (Anastasiadou et al. 2018). These non-coding elements together with coding regions of the genome form a complex network regulating nearly all cellular functions. Here, we focus on miRNAs and lncRNAs because their role (especially that of miRNAs) in cancer is the best characterized. miRNAs can act both as TSGs and proto-oncogenes and play a role in a plethora of cellular networks associated with cancer hallmarks (Ruan et al. 2009; Budakoti et al. 2021). The machinery required for miRNA biogenesis can be altered in cancer as well (Budakoti et al. 2021). In Table 1.4, we give functional examples of miRNAs and lncRNAs in carcinogenesis.

1.4 Conclusion

The detailed nature and importance of epigenetic modifications characteristic of normal and cancer cells have gradually unfolded over several decades, and along with methodological advances, more and more information is becoming available. Discoveries made in individual research laboratories (many cited above) together with coordinated large-scale efforts, such as the recently completed Epigenome Roadmap (Skipper et al. 2015) have generated valuable epigenomic references for normal and disease states that will facilitate future research.

Table 1.4 Examples of ncRNAs and their association with cancer hallmarks

Cancer hallmark	ncRNA	Functional example	Reference(s)
Sustaining proliferative signaling	miRNA, lncRNA	miRNAS regulate various TSGs and proto-oncogenes. For example, miR-124a, a repressor for CDK6, is silenced in cancer resulting in inactivation of Rb. SRA functions as a lncRNA (as well as a coactivator/corepressor protein) upregulate nuclear receptors	miRNA: Ferreira and Esteller (2018) lncRNA: Gutschner and Diederichs (2012)
Evading growth suppressors	miRNA, lncRNA	Multiple miRNAs target e.g. p53 directly or indirectly. lncRNA, ANRIL, represses a TSG INK4B	miRNA: Liu et al. (2017) lncRNA: Gutschner and Diederichs (2012)
Inducing or accessing vasculature	miRNA, lncRNA	Multiple miRNAs target angiogenesis-associated genes. For example, activation of miR-17-92 promotes angiogenesis. lncRNA, aHIF, is a negative regulator of HIF1α, an important angiogenesis-associated regulator	miRNA: Budakoti et al. (2021) lncRNA: Gutschner and Diederichs (2012)
Activating invasion and metastasis	miRNA, lncRNA	Multiple miRNAs target TSGs or proto-oncogenes regulating invasion. EMT is regulated by miRNAs; e.g., miR-200 family targets EMT-associated transcription factors ZEB1/2. lncRNA HOTAIR targets histone-modifying enzymes PRC2 and LSD1, which suppress HOX genes and metastasis suppressors	miRNA: Budakoti et al. (2021) lncRNA: Sun and Fang (2016)
Enabling replicative immortality	miRNA, lncRNA	Multiple miRNAs target hTERT. lncRNAs called TERRA are involved in telomere heterochromatin formation, and cancer cells with active telomerase show low levels of TERRA	miRNA: Lewis and Tollefsbol (2016) lncRNA: Gutschner and Diederichs (2012)
Resisting cell death	miRNA, lncRNA	Multiple miRNAs target p53 directly or indirectly; e.g., the expression of mir-34 family correlates with p53 expression, and its downregulation is associated with attenuated 53-mediated apoptosis. lncRNA, PCGEM1 is a prostate cancer-associated lncRNA with antiapoptotic function	miRNA: Liu et al. (2017) lncRNA: Gutschner and Diederichs (2012)
Avoiding immune destruction	miRNA, lncRNA	Multiple miRNAs contribute to evasion of immunosurveillance, e.g., by downregulating MICA and MICB proteins. lncRNA HOTAIR	miRNA: Ruan et al. (2009) lncRNA: Ferreira and Esteller (2018)

(continued)

1 An Overview of Epigenetics Modifications in Normal and Cancer Cell

Table 1.4 (continued)

Cancer hallmark	ncRNA	Functional example	Reference(s)
		(by inhibiting miR-152) upregulates HLA-G	
Tumor-promoting inflammation	miRNA	miRNAs regulate multiple inflammation-associated genes, and TAMs contribute to pro-tumor environment. For example, an upregulated miRNA-155 in ulcerative colitis regulates toll-like receptors leading to alterations in innate immune response, increase the levels of IL-13 and decrease E-cadherin expressions, which together promote epithelial instability and metastatic risk	Larionova et al. (2020), Budakoti et al. (2021), Krishnachaitanya et al. (2022)
Deregulating cellular metabolism	miRNA	Multiple miRNAs contributing to metabolic reprogramming, e.g., glucose transporter GLUT1 is suppressed by miR-132 which is downregulated in many cancers leading to GLUT1 overexpression	Subramaniam et al. (2019)
Genome instability and mutation	miRNA, lncRNA	Genomic instability itself can cause deregulation of miRNAs. Additionally, for example, miR-17 and miR-20a, both regulated by c-Myc, promote the transition to G1 checkpoint, and inhibition of those miRNAs leads to increase of DNA double-strand breaks. Many lncRNAs regulate proteins involved in, e.g., double-strand break repair	miRNA: Ruan et al. (2009) lncRNA: Guo et al. (2015)
Non-mutational epigenetic reprogramming	miRNA	Multiple miRNAs target e.g. TET enzymes and interact with histone-modifying complexes	Ferreira and Esteller (2018)
Polymorphic microbiomes	miRNA	Some mouse studies indicate the connection between certain miRNAs and altered gut microbiome	Allen and Sears (2019)
Unlocking phenotypic plasticity	miRNA	Multiple miRNAs target HOXA5 and SMAD4 which may result in dedifferentiation and WNT-driven hyperproliferation in colorectal tumorigenesis	Liu et al. (2013), Ordóñez-Morán et al. (2015), Mo et al. (2019)
Senescent cells	miRNA, lncRNA	Multiple miRNAs target p53 directly or indirectly, and other regulators of senescence. Multiple lncRNAs regulate senescence-associated pathways	miRNA: Liu et al. (2017), Budakoti et al. (2021) lncRNA: Puvvula (2019)

For the construction of comprehensive epigenomic landscapes, modern technologies to integrate information from the different epigenetic regulators are necessary (Li 2021). Omics studies combining data from epigenome, genome, transcriptome, and proteome are a key. Single-cell methods can shed light on the role of epigenetic modifications as driver or passenger events in tumorigenesis and serve as tools to observe intercellular differences (Clark et al. 2016). Additionally, 3D models can help to define the molecular details between cancer cells and their microenvironment (Rodrigues et al. 2021). Advances in epigenetic bioinformatics are crucial in parallel to enable insightful handling of multi-omics data derived from combinatorial analysis (Li 2021).

There is an obvious need to exploit all this acquired knowledge in the clinical setting for the ultimate benefit of the patient. Diagnostic, prognostic, and predictive epigenetic biomarkers are available for different cancer types (Berdasco and Esteller 2019). The possibility of reprogramming the epigenetic landscape of cancer is a promising approach to treatment, and it may help overcome drug resistance (Miranda Furtado et al. 2019). Many currently available "epidrugs" targeting epigenetic enzymes are not site-specific and target the whole epigenome, which may cause problems. Novel epidrugs targeting mutated epigenetic molecules and the progress made in site-specific epigenome editing may avoid that problem (Altucci and Rots 2016; Yang et al. 2021). Even with current limitations that need improvement, CRISPR-dCas9-based epigenetic editing is a powerful tool for advancing the knowledge of epigenetic regulatory mechanisms and facilitating future clinical applications (Gjaltema and Rots 2020).

Compliance with Ethical Standards Funding: The Finnish Academy (grant number 331284 to SM-N, and grant number 330606 to PP), Jane and Aatos Erkko Foundation, Cancer Foundation Finland, and Sigrid Juselius Foundation supported this work.

Conflict of Interest: Authors declare no conflicts of interest.

Ethical Approval: This chapter does not contain any studies with human participants performed by the authors.

References

Alfert A, Moreno N, Kerl K (2019) The BAF complex in development and disease. Epigenetics Chromatin 12(1):19

Allen J, Sears CL (2019) Impact of the gut microbiome on the genome and epigenome of colon epithelial cells: contributions to colorectal cancer development. Genome Med 11(1):11

Allen HF, Wade PA, Kutateladze TG (2013) The NuRD architecture. Cell Mol Life Sci 70(19): 3513–3524

Alles J, Fehlmann T, Fischer U, Backes C, Galata V, Minet M et al (2019) An estimate of the total number of true human miRNAs. Nucleic Acids Res 47(7):3353–3364

Allis CD, Jenuwein T (2016) The molecular hallmarks of epigenetic control. Nat Rev Genet 17(8): 487–500

Altucci L, Rots MG (2016) Epigenetic drugs: from chemistry via biology to medicine and back. Clin Epigenetics 8:56

Anastasiadou E, Jacob LS, Slack FJ (2018) Non-coding RNA networks in cancer. Nat Rev Cancer 18(1):5–18

Arif KMT, Elliott EK, Haupt LM, Griffiths LR (2020) Regulatory mechanisms of epigenetic miRNA relationships in human cancer and potential as therapeutic targets. Cancers (Basel) 12(10)

Bannister AJ, Kouzarides T (2011) Regulation of chromatin by histone modifications. Cell Res 21(3):381–395

Basta J, Rauchman M (2015) The nucleosome remodeling and deacetylase complex in development and disease. Transl Res 165(1):36–47

Benboubker V, Boivin F, Dalle S, Caramel J (2022) Cancer cell phenotype plasticity as a driver of immune escape in melanoma. Front Immunol 13:873116

Berdasco M, Esteller M (2010) Aberrant epigenetic landscape in cancer: how cellular identity goes awry. Dev Cell 19(5):698–711

Berdasco M, Esteller M (2019) Clinical epigenetics: seizing opportunities for translation. Nat Rev Genet 20(2):109–127

Berglund A, Mills M, Putney RM, Hamaidi I, Mulé J, Kim S (2020) Methylation of immune synapse genes modulates tumor immunogenicity. J Clin Invest 130(2):974–980

Bibikova M, Fan JB (2009) GoldenGate assay for DNA methylation profiling. Methods Mol Biol 507:149–163

Bibikova M, Le J, Barnes B, Saedinia-Melnyk S, Zhou L, Shen R et al (2009) Genome-wide DNA methylation profiling using Infinium® assay. Epigenomics 1(1):177–200

Bibikova M, Barnes B, Tsan C, Ho V, Klotzle B, Le JM et al (2011) High density DNA methylation array with single CpG site resolution. Genomics 98(4):288–295

Bitman-Lotan E, Orian A (2021) Nuclear organization and regulation of the differentiated state. Cell Mol Life Sci 78(7):3141–3158

Boland CR, Goel A (2010) Microsatellite instability in colorectal cancer. Gastroenterology 138(6): 2073–2087

Booth MJ, Branco MR, Ficz G, Oxley D, Krueger F, Reik W et al (2012) Quantitative sequencing of 5-methylcytosine and 5-hydroxymethylcytosine at single-base resolution. Science 336(6083):934–937

Budakoti M, Panwar AS, Molpa D, Singh RK, Busselberg D, Mishra AP et al (2021) Micro-RNA: the darkhorse of cancer. Cell Signal 83:109995

Cano-Rodriguez D, Gjaltema RA, Jilderda LJ, Jellema P, Dokter-Fokkens J, Ruiters MH et al (2016) Writing of H3K4Me3 overcomes epigenetic silencing in a sustained but context-dependent manner. Nat Commun 7:12284

Cao J, Yan Q (2020) Cancer epigenetics, tumor immunity, and immunotherapy. Trends Cancer 6(7):580–592

Chow JC, Yen Z, Ziesche SM, Brown CJ (2005) Silencing of the mammalian X chromosome. Annu Rev Genomics Hum Genet 6:69–92

Clapier CR, Iwasa J, Cairns BR, Peterson CL (2017) Mechanisms of action and regulation of ATP-dependent chromatin-remodelling complexes. Nat Rev Mol Cell Biol 18(7):407–422

Clark SJ, Lee HJ, Smallwood SA, Kelsey G, Reik W (2016) Single-cell epigenomics: powerful new methods for understanding gene regulation and cell identity. Genome Biol 17:72

Crouch J, Shvedova M, Thanapaul R, Botchkarev V, Roh D (2022) Epigenetic regulation of cellular senescence. Cells 11(4):672

Deaton AM, Bird A (2011) CpG islands and the regulation of transcription. Genes Dev 25(10): 1010–1022

Efimova OA, Koltsova AS, Krapivin MI, Tikhonov AV, Pendina AA (2020) Environmental epigenetics and genome flexibility: focus on 5-hydroxymethylcytosine. Int J Mol Sci 21(9): 3223

Engreitz JM, Pandya-Jones A, McDonel P, Shishkin A, Sirokman K, Surka C et al (2013) The Xist lncRNA exploits three-dimensional genome architecture to spread across the X chromosome. Science 341(6147):1237973

Esteller M, Corn PG, Baylin SB, Herman JG (2001) A gene hypermethylation profile of human cancer. Cancer Res 61(8):3225–3229

Ferreira HJ, Esteller M (2018) Non-coding RNAs, epigenetics, and cancer: tying it all together. Cancer Metastasis Rev 37(1):55–73

Flavahan WA, Gaskell E, Bernstein BE (2017) Epigenetic plasticity and the hallmarks of cancer. Science 357(6348)

Gaffney DJ, McVicker G, Pai AA, Fondufe-Mittendorf YN, Lewellen N, Michelini K et al (2012) Controls of nucleosome positioning in the human genome. PLoS Genet 8(11):e1003036

Gjaltema RAF, Rots MG (2020) Advances of epigenetic editing. Curr Opin Chem Biol 57:75–81

Goldberg AD, Allis CD, Bernstein E (2007) Epigenetics: a landscape takes shape. Cell 128(4): 635–638

Greenberg MVC, Bourc'his D (2019) The diverse roles of DNA methylation in mammalian development and disease. Nat Rev Mol Cell Biol 20(10):590–607

Grillone K, Riillo C, Scionti F, Rocca R, Tradigo G, Guzzi PH et al (2020) Non-coding RNAs in cancer: platforms and strategies for investigating the genomic "dark matter". J Exp Clin Cancer Res 39(1):117

Guo X, Wang L, Li J, Ding Z, Xiao J, Yin X et al (2015) Structural insight into autoinhibition and histone H3-induced activation of DNMT3A. Nature 517(7536):640–644

Gutschner T, Diederichs S (2012) The hallmarks of cancer: a long non-coding RNA point of view. RNA Biol 9(6):703–719

Hanahan D (2022) Hallmarks of cancer: new dimensions. Cancer Discov 12(1):31–46

Hanahan D, Weinberg RA (2000) The hallmarks of cancer. Cell 100(1):57–70

Hanahan D, Weinberg RA (2011) Hallmarks of cancer: the next generation. Cell 144(5):646–674

Hartnett L, Egan LJ (2012) Inflammation, DNA methylation and colitis-associated cancer. Carcinogenesis 33(4):723–731

Hughes AL, Kelley JR, Klose RJ (2020) Understanding the interplay between CpG Island-associated gene promoters and H3K4 methylation. Biochim Biophys Acta Gene Regul Mech 1863(8):194567

Ito S, D'Alessio AC, Taranova OV, Hong K, Sowers LC, Zhang Y (2010) Role of Tet proteins in 5mC to 5hmC conversion, ES-cell self-renewal and inner cell mass specification. Nature 466(7310):1129–1133

Ito S, Shen L, Dai Q, Wu SC, Collins LB, Swenberg JA et al (2011) Tet proteins can convert 5-methylcytosine to 5-formylcytosine and 5-carboxylcytosine. Science 333(6047):1300–1303

Kadoch C, Crabtree GR (2015) Mammalian SWI/SNF chromatin remodeling complexes and cancer: mechanistic insights gained from human genomics. Sci Adv 1(5):e1500447

Kaelin WG (2005) The von Hippel-Lindau tumor suppressor protein: roles in cancer and oxygen sensing. Cold Spring Harb Symp Quant Biol 70:159–166

Karch KR, Denizio JE, Black BE, Garcia BA (2013) Identification and interrogation of combinatorial histone modifications. Front Genet 4:264

Kriaucionis S, Heintz N (2009) The nuclear DNA base 5-hydroxymethylcytosine is present in Purkinje neurons and the brain. Science 324(5929):929–930

Krishnachaitanya SS, Liu M, Fujise K, Li Q (2022) MicroRNAs in inflammatory bowel disease and its complications. Int J Mol Sci 23(15)

Kurumizaka H, Kujirai T, Takizawa Y (2021) Contributions of histone variants in nucleosome structure and function. J Mol Biol 433(6):166678

Lander ES (2011) Initial impact of the sequencing of the human genome. Nature 470(7333): 187–197

Lander ES, Linton LM, Birren B, Nusbaum C, Zody MC, Baldwin J et al (2001) Initial sequencing and analysis of the human genome. Nature 409(6822):860–921

Larionova I, Kazakova E, Patysheva M, Kzhyshkowska J (2020) Transcriptional, epigenetic and metabolic programming of tumor-associated macrophages. Cancers (Basel) 12(6)

Lewis KA, Tollefsbol TO (2016) Regulation of the telomerase reverse transcriptase subunit through epigenetic mechanisms. Front Genet 7:83

Li Y (2021) Modern epigenetics methods in biological research. Methods 187:104–113

Li W, Liu M (2011) Distribution of 5-hydroxymethylcytosine in different human tissues. J Nucleic Acids 2011:870726

Li S, Tollefsbol TO (2021) DNA methylation methods: global DNA methylation and methylomic analyses. Methods 187:28–43

Lismer A, Dumeaux V, Lafleur C, Lambrot R, Brind'Amour J, Lorincz MC et al (2021) Histone H3 lysine 4 trimethylation in sperm is transmitted to the embryo and associated with diet-induced phenotypes in the offspring. Dev Cell 56(5):671–686.e676

Liu L, Nie J, Chen L, Dong G, Du X, Wu X et al (2013) The oncogenic role of microRNA-130a/301a/454 in human colorectal cancer via targeting Smad4 expression. PLoS One 8(2):e55532

Liu J, Zhang C, Zhao Y, Feng Z (2017) MicroRNA control of p53. J Cell Biochem 118(1):7–14

Llinàs-Arias P, Esteller M (2017) Epigenetic inactivation of tumour suppressor coding and non-coding genes in human cancer: an update. Open Biol 7(9):170152

Ma L, Cao J, Liu L, Du Q, Li Z, Zou D et al (2019) LncBook: a curated knowledgebase of human long non-coding RNAs. Nucleic Acids Res 47(5):2699

Mäki-Nevala S, Ukwattage S, Wirta EV, Ahtiainen M, Ristimäki A, Seppälä TT et al (2021) Immunoprofiles and DNA methylation of inflammatory marker genes in ulcerative colitis-associated colorectal tumorigenesis. Biomol Ther 11(10):1440

Meissner A, Gnirke A, Bell GW, Ramsahoye B, Lander ES, Jaenisch R (2005) Reduced representation bisulfite sequencing for comparative high-resolution DNA methylation analysis. Nucleic Acids Res 33(18):5868–5877

Millán-Zambrano G, Burton A, Bannister AJ, Schneider R (2022) Histone post-translational modifications - cause and consequence of genome function. Nat Rev Genet 23(9):563–580

Miranda Furtado CL, Dos Santos Luciano MC, Silva Santos RD, Furtado GP, Moraes MO, Pessoa C (2019) Epidrugs: targeting epigenetic marks in cancer treatment. Epigenetics 14(12):1164–1176

Miranda-Gonçalves V, Lameirinhas A, Henrique R, Jerónimo C (2018) Metabolism and epigenetic interplay in cancer: regulation and putative therapeutic targets. Front Genet 9:427

Mizuguchi G, Shen X, Landry J, Wu WH, Sen S, Wu C (2004) ATP-driven exchange of histone H2AZ variant catalyzed by SWR1 chromatin remodeling complex. Science 303(5656):343–348

Mo JS, Park YR, Chae SC (2019) MicroRNA 196B regulates HOXA5, HOXB6 and GLTP expression levels in colorectal cancer cells. Pathol Oncol Res 25(3):953–959

Münzel M, Globisch D, Trindler C, Carell T (2010) Efficient synthesis of 5-hydroxymethylcytosine containing DNA. Org Lett 12(24):5671–5673

Nebbioso A, Tambaro FP, Dell'Aversana C, Altucci L (2018) Cancer epigenetics: moving forward. PLoS Genet 14(6):e1007362

Nieto MA, Huang RY, Jackson RA, Thiery JP (2016) EMT: 2016. Cell 166(1):21–45

Noguera-Castells A, García-Prieto CA, Álvarez-Errico D, Esteller M (2023) Validation of the new EPIC DNA methylation microarray (900K EPIC v2) for high-throughput profiling of the human DNA methylome. Epigenetics 18(1):2185742. https://doi.org/10.1080/15592294.2023.2185742

Nozawa RS, Nagao K, Igami KT, Shibata S, Shirai N, Nozaki N et al (2013) Human inactive X chromosome is compacted through a PRC2-independent SMCHD1-HBiX1 pathway. Nat Struct Mol Biol 20(5):566–573

Okano M, Bell DW, Haber DA, Li E (1999) DNA methyltransferases Dnmt3a and Dnmt3b are essential for de novo methylation and mammalian development. Cell 99(3):247–257

Ordóñez-Morán P, Dafflon C, Imajo M, Nishida E, Huelsken J (2015) HOXA5 counteracts stem cell traits by inhibiting Wnt signaling in colorectal cancer. Cancer Cell 28(6):815–829

Otlu BD-GM, Vermes I, Bergstrom EN, Barnes M, Alexandrov LB (2023) Topography of mutational signatures in human cancer. Cell Reports 42(8):112930. https://doi.org/10.1016/j.celrep.2023.112930

Pan Y, Yu Y, Wang X, Zhang T (2020) Tumor-associated macrophages in tumor immunity. Front Immunol 11:583084

Pastor WA, Pape UJ, Huang Y, Henderson HR, Lister R, Ko M et al (2011) Genome-wide mapping of 5-hydroxymethylcytosine in embryonic stem cells. Nature 473(7347):394–397

Penn NW, Suwalski R, O'Riley C, Bojanowski K, Yura R (1972) The presence of 5-hydroxymethylcytosine in animal deoxyribonucleic acid. Biochem J 126(4):781–790

Pessia E, Makino T, Bailly-Bechet M, McLysaght A, Marais GA (2012) Mammalian X chromosome inactivation evolved as a dosage-compensation mechanism for dosage-sensitive genes on the X chromosome. Proc Natl Acad Sci U S A 109(14):5346–5351

Pidsley R, Zotenko E, Peters TJ, Lawrence MG, Risbridger GP, Molloy P et al (2016) Critical evaluation of the illumina methylation EPIC BeadChip microarray for whole-genome DNA methylation profiling. Genome Biol 17(1):208

Puvvula PK (2019) LncRNAs regulatory networks in cellular senescence. Int J Mol Sci 20(11):2615

Rasmussen KD, Helin K (2016) Role of TET enzymes in DNA methylation, development, and cancer. Genes Dev 30(7):733–750

Rodrigues J, Heinrich MA, Teixeira LM, Prakash J (2021) 3D in vitro model (R)evolution: unveiling tumor-stroma interactions. Trends Cancer 7(3):249–264

Rošić S, Amouroux R, Requena CE, Gomes A, Emperle M, Beltran T et al (2018) Evolutionary analysis indicates that DNA alkylation damage is a byproduct of cytosine DNA methyltransferase activity. Nat Genet 50(3):452–459

Ruan K, Fang X, Ouyang G (2009) MicroRNAs: novel regulators in the hallmarks of human cancer. Cancer Lett 285(2):116–126

Sahnane N, Magnoli F, Bernasconi B, Tibiletti MG, Romualdi C, Pedroni M et al (2015) Aberrant DNA methylation profiles of inherited and sporadic colorectal cancer. Clin Epigenetics 7:131

Saldaña-Meyer R, Recillas-Targa F (2011) Transcriptional and epigenetic regulation of the p53 tumor suppressor gene. Epigenetics 6(9):1068–1077

Simpson JT, Workman RE, Zuzarte PC, David M, Dursi LJ, Timp W (2017) Detecting DNA cytosine methylation using nanopore sequencing. Nat Methods 14(4):407–410

Skipper M, Eccleston A, Gray N, Heemels T, Le Bot N, Marte B et al (2015) Presenting the epigenome roadmap. Nature 518(7539):313

Skvortsova K, Stirzaker C, Taberlay P (2019) The DNA methylation landscape in cancer. Essays Biochem 63(6):797–811

Song CX, Szulwach KE, Fu Y, Dai Q, Yi C, Li X et al (2011) Selective chemical labeling reveals the genome-wide distribution of 5-hydroxymethylcytosine. Nat Biotechnol 29(1):68–72

Stewart SK, Morris TJ, Guilhamon P, Bulstrode H, Bachman M, Balasubramanian S et al (2015) oxBS-450K: a method for analysing hydroxymethylation using 450K BeadChips. Methods 72:9–15

Subramaniam S, Jeet V, Clements JA, Gunter JH, Batra J (2019) Emergence of MicroRNAs as key players in cancer cell metabolism. Clin Chem 65(9):1090–1101

Sun YM, Chen YQ (2020) Principles and innovative technologies for decrypting noncoding RNAs: from discovery and functional prediction to clinical application. J Hematol Oncol 13(1):109

Sun L, Fang J (2016) Epigenetic regulation of epithelial-mesenchymal transition. Cell Mol Life Sci 73(23):4493–4515

Tahiliani M, Koh KP, Shen Y, Pastor WA, Bandukwala H, Brudno Y et al (2009) Conversion of 5-methylcytosine to 5-hydroxymethylcytosine in mammalian DNA by MLL partner TET1. Science 324(5929):930–935

Talbert PB, Henikoff S (2021) The Yin and Yang of histone marks in transcription. Annu Rev Genomics Hum Genet 22:147–170

Tan L, Xiong L, Xu W, Wu F, Huang N, Xu Y et al (2013) Genome-wide comparison of DNA hydroxymethylation in mouse embryonic stem cells and neural progenitor cells by a new comparative hMeDIP-seq method. Nucleic Acids Res 41(7):e84

Terry S, Savagner P, Ortiz-Cuaran S, Mahjoubi L, Saintigny P, Thiery JP et al (2017) New insights into the role of EMT in tumor immune escape. Mol Oncol 11(7):824–846

Thienpont B, Steinbacher J, Zhao H, D'Anna F, Kuchnio A, Ploumakis A et al (2016) Tumour hypoxia causes DNA hypermethylation by reducing TET activity. Nature 537(7618):63–68

Urich MA, Nery JR, Lister R, Schmitz RJ, Ecker JR (2015) MethylC-seq library preparation for base-resolution whole-genome bisulfite sequencing. Nat Protoc 10(3):475–483

Waddington CH (2012) The epigenotype. 1942. Int J Epidemiol 41(1):10–13

Wagner I, Capesius I (1981) Determination of 5-methylcytosine from plant DNA by high-performance liquid chromatography. Biochim Biophys Acta 654(1):52–56

Warburg O, Wind F, Negelein E (1927) The metabolism of tumors in the body. J Gen Physiol 8(6):519–530

Wei JW, Huang K, Yang C, Kang CS (2017) Non-coding RNAs as regulators in epigenetics (review). Oncol Rep 37(1):3–9

Wu Q, Ni X (2015) ROS-mediated DNA methylation pattern alterations in carcinogenesis. Curr Drug Targets 16(1):13–19

Wu X, Zhang Y (2017) TET-mediated active DNA demethylation: mechanism, function and beyond. Nat Rev Genet 18(9):517–534

Xu T, Gao H (2020) Hydroxymethylation and tumors: can 5-hydroxymethylation be used as a marker for tumor diagnosis and treatment? Hum Genomics 14(1):15

Yang T, Yang Y, Wang Y (2021) Predictive biomarkers and potential drug combinations of epi-drugs in cancer therapy. Clin Epigenetics 13(1):113

Yates J, Boeva V (2022) Deciphering the etiology and role in oncogenic transformation of the CpG Island methylator phenotype: a pan-cancer analysis. Brief Bioinform 23(2):bbab610

Yin Y, Morgunova E, Jolma A, Kaasinen E, Sahu B, Khund-Sayeed S et al (2017) Impact of cytosine methylation on DNA binding specificities of human transcription factors. Science 356(6337):eaaj2239

Yu M, Hon GC, Szulwach KE, Song CX, Jin P, Ren B et al (2012) Tet-assisted bisulfite sequencing of 5-hydroxymethylcytosine. Nat Protoc 7(12):2159–2170

Yuan S, Norgard RJ, Stanger BZ (2019) Cellular plasticity in cancer. Cancer Discov 9(7):837–851

Zeng Y, Chen T (2019) DNA methylation reprogramming during mammalian development. Genes (Basel) 10(4):257

Zhang Y, Jurkowska R, Soeroes S, Rajavelu A, Dhayalan A, Bock I et al (2010) Chromatin methylation activity of Dnmt3a and Dnmt3a/3L is guided by interaction of the ADD domain with the histone H3 tail. Nucleic Acids Res 38(13):4246–4253

Zhao Y, Wang C, Goel A (2021) Role of gut microbiota in epigenetic regulation of colorectal cancer. Biochim Biophys Acta Rev Cancer 1875(1):188490

Zhu X, Asa SL, Ezzat S (2009) Histone-acetylated control of fibroblast growth factor receptor 2 intron 2 polymorphisms and isoform splicing in breast cancer. Mol Endocrinol 23(9):1397–1405

Chapter 2
Epigenetic Enzymes and Their Mutations in Cancer

Aysegul Dalmizrak and Ozlem Dalmizrak

Abstract Epigenetic mechanisms are crucial for normal development and maintenance of tissue-specific gene expression patterns in mammals. Impaired epigenetic processes can cause alterations in gene function and malignant cellular transformation. It is now known that epigenetic abnormalities, together with genetic changes, have a role in the onset and progression of cancer, which was once thought to be a genetic disease. Recent developments in the field of cancer epigenetics have demonstrated substantial reprogramming of all elements of the epigenetic machinery in cancer, including DNA methylation, histone modifications, nucleosome positioning, and non-coding RNAs. DNA methyltransferases, histone acetyltransferases, and histone deacetylases are a few examples of epigenetic regulatory enzymes that are involved in epigenetic modification. In recent years, an increasing number of studies have demonstrated that mutations in epigenetic regulatory enzymes occur in various cancer types and are closely associated with the malignant phenotype. Hence, research on inhibitors that target these mutant enzymes has gradually shifted into preclinical and clinical stages. In this chapter, we first discuss the epigenetic regulatory enzymes and then how their mutations are associated with carcinogenesis.

Keywords DNA methylation · Histone modifications · DNA methyltransferases · Histone acetyltransferases · Histone deacetylases · Mutations of epigenetic enzymes

Abbreviations

5′-C	5′-Residue of cytosine
5caC	5-Carboxylcytosine

A. Dalmizrak
Department of Medical Biology, Faculty of Medicine, Balıkesir University, Balıkesir, Turkey

O. Dalmizrak (✉)
Department of Medical Biochemistry, Faculty of Medicine, Near East University, Nicosia, Turkey
e-mail: ozlem.dalmizrak@neu.edu.tr

5fC	5-Formylcytosine
5hmC	5-Hydroxymethylcytosine
5hmU	5-Hydroxymethyluracil
α-KG	α-Ketoglutarate
ADD	ATRX-DNMT3-DNMT3L domain
ALL	Acute lymphoblastic leukemia
AML	Acute myeloid leukemia
ATC	Anaplastic thyroid carcinoma
ATM	Ataxia telangiectasia mutant
BAH1 and BAH2	Bromo-adjacent homology 1 and 2
BER	Base excision repair
BRCA1	Breast cancer gene-1
BRD	Bromodomain
CARM1	Cofactor-associated arginine methyltransferase
CBP	CREB-binding protein
CDC6	Cell division cycle-6
CDK9	Cyclin-dependent kinase-9
CHIP	Clonal hematopoiesis of indeterminate potential
CLL	Chronic lymphocytic leukemia
CN-AML	Cytogenetically normal acute myeloid leukemia
CR	Complete response
CRC	Colorectal cancer
CREB	cAMP-response element binding protein
DFS	Disease-free survival
DLBCL	Diffuse large B-cell lymphoma
DMAP1	DNA methyltransferase associated protein 1
DNMT1	DNA methyltransferase 1
DNMT3	DNA methyltransferase 3
DNMTs	DNA methyltransferase enzymes
DSBH	Double-stranded helix
EFS	Event-free survival
EGFR	Epidermal growth factor receptor
EMT	Epithelial–mesenchymal transition
ES	Embryonic stem
ESCC	Esophageal squamous cell carcinoma
ESCs	Embryonic stem cells
EZH	Enhancer of Zeste Homolog
GBM	Glioblastoma multiforme
GCB	Germinal center B-cell-like
GNAT	GCN5-related N-acetyltransferase
HATs	Histone acetyltransferases
HBO1	Lysine acetyltransferase 7 or Histone acetyltransferase bound to origin recognition complex 1
HDAC	Histone deacetylase

HDLP	Histone deacetylase-like protein
HDMT	Histone demethylase
HMTs	Histone methyltransferases
HPV	Human papilloma virüs
HSCT	Hematopoietic stem cell transplant
ICI	Immune checkpoint inhibitor
IDH	Isocitrate dehydrogenase
ITD	Internal tandem duplication
JmJC	Jumonji C
KAT	Lysine acetyltransferase
KDM1	Lysine demethylase 1
KMT	Lysine methyltransferases
KMT2C	Histone lysine methyltransferase 2C
LFS	Leukemia-free survival
LSD1/KDM1A	Lysine-specific demethylase 1A
MDS	Myelodysplastic syndrome
MIBC	Muscle-invasive bladder cancer
MLL	Mixed lineage leukemia
MORF	MOZ-related factor
MOZ	Monocytic leukemic zinc finger protein
ncRNA	Non-coding RNA
NLS	A nuclear localization sequence
ORC	Origin recognition complex
OS	Overall survival
PADI4	Peptidyl arginine deiminase 4
PBD	PCNA binding domain
PBHD	Polybromo homology domain
PCFBCL	Primary cutaneous follicular B-cell lymphoma
PCNA	Proliferating cell nuclear antigen
PD-L1	Programmed death-ligand 1
PDTC	Poorly differentiated thyroid carcinoma
PFS	Progression-free survival
PHD	Plant homeodomain
PRMT	Arginine methyltransferases
QM/MM MD	Quantum mechanics/molecular mechanics molecular dynamics
RFT	Replication foci targeting domain
SAM	S-adenosyl-l-methionine
T-ALL	T-cell acute lymphoblastic leukemia
T-dCyd	4′-Thio-2′ deoxycytidine
TAF	TATA box-binding protein (TBP)-associated factors
TBP	TATA box-binding protein (TBP)
TDG	Thymine DNA glycosylase
TEs	Transposable elements
TET	Ten eleven translocation

TMB	Tumor mutation burden
TNBC	Triple-negative breast cancer
TRD	Target recognition domain
TS	Targeting sequence
VAF	Variant allele frequency
VEGF	Vascular endothelial growth factor

2.1 Introduction

The concept of epigenetics was first described by Conrad Hal Waddington in 1942 as "the field of biology which examines the causal interaction between genes and their products, which give rise to the phenotype" (Waddington 1942). Waddington's concept originally focused on the function of epigenetics in embryonic development; however, as epigenetics has come to be associated with a wide range of biological processes, the definition has changed through time. Currently, the word "epigenetics" refers to heritable changes in gene expression that occur during mitosis and/or meiosis without altering the DNA sequence. The majority of these genetic alterations are generated during differentiation and are stable enough to last through successive cycles of cell division, allowing cells to have diverse identities while still sharing the same genetic material. This heritability of gene expression patterns is mediated by epigenetic modifications including DNA methylation, a wide range of covalent histone modifications, nucleosome localization, non-coding RNA (ncRNA) expression, and chromatin 3D structure (Allis and Jenuwein 2016). The epigenome which is the sum of these alterations provides a mechanism for cellular diversity by controlling which genetic information can be accessible by cellular machinery. Epigenetic regulatory enzymes are important for regulating chromatin structure and gene expression and studies have shown that the dysregulation caused by changes in the amino acid sequence of these enzymes is closely correlated with tumor onset and progression. Failure of the proper maintenance of heritable epigenetic marks can result in inappropriate activation or inhibition of various signaling pathways and lead to diseases such as cancer (Jones and Baylin 2002; Egger et al. 2004). In particular, abnormal expression of cancer-related genes, tumor suppressor genes, or oncogenes through dysregulated epigenetic regulatory enzymes can cause carcinogenesis by altering basic processes including DNA repair, cell proliferation, and mortality (Miranda Furtado et al. 2019; Park and Han 2019). Throughout this chapter, we will summarize the current knowledge about the epigenetic enzymes and their mutations implicated in cancer.

2.1.1 DNA Methylation

DNA methylation is a crucial epigenetic sign that all living organisms have utilized to survive in various environmental conditions. Prokaryotes, for instance, employ it to distinguish their own DNA from foreign DNA and prevent endoreduplication (Oliveira and Fang 2021). In eukaryotes, DNA methylation is employed to silence DNA fragments and entire chromosomes, control cell differentiation, and prevent DNA cell segregation errors (Ponger and Li 2005). In mammals, DNA methylation occurs almost exclusively at cytosine residues at C5 positions in CpG sequences. These dinucleotides are distributed unevenly throughout the genome, and most are extensively methylated. However, only a portion of CG sites is methylated, leading to a pattern of methylation that is unique to different tissues and cell types. The human genome has 56 million CG sites, of which 60–80% are methylated and account for 4–6% of all cytosines (Jurkowska et al. 2011). Methylation levels and patterns vary depending on the type of cell, with embryonic stem (ES) cells displaying the greatest variations (Lister et al. 2009). Furthermore, 5-hydroxymethylcytosine (5hmC), which arises from the oxidation of the methyl group of 5-methylcytosine, has been found in mammalian DNA (Kriaucionis and Heintz 2009).

DNA methylation is achieved by DNA methyltransferase enzymes (DNMTs) by transferring the methyl group of S-adenosyl-L-methionine (SAM) to the 5′-residue of cytosine (5′-C) in DNA. The DNMT family includes DNMT1, DNMT2, DNMT3A, DNMT3B, and DNMT3L, which differ based on their structural characteristics and functional domains (Del Castillo Falconi et al. 2022). The addition of methylation marks to genomic DNA is catalyzed by the canonical cytosine-5 DNMTs, DNMT1, DNMT3A, and DNMT3B. As they lack catalytic DNMT activity, DNMT2 and DNMT3L are non-canonical family members (Lyko 2018). Mammalian DNMTs contain two regions: a large multidomain N-terminal regulatory region and a catalytic C-terminal region. The N-terminal region of the enzymes directs their nuclear localization and facilitates their interactions with other proteins, DNA, and chromatin. The active center of the enzyme is located in the smaller C-terminal region, which is conserved among both eukaryotic and prokaryotic DNMTs. This region also contains ten amino acids that are unique to all C5 cytosine DNMTs (Cheng 1995). All DNMTs have a common core structure in their catalytic domains known as the "AdoMet-dependent MTase fold," which is composed of a mixed seven-stranded β-sheet made up of six parallel β strands and a seventh strand inserted between strands 5 and 6 in an antiparallel orientation. The central β sheet is encircled by six helices (Cheng and Blumenthal 2008). This domain participates in catalysis (motifs IV, VI, and VIII) as well as cofactor binding (motifs I and X). DNA recognition and specificity are mediated by the so-called target recognition domain (TRD), a non-conserved region located between motifs VIII and IX (Cheng 1995; Jeltsch 2002) (Fig. 2.1). DNMTs have a common mechanistic property in addition to their conserved structures. All of them remove their target base from the DNA helix and bury it within a hydrophobic cavity of the active center. The catalytic cysteine

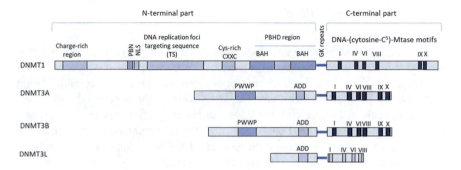

Fig. 2.1 Functional domains of mammalian DNA methyltransferases

residue in a PCQ motif (motif IV) is involved in the nucleophilic attack of the enzyme on the cytosine in the sixth position which results in the formation of a covalent bond between the enzyme and the substrate base. This reaction causes an increase in the negative charge density at the C5 atom of the cytosine, which attacks the methyl group bound to SAM. A transient protonation of the cytosine ring at the endocyclic nitrogen atom (N3) by an acid derived from an enzyme has been proposed as a possible mechanism for how the nucleophilic attack of the cysteine may be facilitated. It has been suggested that the conserved glutamate residue from the ENV motif is responsible for carrying out this reaction (motif V). Moreover, this residue stabilizes the flipped base by making contact with the exocyclic N4 amino group. The arrangement of the glutamate and the flipped cytosine base may also be influenced by the arginine residue from the RXR motif (motif VIII). The covalent link between the enzyme and DNA is broken as a result of the addition of the methyl group to the cytosine base and the subsequent deprotonation at C5 (Jeltsch 2002).

The most prominent and earliest recognized change in DNA methylation patterns in cancer cells is DNA hypomethylation. Albeit all of the effects of these losses remain to be fully understood, DNA demethylation may be a factor in genomic instability and an increase in aneuploidy which are the hallmarks of cancer (Ehrlich and Lacey 2013). Indeed, increased mutation rates, aneuploidies, and tumor induction caused by the deletion or decrease of the Dnmt1 is conclusive proof that DNA hypomethylation actively contributes to the increase in chromosomal fragility (Gaudet et al. 2003). Loss of DNA methylation may be accompanied by the activation of transcription, allowing the transcription of repeats, transposable elements (TEs), and oncogenes (Ehrlich and Lacey 2013; Hur et al. 2014).

A well-documented DNA methylation alteration in cancer is abnormal hypermethylation of CpG islands in the 5′ regions of cancer-related genes. This alteration can be directly linked to transcriptional silencing of genes with tumor suppressor function (Jones and Baylin 2002) (Fig. 2.2). Considering that 60% of all gene promoters contain CpG islands, the majority of which are never methylated during normal development or in adult cell renewal mechanisms. The more open chromatin states and active expression status of these genes are fundamentally

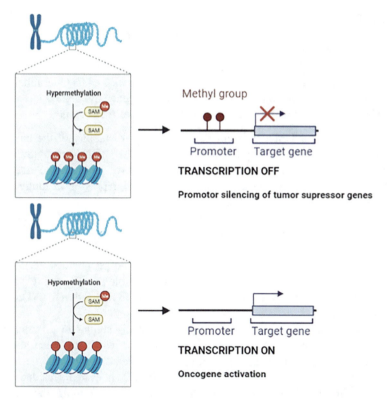

Fig. 2.2 Epigenetic changes related to DNA methylation can contribute to cancer through diverse mechanisms. Hypermethylation at gene promoters can result in the heritable silencing and subsequent inactivation of tumor suppressors and other genes. On the other hand, reduced DNA cytosine methylation can lead to genomic instability and oncogene activation

dependent on this lack of methylation (Baylin and Jones 2011). Because methylated CpG island promoters are so common in malignancies and are known to directly promote carcinogenesis, epigenetic treatment, in which epigenetic modifications are targeted for therapeutic reversal, has significant potential (Azad et al. 2013). It should be emphasized that 5mC frequently appears in the gene body of active genes, and its effects here may frequently be contrary to those of its presence in promoters. Hence, rather than being linked to transcriptional repression, DNA methylation on gene body may promote transcriptional elongation and increase gene expression (Jones 2012).

2.1.1.1 DNA Methyltransferase 1

The first mammalian DNA methyltransferase enzyme to be identified biochemically and cloned was DNA methyltransferase 1 (DNMT1) (Bestor et al. 1988). In somatic tissues, DNMT1 is ubiquitously expressed in dividing cells, constituting the majority of DNMT activity. It is only moderately expressed in non-dividing cells (Robertson et al. 1999). The cell cycle affects the variation in DNMT1 mRNA expression, which rises in the S phase (Lee et al. 1996).

The human DNMT1 protein consists of 1616 amino acids, with an N-terminal regulatory region that makes up two-thirds of the sequence and a C-terminal region that is separated by a highly conserved (GK)n repeat (Lyko 2018). There are multiple functional domains in the N-terminal region of the enzyme:

(a) A charge-rich or DMAP1 interaction domain: The stability of the enzyme is thought to be influenced by a charge-rich or DMAP1 (DNA methyltransferase associated protein 1) interaction domain that participates in the interaction between DNMT1 and DMAP1, a transcriptional repressor (Rountree et al. 2000; Ding and Chaillet 2002).
(b) A PBD (PCNA (proliferating cell nuclear antigen) binding domain) mediates the interaction of DNMT1 with PCNA which also plays a role in directing DNMT1 to replication foci (Chuang et al. 1997).
(c) A nuclear localization sequence (NLS) (Cardoso and Leonhardt 1999).
(d) A TS (targeting sequence) domain that is responsible for targeting DNMT1 to centromeric chromatin (Easwaran et al. 2004) and to replication foci (Leonhardt et al. 1992).
(e) A zinc domain, also known as CXXC domain, is similar to a cysteine-rich domain found in other chromatin-associated proteins. It has been shown to bind unmethylated CGs in vitro and is closely related to the catalytic function of DNMT1 (Lee et al. 2001; Pradhan et al. 2008). Eight cysteine residues form a cluster in the CXXC domain of human DNMT1, forming the sequence $651CX_2CX_2CX_4CX_2CX_2CX_{15}CX_4C697$. The area has been proven to bind radioactive zinc in previous research, and it is situated between amino acids 580 and 697 (Bestor 1992).
(f) The BAH1 and BAH2 (Bromo-adjacent homology 1 and 2) domains are part of the so-called PBHD domain (polybromo homology domain). Nearly, all DNMT1 homologs share two BAH domains with an undefined function. About 20 mammalian proteins with different functions have BAH domains. According to reports, some BAH domains attach to histone tails in a modification-dependent manner (Yang and Xu 2013). BAH domains can be divided into two categories: ORC1-like and SIR3-like. In DNMT1, BAH1 belongs to the ORC1-like group, while BAH2 to the SIR3-like group. The N-terminal regulatory region and the C-terminal catalytic region of BAH2 are joined by a sequence of alternating lysine and glycine residues known as the GK repeats. Studies revealed that the interaction of DNMT1 with replication foci during S phase is mediated by the BAH domains (Yarychkivska et al. 2018).
(g) The GK repeats are located between N- and C-terminal parts of the enzyme (Jurkowska et al. 2011).
(h) The C-terminal domain has the catalytic center of the enzyme (Jurkowska et al. 2011).

Subnuclear localization of DNMT1 fluctuates dynamically throughout the cell cycle; during interphase, when cells are not replicating, it is distributed in the nucleus. However, it localizes to the replication foci in cells actively synthesizing DNA during the early and middle stages of the S phase, resulting in a distinctive punctate pattern (Leonhardt et al. 1992). The PBD domain (Chuang et al. 1997), the RFT domain (replication foci targeting domain, which is part of the TS domain) (Liu et al. 1998), and the PBHD domain of DNMT1 have all been linked to the targeting of the enzyme to replication foci during the S phase (Liu et al. 1998). However, the delivery of DNMT1 to the replication fork was unaffected by the deletion of RFT or PBHD, indicating that the PDB domain is crucial to this process (Easwaran et al. 2004).

DNMT1 exhibits a preference for hemimethylated DNA over unmethylated substrate, supporting its function as a maintenance MTs (Fatemi et al. 2001). After DNA replication, DNMT1 is responsible for re-establishing DNA methylation. The enzyme resides at the replication fork, where it functions as a molecular copier. It rapidly methylates the hemimethylated CG dinucleotides to return the methylation pattern to normal. DNMT1 has a high processing capacity and is able to methylate long DNA regions without dissociation (Goyal et al. 2006). The fact that processive methylation can only occur in one strand of DNA is intriguing and suggests that DNMT1 does not switch the target strand as it moves along the substrate. These characteristics demonstrate that DNMT1 methylates the CG sites on one strand of DNA while maintaining its orientation with regard to the DNA (Hermann et al. 2004).

DNMT1 activity is essential for de novo DNA methylation in addition to its important function as a maintenance DNMT (Jair et al. 2006). As it occurs in vitro, DNMT1 may help DNMT3A and DNMT3B by utilizing the hemimethylated CG sites produced by the DNMT3 enzymes (Fatemi et al. 2002).

Three independent DNA-binding sites (NLS-containing domain, Zn-binding domain, and catalytic domain) have been shown to exist in DNMT1. An enzyme with a strong preference for the methylation of hemimethylated target sites is necessary for the accurate transfer of the methylation pattern. This specificity in DNMT1 is produced by a combination of an intrinsic preference of the catalytic domain for hemimethylated substrates and an allosteric activation of the enzyme that takes place if methylated DNA binds to the Zn-binding domain in the N-terminal region of the protein. It is noteworthy how this allosteric activation mechanism causes DNMT1 to behave in an all-or-none manner, which means that only unmethylated DNA remains unmethylated, whereas partially modified DNA tends to become fully methylated (Fatemi et al. 2001). Moreover, unmethylated substrates have been shown in various studies to have an inhibitory impact, indicating that binding of unmethylated DNA to the N-terminal region of DNMT1 results in the inhibition of the enzyme activity on hemimethylated DNA (Zhang et al. 2015).

2.1.1.2 DNA Methyltransferase 3

The mammalian DNA methyltransferase 3 (DNMT3) family has three members: DNMT3A, DNMT3B, and DNMT3L. In germ cells and at an early stage of mammalian development, the active DNMTs, DNMT3A, and DNMT3B, establish DNA methylation patterns. In germ cells, DNMT3L serves as a regulatory factor even though it is catalytically inactive. Due to the fact that DNMT3A and DNMT3B do not show any preference between hemimethylated and unmethylated DNA, they have been referred to as de novo DNMTs (Okano et al. 1998; Gowher and Jeltsch 2001). Moreover, they contribute to the preservation of DNA methylation in heterochromatic areas (Liang et al. 2002). The enzymatic activities of these two de novo DNMTs are allosterically stimulated by a catalytically inactive family member, i.e., DNMT3L (Chedin et al. 2002).

The mammalian DNMT3A and DNMT3B exhibit distinctive physiological and pathological roles as well as different enzymatic properties, while sharing a similar domain organization and high sequence identity (85%) in the methyltransferase catalytic domain (Okano et al. 1999; Gowher and Jeltsch 2002; Suetake et al. 2003). These two proteins methylate distinct as well as overlapping targets at various developmental stages. DNMT3A is necessary for the methylation of imprinted genes and distributed repeated elements, while centromeric minor satellite repeats and actively transcribed genes within the gene body are methylated by DNMT3B to prevent erroneous transcription initiation (Okano et al. 1999; Li and Zhang 2014).

Embryonic tissues and undifferentiated ES cells highly express DNMT3A and DNMT3B, whereas they are downregulated in differentiated cells. In contrast to its shorter isoform, DNMT3A2, whose expression is tightly controlled, DNMT3A is ubiquitously expressed and found in the majority of organs. DNMT3A2 predominates in embryonic stem cells, germ cells, and embryonal carcinoma cells, as well as in the spleen and the thymus; however, it is silenced in adult tissues (Chen et al. 2002). Similar to this, DNMT3B isoforms have diverse expression profiles and localization patterns during development, raising the potential that they may facilitate the methylation of various sets of genomic sequences. Intriguingly, ES cells exclusively express variants that contain the conserved exons 10 and 11, but murine DNMT3 proteins identified in somatic lines are characterized by the absence of these exons. The existence of exons 10 and 11 in ES cells raises the possibility that these areas are crucial for proper embryonic development or may only occur in undifferentiated cells (Weisenberger et al. 2004). DNMT3L is expressed particularly in germ cells throughout the gametogenesis and embryonic stages (Hata et al. 2002).

Both DNMT3A and DNMT3B permanently interact with chromatin harboring methylated DNA (Jeong et al. 2009), including mitotic chromosomes, and localize to pericentromeric heterochromatin (Bachman et al. 2001; Chen et al. 2004). The PWWP domain, which is found in the N-terminal regions of DNMT3A and DNMT3B, is necessary for the enzymes to target chromatin (Chen et al. 2004). Interaction of DNMT3L with DNMT3A or DNMT3B determines its nuclear and subnuclear localization. DNMT3L is distributed throughout the cytoplasm and nucleus in the absence of DNMT3A and DNMT3B; it was found that only after binding to DNMT3A, DNMT3L was localized in chromatin foci (Nimura et al. 2006).

The DNMT3 enzymes are similar to DNMT1 in that they include an N-terminal regulatory domain and a C-terminal catalytic domain that contains the conserved C5 DNMTs motifs. The catalytic domains of DNMT1 and DNMT3A/3B are similar, although their N-terminal domains are different. N-terminal domain of DNMT3A and DNMT3B has two distinct domains: a cysteine-rich area known as the ADD (ATRX-DNMT3-DNMT3L) domain, also termed as PHD (plant homeodomain) domain, and a PWWP domain. DNMT3L lacks the PWWP domain, as well as the MTs motifs IX and X, and all significant catalytic residues in its C-terminal domain (Chen and Chan 2014). The ADD domain binds zinc ions and is a site for different protein–protein interactions. It has been demonstrated that it mediates the interaction of numerous proteins with DNMT3A (Fuks et al. 2001; Brenner et al. 2005). It has been discovered that the ADD domains of DNMT3A, DNMT3B, and DNMT3L particularly contact with the N-terminal region of histone H3 tails that are not methylated at lysine 4, methylation of H3 at K4 destroyed the interaction (Ooi et al. 2007; Otani et al. 2009). Moreover, it has been demonstrated that the interaction of the ADD domain with the H3 histone that is unmethylated at K4 promotes DNMT3A to methylate chromatin-linked DNA in vitro (Zhang et al. 2010). These findings show that the ADD domain of DNMTs can direct DNA methylation in response to particular histone modifications and give proof that DNMTs might be directed to chromatin that carries particular marks. The PWWP domain of DNMT3A and DNMT3B has 100–150 amino acids. This poorly conserved region contains a proline-tryptophan motif and is essential for the targeting of the MTs to pericentromeric chromatin (Chen et al. 2004).

The C-terminal domains of DNMT3A and DNMT3L form long heterotetrameric complexes consisting of two DNMT3A (in the center) and two DNMT3L (on the edges) molecules. The DNMT3A C-terminal domain presents two sites for protein–protein interactions: one polar RD interface (characterized by a hydrogen bonding network between arginine and aspartate residues) and one hydrophobic FF interface (characterized by the stacking interaction of two phenylalanine residues). DNMT3L also has an FF interface and it makes DNMT3A/3L contact possible. In contrast, DNMT3L lacks RD interface. The α helices C, D, and E of DNMT3A are likely to be affected by the interaction of DNMT3A with DNMT3L through the FF interface. These helices' residues directly engage with the critical catalytic or SAM-binding residues, which may help to explain how DNMT3L stimulates DNMT3A to bind SAM and make the catalysis. One turn of the DNA helix separates the active sites of the two central DNMT3A subunits (approximately 10 bps) which implies that two CG sites on opposing strands might be methylated by DNMT3A in a single binding event. In fact, in vitro methylation studies showed a correlation of methylation between two sites located 10 bps apart in both the same strand and the opposite strand (Jurkowska et al. 2008).

2.1.1.3 Mutations of DNA Methyltransferases

Among the DNMT mutations that play a role in cancer development, the most prominent mutations are those found in the *DNMT1* and *DNMT3A* genes.

Mutations in *DNMT1* have been identified in a variety of human cancers. These mutations have been shown to alter the enzymatic activity of DNMT1, leading to abnormal DNA methylation patterns in cancer cells. In a study based on the clinical and genetic data of colorectal cancer (CRC) patients receiving immune checkpoint inhibitors (ICIs) therapy, it was determined that the overall survival (OS) was longer and better in male patients over 65 years of age with a mutation count more than 11 (tumor mutation burden (TMB)-high). It has been emphasized that DNMT1 may be a protective predictive biomarker (Lin et al. 2020). In the pan-cancer study, which examined both the expression and mutations of DNMT1 in head and neck squamous cell carcinoma, it was found that DNMT1 was overexpressed in male patients over 60 years of age, Caucasians, advanced-level tumors, human papilloma virus (HPV)-positive patients, and was associated with a poor prognosis. In addition, eight somatic mutations (P1330S, P1325S, E912Q, S1352G, P692S, H370Y, T616M, and R325L), all missense mutations, and eight genes (*CLSPN, UHRF1, BRCA1, ATAD5, TIMELESS, CIT, KIF4B,* and *DTL*) have been identified. It was concluded that DNMT1 could be a new diagnostic biomarker and a therapeutic target (Cui et al. 2021a). In another study, *DNMT1* mutation (c.358G > C, p.Val120Leu) was found to be the fifth most common mutation in papillary thyroid cancer patients who underwent thyroidectomy, after mutations in *BRAF, BCR, CREB3L2,* and *IRS2* genes (Qi et al. 2021).

Mutations in *DNMT2* have been reported in some cancers, although they are less common than mutations in *DNMT1*. In order to determine the effect of *DNMT2* somatic mutations on enzyme activity in cancer tissues, *DNMT2* variants were created in a study using COSMIC in spring 2014 data. It was determined that the E63K mutation caused an increase in enzyme activity, while the G155S and L257V mutations caused a decrease. R371H and G155V mutations were also found to have inhibitory effects. It was concluded that these somatic mutations may have a functional effect on tumorigenesis (Elhardt et al. 2015).

Mutations in *DNMT3A* have been identified in a variety of human cancers, including acute myeloid leukemia (AML), myelodysplastic syndrome (MDS), and lymphoid malignancies. The most prominent mutation for *DNMT3A* is R882 missense mutation and it is observed quite frequently. Comprehensive genetic and clinical-biological analyses of T-cell acute lymphoblastic leukemia (T-ALL) patients with *DNMT3A* missense (L373V, P385L, G543C, G543V, M548T, C549R, V563M, R635W, R729W, R866W, R882C, R882H), nonsense (Q249*, W306*), and frameshift (V563GfsX14, P718LfsX61, W795GfsX7) mutations treated during the GRAALL-2003 and -2005 studies showed that mutations are associated with older age, immature T-cell receptor genotype, lower remission rates, worse clinical outcome, higher cumulative incidence of relapse, poorer event-free survival (EFS) and OS. Therefore, it can be concluded that the DNMT3A genotype may be a predictor of aggressive T-ALL biology (Bond et al. 2019). In Chinese AML patients, *DNMT3A* R882 mutation is associated with a worse prognosis. However, the effect is dependent on the DNMT3A R882 mutant allele ratio, and patients with a higher allele ratio have a shorter OS as compared with the lower allele ratio group (Yuan et al. 2019). In cytogenetically normal AML (CN-AML) patients, *DNMT3A* missense (C.2645G > A/R882H, C.2644C > T/R882C, C.2645G > C/R882P), TET2 missense (G1933T/R534I, G1285A/G429R, C817T/Q273X), and frameshift (A3023ins/K1008X) mutations occur separately, also both *DNMT3A* and *TET2* mutations (C.2645G > A& G1933T/R882H& R534I, C.2645G > A& G1285A/ R882H& G429R) coexist. Mutated *TET2* or *DNMT3A* genes were significantly associated with failure of complete remission, higher mortality rate, shorter OS, and disease-free survival (DFS) (Aref et al. 2022). According to in vitro study, AML cells with a *DNMT3A* R882H mutation proliferate at a high rate and do not undergo apoptosis. However, they are less sensitive to daunorubicin and have higher NRF2 expression. The NRF2/NQO1 pathway is active in mutant cells in response to daunorubicin treatment. The *DNMT3A* R882H mutation regulates NRF2 expression by affecting protein stability rather than reducing methylation of the NRF2 promoter. Inhibition of the NRF2/NQO1 pathway significantly increases the daunorubicin sensitivity of mutant cells. Therefore, targeting NFR2 is considered a new therapeutic approach in AML patients with the *DNMT3A* R882H mutation (Chu et al. 2022). Another issue about *DNMT3A* mutations is the coexistence of the mutation with *NPM1* and *FLT3* gene mutations. *DNMT3A* and *NPM1* mutations are the most common mutations in Chinese AML patients. In cytogenetically normal AML patients, *DNMT3A* mutation tends to co-occur with NPM1 and FLT3-internal tandem duplication (FLT3-ITD) mutations (Lit et al. 2022). *FLT3* and *DNMT3A* R882 mutations negatively affect the complete response (CR) rates in Egyptian AML patients. FLT3-ITD mutation is associated with lower OS in advanced age and DNMT3A/FLT3 combined mutant genotypes, while mutant NPM1/wild FLT3, wild DNMT3A/FLT3, and mutant NPM1A/wild DNMT3A combinations are associated with high CR rates (El Gammal et al. 2019). The clinical outcome in patients carrying the mutant form of all three genes is worse than those carrying the mutations individually or in binary combinations. This is associated with the adverse prognostic effect of the *DNMT3A* mutation (Elrhman et al. 2021). On the other hand, cytogenetically normal Syrian AML patients with FLT3-ITD and *NPM1* mutations have the worst prognosis, and the presence of these mutations is significantly associated with OS and EFS. However, DNMT3A as an independent factor does not have an extremely poor prognostic effect (Moualla et al. 2022). In a study in which *DNMT3A* mutation was found to be associated with age, percentage of blasts in peripheral blood and FLT3 mutation, it was determined that the affected gene expressions were associated with neutrophil degranulation, myeloid cell differentiation, stem cell proliferation, positive regulation of system process, leukocyte migration, and tissue morphogenesis. Seven key genes (*BMP4, MPO, THBS1, APP, ELANE, HOXA7,* and *VWF*) have been also identified (Chen et al. 2020). The **DNMT3A** R882H variant is most common in AML patients with a normal karyotype. This mutation is followed by mutations in *NPM1, FLT3, TET2,* and isocitrate dehydrogenase (*IDH*) 1 and 2 genes. Patients with a high *DNMT3A* VAF (variant allele frequency, DNMT^{high}) mutation have leukocytosis, a high number of blasts in the bone marrow and blood. However, compared to DNMT3A^{low}, DNMT3A^{high} is associated with much shorter EFS and OS (Narayanan et al. 2021). In AML patients with *DNMT3A* frameshift, missense, nonsense, and splice site mutations, FLT3 and/or NPM1 mutations cause differences in patient survival. In those with a shorter lifespan, either one or both of the genes may be mutated. However, p53, vascular endothelial growth factor (VEGF), and DNA replication pathway genes are upregulated and PI3K-Akt pathway genes are downregulated in this group. Also, in the same group, miRNAs are downregulated (miR-153-2, miR-3065, miR-95, miR-6718), which are thought to be important for AML prognosis but have not been reported so far (Lauber et al. 2020). *DNMT3A* mutations also have effects on different cancer types besides leukemia. In papillary thyroid cancer in the Middle East population, missense (c.2312G > A/p.Arg771Gln, c.2239G > A/p.Asp747Asn, c.2191 T > C/p.Phe731Leu, c.2186G > A/p.Arg729Gln, c.2161A > G/p.Lys721Glu, c.2114 T > C/p.Ile705Thr, c.2063G > T/p.Arg688Leu, c.1984G > T/p.Ala662Ser, c.1976G > A/p.Arg659His, c.892G > A/p.Gly298Arg, c.T1408C/p.Ile470Val, c.C700T/p.Gly234Arg) and frameshift (c.2266G > T p.Glu756Stop) mutations of DNMT3A are associated with aggressive clinical parameters and poor outcome (Siraj et al. 2019). *DNMT3A* mutation is significantly associated with short life expectancy in patients with poorly differentiated and anaplastic thyroid carcinoma (PDTC and ATC). The mutation is being evaluated as a potential predictive biomarker or therapeutic target for the prognosis and treatment of thyroid cancer (Guo et al. 2019). In in vitro and in vivo studies with different cancer cell lines and multiple xenograft models, it was found that after 4′-thio-2′ deoxycytidine (T-dCyd) treatment survival was suppressed in breast, lung, melanoma, and renal cancer cell lines with deleterious *TET2* and non-synonymous *DNMT3A* mutations. In addition, it was determined that p21 was upregulated and cell cycle was stimulated in the lung cancer cell line and tumor growth was suppressed in the xenograft model. In the xenograft model carrying both mutations, a significant increase in p21 and almost destruction of tumor cells were determined. In the lung cancer cell line, the *TET2* c.5162 T > G p.L1721W missense mutation was first detected (Yang et al. 2021).

2.1.1.4 Ten Eleven Translocation Proteins

DNA methylation is a stable and highly conserved epigenetic signature present in many organisms. It has a substantial impact on a variety of biological processes, including genomic imprinting, X-chromosome inactivation, and the suppression of transposons (Feng et al. 2010). However until the role of the three ten eleven translocation (TET)-family 5mC oxidases, TET1, TET2, and TET3, was discovered, it was unclear how methyl groups are lost independently of DNA replication (Tahiliani et al. 2009). TET1 was the first member discovered in patients having a ten eleven chromosomal translocation t(10;11)(q22;q23) as a fusion partner of the mixed lineage leukemia (MLL) gene (Lorsbach et al. 2003). Two other TET genes, *TET2* and *TET3*, were found in the human genome based on sequence homology. It has been established that TET1 oxidizes 5mC to 5-hydroxymethylcytosine (5hmC) (Tahiliani et al. 2009), and further studies have revealed that TET2 and TET3 can also catalyze 5mC oxidation. Each of the three TETs is capable of oxidizing 5hmC further to produce 5-formylcytosine (5fC) and 5-carboxylcytosine (5caC) (Ito et al. 2011). The catalytic function of TETs depends on iron (Fe^{2+}) and α-ketoglutarate (α-KG, also known as 2-oxoglutarate) (Tahiliani et al. 2009). The catalytic domain of TETs contains both a double-stranded helix (DSBH) domain and a cysteine-rich domain. The cysteine-rich domain has the function of stabilizing the interaction between TET and DNA. The CXXC domain identifies and binds to unmethylated CpG sites. TET1 and TET3 both have a CXXC-type zinc finger domain at the N-terminus, whereas TET2 does not. A Fe (II) binding domain is present in the catalytic domain of the TETs (Melamed et al. 2018).

The replication-dependent "passive" DNA demethylation can take place when the parental strand contains 5hmC throughout DNA replication cycles due to the low affinity of DNMT1 for the hemi-5hmC site relative to the hemi-5C site. As a result, repeated DNA replication cycles result in gradual dilution of cytosine methylation (Seiler et al. 2018). The DNA repair enzyme thymine DNA glycosylase (TDG) can also remove 5fC and 5caC through base excision repair (BER), and then replace them with an unmodified cytosine which is known "active" demethylation (He et al. 2011). It has been demonstrated that 5hmC undergoes deamination to form 5-hydroxymethyluracil (5hmU), which is then converted to cytosine via the TDG/BER pathway. Also, it was demonstrated that TDG targets 5fC:G and 5caC: G more effectively than thymidine or 5hmC:G mismatches. It has been discovered that TDG deficiency increased 5fC and 5caC levels up to ten-fold in embryonic stem cells (ESCs) (Shen et al. 2013). Moreover, other DNA damage and BER pathway members like p53, PARP, GADD45, and NEIL1/2 help to maintain methylation homeostasis by blocking DNA hypermethylation (Li et al. 2015; Tovy et al. 2017). TET proteins were reported to oxidize thymine to 5hmU, resulting in mismatched 5hmU:A and triggering the indirect removal of 5mC from the genome by a subsequent long patch BER or non-standard mismatch repair mechanism (Olinski et al. 2016). These findings show that the combined effects of DNMTs and TET enzymes continuously regulate the balance between DNA methylation and demethylation in mammals.

2.1.1.5 Mutations of Ten Eleven Translocation Proteins

TET1 mutations have been identified in a variety of hematological malignancies and solid tumors. These mutations are thought to contribute to cancer development by altering DNA methylation patterns and gene expressions. *TET1, TET3,* and *ASXL2* loss-of-function mutations are rarely seen in patients with MDS/MPN overlap syndrome. In chronic myelomonocytic leukemia, *TET1* and *TET3* mutations coexist independently of *TET2* mutation (Lasho et al. 2018). On the other hand, in a clinical cohort study investigating the *TET1* mutation in patients with different cancer types responding to ICI treatment, it was found that the mutation occurred more frequently in skin, lung, gastrointestinal, and urogenital cancers. In addition, it was concluded that the *TET1* mutation is associated with a higher objective response rate, better durable clinical benefit, longer progression-free survival (PFS), and improved OS in patients receiving ICI therapy. Therefore, it is thought that the mutation may serve as a new predictive biomarker for immune checkpoint blockade in multiple cancer types (Wu et al. 2019). In colon adenocarcinoma, the OS of *TET1* mutant patients receiving ICI treatment is significantly longer than those without having the mutations. Compared with wild-type patients, patients with TET1 mutations have higher TMB and neoantigen load, abundance of tumor infiltrating immune cells, increased expression of immune-related genes, and mutation number of DNA damage repair pathways. In addition, patients with *TET1* mutations are more sensitive to lapatinib and 5-fluorouracil (Qiu et al. 2022). In glioblastoma multiforme (GBM), TET1 deletion rate is higher in IDH wild-type patients than in *IDH* mutant patients. Biallelic *TET1* deletions often occur with epidermal growth factor receptor (*EGFR*) amplification and are associated with low levels of TET1 mRNA expression, indicating loss of TET1 activity. Focal amplification of *EGFR* correlates positively with overall mutational burden, tumor size, and poor long-term survival. Although biallelic TET1 deletions are not an independent prognostic factor, they are associated with poor outcomes in IDH-wt GBM with concomitant *EGFR* amplification (Stasik et al. 2020).

TET2 mutations have been associated with a poorer prognosis in patients with MDS, a group of blood disorders that can progress to AML. *TET2* mutations have also been found in other blood cancers, such as chronic lymphocytic leukemia (CLL) and acute lymphoblastic leukemia (ALL). *TET2* is frequently mutated in Chinese cytogenetically normal AML patients with elderly age (Wang et al. 2018). In MDS, the presence of TP53 mutations but the absence of *TET2, DNMT3A*, or *ASXL1* mutations is significantly associated with shorter OS (Du et al. 2020). The prevalence of *DNMT3A* (R882A) and *TET2* (mutations in exons 6–10) mutations in Mexican AML patients was 2.7% and 11.8%, respectively. Mutations in *DNMT3A* and *TET2* cause irregular DNA methylation patterns and transcriptional expression levels in genes known to be involved in the pathogenesis of AML. Therefore, it is thought that alterations in *DNMT3A* and *TET2* genes may be associated with AML prognosis (Ponciano-Gómez et al. 2017). Four new TET2 variants not included in the database were identified in Pakistani AML patients. These variants are frameshift deletion (p.T395fs), frameshift insertion (p.G494fs) and nonsense (p.G898X, p.Q1191X) mutations. Especially, the majority of mutations in exon 3 are seen in patients diagnosed with mature AML (Shaikh et al. 2021). Among the Nordic population, the most frequently mutated genes in chronic myelomonocytic leukemia patients who underwent allogeneic hematopoietic stem cell transplantation are *ASXL1, TET2, RUNX1, SRSF2*, and *NRAS*, and *TET2* mutations are associated with significantly higher 3-year OS (Wedge et al. 2021). In AML with t(9;22)(q34;q11) among the Swedish population, interestingly, there are no mutations in *NPM1, FLT3* or *DNMT3A*, the three genes that are frequently mutated in AML.

Instead, RUNX1 is the most frequently mutated gene. Less frequently, mutations are found in the *IDH2, NRAS, TET2*, and TP53 genes (Orsmark-Pietras et al. 2021). The most common mutations in myeloproliferative neoplasm patients among the southern Iran population are JAK2V617 and TET2 mutations. The highest rate of JAK2V617 mutations is found in Polycythemia Vera. The heterozygous form of the *TET2* mutation has a high prevalence, especially among the elderly. There is no correlation between *JAK2* and *TET2* mutations, although both are more common in people aged 60 years and older (Abedi et al. 2021). In elderly Korean patients (aged ≥60 years) whose CHIP (clonal hematopoiesis of indeterminate potential) mutations were investigated, the prevalence of CHIP increased with age, and *DNMT3A* and *TET2* loss-of-function mutations were found to be the most common mutations (Moon et al. 2023). Apart from population studies, in a study investigating the mutational landscape of chronic myelomonocytic leukemia, *NRAS, ASXL1, TET2, SRSF2, RUNX1, KRAS,* and *SETBP1* genes were found to be the most commonly mutated. It was also found that patients aged 60 years and older had more frequent mutations in *TET2* and *ASXL1* than patients younger than 60 years (Han et al. 2022). In a study investigating the interaction of *ASXL1* and *TET2* gene mutations in the same leukemia type, it was reported that the presence of *TET2* mutant and *ASXL1* wild-type genotypes was the most beneficial genotype for the survival of patients (Zhao et al. 2022). Finally, the most common somatic mutations in chronic myelomonocytic leukemia patients who underwent allogeneic hematopoietic cell transplantation are *ASXL1, TET2, KRAS/NRAS,* and *SRSF2* gene mutations. *DNMT3A* and TP53 mutations are associated with decreased OS, while *DNMT3A, JAK2,* and TP53 mutations are associated with decreased disease-free survival. The only mutation associated with increased relapse is TP53 gene (Mei et al. 2023).

2.1.2 Histone Modifications

Histone proteins are important components of nucleosomes, and their post-translational modifications are related to chromatin structure. There are six types of histones: H1, H2A, H2B, H3, H4, and H5. They are highly rich in positively charged amino acids—lysine and arginine. Nucleosomal histones are extremely conserved proteins. They show almost 100% amino acid homology in all eukaryotic organisms. The core of the nucleosome, which is made up of two tetrameric complexes of the repeating histones H2A, H2B, H3, and H4, is the essential component of chromatin. The octameric units are packaged in 30 nm fibrils and joined by a linker formed of histone H1 or, in rare instances, H5. The fibrils are then arranged into chromatids, loops, sockets, and helixes (Onufriev and Schiessel 2019). Higher-order structures are made possible by the stabilization of chromosomes by the linker histones H1 and H5 (Fyodorov et al. 2018). Histone modification is one of the most critical and essential regulatory epigenetic mechanisms in cancer development (Qin et al. 2020). Acetylation, ubiquitination, phosphorylation, and methylation are the most important post-translational histone modifications. Methylation and

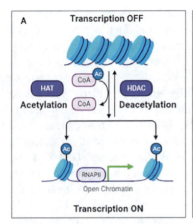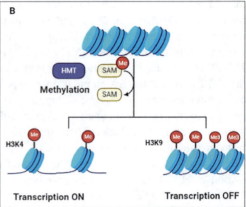

Fig. 2.3 Acetylation and methylation of histone proteins. (a) Histone acetylation takes place at multiple lysine residues located at the N-terminus, catalyzed by enzymes called histone acetyltransferases (HATs) or lysine acetyltransferases (KATs). Through various mechanisms, histone acetylation plays a crucial role in regulating the compactness of chromatin. These mechanisms include neutralizing the positive charge at unmodified lysine residues and promoting active transcription, particularly at gene promoters, enhancers, and the gene body. Additionally, histone acetylation facilitates the recruitment of coregulators and RNA polymerase complexes to specific loci. HATs transfer acetyl groups from acetyl-CoA cofactors to lysine residues on histones, while histone deacetylases (HDACs) perform the opposite function, resulting in the highly reversible nature of histone acetylation. (b) Histone lysine methylation can occur in three different states: mono-, di-, or trimethylation. In particular, di- and trimethylation at specific sites such as H3K4, H3K36, and H3K79 are typically associated with gene activation. For instance, trimethylation at H3K4 (known as H3K4me3) acts as a marker for promoters, while H3K36 and H3K79 methylations predominantly occur across gene bodies. On the other hand, mono-methylation of H3K4 serves as an activating mark specific to enhancers. Conversely, methylations at H3K9 and H3K27 are generally associated with gene repression

acetylation are the most common types of histone modifications (Kouzarides 2007) (Fig. 2.3.A). These changes can modulate chromatin structure by altering non-covalent interactions within and between nucleosomes. They also act as docking sites for specialized proteins with distinctive domains that identify these alterations specifically. Enzymes responsible for these post-translational histone modifications include histone acetyltransferase (HAT), histone deacetylase (HDAC), histone methyl transferase (HMT), histone demethylase (HDMT), kinases, E3-ubiquitin (Prachayasittikul et al. 2017).

2.1.2.1 Histone Acetyltransferases

Histone acetyltransferases (HATs) catalyze the transfer of acetyl group from acetyl-CoA to the ε-amino group of the internal lysine residue located close proximity to the amino termini of the histone proteins. The addition of an acetyl group removes the positive charge of lysine destroying the electrostatic interaction between histones and DNA which leads to a relief in the chromatin structure, affects the gene assembly, and then alters the transcription process (Di Martile et al. 2016). Euchromatin, the open and active conformation of chromatin, is related to histone acetylation, while condensed and inactive chromatin is typically associated with histone deacetylation (i.e., heterochromatin). All HAT members have the ability to activate transcription, which is their primary function, and as a result, they are essential for a variety of cellular processes.

HATs are divided into type A and type B groups according to their cellular localization. Type B is located in the cytosol and has five members: HAT1, HAT2, HATB3.1, Rtt109, and HAT4. They modify newly synthesized free histones, upon which they are transported to the nucleus and associated with the DNA (Trisciuoglio et al. 2018). Type A is categorized into five major families and primarily displays nuclear localization. Family I has two members with similar structures and functions, CBP (CREB-binding protein, CREBBP) and its paralog p300 (or EP300). Both CBP and p300 have the HAT domain, the bromodomain (BRD), and three cysteine and histidine-rich domains (TAZ, PHD, and ZZ) that are used for protein-protein interaction (Dancy and Cole, 2015). Family II which is also known as GCN5-related N-acetyltransferase (GNAT), contains 12 members and they acetylate both histone and non-histone proteins. They also have the HAT domain and conserved BRD at the C-terminus, which identifies and binds to acetyl-lysine residues (Salah Ud-Din et al. 2016). Family III has highly conserved MYST domain and other protein recognition domains (Sapountzi and Côté 2011). Family IV—nuclear receptor coactivator-related HAT—consists of steroid receptor coactivators (SCR1, SCR2, and SCR3) that are involved in chromatin remodeling and the recruitment/stabilization of common transcription factors (Wang and Dent 2014). The TATA box-binding protein (TBP)-associated factors (TAF), TAFII250 and TFIIIC are members of the transcription factor-related HAT V family (Hsieh et al. 1999).

The transcriptional coactivator CBP (also known as KAT3A) and its paralog p300 (KAT3B) are both members of the p300/CBP family. The human CBP locus is found at 16p13.3 on chromosome and shares similarities with the p300 gene at chromosome 22q13. Moreover, they are structurally identical and exhibit 63% and 86% amino acid and KAT domain sequence similarity, respectively (Wang et al. 2013). CBP and p300 have alternate functions to maintain cellular homeostasis. They both operate as transcriptional coactivators of different sequence-specific transcription factors, which are involved in a variety of biological processes including DNA repair, cell proliferation, senescence, differentiation, and apoptosis (Kalkhoven 2004). p300 also regulates the expression and function of tumor-related genes including androgen receptor (Zhong et al. 2014), p53 (Teufel et al. 2007), c-myc (Vervoorts et al. 2003), and breast cancer gene-1 (BRCA1) (Pao et al. 2000).

GCN5 and PCAF, the two main members of GNAT family, are related proteins. Whereas the latter only occurs in higher eukaryotes, the former has homologs in both yeast and humans. GNATs generally contribute to cellular proliferation and are crucial for controlling the cell cycle. For instance, cell division cycle-6 (CDC6) is specifically acetylated by GCN5 at three lysine residues on either side of its cyclin-docking motif. This alteration is necessary for the protein to be subsequently phosphorylated by cyclin A-cyclin-dependent kinase (CDKs) at a particular position near the acetylation site. The relocalization of the protein to the cell cytoplasm during the S phase as well as the control of its stability depend on GCN5-mediated acetylation and site-specific phosphorylation of CDC6 (Paolinelli et al. 2009). Both GCN5 and PCAF exclusively acetylate the catalytic core of cyclin-dependent kinase-9 (CDK9) to control its activity. This alteration moves the enzyme to the insoluble nuclear matrix compartment and significantly inhibits CDK9's transcriptional and catalytic activity (Sabò et al. 2008).

The MYST family consists of five members: Tip60 (HIV1 TAT interacting 60 kDa protein), MOF, MOZ (monocytic leukemia zinc finger protein), MORF (MOZ related factor), HBO1 (histone acetyltransferase bound to origin recognition complex (ORC)). All have conserved MYST domain containing an acetyl-CoA binding and zinc finger motifs (Sapountzi and Côté 2011). They also have additional domains to recognize other proteins facilitating their function in the regulation of transcription, DNA damage response, cell growth, and survival (Avvakumov and Côté 2007). Tip60, a member of the MYST family, is associated with a variety of cellular processes, including transcription, DNA damage-induced checkpoint activation, and apoptosis. It is a crucial enzyme for DNA repair and restoring normal cellular function because it controls the ataxia telangiectasia mutant (ATM) protein kinase, which phosphorylates and activates proteins involved in DNA repair. ATM protein kinase, however, needs to be acetylated by the Tip60 protein in order to be active. Absence of Tip60 inhibits ATM protein kinase activity and lowers the capacity of cells to repair DNA (Sun et al. 2005). Also reported, the acetylation of p53 by Tip60 at K120 was shown to be crucial for p53-induced cell death (Sykes et al. 2006).

The substrate specificity of human MOF is substantially preserved from fly to human. It is responsible for the acetylation of histone H4 at lysine 16 in human cells, which has clear connections to cancer (Taipale et al. 2005). It has long been understood that MOF depletion can affect a variety of intracellular biological processes, including chromatin integrity, cell cycle, gene transcription, DNA damage repair, and early embryonic development (Su et al. 2016). It has been demonstrated that MOF contributes to the regeneration of embryonic stem cells. Particularly, MOF is a crucial member of the embryonic stem cell core transcriptional network and primes genes for a variety of developmental programs, making it an essential factor in both normal physiology and illness (Li et al. 2012). In animal cells, MOF reduction can lead to aberrant gene transcription, particularly resulting in abnormal expression of specific tumor suppressors or oncogenes (Gupta et al. 2008).

The stable multisubunit complexes that human MOZ and MORF generate are in charge of acetylating a considerable portion of histone H3. Normal developmental programs as well as the control of several genes, particularly the Hox family, depend on the acetylation activity of MOZ/MORF complexes (Yang and Ullah 2007; Perez-Campo et al. 2013).

HBO1 was identified as a fifth human MYST protein through a two-hybrid screen as a result of its interaction with the ORC1 subunit of the ORC origin recognition complex. HBO1 appears to be crucial for DNA replication, therefore it seems reasonable that abnormalities in its action should have a serious negative effect on the cell and promote oncogenesis (Iizuka and Stillman 1999).

2.1.2.2 Mutations of Histone Acetyltransferases

Mutations of HATs can have various effects on cancer. For example, mutations in the CREBBP (CBP) and EP300 genes, which encode two HAT enzymes, have been identified in various types of cancer, including leukemia, lymphoma, and solid tumors (Ojesina et al. 2014; Nann et al. 2020; Zhu et al. 2022; Michot et al. 2023; Xu et al. 2023). These mutations can result in decreased HAT activity and altered gene expression, leading to increased cell proliferation, resistance to apoptosis, and other hallmarks of cancer. In a study of Norwegian triple-negative breast cancer (TNBC) patients, the G211S point mutation detected in the EP300 gene was found to be significantly associated with this TNBC subset and highly reduced the probability of other pathological somatic mutations. Interestingly, the EP300-G211S mutation results in a lower risk of recurrence and breast cancer-specific mortality during long-term follow-up of patients (Bemanian et al. 2018). In Chinese esophageal squamous cell carcinoma (ESCC) patients, EP300 mutations are associated with tumor grade, pathological T stage, and lymph node metastasis. Nonsense, missense, frameshift, and splicing mutations correlate with poor prognosis, and their deletion suppresses angiogenesis, hypoxia, and the epithelial–mesenchymal transition (EMT) process. The most common type of mutation is the missense mutation. The majority of these mutations are c.G4195/p.D1399, c.4241A/p.Y1414, and c.4540G/p.E1514 (Bi et al. 2019). On the other hand, in bladder cancer, EP300 mutations are associated with higher TMB and favorable clinical prognosis. The mutation upregulates the signaling pathways in the immune system and increases the antitumor immune response (Zhu et al. 2020). However, in a pan-cancer study using datasets from eleven different cancer types, it was determined that EP300 mutations were associated with genome instability (increased TMB), increased antitumor immunity, and programmed death-ligand 1 (PD-L1) expression. Because of higher TMB and PD-L1 expressions are associated with a more active response to ICIs, EP300 mutated cancers respond better to ICIs. Also, cancers with EP300 mutations are more sensitive to several cell cycle inhibitors, including AZD7762, Wee1 inhibitor, RO-3306, palbociclib, BI-2536, MK-1775, dinaciclib, ribociclib, and MK-8776 (Chen et al. 2021a). In an in vitro study of bladder cancer, it was determined that among the missense mutations (H1451L, D1485V, E1521Q, K1554N, R1627W, and Q2295K), the EP300-R1627W mutation impairs EP300 transactivation activity in both p21 and p16 promoters. In addition, the mutation has been found to have a more aggressive effect on growth and invasion in vitro and in vivo, and it has been reported that there is a driver mutation in the development and progression of bladder cancer (Luo et al. 2023). On the other hand, in Chinese urothelial bladder carcinoma patients, CREBBP mutations are frequently found in muscle-invasive bladder cancer (MIBC) patients (Wang et al. 2020). In CRC, NOTCH3, histone lysine methyltransferase 2C (KMT2C), and CREBBP are associated with tumor location, stage, and PFS, respectively, and can be considered as prospective biomarkers for diagnosis and prognosis (Liu et al. 2021). In primary cutaneous follicular B-cell lymphoma (PCFBCL), a rare lymphoma subtype of the skin, somatic mutations in the *CREBBP, TNFRSF14, STAT6,* and TP53 genes are among the most frequently identified oncogenic changes. Identification of such genetic alterations helps differentiate PCFBCL from cutaneous pseudo-lymphoma and thus provides an additional diagnostic tool in difficult-to-diagnose cases (Wobser et al. 2022). There are also studies in the literature evaluating EP300 and CREBBP co-mutations. For example, in SAKK 38/07 prospective clinical trial cohort study that performed mutational analysis in diffuse large B-cell lymphoma (DLBCL), it has been determined that CREBBP and EP300 mutations have negative effects on OS, PFS, and EFS in patients equally treated with six courses of R-CHOP followed by two courses of R (R-CHOP-14) (Juskevicius et al. 2017). In the phase II study of the use of tucidinostat (CR-CHOP) in addition to R-CHOP in the treatment of newly diagnosed advanced age DLBCL patients, it was stated that CR-CHOP mitigates the negative prognostic effect of CREBBP/EP300 mutations and also effective and safe in the treatment (Zhang et al. 2020a).

The MYST subgroup also has an effect on cancer development. Especially KAT6A and KAT6B contribute to the development of cancer with chromosomal rearrangements. In a retroperitoneal leiomyoma patient with t(10;17)(q22;q21), which resulted in the formation of the KAT6B-KANSL1 fusion gene, a fusion transcript was not found and this was interpreted as either absent or unexpressed (Panagopoulos et al. 2015). In adult AML patients with t(8;16)(p11.2;p13.3) resulting in the formation of the KAT6A-CREBBP fusion gene, translocation exhibits monoblastic or myelomonocytic differentiation, and arises in patients with a history of cancer treated with cytotoxic therapies. It is also associated with the good outcome of de novo AML and t-AML patients without adverse prognostic factors (Xie et al. 2019). On the other hand, in a study showing the effect of KAT6A amplification in endometrial serous carcinoma, it was determined that amplification occurs more frequently in younger patients and is associated with short PFS and OS (Saglam et al. 2020).

2.1.2.3 Histone Deacetylases

Deacetylation is the opposite of acetylation, whereby histone proteins move toward one another as a result of the formation of nucleosome's compact structure, which prevents the activation of gene transcription. Histone deacetylase (HDAC) enzymes catalyze the cleavage of the acetyl group of lysine (Di Martile et al. 2016) and are essential for controlling transcription (Huang et al. 2019). So far, 18 members of the human HDAC family have been discovered and are categorized into four classes (Seto and Yoshida 2014). Class I proteins (HDAC1, 2, 3, and 8) that are located in the nucleus have highly conserved deacetylase domains flanked by short amino acid and carboxy-terminal members (Yang and Seto 2008). The members of class II are separated into two subclasses, IIa (HDAC4, 5, 7, and 9) and IIb (HDAC6 and 10), which are found in both the cytoplasm and nucleus. They have a regulatory N-terminal domain, which ensures their interaction with corepressors and transcription factors specific to different tissues, in addition to the conservative deacetylase domain (Parra and Verdin 2010). Class III, known as SIRT-like enzymes, has seven members with various cellular localization and have NAD-dependent protein deacetylase and/or ADP ribosylase activities (Hallows et al. 2008). Class IV only has one member, HDAC11, which has conservative residues in the catalytic core regions that are identical to those of class I and II HDAC (Gao et al. 2002).

The classical HDAC family of enzymes (Class I, II, and IV) require zinc ions for their catalytic function. Histone deacetylase-like protein (HDLP), produced by the hyperthermophilic bacterium *Aquifex aeolicus*, was the first classical HDAC family protein to have its X-ray crystal structure characterized (Finnin et al. 1999). The catalytic domain structure of mammalian HDACs as determined by X-ray crystallography is virtually the same as that of HDLP, with the HDAC family sharing the same active site and contact inhibitor residues. However, structural studies on HDAC8 and its mutations suggested a different model. According to this concept, one of the histidine residues—H143—acts as the general base while the other— H142—acts as a general electrostatic catalyst. In line with the proposed model of action, the HDAC8 H143A mutant has essentially no activity compared to an H142A mutant's residual activity (Gantt et al. 2010). Moreover, it is suggested by quantum mechanics/molecular mechanics molecular dynamics (QM/MM MD) simulations that a neutral H143 first acts as the general base to accept a proton from the zinc-bound water molecule in the initial rate-determining nucleophilic attack step before transferring it to the amide nitrogen atom to aid in the cleavage of the amide bond (Wu et al. 2011).

In contrast to Class I, II, and IV enzymes, which depend on zinc for catalysis, Class III HDACs require NAD^+ as a cofactor. Structural analyses of archaeal, yeast, and human homologs of Sir2 have revealed that the catalytic domain of sirtuins is located in a cleft between a large domain with a Rossmann-fold and a small zinc-binding domain. The sirtuin family shares the same amino acid residues in the cleft, forming a protein tunnel where the substrate binds with NAD^+ (Finnin et al. 2001). The proposed mechanism is based on the nucleophilic attack of the acetamide oxygen to the C1′ position of the nicotinamide ribose resulting C1′-O-alkylamidate intermediate and free nicotinamide. Then the C1′-O-alkylamidate intermediate is converted to a 1′, 2′-cyclic intermediate, which eventually releases lysine and 2′-O-acetyl-ADP ribose (Avalos et al. 2004).

2.1.2.4 Mutations of Histone Deacetylases

Mutations or alterations in HDAC genes have been linked to the development and progression of several types of cancer. For instance, HDAC1, HDAC2, HDAC3, and HDAC6 are frequently overexpressed in various cancer types, and this overexpression has been associated with poor prognosis and resistance to therapy. Additionally, mutations in HDAC genes have been reported in some cancers. These mutations can lead to changes in the enzymatic activity of HDACs, resulting in abnormal gene expression and contributing to the development and progression of cancer. For example, in a study investigating the sensitivity of cisplatin in gastric cancer, it was determined that there are point mutations, frameshift deletion, or deep deletion that may affect the function of the *HDAC4* gene. These changes correlate with good prognosis (Spaety et al. 2019). In addition, in a study investigating gene mutations in ovarian lymphoma, *HDAC4* mutations were detected only in ovarian DLBCL, but not in conventional DLBCL. *NOTCH3* and *HDAC4* mutations are found in the germinal center B-cell-like (GCB) subtype (Xu et al. 2020). In a study investigating SIRT1 mutations in 41 breast and cervical cancer cell lines, a total of 31 sequence variants were identified. Although 6 of them have not been known or reported before, 4 variants detected in breast cancer are in the coding region and are missense mutations. Two of them (2244A > G (I731V) and 2268G > T (D739Y)) were detected for the first time. The R65_A72del mutation was detected in the cervical cancer cell line. However, it has been shown that these mutations do not alter SIRT1 deacetylase activity or telomerase activity (Han et al. 2013). In a cohort study investigating *SIRT1* polymorphisms (rs10997870 and rs12778366) in CRC, it was found that the rs12778366 TC/CC versus TT genotype was inversely related to microsatellite instable CRC and not associated with microsatellite stable tumors (Hrzic et al. 2020).

2.1.2.5 Histone Methyltransferases

Histone methylation particularly occurs in specific lysine and arginine residues at the amino terminal ends of histones core by the action of histone methyltransferases (HMTs) (Greer and Shi, 2012). Each lysine can be mono-, di-, or tri-methylated covalently on the amino group of lysine. However, post-methylation of arginine can occur in mono-, di-, symmetrical, and asymmetrical forms with methylation of the molecule's guanidyl group (Bedford and Richard 2005). Whether histone methylation activates or represses transcription is correlated with differences in residue methylation and modification states (Li et al. 2007). For instance, H3 lysine 4 (H3K4), H3K36, and H3K79 lysine methylation are linked to transcriptional activation. In contrast, gene suppression is associated with methylation at the H3K9, H3K27, and H4K20 (Sims et al. 2003) (Fig. 2.3.B). In humans, more than 50 lysine methyltransferases (KMTs) have been identified so far. KMTs are further divided into two groups based on the catalytic domain sequence: SET domain-containing KMTs, such as Su(var)3–9, Enhancer of Zeste Homolog (EZH), and Trithorax, and non-SET domain-containing KMTs, such as the DOT1-like proteins (Feng et al. 2002; Herz et al. 2013). SET methyltransferase structure has pre-SET, SET, and post-SET domains. The SET methyltransferases are further divided into various families. In the SET1 family, the SET domain is followed by a post-SET domain. This family includes the well-known EZH1 and EZH2 methyltransferases, despite the fact that they lack the post-SET domain. The nuclear receptor binds to the SET domain of the SET2 family of proteins, which includes the NSD1–3, SETD2, and SMYD family of proteins, which is usually accompanied by a post-SET and an AWS domain. SUV39H1, SUV39H2, G9a, GLP, ESET, and CLLL8 are only a few of the SUV39 family members that all exhibit a pre-SET domain (Rea et al. 2000). Other SET domain-containing methyltransferases, such as SET7/9, SET8, SUV4-20H1, and SUV4-20H2, have not been divided into distinct subgroups (Dillon et al. 2005). The human DOT1-like (DOT1L) protein is a methyltransferase that does not possess a SET domain and methylates a lysine residue in the histone's globular core (Wood and Shilatifard 2004).

Based on various arginine binding pockets, the arginine methyltransferases (PRMTs) have three different types of methylation patterns. Monomethyl arginine and asymmetric dimethylarginine can be produced by the first class of PRMTs, which includes PRMT1, PRMT3, and cofactor-associated arginine methyltransferase (CARM1) (Chen et al. 1999; McBride et al. 2000). The monomethyl or symmetric dimethylarginine can be produced by the second class of PRMTs, which includes PRMT5 (Branscombe et al. 2001). The only product of the third class of PRMTs, which includes PRMT7, is monomethylated arginine (Blanc and Richard 2017).

2.1.2.6 Mutations of Histone Methyltransferases

It is well recognized that HMT gene mutations play a significant role in carcinogenesis, particularly in solid tumors and hematological cancers. Gene mutations in the *EZH2, KMT2A, NSD*, and *SET* are important in particular.

Regarding solid tumors, the c.2201G > C mutation of *EZH2* in CRC and the c.1544A > G mutation in liver cancer are thought to be used as biomarkers (Mahasneh et al. 2021; Cui et al. 2021b). In melanoma, *EZH2* Y641F (activating point mutation) upregulates interferon-related genes. Upregulation of these genes is not a direct effect of changes in H3K27me3 but through a non-canonical interaction between EZH2 and STAT3. Together, EZH2 and STAT3 function as transcriptional activators to mediate gene activation of numerous genes, including MHC Class 1b antigen processing genes. Furthermore, expression of STAT3 is required to maintain the antitumor immune response and to prevent melanoma progression and recurrence in EZH2 Y641F melanomas (Zimmerman et al. 2022). With regard to hematological malignancies, the pathogenic *EZH2* mutations tend to co-occur ASXL1 in MDS. If mutations are alone or co-presence with *ASXL1, RUNX1* mutations, and chromosome 7 abnormalities (del(7q) and monosomy 7), they are associated with poor OS (Ball et al. 2023). On the other hand, EZH2 dysregulation caused by mutation and underexpression defines specific subtypes of AML. Patients with EZH2 mutation have shorter OS and leukemia-free survival (LFS) after receiving autologous or allogeneic hematopoietic stem cell transplant (HSCT) than patients without EHZ2 mutation. However, EZH2 expression has no effect on OS and LFS of AML patients. Notably, in the low EZH2 expression group, patients undergoing HSCT had significantly better OS and LFS compared to patients receiving chemotherapy alone, while there was no significant difference in OS and LFS between chemotherapy and HSCT patients in the high EZH2 expression group. EZH2 dysregulation may serve as potential biomarkers that predict prognosis and guide treatment choice between transplantation and chemotherapy (Chu et al. 2020). The *EZH2* loss-of-function mutation in AML provides resistance to cytarabine. This resistance is the result of the upregulation of EZH2 target genes responsible for apoptosis, proliferation, and transport (Kempf et al. 2021). In the Mexican-Mestizo DLBCL population, Tyr641His and Tyr641Ser mutations of EHZ2 exon-16 are negatively associated with relapse/progression and tend to lack complete response (Oñate-Ocaña et al. 2021). In follicular lymphoma, on the other hand, non-mutated patients receiving R-CHOP have significantly more relapses than patients receiving R-Bendamustine. Furthermore, mutated EZH2 patients treated with R-CHOP show a lower incidence of relapse, higher PFS, and higher OS compared to those treated with R-Bendamustine. Therefore, R-CHOP for mutated patients and R-Bendamustine for non-mutated patients is a more appropriate treatment option (Martínez-Laperche et al. 2022). There are also pharmacological studies on EZH2 mutations in the literature. For example, in a phase II study investigating the effect of Tazemetostat in EZH2 mutant relapsed or refractory follicular lymphoma patients, it was determined that the agent showed clinically significant and durable responses. Moreover, it has been found to be generally well tolerated in patients who have received intensive pretreatment. (Morschhauser et al. 2020). In another phase II study, it was reported that Tazemetostat is effective, safe, and can be used in the treatment of EZH2 mutant relapsed or refractory B-cell non-Hodgkin lymphoma (Izutsu et al. 2021).

The gene most known to be effective in cancer development in the KMT family is KMT2A. The effect of KMT2A is mostly seen as rearrangements. The KMT2A-ARHGEF12 fusion gene generated as a result of a 1.95 Mb interstitial deletion in the long arm of chromosome 11, joining exon 10 of the KMT2A gene to exon 12 of the ARHGEF12 gene, was detected in high-grade B-cell lymphoma (Jung et al. 2020). In AML, patients with 11q23/KMT2A rearrangements have a low number of additional gene mutations involving the RAS pathway (*KRAS, NRAS*, and *PTPN11*). KRAS mutations occur more frequently in patients with t(6;11)(q27; q23)/KMT2A-AFDN compared to patients with other 11q23/KMT2A subsets. Younger (age < 60 years) patients with t(9;11) (p22;q23)/KMT2A-MLLT3 have better outcomes than patients with other 11q23/KMT2A rearrangements. On the other hand, elderly patients (age ≥ 60 years) with the same translocation have poor outcomes (Bill et al. 2020). RAS pathway (*KRAS, NRAS*, and *PTPN11*) and SETD2 mutations are frequently seen in pediatric 11q23/KMT2A-rearranged AML patients. KRAS mutations correlate with worse 5-year EFS and 5-year OS. The presence of SETD2 mutations increases the 5-year relapse rate. KRAS mutations in 11q23/KMT2A-rearranged AML are thought to be an independent predictor for poor EFS (Yuen et al. 2023). In a case report, it was stated that in a patient with KMT2A-MLLT3 rearranged AML who reached remission, the disease relapsed as KMT2A-MLLT3 rearranged ALL after a while. Exome analysis of the relapse sample revealed two somatic mutations of PAX5 (p.Ser285X and p.Gly30Lys). It has been suggested that these two PAX5 alterations cause loss of function, thus playing a role in the transition from acute monocytic leukemia to acute lymphocytic leukemia (Nakajima et al. 2022). In the pan-sarcoma genomic analysis, YAP1–KMT2A–YAP1 and VIM–KMT2A fusions were detected. YAP1–KMT2A fusion-positive sarcomas show a sclerosing epithelioid fibrosarcoma-like histology, while VIM–KMT2A sarcomas have spindle-to-round cell morphology (Massoth et al. 2020). KMT2A mutations in CRC are associated with enhanced genomic instability, including a high level of microsatellite instability and TMB. Mutant cancers also have co-occurring gene mutations within Wnt signaling, ERBB2/4, TGF-β superfamily pathway, and PI-3-kinase pathway. Therefore, it is stated that KMT2A mutations may be a predictive biomarker for better overall survival in metastatic CRC (Liao et al. 2022).

The most important rearrangement of the *NSD1* gene from the NSD family is the formation of the t(5;11)(q35;p15.5) (NUP98-NSD1) gene fusion. NUP98-NSD1 fusion is associated with poor outcome in AML (Shiba et al. 2013). However, in pediatric AML, in addition to the fusion gene, in the presence of FLT3-ITD and the absence of *NPM1* and *CEBPA* mutations, the response to treatment is poor (Akiki et al. 2013). On the other hand, *NSD2* mutation, which is the other member of the family, is frequently found in pediatric hematological malignancies (Huether et al. 2014). The p.E1099K NSD2 mutation has been described in pediatric ALL, and ectopic expression of the variant induces a chromatin signature feature of NSD2 hyperactivation and promotes transformation (Jaffe et al. 2013, Oyer et al. 2014). Cell lines harboring the E1099K mutation exhibit increased H3K36 dimethylation and decreased H3K27 trimethylation, particularly on histone H3.1-containing nucleosomes. The mutation is associated with reduced apoptosis and enhanced proliferation, clonogenicity, adhesion, and migration. Also, in mouse xenografts, mutant NSD2 cells are more lethal and brain invasive than wild-type cells (Swaroop et al. 2019).

The SET family member, SETD2, is an important gene involved in cancer development. Although *SETD2* mutations (frameshift/truncating mutations or point mutations at high allele frequencies) are most common in high-grade gliomas of the cerebral hemispheres, they can also be found in various primary central nervous system tumors (Viaene et al. 2018). In metastatic non-small cell lung cancer, SETD2 mutation (missense mutation p.T1171K) and CREB1 inactivation contribute to cisplatin cytotoxicity through regulation of ERK signaling pathway, and their inactivation may lead to cisplatin resistance (Kim et al. 2019). In CRC, SETD2 mutation is associated with co-occurring p53 mutations and abnormal beta-catenin expression (Bushara et al. 2023). Mutations in *SETD2, PBRM1, BAP1*, and *KDM5C* are the most common mutations in clear cell renal cell carcinoma. Mutant SETD2 increases the malignancy of clear cell renal carcinoma with PBRM1 mutation and can develop local or distant metastases (Liu et al. 2023).

2.1.2.7 Histone Demethylases

While methyltransferases are in control of generating methylation patterns, demethylases have the ability to remove methyl groups from proteins besides histones (Nicholson and Chen 2009). The understanding of the role of epigenetics in carcinogenesis has been substantially enhanced by the discovery of histone demethylases and their function in the regulation of post-translational modifications of chromatin, which may present new therapeutic targets for the treatment of cancer (Højfeldt et al. 2013).

Peptidyl arginine deiminase 4 (PADI4) demethylates arginine residues by converting them to citrulline in order to reverse methylation. However, because a methyl group is lost during the conversion of arginine to citrulline rather than a free arginine, this alteration is not regarded as demethylation (Cuthbert et al. 2004). Amine oxidase homolog lysine demethylase 1 (KDM1) and JmjC domain-containing histone demethylases are two families of actual histone demethylases that are capable of removing the methyl groups attached to lysine amino acids of histone proteins. KDM1A and KDM1B are the two members of the KDM1 family. Shi et al. described KDM1A, also known as lysine-specific demethylase 1 (LSD1) for the first time in 2004. This enzyme is a highly conserved flavin-containing amine oxidase homolog and eliminates mono- and di-methylated lysines at lysine 4 or lysine 9 of H3 (Shi et al. 2004). KDM1A produces formaldehyde when it removes methyl groups through oxidation (Shi et al. 2004). According to numerous studies, KDM1A preferentially demethylates H3K4me1 and H3K4me2 through an interaction between its tower domain and CoREST, which results in transcriptional inactivation (Lee et al. 2005). Nevertheless, KDM1A demethylates H3K9me1 and H3K9me2 when it complexes with androgen receptors causing transcriptional activation (Wissmann et al. 2007). KDM1A demethylates K370me2 to prevent p53 from interacting with 53BP1, which inhibits the functions of p53, including the stimulation of apoptosis (Huang et al. 2007). Moreover, it has been shown that DNA cytosine-5-methyltransferase 1 (DNMT1) is demethylated by KDM1A, which stabilizes DNMT1 and enables it to retain DNA methylation patterns in embryonic stem cells (Nicholson and Chen 2009). Similar to KDM1A, KDM1B is a homolog of the FAD-dependent amine oxidase that targets H3K4me1 and H3K4me2 specifically (Karytinos et al. 2009). However, due to the absence of a tower domain, KDM1B is unable to assemble a complex with CoREST (Karytinos et al. 2009). Although more recent research have indicated that KDM1B plays a role in maternal imprinting in oocytes and may be involved in activating NF-κB, the regulatory activities of KDM1B are currently being explored (Ciccone et al. 2009).

The Jumonji C (JmJC) domain-containing histone demethylases are the second and largest subclass of these enzymes. Around 20 JmJC domain proteins that have been discovered are thought to be lysine-specific demethylases (Højfeldt et al. 2013). These enzymes are the members of the 2-oxoglutarate-dependent dioxygenases and need Fe^{2+} and oxygen to conduct the hydroxylation required to remove methyl groups. This family of enzymes can remove trimethylations, in contrast to KDM1

(Cloos et al. 2008). The JmjC KDMs have been grouped into a variety of distinct subfamilies, including KDM2, KDM3, KDM4, KDM5, KDM6, and others. KDM2/FBXL subfamily has KDM2A and KDM2B. While both of these enzymes have the ability to exclusively demethylate H3K36me2 (He et al. 2008; Kottakis et al. 2011), KDM2B also has the capacity to demethylate H3K4me3 (Frescas et al. 2007). KDM2B controls p15Ink4b by demethylating H3K36me2, which causes repression at that locus (He et al. 2008). It has been demonstrated that KDM2B knockdown causes cellular senescence to be induced in a p53 and RB-dependent way (He et al. 2008). One study suggested KDM2B might prevent oxidative stress by blocking ROS-mediated signaling, while another suggested KDM2A might inhibit the NF-κB pathway (Polytarchou et al. 2008; Lu et al. 2010). The KDM3/JMJD1C subfamily consists of KDM3A and its two human homologs, KDM3B and JMJD1C. KDM3A and KDM3B are specialized for demethylating H3K9me1 and H3K9me2, whereas JMJD1C lacks histone demethylase activity. KDM3A has been demonstrated to act on androgen receptors in a ligand-dependent way and to have a role in spermatogenesis and metabolism (Wilson et al. 2017) and KDM3B appears to be involved in spermatogenesis (Liu et al. 2015). The KDM4 subfamily catalyzes the specific demethylation of H3K9me2, H3K9me3, H3K36me2, and H3K36me3. KDM4C has been demonstrated to transcriptionally activate amino acid biosynthesis and transport, contributing the intracellular amino acid levels (Zhao et al. 2016). Recently, it has been proven that N-Myc and KDM4B interact directly (Yang et al. 2015). It's interesting to note that KDM4A controls protein synthesis and has been linked to translational machinery (Van Rechem et al. 2015). KDM5 subfamily exhibits catalytic activity on H3K4me2 and H3K4me3. Through its direct contact with RBP-J, KDM5A has been linked to the Notch/RBP-J complex gene silencing (Liefke et al. 2010). Its function in gene regulation via the PRC2 complex has been highlighted in several studies (Pasini et al. 2008). KDM6A/UTX controls the cell cycle in an RB-dependent way and blocks growth signals by keeping RB-binding proteins active to cause cell cycle arrest (Wang et al. 2010). KDM6B/JMJD3 has been demonstrated to not only interact with p53 but also to be attracted to the promoter and enhancer regions of 263 of p53 target genes (Williams et al. 2014). Moreover, it has been demonstrated that KDM6B increases the expression of p16INK4A and p14ARF, which are located at the INK4A-ARF locus, as well as stabilizes nuclear p53 by direct contact (Agger et al. 2009; Ene et al. 2012).

2.1.2.8 Mutations of Histone Demethylases

One well-known example is the mutations in the histone demethylase enzyme called lysine-specific demethylase 1A (LSD1/KDM1A). Studies have shown that mutations in LSD1 are associated with several types of cancers. These mutations are thought to contribute to the development and progression of cancer by altering the expression of genes involved in cell growth and proliferation. For example, somatic mutations in epigenetic regulators (ASXL1, TET2, TET3, KDM1A, and MSH6) associated with cell signaling and cell division pathways have been detected in most patients with newly diagnosed chronic phase chronic myeloid leukemia (Togasaki et al. 2017). In CRC, LSD1 gene deletion is associated with lymph node metastasis and advanced stages of cancer. For this reason, it is thought to be a biomarker with prognostic value (Ramírez-Ramírez et al. 2020). On the other hand, breast cancer patients with the LSD1 mutation show significantly worse outcomes than those without the LSD1 mutation. LSD1 R251Q mutation increases the invasion and migration of luminal breast cancer cells. It also alters the expression of genes that modulate the EMT. In addition, the R251Q mutation disrupts the H3K4me2 demethylation activity of LSD1, abolishing the interaction between LSD1 and CoREST, leading to increased expression of TRIM37, a histone H2A ubiquitin ligase that regulates E-cadherin expression (Zhang et al. 2020b). The E239K mutation eliminates the suppressive function of LSD1 on the migration and invasion of breast cancer cells by disrupting the interaction between LSD1 and GATA3 (Zhang et al. 2022). In gastric cancer, LSD1 deletion suppresses gastric cancer migration by upregulating CD9 via reducing intracellular miR-142−5p (Zhao et al. 2020).

In addition to LSD1, mutations in other histone demethylases, such as lysine-specific demethylase 6A (UTX/KDM6A), have also been implicated in cancer. Inactivating mutations in KDM6A include homozygous or hemizygous large deletions, nonsense mutations, small frame-shifting insertion/deletions, and consensus splice site mutations which lead to aberrant splicing and premature termination codons (van Haaften et al. 2009). The KDM6A mutation is particularly common in bladder cancer. It has been determined that mutations in low-grade non-muscle-invasive bladder cancer are more common in women than in men (Hurst et al. 2017, Nassar et al. 2019). However, in vitro and in vivo experiments examining KDM6A depletion and overexpression in tumor cells support the role of KDM6A as a suppressor for tumor growth and cell migration, thus highlighting its prognostic value (Nickerson et al. 2014). In a study investigating the potential role of KDM6A in the regulation of the antitumor immune response, it was determined that the KDM6A mutation was associated with a lower number of tumor-infiltrating immune cells. KDM6A mutation is associated with lower KDM6A mRNA levels compared to samples carrying the wild-type gene. Patients with low KDM6A expression have a worse prognosis than patients with high KDM6A expression. In addition, the KDM6A mutation downregulates nine signaling pathways (intestinal immune network for IgA production, chemokine signaling pathway, natural killer cell-mediated cytotoxicity, B-cell receptor signaling pathway, T-cell receptor signaling pathway, Fc epsilon Ri signaling pathway, Fc gamma R-mediated phagocytosis, primary immunodeficiency, and the Toll-like receptor signal pathway) involved in the immune system and attenuates the tumor immune response (Chen et al. 2021b).

2.2 Conclusion

In eukaryotes, prominent epigenetic alterations include DNA methylation and histone modification. Dysregulation of epigenetic regulatory enzymes is strongly associated with the development and progression of different types of cancer. Changes in the activity of epigenetic enzymes may result from mutations, whereby mutant epigenetic regulatory enzymes alter epigenetic modifications and facilitate the proliferation, migration, and colony formation of cancer cells. However, in many cases, the mechanisms by which mutations alter the activity or function of epigenetic regulatory enzymes are not fully understood. A better understanding of these pathways would enable us to comprehend the properties of various tumor types more effectively. The individual-based treatment of these malignancies could be accelerated by further research into medications that target these mutated enzymes.

Compliance with Ethical Standards Conflict of Interest: The authors declare that they have no conflict of interest.

Ethical Approval: This chapter does not contain any studies with human participants or animals performed by any of the authors.

Funding: No funding was received to assist with the preparation of this manuscript.

References

Abedi E, Ramzi M, Karimi M, Yaghobi R, Mohammadi H, Bayat E, Moghadam M, Farokhian F, Dehghani M, Golafshan HA, Haghpanah S (2021) TET2, DNMT3A, IDH1, and JAK2 mutation in myeloproliferative neoplasms in Southern Iran. Int J Organ Transplant Med 12(3):12–20

Agger K, Cloos PA, Rudkjaer L, Williams K, Andersen G, Christensen J, Helin K (2009) The H3K27me3 demethylase JMJD3 contributes to the activation of the INK4A-ARF locus in response to oncogene- and stress-induced senescence. Genes Dev 23(10):1171–1176. https://doi.org/10.1101/gad.510809

Akiki S, Dyer SA, Grimwade D, Ivey A, Abou-Zeid N, Borrow J, Jeffries S, Caddick J, Newell H, Begum S, Tawana K, Mason J, Velangi M, Griffiths M (2013) NUP98-NSD1 fusion in association with FLT3-ITD mutation identifies a prognostically relevant subgroup of pediatric acute myeloid leukemia patients suitable for monitoring by real time quantitative PCR. Genes Chromosomes Cancer 52(11):1053–1064. https://doi.org/10.1002/gcc.22100

Allis CD, Jenuwein T (2016) The molecular hallmarks of epigenetic control. Nat Rev Genet 17(8): 487–500. https://doi.org/10.1038/nrg.2016.59

Aref S, Sallam N, Abd Elaziz S, Salama O, Al Ashwah S, Ayed M (2022) Clinical implication of DNMT3A and TET2 genes mutations in cytogenetically normal acute myeloid leukemia. Asian Pac J Cancer Prev 23(12):4299–4305. https://doi.org/10.31557/APJCP.2022.23.12.4299

Avalos JL, Boeke JD, Wolberger C (2004) Structural basis for the mechanism and regulation of Sir2 enzymes. Mol Cell 13(5):639–648. https://doi.org/10.1016/s1097-2765(04)00082-6

Avvakumov N, Côté J (2007) The MYST family of histone acetyltransferases and their intimate links to cancer. Oncogene 26(37):5395–5407. https://doi.org/10.1038/sj.onc.1210608

Azad N, Zahnow CA, Rudin CM, Baylin SB (2013) The future of epigenetic therapy in solid tumours--lessons from the past. Nat Rev Clin Oncol 10(5):256–266. https://doi.org/10.1038/nrclinonc.2013.42

Bachman KE, Rountree MR, Baylin SB (2001) Dnmt3a and Dnmt3b are transcriptional repressors that exhibit unique localization properties to heterochromatin. J Biol Chem 276(34): 32282–32287. https://doi.org/10.1074/jbc.M104661200

Ball S, Aguirre LE, Jain AG, Ali NA, Tinsley SM, Chan O, Kuykendall AT, Sweet K, Lancet JE, Sallman DA, Hussaini MO, Padron E, Komrokji RS (2023) Clinical characteristics and outcomes of EZH2-mutant MDS: a large single institution analysis of 1774 patients. Leuk Res 124: 106999. https://doi.org/10.1016/j.leukres.2022.106999

Baylin SB, Jones PA (2011) A decade of exploring the cancer epigenome – biological and translational implications. Nat Rev Cancer 11(10):726–734. https://doi.org/10.1038/nrc3130

Bedford MT, Richard S (2005) Arginine methylation an emerging regulator of protein function. Mol Cell 18(3):263–272. https://doi.org/10.1016/j.molcel.2005.04.003

Bemanian V, Noone JC, Sauer T, Touma J, Vetvik K, Søderberg-Naucler C, Lindstrøm JC, Bukholm IR, Kristensen VN, Geisler J (2018) Somatic EP300-G211S mutations are associated with overall somatic mutational patterns and breast cancer specific survival in triple-negative breast cancer. Breast Cancer Res Treat 172(2):339–351. https://doi.org/10.1007/s10549-018-4927-3

Bestor TH (1992) Activation of mammalian DNA methyltransferase by cleavage of a Zn binding regulatory domain. EMBO J 11(7):2611–2617. https://doi.org/10.1002/j.1460-2075.1992.tb05326.x

Bestor T, Laudano A, Mattaliano R, Ingram V (1988) Cloning and sequencing of a cDNA encoding DNA methyltransferase of mouse cells. The3 carboxyl-terminal domain of the mammalian enzymes is related to bacterial restriction methyltransferases. J Mol Biol 203(4):971–983. https://doi.org/10.1016/0022-2836(88)90122-2

Bi Y, Kong P, Zhang L, Cui H, Xu X, Chang F, Yan T, Li J, Cheng C, Song B, Niu X, Liu X, Liu X, Xu E, Hu X, Qian Y, Wang F, Li H, Ma Y, Yang J, Cheng X (2019) EP300 as an oncogene correlates with poor prognosis in esophageal squamous carcinoma. J Cancer 10(22):5413–5426. https://doi.org/10.7150/jca.34261

Bill M, Mrózek K, Kohlschmidt J, Eisfeld AK, Walker CJ, Nicolet D, Papaioannou D, Blachly JS, Orwick S, Carroll AJ, Kolitz JE, Powell BL, Stone RM, de la Chapelle A, Byrd JC, Bloomfield CD (2020) Mutational landscape and clinical outcome of patients with de novo acute myeloid leukemia and rearrangements involving 11q23/KMT2A. Proc Natl Acad Sci U S A 117(42): 26340–26346. https://doi.org/10.1073/pnas.2014732117

Blanc RS, Richard S (2017) Arginine methylation: the coming of age. Mol Cell 65(1):8–24. https://doi.org/10.1016/j.molcel.2016.11.003

Bond J, Touzart A, Leprêtre S, Graux C, Bargetzi M, Lhermitte L, Hypolite G, Leguay T, Hicheri Y, Guillerm G, Bilger K, Lhéritier V, Hunault M, Huguet F, Chalandon Y, Ifrah N, Macintyre E, Dombret H, Asnafi V, Boissel N (2019) DNMT3A mutation is associated with increased age and adverse outcome in adult T-cell acute lymphoblastic leukemia. Haematologica 104(8):1617–1625. https://doi.org/10.3324/haematol.2018.197848

Branscombe TL, Frankel A, Lee JH, Cook JR, Yang Z, Pestka S, Clarke S (2001) PRMT5 (Janus kinase-binding protein 1) catalyzes the formation of symmetric dimethylarginine residues in proteins. J Biol Chem 276(35):32971–32976. https://doi.org/10.1074/jbc.M105412200

Brenner C, Deplus R, Didelot C, Loriot A, Viré E, De Smet C, Gutierrez A, Danovi D, Bernard D, Boon T, Pelicci PG, Amati B, Kouzarides T, de Launoit Y, Di Croce L, Fuks F (2005) Myc represses transcription through recruitment of DNA methyltransferase corepressor. EMBO J 24(2):336–346. https://doi.org/10.1038/sj.emboj.7600509

Bushara O, Wester JR, Jacobsen D, Sun L, Weinberg S, Gao J, Jennings LJ, Wang L, Lauberth SM, Yue F, Liao J, Yang GY (2023) Clinical and histopathologic characterization of SETD2-mutated colorectal cancer. Hum Pathol 131:9–16. https://doi.org/10.1016/j.humpath.2022.12.001

Cardoso MC, Leonhardt H (1999) DNA methyltransferase is actively retained in the cytoplasm during early development. J Cell Biol 147(1):25–32. https://doi.org/10.1083/jcb.147.1.25

Chedin F, Lieber MR, Hsieh CL (2002) The DNA methyltransferase-like protein DNMT3L stimulates de novo methylation by Dnmt3a. Proc Natl Acad Sci U S A 99(26):16916–16921. https://doi.org/10.1073/pnas.262443999

Chen BF, Chan WY (2014) The de novo DNA methyltransferase DNMT3A in development and cancer. Epigenetics 9(5):669–677. https://doi.org/10.4161/epi.28324

Chen D, Ma H, Hong H, Koh SS, Huang SM, Schurter BT, Aswad DW, Stallcup MR (1999) Regulation of transcription by a protein methyltransferase. Science 284(5423):2174–2177. https://doi.org/10.1126/science.284.5423.2174

Chen T, Ueda Y, Xie S, Li E (2002) A novel Dnmt3a isoform produced from an alternative promoter localizes to euchromatin and its expression correlates with active de novo methylation. J Biol Chem 277(41):38746–38754. https://doi.org/10.1074/jbc.M205312200

Chen T, Tsujimoto N, Li E (2004) The PWWP domain of Dnmt3a and Dnmt3b is required for directing DNA methylation to the major satellite repeats at pericentric heterochromatin. Mol Cell Biol 24(20):9048–9058. https://doi.org/10.1128/MCB.24.20.9048-9058.2004

Chen S, Chen Y, Lu J, Yuan D, He L, Tan H, Xu L (2020) Bioinformatics analysis identifies key genes and pathways in acute myeloid leukemia associated with DNMT3A mutation. Biomed Res Int 2020:9321630. https://doi.org/10.1155/2020/9321630

Chen Z, Chen C, Li L, Zhang T, Wang X (2021a) Pan-Cancer analysis reveals that E1A binding protein p300 mutations increase genome instability and antitumor immunity. Front Cell Dev Biol 9:729927. https://doi.org/10.3389/fcell.2021.729927

Chen X, Lin X, Pang G, Deng J, Xie Q, Zhang Z (2021b) Significance of KDM6A mutation in bladder cancer immune escape. BMC Cancer 21(1):635. https://doi.org/10.1186/s12885-021-08372-9

Cheng X (1995) Structure and function of DNA methyltransferases. Annu Rev Biophys Biomol Struct 24:293–318. https://doi.org/10.1146/annurev.bb.24.060195.001453

Cheng X, Blumenthal RM (2008) Mammalian DNA methyltransferases: a structural perspective. Structure 16(3):341–350. https://doi.org/10.1016/j.str.2008.01.004

Chu MQ, Zhang TJ, Xu ZJ, Gu Y, Ma JC, Zhang W, Wen XM, Lin J, Qian J, Zhou JD (2020) EZH2 dysregulation: potential biomarkers predicting prognosis and guiding treatment choice in acute myeloid leukaemia. J Cell Mol Med 24(2):1640–1649. https://doi.org/10.1111/jcmm.14855

Chu X, Zhong L, Dan W, Wang X, Zhang Z, Liu Z, Lu Y, Shao X, Zhou Z, Chen S, Liu B (2022) DNMT3A R882H mutation drives daunorubicin resistance in acute myeloid leukemia via regulating NRF2/NQO1 pathway. Cell Commun Signal 20(1):168. https://doi.org/10.1186/s12964-022-00978-1

Chuang LS, Ian HI, Koh TW, Ng HH, Xu G, Li BF (1997) Human DNA-(cytosine-5) methyltransferase-PCNA complex as a target for p21WAF1. Science 277(5334):1996–2000. https://doi.org/10.1126/science.277.5334.1996

Ciccone DN, Su H, Hevi S, Gay F, Lei H, Bajko J, Xu G, Li E, Chen T (2009) KDM1B is a histone H3K4 demethylase required to establish maternal genomic imprints. Nature 461(7262): 415–418. https://doi.org/10.1038/nature08315

Cloos PA, Christensen J, Agger K, Helin K (2008) Erasing the methyl mark: histone demethylases at the center of cellular differentiation and disease. Genes Dev 22(9):1115–1140. https://doi.org/10.1101/gad.1652908

Cui J, Zheng L, Zhang Y, Xue M (2021a) Bioinformatics analysis of DNMT1 expression and its role in head and neck squamous cell carcinoma prognosis. Sci Rep 11(1):2267. https://doi.org/10.1038/s41598-021-81971-5

Cui Y, Li H, Zhan H, Han T, Dong Y, Tian C, Guo Y, Yan F, Dai D, Liu P (2021b) Identification of potential biomarkers for liver cancer through gene mutation and clinical characteristics. Front Oncol 11:733478. https://doi.org/10.3389/fonc.2021.733478

Cuthbert GL, Daujat S, Snowden AW, Erdjument-Bromage H, Hagiwara T, Yamada M, Schneider R, Gregory PD, Tempst P, Bannister AJ, Kouzarides T (2004) Histone deimination antagonizes arginine methylation. Cell 118(5):545–553. https://doi.org/10.1016/j.cell.2004.08.020

Dancy BM, Cole PA (2015) Protein lysine acetylation by p300/CBP. Chem Rev 115(6): 2419–2452. https://doi.org/10.1021/cr500452k

Del Castillo Falconi VM, Torres-Arciga K, Matus-Ortega G, Díaz-Chávez J, Herrera LA (2022) DNA methyltransferases: From evolution to clinical applications. Int J Mol Sci 23(16):8994. https://doi.org/10.3390/ijms23168994

Di Martile M, Del Bufalo D, Trisciuoglio D (2016) The multifaceted role of lysine acetylation in cancer: prognostic biomarker and therapeutic target. Oncotarget 7(34):55789–55810. https://doi.org/10.18632/oncotarget.10048

Dillon SC, Zhang X, Trievel RC, Cheng X (2005) The SET-domain protein superfamily: protein lysine methyltransferases. Genome Biol 6(8):227. https://doi.org/10.1186/gb-2005-6-8-227

Ding F, Chaillet JR (2002) In vivo stabilization of the Dnmt1 (cytosine-5)-methyltransferase protein. Proc Natl Acad Sci U S A 99(23):14861–14866. https://doi.org/10.1073/pnas.232565599

Du MY, Xu M, Deng J, Liu L, Guo T, Xia LH, Hu Y, Mei H (2020) Evaluation of different scoring systems and gene mutations for the prognosis of myelodysplastic syndrome (MDS) in Chinese population. J Cancer 11(2):508–519. https://doi.org/10.7150/jca.30363

Easwaran HP, Schermelleh L, Leonhardt H, Cardoso MC (2004) Replication-independent chromatin loading of Dnmt1 during G2 and M phases. EMBO Rep 5(12):1181–1186. https://doi.org/10.1038/sj.embor.7400295

Egger G, Liang G, Aparicio A, Jones PA (2004) Epigenetics in human disease and prospects for epigenetic therapy. Nature 429(6990):457–463. https://doi.org/10.1038/nature02625

Ehrlich M, Lacey M (2013) DNA hypomethylation and hemimethylation in cancer. Adv Exp Med Biol 754:31–56. https://doi.org/10.1007/978-1-4419-9967-2_2

El Gammal MM, Ebid GT, Madney YM, Abo-Elazm OM, Kelany AK, Torra OS, Radich JP (2019) Clinical effect of combined mutations in DNMT3A, FLT3-ITD, and NPM1 among Egyptian acute myeloid leukemia patients. Clin Lymphoma Myeloma Leuk 19(6):e281–e290. https://doi.org/10.1016/j.clml.2019.02.001

Elhardt W, Shanmugam R, Jurkowski TP, Jeltsch A (2015) Somatic cancer mutations in the DNMT2 tRNA methyltransferase alter its catalytic properties. Biochimie 112:66–72. https://doi.org/10.1016/j.biochi.2015.02.022

Elrhman HAEA, El-Meligui YM, Elalawi SM (2021) Prognostic impact of concurrent DNMT3A, FLT3 and NPM1 gene mutations in acute myeloid leukemia patients. Clin Lymphoma Myeloma Leuk 21(12):e960–e969. https://doi.org/10.1016/j.clml.2021.07.011

Ene CI, Edwards L, Riddick G, Baysan M, Woolard K, Kotliarova S, Lai C, Belova G, Cam M, Walling J, Zhou M, Stevenson H, Kim HS, Killian K, Veenstra T, Bailey R, Song H, Zhang W, Fine HA (2012) Histone demethylase Jumonji D3 (JMJD3) as a tumor suppressor by regulating p53 protein nuclear stabilization. PLoS One 7(12):e51407. https://doi.org/10.1371/journal.pone.0051407

Fatemi M, Hermann A, Pradhan S, Jeltsch A (2001) The activity of the murine DNA methyltransferase Dnmt1 is controlled by interaction of the catalytic domain with the N-terminal part of the enzyme leading to an allosteric activation of the enzyme after binding to methylated DNA. J Mol Biol 309(5):1189–1199. https://doi.org/10.1006/jmbi.2001.4709

Fatemi M, Hermann A, Gowher H, Jeltsch A (2002) Dnmt3a and Dnmt1 functionally cooperate during de novo methylation of DNA. Eur J Biochem 269(20):4981–4984. https://doi.org/10.1046/j.1432-1033.2002.03198.x

Feng Q, Wang H, Ng HH, Erdjument-Bromage H, Tempst P, Struhl K, Zhang Y (2002) Methylation of H3-lysine 79 is mediated by a new family of HMTases without a SET domain. Curr Biol 12(12):1052–1058. https://doi.org/10.1016/s0960-9822(02)00901-6

Feng S, Jacobsen SE, Reik W (2010) Epigenetic reprogramming in plant and animal development. Science 330(6004):622–627. https://doi.org/10.1126/science.1190614

Finnin MS, Donigian JR, Cohen A, Richon VM, Rifkind RA, Marks PA, Breslow R, Pavletich NP (1999) Structures of a histone deacetylase homologue bound to the TSA and SAHA inhibitors. Nature 401(6749):188–193. https://doi.org/10.1038/43710

Finnin MS, Donigian JR, Pavletich NP (2001) Structure of the histone deacetylase SIRT2. Nat Struct Biol 8(7):621–625. https://doi.org/10.1038/89668

Frescas D, Guardavaccaro D, Bassermann F, Koyama-Nasu R, Pagano M (2007) JHDM1B/ FBXL10 is a nucleolar protein that represses transcription of ribosomal RNA genes. Nature 450(7167):309–313. https://doi.org/10.1038/nature06255

Fuks F, Burgers WA, Godin N, Kasai M, Kouzarides T (2001) Dnmt3a binds deacetylases and is recruited by a sequence-specific repressor to silence transcription. EMBO J 20(10):2536–2544. https://doi.org/10.1093/emboj/20.10.2536

Fyodorov DV, Zhou BR, Skoultchi AI, Bai Y (2018) Emerging roles of linker histones in regulating chromatin structure and function. Nat Rev Mol Cell Biol 19(3):192–206. https://doi.org/10.1038/nrm.2017.94

Gantt SL, Joseph CG, Fierke CA (2010) Activation and inhibition of histone deacetylase 8 by monovalent cations. J Biol Chem 285(9):6036–6043. https://doi.org/10.1074/jbc.M109.033399

Gao L, Cueto MA, Asselbergs F, Atadja P (2002) Cloning and functional characterization of HDAC11, a novel member of the human histone deacetylase family. J Biol Chem 277(28): 25748–25755. https://doi.org/10.1074/jbc.M111871200

Gaudet F, Hodgson JG, Eden A, Jackson-Grusby L, Dausman J, Gray JW, Leonhardt H, Jaenisch R (2003) Induction of tumors in mice by genomic hypomethylation. Science 300(5618):489–492. https://doi.org/10.1126/science.1083558

Gowher H, Jeltsch A (2001) Enzymatic properties of recombinant Dnmt3a DNA methyltransferase from mouse: the enzyme modifies DNA in a non-processive manner and also methylates non-CpG [correction of non-CpA] sites. J Mol Biol 309(5):1201–1208. https://doi.org/10.1006/jmbi.2001.4710

Gowher H, Jeltsch A (2002) Molecular enzymology of the catalytic domains of the Dnmt3a and Dnmt3b DNA methyltransferases. J Biol Chem 277(23):20409–20414. https://doi.org/10.1074/jbc.M202148200

Goyal R, Reinhardt R, Jeltsch A (2006) Accuracy of DNA methylation pattern preservation by the Dnmt1 methyltransferase. Nucleic Acids Res 34(4):1182–1188. https://doi.org/10.1093/nar/gkl002

Greer EL, Shi Y (2012) Histone methylation: a dynamic mark in health, disease and inheritance. Nat Rev Genet 13(5):343–357. https://doi.org/10.1038/nrg3173

Guo LC, Zhu WD, Ma XY, Ni H, Zhong EJ, Shao YW, Yu J, Gu DM, Ji SD, Xu HD, Ji C, Yang JM, Zhang Y (2019) Mutations of genes including DNMT3A detected by next-generation sequencing in thyroid cancer. Cancer Biol Ther 20(3):240–246. https://doi.org/10.1080/15384047.2018.1523856

Gupta A, Guerin-Peyrou TG, Sharma GG, Park C, Agarwal M, Ganju RK, Pandita S, Choi K, Sukumar S, Pandita RK, Ludwig T, Pandita TK (2008) The mammalian ortholog of Drosophila MOF that acetylates histone H4 lysine 16 is essential for embryogenesis and oncogenesis. Mol Cell Biol 28(1):397–409. https://doi.org/10.1128/MCB.01045-07

Hallows WC, Albaugh BN, Denu JM (2008) Where in the cell is SIRT3?--functional localization of an NAD+-dependent protein deacetylase. Biochem J 411(2):e11–e13. https://doi.org/10.1042/BJ20080336

Han J, Hubbard BP, Lee J, Montagna C, Lee HW, Sinclair DA, Suh Y (2013) Analysis of 41 cancer cell lines reveals excessive allelic loss and novel mutations in the SIRT1 gene. Cell Cycle 12(2): 263–270. https://doi.org/10.4161/cc.23056

Han W, Zhou F, Wang Z, Hua H, Qin W, Jia Z, Cai X, Chen M, Liu J, Chao H, Lu X (2022) Mutational landscape of chronic myelomonocytic leukemia and its potential clinical significance. Int J Hematol 115(1):21–32. https://doi.org/10.1007/s12185-021-03210-x

Hata K, Okano M, Lei H, Li E (2002) Dnmt3L cooperates with the Dnmt3 family of de novo DNA methyltransferases to establish maternal imprints in mice. Development 129(8):1983–1993. https://doi.org/10.1242/dev.129.8.1983

He J, Kallin EM, Tsukada Y, Zhang Y (2008) The H3K36 demethylase Jhdm1b/Kdm2b regulates cell proliferation and senescence through p15(Ink4b). Nat Struct Mol Biol 15(11):1169–1175. https://doi.org/10.1038/nsmb.1499

He YF, Li BZ, Li Z, Liu P, Wang Y, Tang Q, Ding J, Jia Y, Chen Z, Li L, Sun Y, Li X, Dai Q, Song CX, Zhang K, He C, Xu GL (2011) Tet-mediated formation of 5-carboxylcytosine and its excision by TDG in mammalian DNA. Science 333(6047):1303–1307. https://doi.org/10.1126/science.1210944

Hermann A, Goyal R, Jeltsch A (2004) The Dnmt1 DNA-(cytosine-C5)-methyltransferase methylates DNA processively with high preference for hemimethylated target sites. J Biol Chem 279(46):48350–48359. https://doi.org/10.1074/jbc.M403427200

Herz HM, Garruss A, Shilatifard A (2013) SET for life: biochemical activities and biological functions of SET domain-containing proteins. Trends Biochem Sci 38(12):621–639. https://doi.org/10.1016/j.tibs.2013.09.004

Højfeldt JW, Agger K, Helin K (2013) Histone lysine demethylases as targets for anticancer therapy. Nat Rev Drug Discov 12(12):917–930. https://doi.org/10.1038/nrd4154

Hrzic R, Simons CCJM, Schouten LJ, van Engeland M, Brandt PVD, Weijenberg MP (2020) Investigation of sirtuin 1 polymorphisms in relation to the risk of colorectal cancer by molecular subtype. Sci Rep 10(1):3359. https://doi.org/10.1038/s41598-020-60300-2

Hsieh YJ, Kundu TK, Wang Z, Kovelman R, Roeder RG (1999) The TFIIIC90 subunit of TFIIIC interacts with multiple components of the RNA polymerase III machinery and contains a histone-specific acetyltransferase activity. Mol Cell Biol 19(11):7697–7704. https://doi.org/10.1128/MCB.19.11.7697

Huang J, Sengupta R, Espejo AB, Lee MG, Dorsey JA, Richter M, Opravil S, Shiekhattar R, Bedford MT, Jenuwein T, Berger SL (2007) p53 is regulated by the lysine demethylase LSD1. Nature 449(7158):105–108. https://doi.org/10.1038/nature06092

Huang M, Zhang J, Yan C, Li X, Zhang J, Ling R (2019) Small molecule HDAC inhibitors: promising agents for breast cancer treatment. Bioorg Chem 91:103184. https://doi.org/10.1016/j.bioorg.2019.103184

Huether R, Dong L, Chen X, Wu G, Parker M, Wei L, Ma J, Edmonson MN, Hedlund EK, Rusch MC, Shurtleff SA, Mulder HL, Boggs K, Vadordaria B, Cheng J, Yergeau D, Song G, Becksfort J, Lemmon G, Weber C, Downing JR (2014) The landscape of somatic mutations in epigenetic regulators across 1,000 paediatric cancer genomes. Nat Commun 5:3630. https://doi.org/10.1038/ncomms4630

Hur K, Cejas P, Feliu J, Moreno-Rubio J, Burgos E, Boland CR, Goel A (2014) Hypomethylation of long interspersed nuclear element-1 (LINE-1) leads to activation of proto-oncogenes in human colorectal cancer metastasis. Gut 63(4):635–646. https://doi.org/10.1136/gutjnl-2012-304219

Hurst CD, Alder O, Platt FM, Droop A, Stead LF, Burns JE, Burghel GJ, Jain S, Klimczak LJ, Lindsay H, Roulson JA, Taylor CF, Thygesen H, Cameron AJ, Ridley AJ, Mott HR, Gordenin DA, Knowles MA (2017) Genomic subtypes of non-invasive bladder cancer with distinct metabolic profile and female gender bias in KDM6A mutation frequency. Cancer Cell 32(5): 701–715.e7. https://doi.org/10.1016/j.ccell.2017.08.005

Iizuka M, Stillman B (1999) Histone acetyltransferase HBO1 interacts with the ORC1 subunit of the human initiator protein. J Biol Chem 274(33):23027–23034. https://doi.org/10.1074/jbc.274.33.23027

Ito S, Shen L, Dai Q, Wu SC, Collins LB, Swenberg JA, He C, Zhang Y (2011) Tet proteins can convert 5-methylcytosine to 5-formylcytosine and 5-carboxylcytosine. Science 333(6047): 1300–1303. https://doi.org/10.1126/science.1210597

Izutsu K, Ando K, Nishikori M, Shibayama H, Teshima T, Kuroda J, Kato K, Imaizumi Y, Nosaka K, Sakai R, Hojo S, Nakanishi T, Rai S (2021) Phase II study of tazemetostat for relapsed or refractory B-cell non-Hodgkin lymphoma with EZH2 mutation in Japan. Cancer Sci 112(9):3627–3635. https://doi.org/10.1111/cas.15040

Jaffe JD, Wang Y, Chan HM, Zhang J, Huether R, Kryukov GV, Bhang HE, Taylor JE, Hu M, Englund NP, Yan F, Wang Z, Robert McDonald E 3rd, Wei L, Ma J, Easton J, Yu Z, de Beaumount R, Gibaja V, Venkatesan K, Stegmeier F (2013) Global chromatin profiling reveals NSD2 mutations in pediatric acute lymphoblastic leukemia. Nat Genet 45(11):1386–1391. https://doi.org/10.1038/ng.2777

Jair KW, Bachman KE, Suzuki H, Ting AH, Rhee I, Yen RW, Baylin SB, Schuebel KE (2006) De novo CpG island methylation in human cancer cells. Cancer Res 66(2):682–692. https://doi.org/10.1158/0008-5472.CAN-05-1980

Jeltsch A (2002) Beyond Watson and Crick: DNA methylation and molecular enzymology of DNA methyltransferases. Chembiochem 3(4):274–293. https://doi.org/10.1002/1439-7633(20020402)3:4<274::AID-CBIC274>3.0.CO;2-S

Jeong S, Liang G, Sharma S, Lin JC, Choi SH, Han H, Yoo CB, Egger G, Yang AS, Jones PA (2009) Selective anchoring of DNA methyltransferases 3A and 3B to nucleosomes containing methylated DNA. Mol Cell Biol 29(19):5366–5376. https://doi.org/10.1128/MCB.00484-09

Jones PA (2012) Functions of DNA methylation: islands, start sites, gene bodies and beyond. Nat Rev Genet 13(7):484–492. https://doi.org/10.1038/nrg3230

Jones PA, Baylin SB (2002) The fundamental role of epigenetic events in cancer. Nat Rev Genet 3(6):415–428. https://doi.org/10.1038/nrg816

Jung HS, Lin F, Wolpaw A, Reilly AF, Margolskee E, Luo M, Wertheim GB, Li MM (2020) A novel KMT2A-ARHGEF12 fusion gene identified in a high-grade B-cell lymphoma. Cancer Genet 246-247:41–43. https://doi.org/10.1016/j.cancergen.2020.08.003

Jurkowska RZ, Anspach N, Urbanke C, Jia D, Reinhardt R, Nellen W, Cheng X, Jeltsch A (2008) Formation of nucleoprotein filaments by mammalian DNA methyltransferase Dnmt3a in complex with regulator Dnmt3L. Nucleic Acids Res 36(21):6656–6663. https://doi.org/10.1093/nar/gkn747

Jurkowska RZ, Jurkowski TP, Jeltsch A (2011) Structure and function of mammalian DNA methyltransferases. Chembiochem 12:206–222. https://doi.org/10.1002/cbic.201000195

Juskevicius D, Jucker D, Klingbiel D, Mamot C, Dirnhofer S, Tzankov A (2017) Mutations of CREBBP and SOCS1 are independent prognostic factors in diffuse large B cell lymphoma: mutational analysis of the SAKK 38/07 prospective clinical trial cohort. J Hematol Oncol 10(1):70. https://doi.org/10.1186/s13045-017-0438-7

Kalkhoven E (2004) CBP and p300: HATs for different occasions. Biochem Pharmacol 68(6):1145–1155. https://doi.org/10.1016/j.bcp.2004.03.045

Karytinos A, Forneris F, Profumo A, Ciossani G, Battaglioli E, Binda C, Mattevi A (2009) A novel mammalian flavin-dependent histone demethylase. J Biol Chem 284(26):17775–17782. https://doi.org/10.1074/jbc.M109.003087

Kempf JM, Weser S, Bartoschek MD, Metzeler KH, Vick B, Herold T, Völse K, Mattes R, Scholz M, Wange LE, Festini M, Ugur E, Roas M, Weigert O, Bultmann S, Leonhardt H, Schotta G, Hiddemann W, Jeremias I, Spiekermann K (2021) Loss-of-function mutations in the histone methyltransferase EZH2 promote chemotherapy resistance in AML. Sci Rep 11(1):5838. https://doi.org/10.1038/s41598-021-84708-6

Kim IK, McCutcheon JN, Rao G, Liu SV, Pommier Y, Skrzypski M, Zhang YW, Giaccone G (2019) Acquired SETD2 mutation and impaired CREB1 activation confer cisplatin resistance in metastatic non-small cell lung cancer. Oncogene 38(2):180–193. https://doi.org/10.1038/s41388-018-0429-3

Kottakis F, Polytarchou C, Foltopoulou P, Sanidas I, Kampranis SC, Tsichlis PN (2011) FGF-2 regulates cell proliferation, migration, and angiogenesis through an NDY1/KDM2B-miR-101-EZH2 pathway. Mol Cell 43(2):285–298. https://doi.org/10.1016/j.molcel.2011.06.020

Kouzarides T (2007) Chromatin modifications and their function. Cell 128(4):693–705. https://doi.org/10.1016/j.cell.2007.02.005

Kriaucionis S, Heintz N (2009) The nuclear DNA base 5-hydroxymethylcytosine is present in Purkinje neurons and the brain. Science 324(5929):929–930. https://doi.org/10.1126/science.1169786

Lasho TL, Vallapureddy R, Finke CM, Mangaonkar A, Gangat N, Ketterling R, Tefferi A, Patnaik MM (2018) Infrequent occurrence of TET1, TET3, and ASXL2 mutations in myelodysplastic/myeloproliferative neoplasms. Blood Cancer J 8(3):32. https://doi.org/10.1038/s41408-018-0057-8

Lauber C, Correia N, Trumpp A, Rieger MA, Dolnik A, Bullinger L, Roeder I, Seifert M (2020) Survival differences and associated molecular signatures of DNMT3A-mutant acute myeloid leukemia patients. Sci Rep 10(1):12761. https://doi.org/10.1038/s41598-020-69691-8

Lee PJ, Washer LL, Law DJ, Boland CR, Horon IL, Feinberg AP (1996) Limited up-regulation of DNA methyltransferase in human colon cancer reflecting increased cell proliferation. Proc Natl Acad Sci U S A 93(19):10366–10370. https://doi.org/10.1073/pnas.93.19.10366

Lee JH, Voo KS, Skalnik DG (2001) Identification and characterization of the DNA binding domain of CpG-binding protein. J Biol Chem 276(48):44669–44676. https://doi.org/10.1074/jbc.M107179200

Lee MG, Wynder C, Cooch N, Shiekhattar R (2005) An essential role for CoREST in nucleosomal histone 3 lysine 4 demethylation. Nature 437(7057):432–435. https://doi.org/10.1038/nature04021

Leonhardt H, Page AW, Weier HU, Bestor TH (1992) A targeting sequence directs DNA methyltransferase to sites of DNA replication in mammalian nuclei. Cell 71(5):865–873. https://doi.org/10.1016/0092-8674(92)90561-p

Li E, Zhang Y (2014) DNA methylation in mammals. Cold Spring Harb Perspect Biol 6(5): a019133. https://doi.org/10.1101/cshperspect.a019133

Li B, Carey M, Workman JL (2007) The role of chromatin during transcription. Cell 128(4): 707–719. https://doi.org/10.1016/j.cell.2007.01.015

Li X, Li L, Pandey R, Byun JS, Gardner K, Qin Z, Dou Y (2012) The histone acetyltransferase MOF is a key regulator of the embryonic stem cell core transcriptional network. Cell Stem Cell 11(2):163–178. https://doi.org/10.1016/j.stem.2012.04.023

Li Z, Gu TP, Weber AR, Shen JZ, Li BZ, Xie ZG, Yin R, Guo F, Liu X, Tang F, Wang H, Schär P, Xu GL (2015) Gadd45a promotes DNA demethylation through TDG. Nucleic Acids Res 43(8): 3986–3997. https://doi.org/10.1093/nar/gkv283

Liang G, Chan MF, Tomigahara Y, Tsai YC, Gonzales FA, Li E, Laird PW, Jones PA (2002) Cooperativity between DNA methyltransferases in the maintenance methylation of repetitive elements. Mol Cell Biol 22(2):480–491. https://doi.org/10.1128/MCB.22.2.480-491.2002

Liao C, Huang W, Lin M, Li H, Zhang Z, Zhang X, Chen R, Huang M, Yu P, Zhang S (2022) Correlation of KMT2 family mutations with molecular characteristics and prognosis in colorectal cancer. Int J Biol Markers 37(2):149–157. https://doi.org/10.1177/03936155221095574

Liefke R, Oswald F, Alvarado C, Ferres-Marco D, Mittler G, Rodriguez P, Dominguez M, Borggrefe T (2010) Histone demethylase KDM5A is an integral part of the core Notch-RBP-J repressor complex. Genes Dev 24(6):590–601. https://doi.org/10.1101/gad.563210

Lin A, Zhang H, Hu X, Chen X, Wu G, Luo P, Zhang J (2020) Age, sex, and specific gene mutations affect the effects of immune checkpoint inhibitors in colorectal cancer. Pharmacol Res 159:105028. https://doi.org/10.1016/j.phrs.2020.105028

Lister R, Pelizzola M, Dowen RH, Hawkins RD, Hon G, Tonti-Filippini J, Nery JR, Lee L, Ye Z, Ngo QM, Edsall L, Antosiewicz-Bourget J, Stewart R, Ruotti V, Millar AH, Thomson JA, Ren B, Ecker JR (2009) Human DNA methylomes at base resolution show widespread epigenomic differences. Nature 462(7271):315–322. https://doi.org/10.1038/nature08514

Lit BMW, Guo BB, Malherbe JAJ, Kwong YL, Erber WN (2022) Mutation profile of acute myeloid leukaemia in a Chinese cohort by targeted next-generation sequencing. Cancer Rep (Hoboken) 5(10):e1573. https://doi.org/10.1002/cnr2.1573

Liu Y, Oakeley EJ, Sun L, Jost JP (1998) Multiple domains are involved in the targeting of the mouse DNA methyltransferase to the DNA replication foci. Nucleic Acids Res 26(4): 1038–1045. https://doi.org/10.1093/nar/26.4.1038

Liu Z, Oyola MG, Zhou S, Chen X, Liao L, Tien JC, Mani SK, Xu J (2015) Knockout of the histone demethylase Kdm3b decreases spermatogenesis and impairs male sexual behaviors. Int J Biol Sci 11(12):1447–1457. https://doi.org/10.7150/ijbs.13795

Liu K, Wang JF, Zhan Y, Kong DL, Wang C (2021) Prognosis model of colorectal cancer patients based on NOTCH3, KMT2C, and CREBBP mutations. J Gastrointest Oncol 12(1):79–88. https://doi.org/10.21037/jgo-21-28

Liu Y, Li Y, Xu H, Zhou L, Yang X, Wang C (2023) Exploration of morphological features of clear cell renal cell carcinoma with PBRM1, SETD2, BAP1, or KDM5C mutations. Int J Surg Pathol 10668969231157317. https://doi.org/10.1177/10668969231157317

Lorsbach RB, Moore J, Mathew S, Raimondi SC, Mukatira ST, Downing JR (2003) TET1, a member of a novel protein family, is fused to MLL in acute myeloid leukemia containing the t (10;11)(q22;q23). Leukemia 17(3):637–641. https://doi.org/10.1038/sj.leu.2402834

Lu T, Jackson MW, Wang B, Yang M, Chance MR, Miyagi M, Gudkov AV, Stark GR (2010) Regulation of NF-kappaB by NSD1/FBXL11-dependent reversible lysine methylation of p65. Proc Natl Acad Sci U S A 107(1):46–51. https://doi.org/10.1073/pnas.0912493107

Luo M, Zhang Y, Xu Z, Lv S, Wei Q, Dang Q (2023) Experimental analysis of bladder cancer-associated mutations in EP300 identifies EP300-R1627W as a driver mutation. Mol Med 29(1):7. https://doi.org/10.1186/s10020-023-00608-7

Lyko F (2018) The DNA methyltransferase family: a versatile toolkit for epigenetic regulation. Nat Rev Genet 19(2):81–92. https://doi.org/10.1038/nrg.2017.80

Mahasneh AA, Alnegresh FS, Alfaqih MA (2021) Mutational analysis of EZH2 gene in patients with colorectal adenoma reveals a genetic variant associated with risk of malignant transformation. Asian Pac J Cancer Prev 22(12):4085–4094. https://doi.org/10.31557/APJCP.2021.22.12.4085

Martínez-Laperche C, Sanz-Villanueva L, Díaz Crespo FJ, Muñiz P, Martín Rojas R, Carbonell D, Chicano M, Suárez-González J, Menárguez J, Kwon M, Diez Martín JL, Buño I, Bastos Oreiro M (2022) EZH2 mutations at diagnosis in follicular lymphoma: a promising biomarker to guide frontline treatment. BMC Cancer 22(1):982. https://doi.org/10.1186/s12885-022-10070-z

Massoth LR, Hung YP, Nardi V, Nielsen GP, Hasserjian RP, Louissaint A Jr, Fisch AS, Deshpande V, Zukerberg LR, Lennerz JK, Selig M, Glomski K, Patel PJ, Williams KJ, Sokol ES, Alexander BM, Vergilio JA, Ross JS, Pavlick DC, Chebib I, Williams EA (2020) Pan-sarcoma genomic analysis of KMT2A rearrangements reveals distinct subtypes defined by YAP1-KMT2A-YAP1 and VIM-KMT2A fusions. Mod Pathol 33(11):2307–2317. https://doi.org/10.1038/s41379-020-0582-4

McBride AE, Weiss VH, Kim HK, Hogle JM, Silver PA (2000) Analysis of the yeast arginine methyltransferase Hmt1p/Rmt1p and its in vivo function. Cofactor binding and substrate interactions. J Biol Chem 275(5):3128–3136. https://doi.org/10.1074/jbc.275.5.3128

Mei M, Pillai R, Kim S, Estrada-Merly N, Afkhami M, Yang L, Meng Z, Abid MB, Aljurf M, Bacher U, Beitinjaneh A, Bredeson C, Cahn JY, Cerny J, Copelan E, Cutler C, DeFilipp Z, Diaz Perez MA, Farhadfar N, Freytes CO, Nakamura R (2023) The mutational landscape in chronic myelomonocytic leukemia and its impact on allogeneic hematopoietic cell transplantation outcomes: a Center for Blood and Marrow Transplantation Research (CIBMTR) analysis. Haematologica 108(1):150–160. https://doi.org/10.3324/haematol.2021.280203

Melamed P, Yosefzon Y, David C, Tsukerman A, Pnueli L (2018) Tet enzymes, variants, and differential effects on function. Front Cell Dev Biol 6:22. https://doi.org/10.3389/fcell.2018.00022

Michot JM, Quivoron C, Sarkozy C, Danu A, Lazarovici J, Saleh K, El-Dakdouki Y, Goldschmidt V, Bigenwald C, Dragani M, Bahleda R, Baldini C, Arfi-Rouche J, Martin-Romano P, Tselikas L, Gazzah A, Hollebecque A, Lacroix L, Ghez D, Vergé V, Ribrag V (2023) Sequence analyses of relapsed or refractory diffuse large B-cell lymphomas unravel three genetic subgroups of patients and the GNA13 mutant as poor prognostic biomarker, results of LNH-EP1 study. Am J Hematol 98(4):645–657. https://doi.org/10.1002/ajh.26835

Miranda Furtado CL, Dos Santos Luciano MC, Silva Santos RD, Furtado GP, Moraes MO, Pessoa C (2019) Epidrugs: targeting epigenetic marks in cancer treatment. Epigenetics 14(12): 1164–1176. https://doi.org/10.1080/15592294.2019.16405464

Moon I, Kong MG, Ji YS, Kim SH, Park SK, Suh J, Jang MA (2023) Clinical, mutational, and transcriptomic characteristics in elderly Korean individuals with clonal hematopoiesis driver mutations. Ann Lab Med 43(2):145–152. https://doi.org/10.3343/alm.2023.43.2.145

Morschhauser F, Tilly H, Chaidos A, McKay P, Phillips T, Assouline S, Batlevi CL, Campbell P, Ribrag V, Damaj GL, Dickinson M, Jurczak W, Kazmierczak M, Opat S, Radford J, Schmitt A, Yang J, Whalen J, Agarwal S, Adib D, Salles G (2020) Tazemetostat for patients with relapsed or refractory follicular lymphoma: an open-label, single-arm, multicentre, phase 2 trial. Lancet Oncol 21(11):1433–1442. https://doi.org/10.1016/S1470-2045(20)30441-1

Moualla Y, Moassass F, Al-Halbi B, Al-Achkar W, Georgeos M, Yazigi H, Khamis A (2022) Prognostic relevance of DNMT3A, FLT3 and NPM1 mutations in Syrian acute myeloid leukemia patients. Asian Pac J Cancer Prev 23(4):1387–1395. https://doi.org/10.31557/APJCP.2022.23.4.1387

Nakajima K, Kubota H, Kato I, Isobe K, Ueno H, Kozuki K, Tanaka K, Kawabata N, Mikami T, Tamefusa K, Nishiuchi R, Saida S, Umeda K, Hiramatsu H, Adachi S, Takita J (2022) PAX5 alterations in an infant case of KMT2A-rearranged leukemia with lineage switch. Cancer Sci 113(7):2472–2476. https://doi.org/10.1111/cas.15380

Nann D, Ramis-Zaldivar JE, Müller I, Gonzalez-Farre B, Schmidt J, Egan C, Salmeron-Villalobos J, Clot G, Mattern S, Otto F, Mankel B, Colomer D, Balagué O, Szablewski V, Lome-Maldonado-C, Leoncini L, Dojcinov S, Chott A, Copie-Bergman C, Bonzheim I, Quintanilla-Martinez L (2020) Follicular lymphoma t(14;18)-negative is genetically a heterogeneous disease. Blood Adv 4(22):5652–5665. https://doi.org/10.1182/bloodadvances.2020002944

Narayanan D, Pozdnyakova O, Hasserjian RP, Patel SS, Weinberg OK (2021) Effect of DNMT3A variant allele frequency and double mutation on clinicopathologic features of patients with de novo AML. Blood Adv 5(11):2539–2549. https://doi.org/10.1182/bloodadvances.2021004250

Nassar AH, Umeton R, Kim J, Lundgren K, Harshman L, Van Allen EM, Preston M, Dong F, Bellmunt J, Mouw KW, Choueiri TK, Sonpavde G, Kwiatkowski DJ (2019) Mutational analysis of 472 urothelial carcinoma across grades and anatomic sites. Clin Cancer Res 25(8): 2458–2470. https://doi.org/10.1158/1078-0432.CCR-18-3147

Nicholson TB, Chen T (2009) LSD1 demethylates histone and non-histone proteins. Epigenetics 4(3):129–132. https://doi.org/10.4161/epi.4.3.8443

Nickerson ML, Dancik GM, Im KM, Edwards MG, Turan S, Brown J, Ruiz-Rodriguez C, Owens C, Costello JC, Guo G, Tsang SX, Li Y, Zhou Q, Cai Z, Moore LE, Lucia MS, Dean M, Theodorescu D (2014) Concurrent alterations in TERT, KDM6A, and the BRCA pathway in bladder cancer. Clin Cancer Res 20(18):4935–4948. https://doi.org/10.1158/1078-0432.CCR-14-0330

Nimura K, Ishida C, Koriyama H, Hata K, Yamanaka S, Li E, Ura K, Kaneda Y (2006) Dnmt3a2 targets endogenous Dnmt3L to ES cell chromatin and induces regional DNA methylation. Genes Cells 11(10):1225–1237. https://doi.org/10.1111/j.1365-2443.2006.01012.x

Ojesina AI, Lichtenstein L, Freeman SS, Pedamallu CS, Imaz-Rosshandler I, Pugh TJ, Cherniack AD, Ambrogio L, Cibulskis K, Bertelsen B, Romero-Cordoba S, Treviño V, Vazquez-Santillan-K, Guadarrama AS, Wright AA, Rosenberg MW, Duke F, Kaplan B, Wang R, Nickerson E, Meyerson M (2014) Landscape of genomic alterations in cervical carcinomas. Nature 506(7488):371–375. https://doi.org/10.1038/nature12881

Okano M, Xie S, Li E (1998) Cloning and characterization of a family of novel mammalian DNA (cytosine-5) methyltransferases. Nat Genet 19(3):219–220. https://doi.org/10.1038/890

Okano M, Bell DW, Haber DA, Li E (1999) DNA methyltransferases Dnmt3a and Dnmt3b are essential for de novo methylation and mammalian development. Cell 99(3):247–257. https://doi.org/10.1016/s0092-8674(00)81656-6

Olinski R, Starczak M, Gackowski D (2016) Enigmatic 5-hydroxymethyluracil: Oxidatively modified base, epigenetic mark or both? Mutat Res Rev Mutat Res 767:59–66. https://doi.org/10.1016/j.mrrev.2016.02.001

Oliveira PH, Fang G (2021) Conserved DNA methyltransferases: A window into fundamental mechanisms of epigenetic regulation in bacteria. Trends Microbiol 29:28–40. https://doi.org/10.1016/j.tim.2020.04.007

Oñate-Ocaña LF, Ponce-Martínez M, Taja-Chayeb L, Gutiérrez-Hernández O, Avilés-Salas A, Cantú-de-León D, Dueñas-González A, Candelaria-Hernández M (2021) A cohort study of the prognostic impact of exon-16 EZH2 mutations in a mexican-mestizo population of patients with diffuse large B-cell lymphoma. Rev Investig Clin 73(6):362–370. https://doi.org/10.24875/RIC.21000070

Onufriev AV, Schiessel H (2019) The nucleosome: from structure to function through physics. Curr Opin Struc Biol 56:119–130. https://doi.org/10.1016/j.sbi.2018.11.003

Ooi SK, Qiu C, Bernstein E, Li K, Jia D, Yang Z, Erdjument-Bromage H, Tempst P, Lin SP, Allis CD, Cheng X, Bestor TH (2007) DNMT3L connects unmethylated lysine 4 of histone H3 to de novo methylation of DNA. Nature 448(7154):714–717. https://doi.org/10.1038/nature05987

Orsmark-Pietras C, Landberg N, Lorenz F, Uggla B, Höglund M, Lehmann S, Derolf Å, Deneberg S, Antunovic P, Cammenga J, Möllgård L, Wennström L, Lilljebjörn H, Rissler M, Fioretos T, Lazarevic VL (2021) Clinical and genomic characterization of patients diagnosed with the provisional entity acute myeloid leukemia with BCR-ABL1, a Swedish population-based study. Genes Chromosomes Cancer 60(6):426–433. https://doi.org/10.1002/gcc.22936

Otani J, Nankumo T, Arita K, Inamoto S, Ariyoshi M, Shirakawa M (2009) Structural basis for recognition of H3K4 methylation status by the DNA methyltransferase 3A ATRX-DNMT3-DNMT3L domain. EMBO Rep 10(11):1235–1241. https://doi.org/10.1038/embor.2009.218

Oyer JA, Huang X, Zheng Y, Shim J, Ezponda T, Carpenter Z, Allegretta M, Okot-Kotber CI, Patel JP, Melnick A, Levine RL, Ferrando A, Mackerell AD Jr, Kelleher NL, Licht JD, Popovic R (2014) Point mutation E1099K in MMSET/NSD2 enhances its methyltranferase activity and leads to altered global chromatin methylation in lymphoid malignancies. Leukemia 28(1): 198–201. https://doi.org/10.1038/leu.2013.204

Panagopoulos I, Gorunova L, Bjerkehagen B, Heim S (2015) Novel KAT6B-KANSL1 fusion gene identified by RNA sequencing in retroperitoneal leiomyoma with t(10;17)(q22;q21). PLoS One 10(1):e0117010. https://doi.org/10.1371/journal.pone.0117010

Pao GM, Janknecht R, Ruffner H, Hunter T, Verma IM (2000) CBP/p300 interact with and function as transcriptional coactivators of BRCA1. Proc Natl Acad Sci U S A 97(3):1020–1025. https://doi.org/10.1073/pnas.97.3.1020

Paolinelli R, Mendoza-Maldonado R, Cereseto A, Giacca M (2009) Acetylation by GCN5 regulates CDC6 phosphorylation in the S phase of the cell cycle. Nat Struct Mol Biol 16(4):412–420. https://doi.org/10.1038/nsmb.1583

Park JW, Han JW (2019) Targeting epigenetics for cancer therapy. Arch Pharm Res 42(2):159–170. https://doi.org/10.1007/s12272-019-01126-z

Parra M, Verdin E (2010) Regulatory signal transduction pathways for class IIa histone deacetylases. Curr Opin Pharmacol 10(4):454–460. https://doi.org/10.1016/j.coph.2010.04.004

Pasini D, Hansen KH, Christensen J, Agger K, Cloos PA, Helin K (2008) Coordinated regulation of transcriptional repression by the RBP2 H3K4 demethylase and Polycomb-Repressive Complex 2. Genes Dev 22(10):1345–1355. https://doi.org/10.1101/gad.470008

Perez-Campo FM, Costa G, Lie-a-Ling M, Kouskoff V, Lacaud G (2013) The MYSTerious MOZ, a histone acetyltransferase with a key role in haematopoiesis. Immunology 139(2):161–165. https://doi.org/10.1111/imm.12072

Polytarchou C, Pfau R, Hatziapostolou M, Tsichlis PN (2008) The JmjC domain histone demethylase Ndy1 regulates redox homeostasis and protects cells from oxidative stress. Mol Cell Biol 28(24):7451–7464. https://doi.org/10.1128/MCB.00688-08

Ponciano-Gómez A, Martínez-Tovar A, Vela-Ojeda J, Olarte-Carrillo I, Centeno-Cruz F, Garrido E (2017) Mutations in TET2 and DNMT3A genes are associated with changes in global and gene-specific methylation in acute myeloid leukemia. Tumour Biol 39(10):1010428317732181. https://doi.org/10.1177/1010428317732181

Ponger L, Li W-H (2005) Evolutionary diversification of DNA methyltransferases in eukaryotic genomes. Mol Biol Evol 22:1119–1128. https://doi.org/10.1093/molbev/msi098

Prachayasittikul V, Prathipati P, Pratiwi R, Phanus-Umporn C, Malik AA, Schaduangrat N, Seenprachawong K, Wongchitrat P, Supokawej A, Prachayasittikul V, Wikberg JE, Nantasenamat C (2017) Exploring the epigenetic drug discovery landscape. Expert Opin Drug Discov 12(4):345–362. https://doi.org/10.1080/17460441.2017.1295954

Pradhan M, Estève PO, Chin HG, Samaranayke M, Kim GD, Pradhan S (2008) CXXC domain of human DNMT1 is essential for enzymatic activity. Biochemistry 47(38):10000–10009. https://doi.org/10.1021/bi8011725

Qi T, Rong X, Feng Q, Sun H, Cao H, Yang Y, Feng H, Zhu L, Wang L, Du Q (2021) Somatic mutation profiling of papillary thyroid carcinomas by whole-exome sequencing and its relationship with clinical characteristics. Int J Med Sci 18(12):2532–2544. https://doi.org/10.7150/ijms.50916

Qin J, Wen B, Liang Y, Yu W, Li H (2020) Histone modifications and their role in colorectal cancer (Review). Pathol Oncol Res 26(4):2023–2033. https://doi.org/10.1007/s12253-019-00663-8

Qiu T, Wang X, Du F, Hu X, Sun F, Song C, Zhao J (2022) TET1 mutations as a predictive biomarker for immune checkpoint inhibitors in colon adenocarcinoma. World J Surg Oncol 20(1):115. https://doi.org/10.1186/s12957-022-02581-7

Ramírez-Ramírez R, Gutiérrez-Angulo M, Peregrina-Sandoval J, Moreno-Ortiz JM, Franco-Topete RA, Cerda-Camacho FJ, Ayala-Madrigal ML (2020) Somatic deletion of KDM1A/LSD1 gene is associated to advanced colorectal cancer stages. J Clin Pathol 73(2):107–111. https://doi.org/10.1136/jclinpath-2019-206128

Rea S, Eisenhaber F, O'Carroll D, Strahl BD, Sun ZW, Schmid M, Opravil S, Mechtler K, Ponting CP, Allis CD, Jenuwein T (2000) Regulation of chromatin structure by site-specific histone H3 methyltransferases. Nature 406(6796):593–599. https://doi.org/10.1038/35020506

Robertson KD, Uzvolgyi E, Liang G, Talmadge C, Sumegi J, Gonzales FA, Jones PA (1999) The human DNA methyltransferases (DNMTs) 1, 3a and 3b: coordinate mRNA expression in normal tissues and overexpression in tumors. Nucleic Acids Res 27(11):2291–2298. https://doi.org/10.1093/nar/27.11.2291

Rountree MR, Bachman KE, Baylin SB (2000) DNMT1 binds HDAC2 and a new co-repressor, DMAP1, to form a complex at replication foci. Nat Genet 25(3):269–277. https://doi.org/10.1038/77023

Sabò A, Lusic M, Cereseto A, Giacca M (2008) Acetylation of conserved lysines in the catalytic core of cyclin-dependent kinase 9 inhibits kinase activity and regulates transcription. Mol Cell Biol 28(7):2201–2212. https://doi.org/10.1128/MCB.01557-07

Saglam O, Tang Z, Tang G, Medeiros LJ, Toruner GA (2020) KAT6A amplifications are associated with shorter progression-free survival and overall survival in patients with endometrial serous carcinoma. PLoS One 15(9):e0238477. https://doi.org/10.1371/journal.pone.0238477

Salah Ud-Din AI, Tikhomirova A, Roujeinikova A (2016) Structure and functional diversity of GCN5-related N-acetyltransferases (GNAT). Int J Mol Sci 17(7):1018. https://doi.org/10.3390/ijms17071018

Sapountzi V, Côté J (2011) MYST-family histone acetyltransferases: beyond chromatin. Cell Mol Life Sci 68(7):1147–1156. https://doi.org/10.1007/s00018-010-0599-9

Seiler CL, Fernandez J, Koerperich Z, Andersen MP, Kotandeniya D, Nguyen ME, Sham YY, Tretyakova NY (2018) Maintenance DNA methyltransferase activity in the presence of oxidized forms of 5-methylcytosine: structural basis for ten eleven translocation-mediated DNA demethylation. Biochemistry 57(42):6061–6069. https://doi.org/10.1021/acs.biochem.8b00683

Seto E, Yoshida M (2014) Erasers of histone acetylation: the histone deacetylase enzymes. Cold Spring Harb Perspect Biol 6(4):a018713. https://doi.org/10.1101/cshperspect.a018713

Shaikh ARK, Ujjan I, Irfan M, Naz A, Shamsi T, Khan MTM, Shakeel M (2021) TET2 mutations in acute myeloid leukemia: a comprehensive study in patients of Sindh, Pakistan. Peer J 9:e10678. https://doi.org/10.7717/peerj.10678

Shen L, Wu H, Diep D, Yamaguchi S, D'Alessio AC, Fung HL, Zhang K, Zhang Y (2013) Genome-wide analysis reveals TET- and TDG-dependent 5-methylcytosine oxidation dynamics. Cell 153(3):692–706. https://doi.org/10.1016/j.cell.2013.04.002

Shi Y, Lan F, Matson C, Mulligan P, Whetstine JR, Cole PA, Casero RA, Shi Y (2004) Histone demethylation mediated by the nuclear amine oxidase homolog LSD1. Cell 119(7):941–953. https://doi.org/10.1016/j.cell.2004.12.012

Shiba N, Ichikawa H, Taki T, Park MJ, Jo A, Mitani S, Kobayashi T, Shimada A, Sotomatsu M, Arakawa H, Adachi S, Tawa A, Horibe K, Tsuchida M, Hanada R, Tsukimoto I, Hayashi Y (2013) NUP98-NSD1 gene fusion and its related gene expression signature are strongly associated with a poor prognosis in pediatric acute myeloid leukemia. Genes Chromosomes Cancer 52(7):683–693. https://doi.org/10.1002/gcc.22064

Sims RJ 3rd, Nishioka K, Reinberg D (2003) Histone lysine methylation: a signature for chromatin function. Trends Genet 19(11):629–639. https://doi.org/10.1016/j.tig.2003.09.007

Siraj AK, Pratheeshkumar P, Parvathareddy SK, Bu R, Masoodi T, Iqbal K, Al-Rasheed M, Al-Dayel F, Al-Sobhi SS, Alzahrani AS, Al-Dawish M, Al-Kuraya KS (2019) Prognostic significance of DNMT3A alterations in Middle Eastern papillary thyroid carcinoma. Eur J Cancer 117:133–144. https://doi.org/10.1016/j.ejca.2019.05.025

Spaety ME, Gries A, Badie A, Venkatasamy A, Romain B, Orvain C, Yanagihara K, Okamoto K, Jung AC, Mellitzer G, Pfeffer S, Gaiddon C (2019) HDAC4 levels control sensibility toward cisplatin in gastric cancer via the p53-p73/BIK pathway. Cancers (Basel) 11(11):1747. https://doi.org/10.3390/cancers11111747

Stasik S, Juratli TA, Petzold A, Richter S, Zolal A, Schackert G, Dahl A, Krex D, Thiede C (2020) Exome sequencing identifies frequent genomic loss of TET1 in IDH-wild-type glioblastoma. Neoplasia 22(12):800–808. https://doi.org/10.1016/j.neo.2020.10.010

Su J, Wang F, Cai Y, Jin J (2016) The functional analysis of histone acetyltransferase MOF in tumorigenesis. Int J Mol Sci 17(1):99. https://doi.org/10.3390/ijms17010099

Suetake I, Miyazaki J, Murakami C, Takeshima H, Tajima S (2003) Distinct enzymatic properties of recombinant mouse DNA methyltransferases Dnmt3a and Dnmt3b. J Biochem 133(6): 737–744. https://doi.org/10.1093/jb/mvg095

Sun Y, Jiang X, Chen S, Fernandes N, Price BD (2005) A role for the Tip60 histone acetyltransferase in the acetylation and activation of ATM. Proc Natl Acad Sci U S A 102(37):13182–13187. https://doi.org/10.1073/pnas.0504211102

Swaroop A, Oyer JA, Will CM, Huang X, Yu W, Troche C, Bulic M, Durham BH, Wen QJ, Crispino JD, MacKerell AD Jr, Bennett RL, Kelleher NL, Licht JD (2019) An activating mutation of the NSD2 histone methyltransferase drives oncogenic reprogramming in acute lymphocytic leukemia. Oncogene 38(5):671–686. https://doi.org/10.1038/s41388-018-0474-y

Sykes SM, Mellert HS, Holbert MA, Li K, Marmorstein R, Lane WS, McMahon SB (2006) Acetylation of the p53 DNA-binding domain regulates apoptosis induction. Mol Cell 24(6): 841–851. https://doi.org/10.1016/j.molcel.2006.11.026

Tahiliani M, Koh KP, Shen Y, Pastor WA, Bandukwala H, Brudno Y, Agarwal S, Iyer LM, Liu DR, Aravind L, Rao A (2009) Conversion of 5-methylcytosine to 5-hydroxymethylcytosine in mammalian DNA by MLL partner TET1. Science 324(5929):930–935. https://doi.org/10.1126/science.1170116

Taipale M, Rea S, Richter K, Vilar A, Lichter P, Imhof A, Akhtar A (2005) hMOF histone acetyltransferase is required for histone H4 lysine 16 acetylation in mammalian cells. Mol Cell Biol 25(15):6798–6810. https://doi.org/10.1128/MCB.25.15.6798-6810.2005

Teufel DP, Freund SM, Bycroft M, Fersht AR (2007) Four domains of p300 each bind tightly to a sequence spanning both transactivation subdomains of p53. Proc Natl Acad Sci U S A 104(17): 7009–7014. https://doi.org/10.1073/pnas.0702010104

Togasaki E, Takeda J, Yoshida K, Shiozawa Y, Takeuchi M, Oshima M, Saraya A, Iwama A, Yokote K, Sakaida E, Hirase C, Takeshita A, Imai K, Okumura H, Morishita Y, Usui N, Takahashi N, Fujisawa S, Shiraishi Y, Chiba K, Naoe T (2017) Frequent somatic mutations in epigenetic regulators in newly diagnosed chronic myeloid leukemia. Blood Cancer J 7(4):e559. https://doi.org/10.1038/bcj.2017.36

Tovy A, Spiro A, McCarthy R, Shipony Z, Aylon Y, Allton K, Ainbinder E, Furth N, Tanay A, Barton M, Oren M (2017) p53 is essential for DNA methylation homeostasis in naïve embryonic stem cells, and its loss promotes clonal heterogeneity. Genes Dev 31(10):959–972. https://doi.org/10.1101/gad.299198.117

Trisciuoglio D, Di Martile M, Del Bufalo D (2018) Emerging role of histone acetyltransferase in stem cells and cancer. Stem Cells Int 2018:8908751. https://doi.org/10.1155/2018/8908751

van Haaften G, Dalgliesh GL, Davies H, Chen L, Bignell G, Greenman C, Edkins S, Hardy C, O'Meara S, Teague J, Butler A, Hinton J, Latimer C, Andrews J, Barthorpe S, Beare D, Buck G, Campbell PJ, Cole J, Forbes S, Futreal PA (2009) Somatic mutations of the histone H3K27 demethylase gene UTX in human cancer. Nat Genet 41(5):521–523. https://doi.org/10.1038/ng.349

Van Rechem C, Black JC, Boukhali M, Aryee MJ, Gräslund S, Haas W, Benes CH, Whetstine JR (2015) Lysine demethylase KDM4A associates with translation machinery and regulates protein synthesis. Cancer Discov 5(3):255–263. https://doi.org/10.1158/2159-8290.CD-14-1326

Vervoorts J, Lüscher-Firzlaff JM, Rottmann S, Lilischkis R, Walsemann G, Dohmann K, Austen M, Lüscher B (2003) Stimulation of c-MYC transcriptional activity and acetylation by recruitment of the cofactor CBP. EMBO Rep 4(5):484–490. https://doi.org/10.1038/sj.embor.embor821

Viaene AN, Santi M, Rosenbaum J, Li MM, Surrey LF, Nasrallah MP (2018) SETD2 mutations in primary central nervous system tumors. Acta Neuropathol Commun 6(1):123. https://doi.org/10.1186/s40478-018-0623-0

Waddington CH (1942) The epigenotype. Endeavour 1:18–20

Wang L, Dent SY (2014) Functions of SAGA in development and disease. Epigenomics 6(3):329–339. https://doi.org/10.2217/epi.14.22

Wang JK, Tsai MC, Poulin G, Adler AS, Chen S, Liu H, Shi Y, Chang HY (2010) The histone demethylase UTX enables RB-dependent cell fate control. Genes Dev 24(4):327–332. https://doi.org/10.1101/gad.1882610

Wang F, Marshall CB, Ikura M (2013) Transcriptional/epigenetic regulator CBP/p300 in tumorigenesis: structural and functional versatility in target recognition. Cell Mol Life Sci 70(21):3989–4008. https://doi.org/10.1007/s00018-012-1254-4

Wang S, Zhang YX, Huang T, Sui JN, Lu J, Chen XJ, Wang KK, Xi XD, Li JM, Huang JY, Chen B (2018) Mutation profile and associated clinical features in Chinese patients with cytogenetically normal acute myeloid leukemia. Int J Lab Hematol 40(4):408–418. https://doi.org/10.1111/ijlh.12802

Wang T, Liu Z, Wang X, Bai P, Sun A, Shao Z, Luo R, Wu Z, Zhang K, Li W, Xiao W, Duan B, Wang Y, Chen B, Xing J (2020) Identification of potential therapeutic targets in urothelial bladder carcinoma of Chinese population by targeted next-generation sequencing. Cancer Biol Ther 21(8):709–716. https://doi.org/10.1080/15384047.2020.1763148

Wedge E, Hansen JW, Dybedal I, Creignou M, Ejerblad E, Lorenz F, Werlenius O, Ungerstedt J, Holm MS, Nilsson L, Kittang AO, Antunovic P, Rohon P, Andersen MK, Papaemmanuil E, Bernard E, Jädersten M, Hellström-Lindberg E, Grønbæk K, Ljungman P, Friis LS (2021) Allogeneic hematopoietic stem cell transplantation for chronic myelomonocytic leukemia: clinical and molecular genetic prognostic factors in a Nordic population. Transplant Cell Ther 27(12):991.e1–991.e9. https://doi.org/10.1016/j.jtct.2021.08.028

Weisenberger DJ, Velicescu M, Cheng JC, Gonzales FA, Liang G, Jones PA (2004) Role of the DNA methyltransferase variant DNMT3b3 in DNA methylation. Mol Cancer Res 2(1):62–72

Williams K, Christensen J, Rappsilber J, Nielsen AL, Johansen JV, Helin K (2014) The histone lysine demethylase JMJD3/KDM6B is recruited to p53 bound promoters and enhancer elements in a p53 dependent manner. PLoS One 9(5):e96545. https://doi.org/10.1371/journal.pone. 0096545

Wilson S, Fan L, Sahgal N, Qi J, Filipp FV (2017) The histone demethylase KDM3A regulates the transcriptional program of the androgen receptor in prostate cancer cells. Oncotarget 8(18): 30328–30343. https://doi.org/10.18632/oncotarget.15681

Wissmann M, Yin N, Müller JM, Greschik H, Fodor BD, Jenuwein T, Vogler C, Schneider R, Günther T, Buettner R, Metzger E, Schüle R (2007) Cooperative demethylation by JMJD2C and LSD1 promotes androgen receptor-dependent gene expression. Nat Cell Biol 9(3):347–353. https://doi.org/10.1038/ncb1546

Wobser M, Schummer P, Appenzeller S, Kneitz H, Roth S, Goebeler M, Geissinger E, Rosenwald A, Maurus K (2022) Panel sequencing of primary cutaneous B-cell lymphoma. Cancers (Basel) 14(21):5274. https://doi.org/10.3390/cancers14215274

Wood A, Shilatifard A (2004) Posttranslational modifications of histones by methylation. Adv Protein Chem 67:201–222. https://doi.org/10.1016/S0065-3233(04)67008-2

Wu R, Lu Z, Cao Z, Zhang Y (2011) Zinc chelation with hydroxamate in histone deacetylases modulated by water access to the linker binding channel. J Am Chem Soc 133(16):6110–6113. https://doi.org/10.1021/ja111104p

Wu HX, Chen YX, Wang ZX, Zhao Q, He MM, Wang YN, Wang F, Xu RH (2019) Alteration in TET1 as potential biomarker for immune checkpoint blockade in multiple cancers. J Immunother Cancer 7(1):264. https://doi.org/10.1186/s40425-019-0737-3

Xie W, Hu S, Xu J, Chen Z, Medeiros LJ, Tang G (2019) Acute myeloid leukemia with t(8;16)(p11.2;p13.3)/KAT6A-CREBBP in adults. Ann Hematol 98(5):1149–1157. doi: https://doi.org/10.1007/s00277-019-03637-7

Xu H, Duan N, Wang Y, Sun N, Ge S, Li H, Jing X, Liang K, Zhang X, Liu L, Xue C, Zhang C (2020) The clinicopathological and genetic features of ovarian diffuse large B-cell lymphoma. Pathology 52(2):206–212. https://doi.org/10.1016/j.pathol.2019.09.014

Xu F, Cui WQ, Liu C, Feng F, Liu R, Zhang J, Sun CG (2023) Prognostic biomarkers correlated with immune infiltration in non-small cell lung cancer. FEBS Open Bio 13(1):72–88. https://doi.org/10.1002/2211-5463.13501

Yang XJ, Seto E (2008) The Rpd3/Hda1 family of lysine deacetylases: from bacteria and yeast to mice and men. Nat Rev Mol Cell Biol 9(3):206–218. https://doi.org/10.1038/nrm2346

Yang XJ, Ullah M (2007) MOZ and MORF, two large MYSTic HATs in normal and cancer stem cells. Oncogene 26(37):5408–5419. https://doi.org/10.1038/sj.onc.1210609

Yang N, Xu RM (2013) Structure and function of the BAH domain in chromatin biology. Crit Rev Biochem Mol Biol 48(3):211–221. https://doi.org/10.3109/10409238.2012.742035

Yang J, AlTahan AM, Hu D, Wang Y, Cheng PH, Morton CL, Qu C, Nathwani AC, Shohet JM, Fotsis T, Koster J, Versteeg R, Okada H, Harris AL, Davidoff AM (2015) The role of histone demethylase KDM4B in Myc signaling in neuroblastoma. J Natl Cancer Inst 107(6):djv080. https://doi.org/10.1093/jnci/djv080

Yang SX, Hollingshead M, Rubinstein L, Nguyen D, Larenjeira ABA, Kinders RJ, Difilippantonio M, Doroshow JH (2021) TET2 and DNMT3A mutations and exceptional response to 4′-thio-2′-deoxycytidine in human solid tumor models. J Hematol Oncol 14(1): 83. https://doi.org/10.1186/s13045-021-01091-5

Yarychkivska O, Shahabuddin Z, Comfort N, Boulard M, Bestor TH (2018) BAH domains and a histone-like motif in DNA methyltransferase 1 (DNMT1) regulate de novo and maintenance methylation in vivo. J Biol Chem 293(50):19466–19475. https://doi.org/10.1074/jbc.RA118.004612

Yuan XQ, Chen P, Du YX, Zhu KW, Zhang DY, Yan H, Liu H, Liu YL, Cao S, Zhou G, Zeng H, Chen SP, Zhao XL, Yang J, Zeng WJ, Chen XP (2019) Influence of DNMT3A R882 mutations on AML prognosis determined by the allele ratio in Chinese patients. J Transl Med 17(1):220. https://doi.org/10.1186/s12967-019-1959-3

Yuen KY, Liu Y, Zhou YZ, Wang Y, Zhou DH, Fang JP, Xu LH (2023) Mutational landscape and clinical outcome of pediatric acute myeloid leukemia with 11q23/KMT2A rearrangements. Cancer Med 12(2):1418–1430. https://doi.org/10.1002/cam4.5026

Zhang Y, Jurkowska R, Soeroes S, Rajavelu A, Dhayalan A, Bock I, Rathert P, Brandt O, Reinhardt R, Fischle W, Jeltsch A (2010) Chromatin methylation activity of Dnmt3a and Dnmt3a/3L is guided by interaction of the ADD domain with the histone H3 tail. Nucleic Acids Res 38(13):4246–4253. https://doi.org/10.1093/nar/gkq147

Zhang ZM, Liu S, Lin K, Luo Y, Perry JJ, Wang Y, Song J (2015) Crystal structure of human DNA methyltransferase 1. J Mol Biol 427(15):2520–2531. https://doi.org/10.1016/j.jmb.2015.06.001

Zhang MC, Fang Y, Wang L, Cheng S, Fu D, He Y, Zhao Y, Wang CF, Jiang XF, Song Q, Xu PP, Zhao WL (2020a) Clinical efficacy and molecular biomarkers in a phase II study of tucidinostat plus R-CHOP in elderly patients with newly diagnosed diffuse large B-cell lymphoma. Clin Epigenetics 12(1):160. https://doi.org/10.1186/s13148-020-00948-9

Zhang Y, Wu T, Wang Y, Zhao X, Zhao B, Zhao X, Zhang Q, Jin Y, Li Z, Hu X (2020b) The R251Q mutation of LSD1 promotes invasion and migration of luminal breast cancer cells. Int J Biol Macromol 164:4000–4009. https://doi.org/10.1016/j.ijbiomac.2020.08.221

Zhang Y, Wu T, Zhao B, Liu Z, Qian R, Zhang J, Shi Y, Wan Y, Li Z, Hu X (2022) E239K mutation abolishes the suppressive effects of lysine-specific demethylase 1 on migration and invasion of MCF7 cells. Cancer Sci 113(2):489–499. https://doi.org/10.1111/cas.15220

Zhao E, Ding J, Xia Y, Liu M, Ye B, Choi JH, Yan C, Dong Z, Huang S, Zha Y, Yang L, Cui H, Ding HF (2016) KDM4C and ATF4 cooperate in transcriptional control of amino acid metabolism. Cell Rep 14(3):506–519. https://doi.org/10.1016/j.celrep.2015.12.053

Zhao LJ, Fan QQ, Li YY, Ren HM, Zhang T, Liu S, Maa M, Zheng YC, Liu HM (2020) LSD1 deletion represses gastric cancer migration by upregulating a novel miR-142-5p target protein CD9. Pharmacol Res 159:104991. https://doi.org/10.1016/j.phrs.2020.104991

Zhao W, Zhang C, Li Y, Li Y, Liu Y, Sun X, Liu M, Shao R (2022) The prognostic value of the interaction between ASXL1 and TET2 gene mutations in patients with chronic myelomonocytic leukemia: a meta-analysis. Hematology 27(1):367–378. https://doi.org/10.1080/16078454.2021.1958486

Zhong J, Ding L, Bohrer LR, Pan Y, Liu P, Zhang J, Sebo TJ, Karnes RJ, Tindall DJ, van Deursen J, Huang H (2014) p300 acetyltransferase regulates androgen receptor degradation and PTEN-deficient prostate tumorigenesis. Cancer Res 74(6):1870–1880. https://doi.org/10.1158/0008-5472.CAN-13-2485

Zhu G, Pei L, Li Y, Gou X (2020) EP300 mutation is associated with tumor mutation burden and promotes antitumor immunity in bladder cancer patients. Aging (Albany NY) 12(3):2132–2141. https://doi.org/10.18632/aging.102728

Zhu Y, Fu D, Shi Q, Shi Z, Dong L, Yi H, Liu Z, Feng Y, Liu Q, Fang H, Cheng S, Wang L, Tian Q, Xu P, Zhao W (2022) Oncogenic mutations and tumor microenvironment alterations of older patients with diffuse large B-cell lymphoma. Front Immunol 13:842439. https://doi.org/10.3389/fimmu.2022.842439

Zimmerman SM, Nixon SJ, Chen PY, Raj L, Smith SR, Paolini RL, Lin PN, Souroullas GP (2022) Ezh2Y641F mutations co-operate with Stat3 to regulate MHC class I antigen processing and alter the tumor immune response in melanoma. Oncogene 41(46):4983–4993. https://doi.org/10.1038/s41388-022-02492-7

Chapter 3
Introduction to Cancer Epigenetics

Ebru Erzurumluoğlu Gökalp, Sevgi Işık, and Sevilhan Artan

Abstract In recent years, many studies have focused on understanding the effects of genetic and epigenetic mechanisms on carcinogenesis, diagnosing the disease at an early stage, and determining personalized treatment strategies. Epigenetic and genetic alterations are effective in the initiation and progression of cancer, the second most common cause of death worldwide. Epigenetics is defined as heritable changes in gene expression without DNA sequence alterations. Epigenetic mechanisms include DNA methylation, histone modifications, and non-coding RNAs. Disruption of the balance in epigenetic processes, which are necessary for the normal maintenance of tissue-specific gene expression, may cause cancer formation and progression. The reversibility of epigenetic abnormalities is a promising feature for epigenetic cancer therapy studies. This chapter aims to summarize information about epigenetic mechanisms, their role in cancer initiation and progression, and their potential use in cancer therapy.

Keywords Epigenetic mechanisms · Cancer · DNA methylation · DNA demethylation · 5-mC · 5-hmC · TET enzymes · histone modifications · non-coding RNAs · miRNAs

Abbreviations

2HG	2hydroxyglutarate
2-OGDD	2-Oxoglutarate-dependent dioxygenases
5caC	5-Carboxylcytosine
5fC	5-Formylcytosine
5mC	5-Methylcytosine
ADGRE2	Adhesion G protein-coupled receptor E2

E. E. Gökalp · S. Işık · S. Artan (✉)
Department of Medical Genetics, Eskisehir Osmangazi Universit, Medical Faculty, Eskisehir, Turkey
e-mail: sartan@ogu.edu.tr

© The Author(s), under exclusive license to Springer Nature Switzerland AG 2023
R. Kalkan (ed.), *Cancer Epigenetics*, Epigenetics and Human Health 11,
https://doi.org/10.1007/978-3-031-42365-9_3

AML	Acute myeloid leukemia
APAF1	Peptidase activating factor 1
BC	Bladder cancer
BER	Base excision repair
CGI	CpG islands
CHK2	Checkpoint kinase 2
CLL	Chronic lymphocytic leukemia
CTCF	CCCTC binding factor
DGCR8	DiGeorge syndrome critical region gene 8
DMRs	Differentially methylated regions
DNMTs	DNA methyltransferases
DSB	Double-strand break
DUSP1	Dual specificity phosphatase 1 gene
E1	Ubiquitin-activating enzyme 1
E2	Ubiquitin-conjugating enzyme 2
EMT	Epithelial-mesenchymal transition
FOXO1	Fork-headed Box Protein O1
GENIE	Genomics Evidence Neoplasia Information Exchange
H2Bub1	H2B monoubiquitination
HATs	Histone acetyltransferases
HDACs	Histone deacetylases
HIF1	Hypoxia-inducible factor 1
HMG	High mobility group
HP1α	Heterochromatin protein 1α
hTERT	Telomerase reverse transcriptase
ICRs	Imprinting control regions
ISWI	Imitation switch
JAK2	Janus kinase 2
KATs	Lysine acetyltransferases
KDMs	Lysine demethylases
KMTs	Lysine methyltransferases
LINEs	Long interspaced nuclear elements
LOI	Loss of imprinting
MAGE	Melanoma antigen gene
MBD	Methyl-CpG binding domains
MBD2	Methyl-CpG-binding domain protein 2
MeCP	Methyl-CpG binding domain protein
MeCP2	Methyl-CpG-binding protein 2
MET	Mesenchymal-to-epithelial transition
MGMT	O6-methylguanine methyltransferase
miRISC	miRNA-induced silencing complex
miRNA	microRNA
MLL1	Mixed lineage leukemia 1
ncRNAs	Non-coding RNAs

PLCD1	Phospholipase C delta1
PMDs	Partially methylated domains
PRMTs	Arginine methyltransferases
PTPRR	ERK phosphatases protein tyrosine phosphatase receptor type R
RAN	Ras-related nuclear protein
S	Serine
SAM	S-adenosyl methionine
SCLC	Small-cell lung cancers
SRA	SET- and RING-associated
SUMO	Small ubiquitin-like modifier
SWI/SNF	Switching defective/sucrosenon-fermenting complex
T	Threonine
TCGA	The Cancer Genome Atlas
TDG	Thymine DNA glycosylase
TET proteins	Ten-eleven translocation methylcytosine dioxygenases
TMZ	Temozolomide
TNBC	Triple-negative breast cancer
Ub	Ubiquitin
USPs	Ubiquitin-specific peptidases
Y	Tyrosine

3.1 Introduction

Cancer is the second leading cause of death in the world behind cardiovascular disease, understanding its etiology and identifying cancer hallmarks is of significant experimental and clinical importance. Although the process of carcinogenesis and the distinguishing features of cancer, mostly based on gene mutations, have been relatively detailed and some treatment approaches have been discovered, the number of cancer-related deaths is still increasing annually (Liang et al. 2019). The underlying reasons for this are the limitations of targeted clinical therapies due to intratumoral heterogeneity, plasticity, epigenomic structure and dormancy in tumor cells, and the inability to overcome the main obstacles to long-term therapeutic efficacy. In addition, the molecular pathologies involved in the metastatic progression of the tumor have yet to be fully elucidated (Marusyk et al. 2020; Hanahan 2022). It has been determined in the last decade that the epigenomic structure is significantly affected by the changes in the tumor microenvironment, leading to deregulation in gene expression control. Moreover, dormant cells are sustained by epigenetic mechanisms (Basu et al. 2021; Robinson et al. 2020). Since dormancy for cancer cells is essential to acquire new mutations, initiate metastasis, adapt to and survive in a new environment, develop resistance to cancer therapy, and avoid immune damage, understanding the mechanisms of dormancy cell cycle arrest is important for developing new targeted therapeutics (Recasens and Munoz 2019). In line with these developments, Hanahan (2022) has expanded cancer hallmarks by including cellular plasticity, non-mutational epigenetic reprogramming, and polymorphic variations in the tissue/organ microbiome. Since the number of cancer-related deaths is increasing annually, each newly discovered cancer feature is vital for understanding cancer development and metastatic progression mechanisms. These developments are also essential because of their potential to reflect on treatment (Liang et al. 2019).

Tumors consist of millions of cancer cells with neoplastic disruptions, which are embedded in a microenvironment. The startling molecular and cellular heterogeneity in tumors and tumor microenvironment heterogeneity are significantly correlated with the progression of the disease and development of resistance to therapy, consequently, clinical outcome.

The heterogeneity of cellular phenotype in tumors is a complicated and multifactorial phenomenon that combines environmental, epigenetic, and genetic features. Even though the genetic heterogeneity aspect of intratumoral heterogeneity has been studied in detail and understood well, there are still inadequacies in its reflection on clinical medicine (McGranahan and Swanton 2017; Marusyk et al. 2020).

In spite of improvements in understanding the complex molecular pathology of cancer, gene mutations continue to be at the center of molecular oncology, and Bert Vogelstein's famous statement would remain valid for many researchers: "The revolution in cancer research can be summed up in a single sentence: cancer is, in essence, a genetic disease" (Vogelstein and Kinzler 2004). The primary goal of cancer research over the past few decades has been identifying tumor-associated genetic alterations and evaluating their functional and clinical implications (Garraway and Lander 2013; Cheng et al. 2021; Marei et al. 2021; Vogelstein et al. 2013). Thanks to molecular technology improvements, DNA sequencing technology has revealed intratumor genetic heterogeneity, surprisingly. In addition, while the morphological and functional features of each normal cell form its own cellular identity, the observation of deviations in cellular identities in tumor cells without DNA-based mutations helped us to understand that not only gene mutations but also changes in epigenetic regulatory mechanisms are common in the process of carcinogenesis (Liang et al. 2019; Klemm et al. 2019).

It is generally accepted that human cancer cells have epigenetic abnormalities, which is the main topic of this chapter, and that global and/or focal epigenetic alterations may play a key role in the initiation and progression of tumorigenesis (Jones and Baylin 2007; Hassler and Egger 2012; Lafave et al. 2022; Bond et al. 2020). Significant changes in different epigenetic regulatory mechanisms characterize the cancer epigenome. In the process of tumor formation, genetic and epigenetic mechanisms are intertwined and mutually benefit from each other. Genetic mutations in epigenetic regulators can cause alterations in the cancer epigenome, while changes in epigenetic processes can result in genetic mutations (You and Jones 2012).

3.1.1 History of Epigenetics

The fundamental concepts of genetics and heredity were established by Mendel's theories in 1865, the isolation of the DNA molecule in 1869, and the discovery of the double helix structure of DNA almost a century later, in 1959. Conrad H. Waddington, a developmental biologist, created the term "epigenetics" to describe a novel biology area focusing on the connections between gene and protein expression (Waddington 2012). In 1957, Waddington put forth the renowned epigenetic landscape, in which a rough surface (which represents extra- and intracellular environmental factors) allows a ball, representing a cell, to travel in various directions (Goldberg et al. 2007). The discovery of the high mobility group (HMG) proteins in the mid-1970s and early 1980s helped us realize that specific proteins, besides the histones, may play an architectural function in chromatin and affect how phenotypes are expressed. Even though the overall structure of DNA was roughly recognized relatively early in the twentieth century, the field of epigenetics could take off until the discovery of specific enzymes acting as writers and erasers of epigenetic marks in the 1990s and 2000s. The well-known markers, including DNA methylation and post-translational histone modifications, were quickly found after understanding the DNA-double helix structure. DNA methylation was first observed in 1965. Histone modifications, such as methylation, acetylation, ubiquitylation, and phosphorylation, were documented from 1962 to 1977 (Peixoto et al. 2020).

Although Waddington's definition initially concerned the interpretation of the involvement of epigenetics in embryonic development and the link between genotype and phenotype, the definition of "epigenetics" has changed accordingly over the last 80 years and has been redefined multiple times. Understanding how a fertilized egg may develop into an organism made up of hundreds of different types of specialized cells, each of which expresses a specific set of genes with the same genetic material, has long been a goal of researchers. It is now widely acknowledged that specific gene expression patterns determine cellular identity. Establishing and maintaining this expression pattern is necessary. The coordinated action of hundreds of transcription factors, which bind to specific DNA sequences to activate or inhibit the transcription of cell lineage genes, is crucial for maintaining the pluripotency of the initial cell and establishing different cell types. The establishment of this phase concerns the mechanisms by which the genotype produces the phenotype during development, similar to Waddington's first definition of epigenetics. In the maintenance phase, non-DNA sequence-specific chromatin cofactors are involved in setting up and maintaining the chromatin states throughout cell division and for extended periods, even in the lack of transcription factors. This stage is similar to Nanney's original definition of epigenetics as the meiotic/mitotic inheritance of alternate chromatin states without changes in DNA sequence. This definition was later expanded upon by Riggs and Holliday and further changed by Bird and others (Felsenfeld 2014; Peixoto et al. 2020; Cavalli and Heard 2019).

3.1.2 Epigenetics and Epigenome

Although all body cells have essentially the same genetic material and hence the same genes, they are categorized into about 200 cell types depending on morphological and functional features. A highly controlled arrangement of DNA into chromatin is necessary to access the fundamental data of the DNA sequence and establish cell type-specific gene expression profiles that are tightly regulated, both temporally and spatially.

It is well known that chromatin, a macromolecular complex made up of DNA and histone proteins, serves as the scaffold for packing the genome into microscopic nuclei. The ability of genes to be silenced or activated is significantly related to the arrangement of the genome into the compact structure. Although there are various factors affecting both local and global chromatin architecture, the covalent modifications of DNA and histones are mainly involved in the coordination of this process. Since specific combinations of genes are expressed in corresponding cell types, cell type has its own distinctive feature known as *cell identity*. Cellular identity is formed during embryogenesis by constraining the developmental potential of embryonic cells toward tissue-specific stem cells and specialized cell types with differentiation programs. These dynamic events take place in cells that have the same genetic information. In normal cells, the genes having roles in the function of a particular cell type are maintained in an accessible state, while the genes without functions are silenced through epigenetic mechanisms.

The epigenetic mechanisms restrict each cell type's potential; thus, the cell's fate depends on the epigenetic regulation of the genetic code. Therefore, epigenetic mechanisms determine each cell type's potential and play vital roles in mammalian development, differentiation, and homeostasis. The complex interplay between these systems is stable during cell division to preserve cellular identity. However, they also respond to intrinsic cellular signals during development or extrinsic ones for adapting to environmental cues through epigenomic features.

The epigenome combines cellular information encoded in the genome with molecular/chemical information of extracellular and environmental origin. The epigenome and the genome establish their unique gene expression program to define the functional identity unique to each cell type, developmental, or disease process. At the same time, the epigenome plays a role in the development of the organism's ability to respond to environmental stimuli in some cases. Therefore, unlike the fixed genome, the epigenome exhibits dynamic and variable behavior in its response to intracellular and extracellular stimuli.

As a result, while epigenetics is concerned with the processes that control when and how specific genes are activated or silenced, epigenomics deals with the analysis of epigenetic alterations across multiple genes in a cell or an entire organism,

3.2 Epigenetic Machinery

Epigenetic modifications provide chromatin organization by creating inherited transcription conditions responsible for maintaining cellular function, i.e., epigenetic regulation occurs through chromatin modifications, which are formed by the packaging of histone and histone-binding proteins with DNA. Epigenetic machinery is composed of four main groups: DNA methylation, histone post-translational modifications, non-coding RNAs (ncRNAs), and chromatin remodeling (Fig. 3.1). However, many subgroups within each main group, together with chromatin rearrangement complexes, regulate gene transcription by controlling chromatin organization. These are cytosine methylation and, recently detailed, hydroxymethylation-induced DNA modifications, ATP-based chromatin rearrangement, and non-coding RNA-mediated pathways, including microRNA and long non-coding RNA.

Previously, these mechanisms have been extensively reviewed elsewhere, we will summarize them in normal cells and then their roles in the carcinogenesis process in detail.

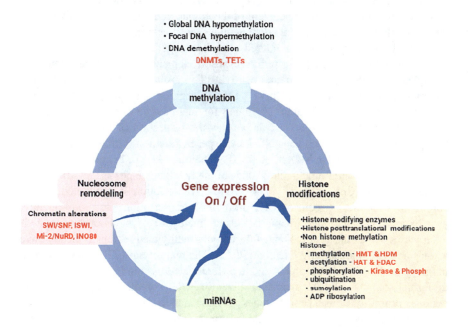

Fig. 3.1 The epigenetic machinery. A collection of related components that work in concert to control both transcriptional and post-transcriptional levels of gene expression make up the epigenetic machinery

3.2.1 DNA Methylation

DNA methylation is the most extensively studied chemical modification in mammals and is now well-known to play a significant regulatory role in the regulation of epigenetic gene expression, developmental processes, cellular differentiation, cell identity establishment, and tissue homeostasis. It alters the functional state of the regulatory areas but has no effect on the cytosine Watson-Crick base pairing rule. Therefore, it exhibits the traditional "epigenetic" signature and has fundamental functions in numerous stable epigenetic suppression mechanisms, including genomic imprinting, X-chromosome inactivation, tissue-specific gene expression, chromosome stability, repression of transposable elements, and aging (Turpin and Salbert 2022; Tucci et al. 2019; Anvar et al. 2021; Cavalli and Heard 2019; Neidhart 2015; Eden et al. 2003; Karpf and Matsui 2005; Smith and Meissner 2013).

The chemical mechanism underlying DNA methylation is the covalent transfer of a methyl (CH3) group from S'Adenosyl methionine to the fifth carbon of the pyrimidine ring of the cytosine (C) base (5-methylcytosine, 5mC) in the CpG dinucleotide under the catalytic action of DNA methyltransferases (DNMTs) (Schübeler 2015; Turpin and Salbert 2022; Ross and Bogdanovic 2019).

However, CpG dinucleotide content of the human genome is not equally distributed throughout the genome. CpG dinucleotides are concentrated in areas with large repetitive genomic sequences scattered all over the genome, such as centromeric repeats, intergenic regions, and retrotransposon elements, and they are generally methylated (70–80%) (Deaton and Bird 2011; Turpin and Salbert 2022). The hypermethylation of large repetitive genomic regions such as pericentromeric, centromeric, and telomeric areas is crucial for maintaining chromosome stability and proper chromosome division, as well as the restriction of the production of transposable elements, such as LINE-1 by hypermethylation (Ortiz-Barahona et al. 2020; Sharma et al. 2010; Roberti et al. 2019; Neidhart 2015). In contrast, less than 10% of total CpGs are found at the 5' ends of many human genes as CpG-rich DNA stretches called "CpG islands" (CGI). While transcription is facilitated by the chromatin structure adjacent to CGI promoters, transcription and, consequently, gene expression is inhibited if CpG islands are methylated. The amount of methylation varies across the genome, and substantially methylated regions typically have lower transcriptional activity (Neidhart 2015). The majority of CGIs usually remain unmethylated during development and in differentiated tissues. Nearly 60% of CGIs in normal somatic cells are mainly localized in gene promoters and the first exon regions, primarily housekeeping genes (Deaton and Bird 2011). However, CGI promoters of some genes that should be transcriptionally silent for a long term during normal development become hypermethylated, such as imprinted genes, the genes located on inactive X-chromosomes, or genes that are exclusively expressed in germ cells but not appropriate to their expressions in somatic cells (Jones and Baylin 2007; Sharma et al. 2010). Besides, CGI hypermethylation in primarily developmentally significant, tissue-specific genes has also been reported (Handy et al. 2011; Roberti et al. 2019).

The genome-wide analyses of the methylome have shown that the methylation position in the transcriptional unit affects gene regulation. Previous studies revealed that although hypermethylation of CGI promoters is blocking the initiation of transcription, gene body methylation may even enhance the elongation of transcription for prevention of the intragenic promoters transcriptions and be involved in alternative splicing regulation (Bond et al. 2020; Neri et al. 2017; Ortiz-Barahona et al. 2020).

On the other hand, DNA methylation alterations occur not only in CGIs and promoters but also in the sequences up to 2 kb from CGIs, which are called CGI "shores." The methylation of CpG shores is associated with transcriptional repression, and methylation patterns in these zones have been reported as tissue-specific, indicating that they play a role in tissue differentiation. Moreover, CGI "shelves," which are located 2 kb upstream and downstream of the CGI shores, have also been identified in the DNA methylation studies. The DNA methylations in different regions and the GC content of these regions have different effects on gene expressions (Nishiyama and Nakanishi 2021; Jones and Baylin 2007).

3.2.1.1 DNA Methyltransferases

During the epigenetic tags incorporation, writers add the marks to chromatin/DNA, whereas readers mediate transcriptional consequences of epigenetic alterations, and finally, erasers remove the added tags.

DNA methyltransferases (DNMTs) are the enzymes responsible for adding the methyl group from S-adenosyl-L-methionine (Ross and Bogdanovic 2019) to cytosine, i.e., DNMTs are DNA methylation "writers." The family comprises five members: DNMT1, DNMT2, DNMT3a, DNMT3b, and DNMT3L. DNA methylation involves three key stages; establishment (de novo methylation), maintenance of methylation, and demethylation. Of DNMT family members, DNMT3A and DNMT3B in combination with DNMT3L are regarded as de novo methylation enzymes targeting unmethylated CpG dinucleotides and establishing new DNA methylation patterns. DNMT3L serves as an accessory partner to the de novo methylation activity of DNMT3A. DNMT3A and DNMT3B play vital roles during early development, and the inactivation of these enzymes results in early embryonic lethality. DNMT1 enzyme recognizes the hemimethylated DNA strands and is responsible for maintaining the methylation process during replication by binding to hemimethylated parental DNA and copying the methylation pattern to fully methylated daughter strands. In the case of aberrant DNA methylation, DNMTs play critical roles. Overexpression of DNMT1, DNMT3a, and DNMT3b has been reported in various solid tumors, such as glioblastoma, gastric, colorectal, pancreatic, hepatic, and lung cancers. In cervical cancers, higher DNMT1 expression was reported in about 70% of the cells, linking to a worse prognosis (Neidhart 2015; Schübeler 2015; Jones and Baylin 2007; Lafave et al. 2022).

3.2.1.2 Methyl-CpG Recognition Proteins

Gene transcription may be impacted by DNA methylation in two different ways: First, DNA methylation itself may physically prevent transcriptional proteins from attaching to the gene. Transcription factors, such as AP-2, c-Myc, E2F, and NF-kB, may be prevented from binding to promoter sites by DNA methylation (Kulis and Esteller 2010). Second, and perhaps more crucially, the established methylated DNA sequences can be read by methyl-CpG binding domain protein (MeCP) families, which then enlist histone deacetylases, a family of enzymes responsible for repressive epigenetic alterations that suppress gene expression and preserve genome integrity (Clouaire and Stancheva 2008; Cheng et al. 2021). MBD1, MBD2, MBD4, and MeCP2 are among the proteins with methyl-CpG binding domains (MBD) and are involved in gene transcription regulation through the cooperation of other proteins. Histone deacetylases and other chromatin remodeling proteins that can change histones are subsequently recruited to the locus by MBDs, resulting in the formation of compact, inactive chromatin known as heterochromatin (Jones and Baylin 2007). It is crucial to understand the relationship between DNA methylation and chromatin structure. Methyl-CpG-binding domain protein 2 (MBD2) regulates the transcriptional silence of hypermethylated genes in cancer, and the lack of methyl-CpG-binding protein 2 (MeCP2) has been linked to Rett syndrome. In contrast to the other four family members, MBD3 attaches to hydroxymethylated DNA rather than methylated DNA (Yildirim et al. 2011). The other family which able to bind 5-mC consists of the ubiquitin-like proteins UHRF1 and UHRF2 (containing PHD and RING fingers domains 1 and 2), which are SET- and RING finger-associated (SRA) domain-containing proteins (Vaughan et al. 2018). Many of these proteins are known to insert repressive histone marks (such as lysine deacetylation and histone lysine/arginine methylation) at their binding sites, either directly or by uptake of proteins that catalyze reactions. Thus, the process of nucleosome remodeling, chromatin compaction, and complex chromatin modifications occur, resulting in transcriptional repression due to the limited access of transcription factors to the promoter.

As previously mentioned, DNMT1 recognizes the hemimethylated DNA for copying the methylated parental DNA strand to form a fully methylated DNA double helix. Therefore, it is responsible for maintaining the methylation process during the replication. The versatile protein UHRF1 is a crucial cofactor for DNMT1 in the process of DNA maintenance methylation (Sharif et al. 2007). The multi-domain protein UHRF1 controls epigenetic changes and mediates between DNA methylation and histone modifications. Through its central SET- and RING-associated (SRA) and C-terminal really fascinating new gene domains, UHRF1 preferentially recognizes hemimethylated DNA and exchanges it by methylating cytosines via its SRA domain at the replication fork. DNMT1 is attracted to its target sites on the freshly synthesized DNA strand by this base-flipping mechanism during the S phase, exposing the unaltered cytosine to DNMT1 (Qin et al. 2015; Berkyurek et al. 2014).

3 Introduction to Cancer Epigenetics

The results of MBD2 inhibition on colon and lung cancer carcinogenesis inhibition seem encouraging. MBD3 interacts with other proteins, including MBD2 and HDAC, to control the methylation process even though it does not directly bind to DNA that has been methylated. MBD4 mutations have been reported in colorectal cancer, endometrial carcinoma, and pancreatic cancers. Additionally, this mutation unexpectedly influences not just CpG sites but also the stability of the entire genome. Because of the interaction between MBD4 and MMR, MBD4 can potentially be crucial for DNA damage repair. In contrast, MeCP2 and the UHRF family seem to stimulate tumor growth when expressed (Mudbhary et al. 2014; Cheng et al. 2021; Cheng et al. 2019).

3.2.1.3 5-Hydroxymethyl Cytosine and TET Enzymes

The enzyme family of 2-oxoglutarate-dependent dioxygenases (2-OGDD) gained a new member in 2009, named Ten-eleven translocation methylcytosine dioxygenases (TET proteins). The ten-eleven translocation (t(10;11)(q22;q23)), which is rarely seen in acute myeloid and lymphocytic leukemia cases, inspired the name of the TET proteins. This structural chromosome aberration caused the fusion of *TET1* gene located on chromosome 10q22 with the mixed lineage leukemia 1 (*MLL1*) gene on chromosome 11q23. TET1 is a Fe(II) and 2-keto glutarate-dependent enzyme involved in the conversion of 5-methyl cytosine dioxygenase to 5-hydroxymethylcytosine (hmC) (Tahiliani et al. 2009). Subsequently, the other members of the TET family, TET2 and TET3, were identified in humans and were shown to possess similar catalytic activity. It is known that the hydroxylation of the 5mC substrate at the CpG dinucleotides to 5hmC can be followed by the sequential oxidation of 5hmC to 5-formyl cytosine (5fC) and to 5-carboxyl cytosine (5caC) by the catalytic activity of the TET enzymes (Ito et al. 2011). For the completion of DNA demethylation, DNA repair enzyme thymine DNA glycosylase (TDG) enzyme recognizes any of these base changes from the genome, which results in the creation of an abasic site. DNA repair mechanisms in the cell (Base excision repair BER) recognize the abasic sites and restore the cytosine in the 5-mC locus (He et al. 2011b).

The TET enzymes are the only recognized "methylation editors" because they catalyze the repetitive oxidation of 5-mC, leading to the demethylation of 5-mC. Because of the demethylation activity of TETs, they can activate transcription and so, they have vital roles in various cellular processes, including embryogenesis, cell differentiation, and tumorigenesis (Ross and Bogdanovic 2019).

3.2.1.4 TET Proteins and DNA Demethylation

As a 2-oxoglutarate/Fe(II)-dependent oxygenase (2OG oxygenase), TETs are iron/ketoglutarate (Fe(II)/KG) dependent dioxygenases. The double-stranded β-helix

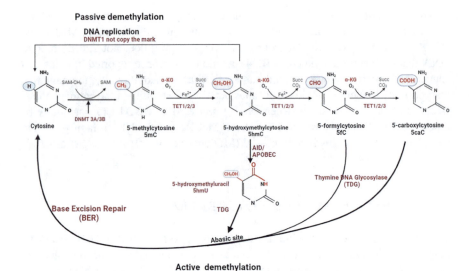

Fig. 3.2 DNA methylation and demethylation mechanisms. DNMT proteins carry out the methylation of cytosines in DNA. The most crucial methylation regulators are DNMT3A/B, while DNMT1 is principally responsible for protecting the 5mC mark during DNA replication. Since DNMT1 does not recognize 5-hmC, it may represent an intermediary in passive demethylation by replication. Two active demethylation mechanisms have recently been identified. The majority of the data points to a route in which TET (Ten-eleven translocation) dioxygenases, which use α-ketoglutarate (α-KG) and iron as cofactors, undergo three consecutive oxidation processes to change 5mC into 5hmC, 5fC, and 5caC. TDG (thymine DNA glycosylase) proteins then identify 5fC and 5caC, activating the base excision repair (BER) process. Additionally, there is evidence for a mechanism in which TDG-mediated BER comes after AID/APOBEC proteins, a group of cytidine deaminases, deaminate 5hmC to 5hmU

(DSBH) and the cysteine-rich domain are at the core catalytic domain at the carboxyl terminus. While the cysteine-rich domain wraps around the DSBH core for stabilizing the overall structure and TET-DNA interaction, the DSBH domain with conserved residues brings Fe(II), KG, and 5mC together for oxidation. Since the methyl group is not involved in the TET–DNA interface, TET can accept various cytosine modifications (Ito et al. 2011; Kao et al. 2016). TET1 and TET3 have a CXXC-type zinc-binding domain, distinguishing methylated and unmethylated DNA at their amino terminus. However, TET2 does not encode a CXXC domain, instead, it is located close to the *IDAX* gene, directly interacting with TET2 and coding a CXCC domain similar to that of other TETs (Pastor et al. 2013).

There are two mechanisms for 5-mC demethylation: passive and active (Fig. 3.2). These mechanisms differ from each other according to whether they are replication dependent or not. *Passive demethylation* is a replication-dependent mechanism in which modified 5mC tags dilute through consecutive cell divisions in the lack of DNMT1-mediated methylation maintenance, and consequently gradually declining degree of methylation. In contrast, the active demethylation mechanism corresponds to a replication-independent mechanism in which methylated Cs are eliminated and replaced with unmodified cytosines through enzymatic activities (Wu and Zhang 2017).

When 5mC is oxidized to 5fC or 5caC, TDG-mediated excision of 5fC or 5caC and BER-dependent repairment of the abasic site can restore unmodified cytosine through the TDG-BER pathway (He et al. 2011b; Wu and Zhang 2017). This process is defined as active modification–active removal (AM–AR) and is independent of DNA replication (Kohli and Zhang 2013). On the other hand, DNA replication can result in the dilution of the oxidized 5mC in restoring the unmodified cytosine pathway; this time, the mechanism is known as active modification-passive dilution. Hemi-modified CpG dyads are produced during DNA replication when unmodified cytosine is integrated into the freshly generated strand. UHRF1 detects a 5mC: C dyad, which aids in bringing DNMT1 to the hemi-5mC location. A CpG site that has been changed with 5hmC, 5fC, or 5caC may become demethylated during several cycles of DNA replication (Wu and Zhang 2017). In regulating the active TET-mediated DNA demethylation, all genes involved can be regulated at the transcriptional, post-transcriptional, and post-translational levels. Moreover, factors belonging to specific genomic regions at which the demethylation process is targeted may also be effective.

The 2-Oxoglutarate (2-OG), also known as α-ketoglutarate (α-KG) and vitamin C, regulates the activity of TET enzymes. In the TET-mediated oxidation processes, oxygen and α-KG are needed as substrates, while Fe(II) is necessary as a cofactor to produce CO_2 and succinate (Kohli and Zhang 2013). Isocitrate dehydrogenase 1 (IDH1), IDH2, and IDH3 are the enzymes responsible for producing α-KG from isocitrate in the Krebs cycle (Losman and Kaelin 2013; Shekhawat et al. 2021). IDH1 or IDH2 overexpression promotes 5hmC synthesis in cells (Waitkus et al. 2015). However, as seen in melanoma and glial tumors, the decreased 5hmC level is linked to IDH2 downregulation (Fig. 3.3). In addition, cancer-related *IDH* mutations cause inhibition of TET activity through the production of 2hydroxyglutarate (2HG) instead of α-KG. The mutant product 2HG is an oncometabolite that challenges α-KG for binding to TET (Xu et al. 2011).

Preimplantation and primordial germ cell development, stem cell differentiation and maintenance, and neuronal functions are biological processes with a global hypomethylation condition that is maintained by 5-hmC through active DNA demethylation. Abnormal DNA demethylation is one of the primary cancer epigenetics subjects and the relation between TET and 5-hmC levels with clinical outcomes in different cancers will be discussed later.

3.2.2 Abnormal Epigenomic Reprogramming in Cancer

Tumor biology is a complex process involving many different mechanisms. Genomic and epigenetic anomalies play a role in the initiation and development of cancer. The genetic and epigenetic basis of cancer has been studied over the past 10 years,

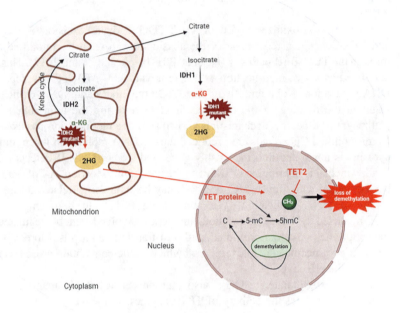

Fig. 3.3 Schematic overview of a cell related to the involvement of *IDH1* and *IDH2* mutations and the resulting loss of TET2 protein demethylation ability in the DNA demethylation process

and the presence of high-frequency changes in numerous epigenetic regulators has clearly demonstrated the crucial role of epigenetic dysregulation in carcinogenesis. During tumorigenesis, the epigenome undergoes many changes, including genome-wide loss of DNA methylation, especially along the repetitive sequences of the genome, regional hypermethylation, mainly in CpG promoter islands of tumor suppressor genes, global changes in histone modification marks, and alterations in networks involving ncRNAs.

Comprehensive investigations of the human cancer genomes have shown that various cancer types have mutations in many key players in the epigenetic control of gene expression, DNA repair, and DNA replication. Cancer initiation and progression frequently result from mutations in epigenetic writers, readers, and editors, as well as components involving chromatin remodeling complex.

3.2.2.1 Cancer-Specific DNA Methylation Alterations

A diagram summarizing the most significant DNA methylation alterations seen in human malignancies is given in Fig. 3.4. These occurrences include DNA hypermethylation at gene promoters, frequently occurring on CpG islands and rendering the afflicted gene silencing. Hypomethylation, or loss of DNA

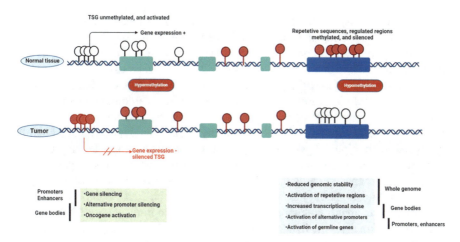

Fig. 3.4 A schematic diagram representing the most significant DNA methylation alterations seen in normal and tumor genomes and genome-scale consequences of methylation alterations. Unmethylated CpG sites are shown by white circles, while methylated CpG sites are shown by red circles. The transcription start location and ongoing loss of transcription following DNA methylation are indicated by the arrows. Exons are demonstrated with green boxes, while the location of repetitive sequences and regulated regions is indicated by the blue rectangle

methylation, affects the entire genome and is frequently found in repeated areas of the genome.

3.2.2.2 Global DNA Hypomethylation

The genome-wide DNA hypomethylation is one of the epigenetics-related hallmarks of cancer and occurs in various genomic regions, including repetitive sequences and regulatory regions. It results in abnormal gene expression, derepression of imprinted genes and retrotransposons, and chromosomal instability (Berdasco and Esteller 2010; Li et al. 2023; Mazloumi et al. 2022; Lozano-Ureña et al. 2021). As already mentioned, hypermethylated pericentromeric, centromeric, and telomeric sequences, preserving chromosomal stability and proper cell division in normal cells, are hypomethylated in tumor cells. Although the majority of CpGs in the genome are known to be 80% methylated, CpG methylation levels in cancer are typically between 40% and 60% (Baylin and Jones 2016). Loss of hypermethylation leads to cell division errors, disrupted chromosome stability, and increased mutation events during multistage carcinogenesis, all classical hallmarks of cancer. The presence of a high frequency of numerical and complex structural chromosome abnormalities are examples seen in tumors (Mazloumi et al. 2022; Pappalardo and Barra 2021). Retrotransposons that are repressed in healthy cells, such as LINEs (long interspaced nuclear elements) and Alu sequences, can be reactivated in cancer cells due to global hypomethylation (Ortiz-Barahona et al. 2020). Studies indicate that up to 50% of cancerous tumors may exhibit retrotransposition activation, which frequently results in structural and copy number changes as well as the induction of oncogene activity. Since the silencing of repetitive genomic regions is through the DNA methylation and repressive chromatin mark, histone H3 lysine 9 (H3K9) methylation, the hypomethylation probably allows the gene expression activation at these repetitive regions (Pfeifer 2018). In most cancer types, including bladder cancer, hepatocellular carcinoma, gastrointestinal stromal tumor, colon cancer, extra-hepatic cholangiocarcinoma, chronic lymphocytic leukemia, ovarian carcinoma, and lung carcinoma, LINE-1 hypomethylation is highly recurrent and tightly correlated with global hypomethylation. It is interesting to note that LINE-1 hypomethylation frequently increases along with the tumor's histological grade and a poor prognosis, particularly in gastrointestinal malignancies (Zheng et al. 2019; Igarashi et al. 2010; Ikeda et al. 2013; Zhang et al. 2020; Baba et al. 2018).

Furthermore, abnormal hypomethylation is also seen in regulatory DNA regions that are normally methylated and repressed. These sequences become hypomethylated in cancer, which can interfere with the repression of normally silenced genes and cellular functions, leading to active transcription of proto-oncogenes, genomic instability, tumorigenesis, and metastasis (Mazloumi et al. 2022).

Through genome-wide sequencing studies, it has been revealed that DNA hypomethylation occurs specifically in DNA blocks called partially methylated domains (PMDs) (Nishiyama and Nakanishi 2021; Hansen et al. 2014). PMDs comprise about half of the genome, usually located in gene-sparse genomic locations, and coincide with nuclear lamina-associated domains and late replication sites (Berman et al. 2012; Hon et al. 2012). They represent a repressive chromatin structure associated with a high somatic mutation rate (Brinkman et al. 2019). Despite this general trend, their location shows some degree of cell type specificity (Schroeder et al. 2011). The enriched genomic regulatory features, which often include promoters and insulators, containing or defined by CTCF regions, are in the boundaries of PMDs (Salhab et al. 2018; Decato et al. 2020).

The gene-specific promoter DNA hypomethylation can also be involved in carcinogenesis. A subset of genes that fall into the germline-specific genes category is activated in cancers as a result of loss of DNA methylation at their promoter regions. Although the information related to the oncogenic potential remains limited, the group of genes, so-called cancer-germline genes, whose expressions are only active during spermatogenesis, can become activated in tumors through promoter hypomethylation. These genes were first identified in melanoma tumors as cytotoxic T lymphocyte antigens, and some of them are known as MAGE (melanoma antigen gene). These genes have an appropriate biomarker potential for malignancy diagnosis and prospective therapeutic targets since they are not expressed in normal somatic tissues but show unique cancer-specific expression patterns. About 250 cancer-germline genes have been identified and although the localizations are dispersed on different chromosomes, X-chromosome hosts many of these genes. The MAGE family, which has more than 50 family members and is evolutionary conserved, is a significant group of these genes. These genes produce ubiquitin ligases, which play a role in reproductive organ germ cell development. Several *MAGE* proteins can bind to and inhibit well-known tumor suppressor proteins such as TP53 and Retinoblastoma (De Souza et al. 2013) (Ladelfa et al. 2012). Activation of *MAGEA11* is frequently observed in prostate cancer and has been associated with increased tumor cell growth. Besides activation of *MAGEB2*, another *MAGE* family member, has been reported in various tumors, such as lung carcinoma, and head and neck carcinoma (Van Tongelen et al. 2017). The BORIS/CTCFL gene family, which codes for a homolog of the insulator protein CCCTC binding factor (CTCF), is one intriguing member of the cancer-testis gene family. The encoded protein BORIS/CTCFL causes an increase in telomerase reverse transcriptase (hTERT) gene expression, encouraging cell immortalization and elevated expression revealed in testicular and ovarian cancers (Renaud et al. 2011).

The overexpression of c-MYC has been determined in various cancer types. The hypomethylated condition of the c-MYC promoter is correlated with its oncogenic potential and resulted from the hypomethylation-related reactivation of the transcriptionally silent retrotransposons (Fatma et al. 2020). The c-MYC promoter hypomethylation and aggressive cancer development correlation has been revealed in about 86,4% of gastric adenocarcinoma samples (De Souza et al. 2013)

Genomic imprinting is an epigenetic marking process that causes the monoallelic gene expression depending on parental origin. As is well known, imprinting patterns vary between tissues. They are regulated by imprinting control regions (ICRs), which are differentially methylated regions (DMRs), to form the parental-specific methylation pattern (Ferguson-Smith 2011). DNA methylation is the most crucial mechanism to govern imprinted gene expression in coordination with other epigenetic mechanisms, including H3K27me3 modification. They play crucial roles in various biological processes, including embryonic and placental growth, fetal development, and adult metabolism. Deletion of these sequences results in loss of imprinting (LOI), which leads to changes in the expression of imprinted genes in the cluster. LOI affects physiological functions and is the cause of the development of imprinting syndromes, including Angelman, Prader-Willi, and Beckwith-Wiedemann syndromes. Furthermore, the dysregulation of the imprinting pattern or the LOI has been described as the most common and early event in different tumors such as esophageal or colorectal cancer, or gliomas, meningiomas, and chronic myeloid leukemia (Jelinic and Shaw 2007). H19, the first reported imprinted gene in humans, and the other IGF2 imprinted gene are both growth regulatory genes that frequently regulate reciprocally. Zhang et al. (2018) and Yang et al. (2021) have reported the role of H19 overexpression in the promotion of leukemogenesis of AML (Zhang et al. 2018; Yang et al. 2021). The loss of the IGF2 imprint gene, related to the Beckwith–Wiedemann syndrome, is also a risk factor for cancer, e.g., colorectal cancer or development of Wilms tumor. The dysregulated expressions of maternally expressed CDKN1C (p57KIP2), H19, MEG3 or paternally expressed *IGF2, PEG3*, Contactin 3 (*CNTN3*), and *DLK1* imprinted genes have been reported as biomarkers associated with the development of high-grade glial tumors and/or prediction of overall survival of patients (Lozano-Urena et al. 2021). Recent studies highlight the potential roles of epigenetic instability of imprinted domains in human cancers and suggest further studies necessary to determine potential use as cancer biomarkers (Bildik et al. 2022; Kim et al. 2015).

3.2.2.2.1 DNA Methyltransferases (DNMTs) and DNA Methylation

The hypomethylation of CpG sites of the genome typically results in the activation of gene expression, whereas the hypermethylation of the sites in enhancers or promoters results in transcriptional silencing (Morgan et al. 2018). DNA methyltransferases (DNMTs), as was previously discussed, are crucial for DNA methylation in the genome. DNMTs regulate the dynamic DNA methylation patterns of embryonic and adult cells in mammals in conjunction with other factors. On the other hand, cancer is typically identified by the abnormal function of DNMTs. As can be expected, there is a close relationship between the aberrant functions of DNMTs and cancer, as well. Common somatic mutations across tumors have been reported by recent large-scale cancer genomics consortia, including The Cancer Genome Atlas (TCGA) and the Genomics Evidence Neoplasia Information Exchange (GENIE). Although many somatic mutations exist in epigenetic regulators, relatively few mutations have been detected in DNMT enzymes (Han et al. 2019). A limited percentage of colon cancer patients have DNMT1 mutations; contrarily, a significant incidence of DNMT3A somatic mutations is seen in patients with acute myeloid leukemia (AML) (Hájková et al. 2012; Lee and Kim 2021).

Focal increases in DNA methylation associated with extensive hypomethylation are hallmarks of cancer genomes. A recent study by Lopez-Mayodo et al. showed a tight correlation between loss of TET function and cancer, as well as the interaction between *DNMT3A* and *TET2* mutations in hematological malignancies. They emphasized that the distinctive pattern of global hypomethylation paired with localized hypermethylation reported in various cancer genomes may be primarily due to loss of TET function (López-Moyado et al. 2019).

3.2.2.2.2 Focal DNA Hypermethylation and Tumor Suppressor Genes

The aberrant hypermethylation of CpG islands (CGI) in the 5′ regions of cancer-related genes is a well-documented DNA methylation alteration in cancer. An alternate pathway to mutation for the deactivation of genes with tumor suppressor activity is this alteration, which can be intimately linked to transcriptional silencing. Accordingly, 60% of all gene promoters contain CpG islands, most of which are unmethylated throughout healthy development or adult cell renewal processes. Therefore, the more open chromatin states and active or ready to be activated, the expression status of these genes is fundamentally dependent on this unmethylated status. Contrarily, methylated CpG island promoters are so common in malignancies (5–10% of CGI genes) and are known to contribute to carcinogenesis directly. These cancer-specific features of the genes have opened up new options for epigenetic therapy, which targets epigenetic modifications for therapeutic reversal (Baylin and Jones 2016).

In order for malignant cells to maintain their uncontrolled development, cancer-related hypermethylation of CpG islands at promoter regions affects genes implicated in all regulatory circuits that control cell proliferation and homeostasis. At every stage of cancer development, hypermethylation events can occur and interact with both other epigenetic mechanisms and genomic abnormalities. Tumor-associated epigenetic lesions are far more common than genetic mutations, according to studies of DNA sequencing and genome-wide methylation data (Vogelstein et al. 2013). Between 5 and 10% of CpG island-containing promoters may be hypermethylated due to cancer.

Genome-wide CGI hypermethylation is evident not only in the majority of primary and metastatic tumors (Costello et al. 2000). However, it is also present in premalignant lesions, such as actinic keratosis lesions of the skin (Rodríguez-Paredes et al. 2018) and early stages of lung cancer (Vrba and Futscher 2019). It makes the most sense to explain a tumor-causing role for a hypermethylated gene in cancer when the methylation event impacts regulatory gene sequences like enhancers or promoter regions. The role of DNA methylation in these situations is typically blocking the related gene expression.

It should be emphasized that 5mC frequently exists in the gene body of active genes, and its effects here may frequently be the opposite of those they have in promoters. At least on a global scale, gene body or transcribed region hypermethylation is linked to increased gene expression levels, and it may encourage carcinogenesis by activating oncogenes if this condition occurs in genes with oncogenic characteristics (Liang and Weisenberger 2017). Nevertheless, CpG island hypermethylation more frequently will result in gene silencing when it affects promoters. If the impacted genes are involved in functional pathways, including cell proliferation control, genomic stability, activation of apoptosis or senescence, DNA repairing, and invasion and metastasis, then methylation-induced silencing events may have a tumor-promoting effect (Pfeifer 2018).

The role of promoter hypermethylation in the repression of gene expression was initially discovered in the retinoblastoma tumor suppressor gene (RB1) promoter region in patients with retinoblastoma (Greger et al. 1989), and then several tumor suppressor genes whose gene expression is repressed by DNA hypermethylation have been found in tumor tissues. Similar to germline mutation in familial malignancies, DNA hypermethylation in these genes is in a tissue-specific manner (Li et al. 2021a).

3.2.2.2.3 Roles of DNA Methylation Aberrations in Cell Proliferation

Cells need external stimuli such as growth factors, mitogens, and hormones for proliferation. Compared to normal cells, tumor cells use different ways to maintain these proliferative signals. They can activate proliferative pathways by deregulating downstream mediators, stimulating cells from the tumor microenvironment to provide them with mitogens (paracrine signaling), or producing their own mitogens (autocrine signaling). An essential component of growth control systems is the restriction of signaling pathways that promote proliferative processes. An important family of protein kinases called cyclin-dependent kinases (CDKs) controls the cell cycle. For CDKs to engage in their kinase activity, they need to be bound to the cyclins. In addition to cyclins, CDK inhibitors (CDKi) also control CDK activity. Cyclins and CDKi, together, are responsive to the stimuli through signal transduction pathways for dividing or staying quiescent of cells. Evading antiproliferative signaling at the different cell cycle checkpoints through epigenetic mechanisms is a characteristic feature of cancer cells. For instance, CDK inhibitor protein-coding genes, including cyclin-dependent kinase inhibitor 2A (CDKN2A), also known as p16INK4a, and a related gene *CDKN2B* (p15INK4a), located next to the CDKN2A locus, are involved in the regulation of cell cycle progression. The suppression of these genes by promoter hypermethylation has been reported in various cancer types. An essential mechanism for controlling cell proliferation is cell cycle-promoting kinase inhibition, and it is predicted that inactivating this mechanism may enhance cell growth. Breast, lung, head and neck cancers, gliomas, and melanomas are tumors associated with the inactivation of *CDKN2A* through promoter hypermethylation. Importantly, base substitution mutations, loss of homozygosity, promoter methylation, and other mutually exclusive events can all inactivate CDKN2A (Ortiz-Barahona et al. 2020; Pfeifer 2018).

In the mitogen-activated protein kinase (MAPK) pathway, a serial set of protein kinase cascades are involved, which is activated through the binding of mitogen to membrane receptors. The protein kinase cascades involved in the mitogen-activated protein kinase (MAPK) pathway are triggered by mitogen binding to membrane receptors, which then activate transcription factors to promote gene expression. Both activating mutations in signaling molecules and modifications to membrane receptors have the ability to constitutively activate the MAPK pathway. For example, a valine to glutamic acid alteration (V600E) in the B-RAF (B-Raf serine/threonine) gene gives rise to constitutive kinase activation, and this substitution is primarily seen in melanomas. Additionally, promoter hypermethylation-related inactivation of the PTPRR (ERK phosphatases protein tyrosine phosphatase receptor type R) and DUSP1 (dual specificity phosphatase 1 gene) genes have been reported in colon cancer (Laczmanska et al. 2013) and oral cavity carcinomas, respectively, meaning leading to MAPK cascade activation (Khor et al. 2013).

In the recent study by Xiang et al., they suggested that the tumor-specific reduced protein expression of PLCD1 (phospholipase C delta1) resulting from promoter hypermethylation could be used as a novel biomarker for early detection and prognostic prediction in colorectal cancers. They also reported that the gene plays important roles in proliferation, migration, invasion, cell cycle progression, and epithelial-mesenchymal transition. The PLCD1 is a negative regulator of the phosphatidylinositol 3-kinase (PI3K)-AKT pathway, another example of a dysregulated proliferative pathway in cancer (Xiang et al. 2019).

The familial cancer syndrome adenomatous polyposis coli is linked to germline mutations of the tumor suppressor gene adenomatous polyposis coli (*APC*), which predisposes its carriers to early-onset colorectal cancer. APC is a negative regulator of the Wingless/Int (WNT) signaling pathway. The other growth-promoting module, the WNT pathway, is especially relevant for intestinal stem cells and their malignancies. Epigenetic alterations in this pathway often result in higher β-catenin expression. Not only in colon cancer, but also APC promoter hypermethylation has been reported in breast, pancreatic, lung, and gastric cancers (Liu et al. 2021a; Zhou et al. 2020; Liang et al. 2017).

3.2.2.2.4 Role of DNA Methylation Changes in Evasion of Apoptosis

Success in tumor development depends not only on maintaining active cell proliferation but also on preventing the programmed cell death that would occur if the pathways were to become dysregulated. A high number of proliferative signals, significant DNA damage caused by the proliferation itself, hypoxia, or externally harmful substances can all cause apoptosis. The primary DNA damage sensor, p53 (TP53), directly controls the transcription of growth arrest genes when it activates in response to significant DNA damage. By epigenetically suppressing p53 targets like stratifin (SFN), tumoral cells can continue the cell cycle despite p53 activity. Stratifin is an important G2/M cell cycle checkpoint regulator and is expressed in response to DNA damage stress via a p53-dependent mechanism. SFN promoter hypermethylation is seen in various tumor types, including small-cell lung cancer (SCLC), prostate, endometrial, and breast cancers (Chauhan et al. 2021).

In normal tissues, if cells are unable to repair DNA damage, p53 activates the intrinsic apoptotic pathway, in which the pro- and anti-apoptotic members of the Bcl-2 family of regulatory proteins take roles in regulation. This route results in the release of cytochrome C and the creation of apoptosomes. The suppression of proapoptotic Bcl-2 family members (BCL2-Associated X Protein (BAX)), BIM (BCL2L11), BCL2 Binding Component 3 or PUMA (BBC3) or silencing of apoptotic peptidase activating factor 1 (APAF1) are examples of cancer-associated epigenetic dysregulation that prevents the development of this cascade (Ortiz-Barahona et al. 2020; Neophytou et al. 2021).

One of the hallmarks of cancer is the evasion of apoptosis. Many pro-apoptotic genes have been discovered to be silenced by methylation in malignant tumors. Death-associated protein kinase (DAPK), an example of hypermethylation-related silenced pro-apoptotic genes, has been revealed in many cancer types as well as in B-cell malignancies. Similarly, neuroblastomas and other malignancies have been shown to have methylation of the caspase 8 gene (CASP8), which encodes a cysteine protease controlled in a death-receptor-dependent and independent way. The paralogue of the well-known tumor suppressor TP53, TP73, has the ability to induce apoptosis. The TP73 promoter is methylated in some malignancies, including neuroblastomas and melanomas (Pfeifer 2018; Ortiz-Barahona et al. 2020).

The hippo signaling pathway is a route that manages cell proliferation and death to govern organ growth. The Hippo signaling pathway is important in inducing apoptosis and limiting cell proliferation. This signaling pathway has grown in importance in human cancer research, as unregulated cell division is a hallmark of many malignancies. MST1 and MST2 (Mammalian sterile 20-like kinases 1 and 2) are present in the pathway's core kinase cassette. Soft tissue sarcomas have been shown to have methylated MST1 and MST2 promoters (Pfeifer 2018). The Ras association domain family (RASSF) of proteins is one of the few positive regulators of MST kinases discovered. The hypermethylation of the RASSF family member, RASSF1A, is practically seen in all human cancers and is mostly already methylated in early preneoplastic lesions. Through the MST1/2 kinases, RASSF1A positively regulates the Hippo growth control system, including its pro-apoptotic output (Motavalli et al. 2021; Malpeli et al. 2019).

3.2.2.2.5 Promotion of Genome Instability by DNA Methylation Alterations

As aforementioned, in addition to a global loss of DNA methylation at repeated sequences in the genome resulting in chromosomal instability, impaired genomic maintenance machinery results in the greater mutability of malignant cells. Changes to this machinery could occur at the DNA damage detection level or at the repairing mechanism itself. Any of these inactive levels make identifying and repairing genetic mistakes more difficult, which may speed up cell division and prevent apoptosis. Either inactivating mutations or promoter hypermethylation-related silencing can result in the loss of these functionalities. Consequently, both levels of DNA methylation can exhibit abnormalities. The hypermethylated Ataxia telangiectasia mutated promoter has been discovered in glioma, breast, and colorectal cancers (Begam et al. 2017). The DNA double-strand break (DSB) sensor ATM phosphorylates multiple important proteins in response to damage, which can result in cell cycle arrest, DNA repair, or apoptosis. The checkpoint kinase 2 (CHK2), a serine-threonine kinase, is also hypermethylated and silent in gliomas (Wang et al. 2010). The DNA repair apparatus is extensive and tailored to diverse forms of damage, from recombination mechanisms for double-strand breaks (DSBs) to mechanisms for single base or nucleotide damage

Depending on which repair mechanisms have been impaired, the inactivation of DNA repair function will probably lead to an increase in the frequency of mutations, either at the single base level or the chromosomal level. Tumors have impaired DNA repair mechanisms, most notably because of mutations in the germline. Xeroderma pigmentosum gene variants, for instance, can induce errors in nucleotide excision repair (e.g., XPA, XPC, and XPF). The mutations in DNA mismatch repair genes cause a hypermutator phenotype that frequently shows up as microsatellite instability. Base excision repair impairment is less frequently linked to cancer. Mutations in *BRCA1, BRCA2,* and *RAD51* genes impair DNA double-strand break repair and recombination repair processes. Both sporadic cancers and familial cancer predisposition syndromes, particularly colorectal malignancies with microsatellite instability, have been linked to mutations in DNA mismatch repair genes. Although Lynch syndrome is due to inherited mutations in DNA mismatch repair genes, including *MSH2, MLH1, MSH6,* or *PMS2,* a majority of mismatch repair deficient sporadic colorectal tumors do not contain mutations; instead, the promoter of the *MLH1* gene is frequently hypermethylated, and biallelic methylation-mediated inactivation causes the loss of protein production. The inactivation of MLH1 is a convincing illustration of a driver methylation event in carcinogenesis because of causes the loss of function similar to gene mutation (Keum and Giovannucci 2019).

The *MGMT* (O6-methylguanine methyltransferase) is a DNA repair gene, encoding a DNA repair protein that removes mutagenic and cytotoxic alkyl groups from the O6 position of guanine and restores the guanine to its original state, i.e., repairs O6-alkylated guanine residues in genomic DNA. By pairing thymine instead of cytosine during DNA replication, guanine-O6 methylation creates a methylated nucleotide with impaired base pairing potential, which encourages G:C to A:T mutations. The promoter of the gene is CpG rich and is epigenetically inactivated through DNA methylation, and consequently, methylation silencing of MGMT diminishes its O6-alkylguanine repairing efficiency. The epigenetically inactivated MGMT is seen in colorectal, gastric, non-small-cell lung cancers, head, and neck squamous cell carcinomas, and significantly in gliomas (Uddin et al. 2020). However, alkylating agents such as Temozolomide (TMZ) are among the most used chemotherapeutic drugs in cancer treatment and are known to cause cell cycle arrest at G2/M, which ultimately leads to apoptosis. Adding methyl groups at the N7 and O6 sites on guanines and the O3 site on adenines in genomic DNA is the mechanism through which TMZ causes cytotoxicity. When the O6 site on guanine is alkylated, a thymine rather than a cytosine match opposite the methylguanine during the following DNA replication, and DNA mismatch errors occur. The mismatches of methylated DNA can be repaired by base excision or DNA mismatch repair pathways through the involvement of a DNA glycosylase like alkylpurine-DNA-N-glycosylase (APNG) or a demethylating enzyme like MGMT. Thus, DNA mismatch repair by active MGMT causes the development of a resistance mechanism against TMZ. In contrast, epigenetically silenced MGMT sensitizes the tumor to TMZ. Glioma patients with a methylated MGMT gene have been shown to have a higher survival rate when treated with the alkylating agent TMZ compared to patients with an unmethylated promoter, possibly due to increased cell killing by the chemotherapy agent (Kukreja et al. 2021; Śledzińska et al. 2021).

3.2.3 Histon Modifications in Cancer

Histone proteins are essential for nucleosome components. In eukaryotes, chromatin is organized into nucleosomes, each formed of a histone octamer and a fragment of surrounding DNA. There are six histones: H1, H2A, H2B, H3, H4, and H5, highly rich in lysine and arginine, two positively charged amino acids (Neganova et al. 2022; Zhao et al. 2021). Since Vincent Allfrey's pioneering work in 1964, it has been known that histones are post-translationally modified (PMTs) (Allfrey et al. 1964). Histon proteins' amino and carboxy termini can undergo transcription-regulating changes, including methylation, acetylation, phosphorylation, sumoylation, ubiquitination, and ADP-ribosylation. They may also act as recognition modules for specific binding proteins (Audia and Campbell 2016).

Histone alterations are classified as active or repressive based on their effects on gene expression. The steady-state cell maintains a balance between particular modifications and modifiers to preserve chromatin structure, execute the correct gene expression program, and regulate the biological outcome. Disruption of this balance in the cell may change the phenotype, leading to the disease's formation and progression (Zhao and Shilatifard 2019; Markouli et al. 2021). Deregulation of these mechanisms results in the development and progression of cancer due to the increased activation of oncogenes or the inhibition of tumor suppressor activity.

3.2.3.1 Histone Acetylation

Histone acetyltransferases (HATs) and histone deacetylases (HDACs) regulate acetylation, a reversible modification of the ε-amino group on lysine residues. HATs transfer the acetyl group of acetyl coenzyme A to the terminal of histone amino acid. Acetylation of the histone tails neutralizes the positively charged lysines, disrupting the connection between the tail and the negatively charged nucleosomal DNA to facilitate chromatin opening and enhance active transcription by making DNA accessible to transcription factors. The lysine residues of non-histone proteins are known to be acetylated such as p53, Rb, and MYC. Therefore, these enzymes are also called lysine acetyltransferases (KATs). In contrast, HDACs remove the terminal acetyl group of histone lysine, resulting in a compact chromatin structure that inhibits transcription (Neganova et al. 2022; Audia and Campbell 2016) (Fig. 3.5).

Acetylated lysines might provide a unique signal for regulatory factors or chromatin remodeling complexes to target specific domains. Bromodomains were discovered to function as acetyl-lysine recognition modules, guiding enzymes with

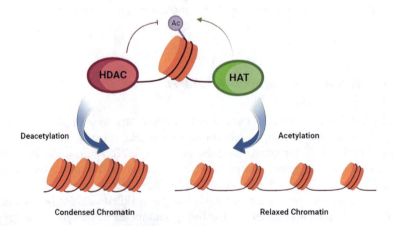

Fig. 3.5 Schematic mechanism of histone acetylation and deacetylation

these domains to specific locations on chromosomes. In addition to transcriptional regulation, new functions for histone acetylation have been identified, including nucleosome assembly, chromatin folding, heterochromatic silencing, DNA damage repair, and replication (Cohen et al. 2011; Zhang et al. 2015).

Numerous studies have shown that aberrant expression or activity of HATs and HDACs significantly affects the cancer acetylome (Li et al. 2019). Depending on the target genes (e.g., tumor suppressor and proto-oncogenes), hyperacetylation and hypoacetylation may disrupt the normal cell cycle, prevent or reverse differentiation, block apoptosis, and enhance cell proliferation, contributing to the formation and metastasis of a cancer phenotype (Di Cerbo and Schneider 2013). Alterations in global histone acetylation, specifically acetylation of H4 at lysine (K)16, have been associated with various cancers and may have predictive significance in some cases (Seligson et al. 2009; Fraga et al. 2005).

Several studies have suggested the dual roles of HATs as oncogenes and tumor suppressors. HAT mutations and altered expression without DNA mutation have been detected in multiple cancers (Chen et al. 2013; Di Cerbo and Schneider 2013).

Well-studied human HAT families are GNAT (HAT1, GCN5, PCAF), MYST (Tip60, MOF, MOZ, MORF, HBO1), and p300/CBP. p300/CBP includes the HAT domain, the bromodomain (BRD), and three cysteine and histidine-rich domains. Germline mutation of CBP causes Rubinstein-Taybi syndrome and increased susceptibility to childhood cancers, probably due to loss of the second allele. p300 has also been linked to hematological malignancies (Cheng et al. 2019; Di Cerbo and Schneider 2013). CBP- and p300-null chimeric mice developed hematological malignancies (Rebel et al. 2002). Several p300 missense mutations have been detected in colorectal adenocarcinoma, gastric adenocarcinoma, and breast cancer (Gayther et al. 2000; Cheng et al. 2019). Small-cell lung cancers and non-Hodgkin B-cell lymphomas have been shown to have mutations close to the HAT catalytic domain that lead to a loss of enzymatic activity (Peifer et al. 2012; Pasqualucci et al. 2011). However, impaired activation of HATs, which are also responsible for the acetylation of tumor suppressor genes such as p53 and Rb, can induce tumorigenesis.

On the other hand, oncogenic effects may result from abnormal activation or localization of p300/CBP. MLL-CBP t(11;16)(q23;p13), MLL-p300 t(11;22)(q23; q13), MOZ-CBP t(8;16)(p11;p13), and MOZ-p300 t(8;22)(p11;q13) have been identified in acute myeloid leukemia (AML), myeloid/lymphoid, or mixed lineage leukemia (MLL) (Cohen et al. 2011). In addition, it has been shown that p300 can modulate some fusion protein activity by acetylation, such as AML1-ETO t(8;21) (q22;q22), which is the most common fusion protein in AMLs. Depletion of p300 impaired its ability to promote leukemic transformation by inhibiting acetylation of AML1-ETO (Wang et al. 2011). The relationship between histone alterations and malignancy in hematological cancers has been broadly studied compared to solid tumors. High p300 expression has been related to poor prognosis in laryngeal squamous cell carcinoma and small-cell lung cancer (Chen et al. 2013; Gao et al. 2014).

Histone acetyltransferase TIP60 regulates apoptosis and DNA damage repair by acetylation of some tumor suppressor genes in addition to histones. Mutations of the human TIP60 gene have been identified in head and neck squamous carcinomas, ductal breast carcinomas, and low-grade B-cell lymphomas (Di Cerbo and Schneider 2013). Low TIP60 mRNA expression was associated with poor overall survival and recurrence-free survival in breast cancer (McGuire et al. 2019). It has also been found that TIP60 can inhibit viability and invasion of lung cancer cells through downregulation of the AKT signaling pathway (Yang et al. 2017). Another acetyltransferase, GCN5, has been shown to regulate gene transcription by catalyzing the acetylation of lysine residues on multiple histones, including H2b, H3, and H4, in addition to transcription factors such as FBP1 and N-Myc. GCN5 mRNA is upregulated in some cancers (Yin et al. 2015).

HDACs are divided into four groups classes I, II, III, and IV. HDAC overexpression has been reported in solid and hematological cancers and is associated with advanced disease and poor patient outcomes. Therefore, HDACs have become promising therapeutic targets (Hosseini and Minucci 2018).

High expression of HDAC1 and 2 is associated with reduced patient survival in colorectal carcinomas. The overexpression of HDAC1, 2, and 6 and HDAC1, 2, and 3 have been described in diffuse large B-cell lymphomas (DLBCL)/peripheral T-cell lymphomas and classical Hodgkin lymphomas, respectively (Dell'Aversana et al. 2012). HDAC6 and HDAC10 have been downregulated in human hepatocellular carcinoma (HCC) tissues and in patients with lung and stomach cancer, respectively, and associated with poor prognosis (Li and Seto 2016). It has been observed that HDAC4 is critical for regulating chromosome structure, while low HDAC4 expression is associated with chromosomal instabilities in high-grade glioma (Cheng et al. 2015). Class III HDACs, known as sirtuins, which play essential roles in regulating gene expression, apoptosis, autophagy, DNA damage repair and, genome stability, have been studied broadly. Increased or decreased class III HDAC expression levels have been detected in myeloid leukemia, prostate and ovarian carcinoma, gliomas, gastric carcinomas, non-melanoma, and melanoma skin cancers (Benedetti et al. 2015).

In addition to alterations in the expression level of HDACs, their enzymatic activity also contributes to cancer development. Some HDACs have been reported to be attracted to target genes by oncogenic proteins such as aberrant HDAC1, 2, or 3 recruitment by AML1-ETO fusion protein. Recruitment of HDACs prevents myeloid differentiation and results in cellular transformation by suppressing AML1 target genes (Falkenberg and Johnstone 2014). Somatic HDAC1 mutations and homozygous HDAC4 deletions have been detected in liposarcomas and melanomas. Also, HDAC2 loss-of-function mutations have been observed in sporadic carcinomas with microsatellite instability and hereditary non-polyposis colorectal cancer syndrome (Hosseini and Minucci 2018; Ropero et al. 2006).

HDACs affect the expression of many cell cycle regulators and also may directly interact with proteins implicated in tumor development, migration, and metastasis. HDAC1 and 2 suppress the expression of the cell cycle inhibitors p21 and p27.

HDAC2 knocked down cells have shown an increase in p21$^{Cip1/WAF1}$ expression independent of p53 in colorectal cancer cells (Huang et al. 2005).

Protein readers play an important role in histone post-translational modifications as well as HATs and HDACs. Readers identify particular locations, attract transcription factors or chromatin-associated protein complexes, and bind to histones to facilitate the localization of enzymes to specific targets (Liu et al. 2021b). The functional protein domains known as bromodomains (BRDS) can identify acetylated lysine residues in histones and other non-histone proteins. Additionally, they can serve as transcription factors and transcriptional coregulators. Another important family, Bromodomain and the extra-terminal domain-containing proteins (BET) include four family members: BRD2, BRD3, BRD4, and BRDT. These proteins play crucial functions as gene transcription activity mediators.

Genetic rearrangements of BRD-containing proteins have been associated with some aggressive tumor types. Nuclear protein midline carcinoma (NMC) of the testis is a highly aggressive tumor associated with translocations involving the NUT protein. BRD4–NUT rearrangements are observed in two-thirds of cases. BRD–NUT blocks cellular differentiation. BRD4–NUT stimulates CBP/p300 HAT activity and inactivation of p53. With recent studies, BET proteins have become potential therapeutic targets against testicular carcinoma, multiple myeloma, lymphoma, lung cancer, and neuroblastoma (Muller et al. 2011; Neganova et al. 2022; Cheng et al. 2019).

The reversible nature of epigenetic modifications has provided the basis for the development of anti-cancer strategies for the regulation of cancer epigenetics. HDAC inhibitors (HDACi) continue to be explored as promising anti-cancer drugs by modulating histone and non-histone proteins, regulating processes such as inhibiting cancer cell invasion, inducing apoptosis, and immunogenicity. Vorinostat, belinostat, Panobinostat, and romidepsin are FDA-approved HDAC inhibitors (Roberti et al. 2019; Karagiannis and Rampias 2021). BET inhibitors (iBETs) that bind reversibly to the bromodomain of BET proteins continue to be studied to suppress oncogenic networks.

3.2.3.2 Histone Methylation

The methylation of histones is a process that occurs mainly at lysines (K) and arginines (R) and plays essential functions in differentiation and development. Dynamic methylation processes require methyl transferases as "writers," demethylases as "erasers," and effector proteins as "readers." Lysine methyltransferases (KMTs) and arginine methyltransferases (PRMTs) are enzymes that transfer methyl groups from S-adenosyl methionine (SAM). Lysine demethylases (KDMs) remove methyl groups from histone lysine residues (Fig. 3.6).

The effects of methylation on histones can be correlated with various gene expression statuses. For instance, methylation of H3K9, H3K27, and H4K20 inhibits gene expression, whereas methylation of H3K4, H3K36, and H3K79 stimulates gene expression but the final effect on chromatin is affected by the interaction of

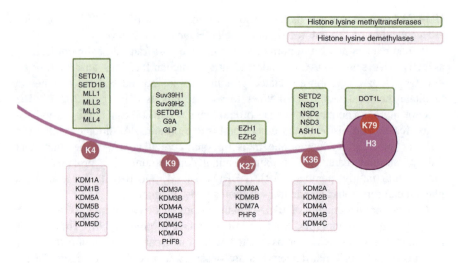

Fig. 3.6 Methylation sites in histone 3 and the enzymes (KMTs and KDMs) involved in process

several histone modifications known as histone crosstalk. The same modification may have distinct functional effects depending on the methylation status (e.g., H3K4me2 and H3K4me3) and chromosomal position (Izzo and Schneider 2010). The involvement of histone methylation in transcriptional regulation is associated with chromatin structure, recruitment of transcriptional factors, interactions with initiation and elongation factors, and effects on RNA processing (Zhao and Shilatifard 2019).

Although methylation and demethylation processes' role in cancer development/progression remains unclear, it is known that abnormalities in the methylation of various lysine residues by histone lysine methyl transferases can alter gene expression specific to certain neoplastic and normal cell types (Neganova et al. 2022). As expected, misregulation of KMTs has been associated with numerous cancers, such as EZH2 overexpression has been detected in breast, bladder, and prostate malignancies, and NSD2 has been associated with tumor aggressiveness and poor prognosis in various types of cancer (Albert and Helin 2010).

All KMTs have SET (Suppressor of variegation, Enhancer of Zeste, Trithorax) domain for their catalytic activity, except disruptor of telomeric silencing 1-like (DOT1L) methyltransferase. The human genome encodes 48 proteins containing SET domains. KMTs also methylate lysines in non-histone proteins. SET7/9, for instance, can stabilize the tumor suppressor p53 by methylating K372 (Chuikov et al. 2004; Cheng et al. 2019; Albert and Helin 2010).

MLL1 (KMT2A), which specifically methylates histone H3 lysine 4, is implicated in various forms of cancer with loss of function and rearrangement. Leukemogenesis can be induced by MLL fusion proteins that alter the proliferation and differentiation of hematopoietic cells. HOXA9 transcriptional regulation is disrupted due to an increase in H3K4me3 elicited by MLL1 translocation in myeloid and lymphoid leukemias. More than 50 MLL fusion proteins have been identified in AML, ALL, and MLLs (Audia and Campbell 2016; Neganova et al. 2022).

Methyltransferase DOT1L catalyzes H3K79 methylation, which occurs in the core of histone H3 rather than on its N-terminal tail and is thought to increase gene expression. H3K79 methylation regulates chromatin structure, transcription, DNA damage response, and cell cycle processes. Misregulation of these mechanisms via aberrant DOT1L function and defects in H3K79 methylation can lead to aneuploidy, telomere elongation, and disturbances in cell proliferation (Ljungman et al. 2019; Guppy et al. 2017). The identification of abnormal upregulation of H3K79 methylation in leukemia led to the development of the DOT1L inhibitor (Zhao and Shilatifard 2019). DOT1L is recruited by MLL fusion partners, resulting in aberrant H3K79 methylation that leads to increased transcription of MLL fusion target genes. DOT1L also has an effect on the development and progression of some solid tumors such as breast, lung, and ovarian cancers (Neganova et al. 2022; Song et al. 2020).

Enhancer of zeste homolog 2 (EZH2), one of the best-studied HMT enzymes involved in oncogenesis, is responsible for the di- and trimethylation of H3K27 (H3K27me2 and -me3). The members of the enhancer of zeste homolog family are the catalytic components of polycomb repressor complexes (PRCs) responsible for gene silencing (Cohen et al. 2011). EZH2 has the potential to function as an oncogene by playing a role in the H3K27me3-mediated aberrant silencing of the promoters of some tumor suppressor genes. EZH2 overexpression and gain-of-function mutations have been associated with many types of cancer. Overexpression of EZH2 has been linked to some solid tumors such as prostate, bladder, colon, and breast cancers and is also associated with aggressive and metastatic disease in prostate cancer (Chase and Cross 2011). B-cell lymphoma cell lines and lymphoma samples with heterozygous $EZH2^{Y641}$ mutations have exhibited elevated H3K27me3 (Yap et al. 2011). Dysregulation of EZH2 in cancer may occur with the effect of multiple microRNAs. For example, targeting EZH2, miR-101 also regulates cell proliferation, invasion, and tumor growth. Loss of miR-101 has been shown in prostate cancer to lead to overexpression of EZH2 (Varambally et al. 2008). EZH2 loss-of-function mutations have also demonstrated a potential tumor suppressor role in hematologic malignancies (Khan et al. 2013).

H3K9 mono-, di-, or trimethylation is associated with different chromatin states, aberrantly regulated in multiple cancers. For example, H3K9me3 correlates with transcriptionally inactive chromatin and acts as a specific binding platform for heterochromatin protein 1 (HP1). The SUV39H1 and SUV39H2 enzymes preferentially trimethylate H3K9 and are crucial in forming constitutive heterochromatin, primarily pericentric heterochromatin (Lachner et al. 2001; Cohen et al. 2011). Dysregulation of members of the H3K9 methyltransferase family has been demonstrated in numerous cancers. KMT1A/SUV39H1 has been overexpressed in breast cancer but has not been correlated with disease progression (Patani et al. 2011).

Histone demethylases can be classified into two groups: The lysine-specific demethylases (LSDs) and Jumonji C (JmjC) domain-containing histone demethylases (KDM2–8) (Cheng et al. 2019). The first reported lysine demethylase specific for residues H3K4 and H3K9 is LSD1 (KDM1A), which has been identified as overexpressed in several cancer types. Non-histone proteins such as p53, E2F1, and HIF-1 are also demethylated by KDM1A (Sterling et al. 2021). For example, LSD1 has been shown to suppress p53 function by inhibiting the interaction of p53 with p53-binding protein 1 (53BP1) (Huang et al. 2007).

KDM2A promotes tumor growth and invasion in lung cancer by increasing ERK1/2 and JNK1/2 activities through H3K36 demethylation at the DUSP3 promoter (Wagner et al. 2013). KDM2B is thought to function as an oncogene and plays a critical role in the development and maintenance of leukemia cells (He et al. 2011a). Similarly, KDM3 enzymes are overexpressed in various tumors and implicated in oncogenic processes. KDM3A has been demonstrated to control the invasion and apoptosis of breast cancer cells and maintain myeloma cells' survival (D'oto et al. 2016). KDM4B and KDM4C catalyze the demethylation of H3K9me3/me2 mark and have been shown that amplified in medulloblastoma, malignant peripheral nerve sheath tumors, and squamous cell carcinoma. KDM4B also plays an important role in the regulation of the N-Myc pathway in neuroblastoma. Glioblastoma stem cells exhibit lower levels of H3K9me3/me2 and H3K27me3/me2 than differentiated cells (Mallm et al. 2020; Yang et al. 2015).

KDM5 subfamily catalyzes only H3K4me3/me2, gene activating marks. KDM5 family members may be involved in the downregulation of tumor suppressors and oncogenes (Sterling et al. 2021). KDM5A is overexpressed in several cancer types. For instance, KDM5A-mediated-H3K4me3 demethylation results in downregulation of the expression of genes encoding the tumor suppressor proteins p16 and p27 in breast cancer (Yang et al. 2019). Furthermore, KDM5B inhibits their oncogenic potential by reducing H3K4me3/me2 on oncogenes such as Hox/Meis in leukemia stem cells (Wong et al. 2015). It has been reported that low KDM5C levels in renal cancer cells trigger genomic instability and are associated with poor prognosis in patients (Rondinelli et al. 2015). Disruption of the histone demethylase KDM6A, the first reported mutation in cancers, leads to cell cycle dysregulation. The roles of KDM6 enzymes appear context-dependent in cancer. Tumor suppressor and oncogenic effects have been observed in different studies (D'oto et al. 2016).

3.2.3.3 Histone Ubiquitination

Ubiquitin (Ub) is a highly conserved, 76-amino acid regulatory protein. In 1977, Gold Knopf et al. identified histone ubiquitination that is involved in many cellular processes, including transcription, DNA repair, and genome stability. Ubiquitination is a modification that tags substrate proteins with Ub and involves a multi-step enzymatic process, including ubiquitin-activating enzyme (E1), ubiquitin-conjugating enzyme (E2), and ubiquitin-protein ligase to attach to the substrate. In this enzymatic process, Ub is adenosine triphosphate-dependently activated and transferred to E2. Finally, a ubiquitin ligase binds ubiquitin to the specific lysine residue. Deubiquitinating enzymes (DUB, also known as ubiquitin-specific peptidases (USPs)) remove ubiquitin (Ub) from target proteins (Fig. 3.7). Considering its cellular functions, it is not surprising that aberrant ubiquitination induces

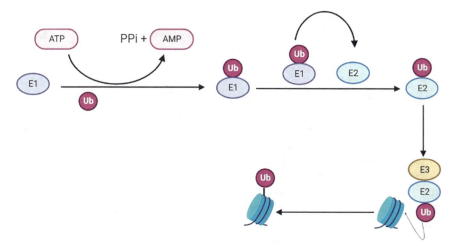

Fig. 3.7 Schematic illustration of the ubiquitination process

oncogenesis by altering the expression of oncogenes and promoting cancer cell proliferation as with other PMTs (Jeusset and McManus 2019; Deng et al. 2020).

Although polyubiquitination of canonical protein is a mark for proteasome-mediated degradation, histone ubiquitination has been associated with controlling various pathways and activities rather than degradation. H2A and H2B are the most abundant ubiquitinated proteins in the nucleus. Although H1, H3, and H4 ubiquitination have been reported, the biological function of these modifications has yet to be fully elucidated (Thompson et al. 2013).

H2A may be either mono- or polyubiquitinated, but H2B is often monoubiquitinated. H2B monoubiquitination (H2Bub1) is a crucial modification for transcriptional activation and tumor suppression. Loss of global H2Bub1 has been reported in breast, lung, and parathyroid cancers and has also been correlated with poor survival in colorectal cancer patients (Cole et al. 2015; Melling et al. 2016). It has been shown that a reduction in H2Bub1 affects the transcriptional mechanism of the ER and may also potentially play a role in estrogen-independent proliferation. H2Bub1 levels have been reported to be decreased in both primary and metastatic breast cancers, although they remain unchanged in benign breast tissue (Prenzel et al. 2011; Wu et al. 2015; Dwane et al. 2017) Depletion E3 ligase RNF20, which is responsible for H2B ubiquitination, has increased cell migration, eliciting transformation and tumorigenesis. RNF20 promoter hypermethylation in primary breast cancer cells and mutation at low frequency in colorectal cancer have been reported (Shema et al. 2008; Marsh and Dickson 2019). Rearrangements of the mixed lineage leukemia proto-oncogene MLL1 initiate aggressive forms of acute leukemia and are associated with poor prognosis. It has been shown that suppression of RNF20, which is required for MLL fusion-mediated leukemogenesis, leads to inhibition of cell proliferation (Wang et al. 2013). H2Bub1 is also required to recruit players in the DNA repair pathways (Moyal et al. 2011). Failure to repair DNA can cause chromosomal instability and contribute to the tumorigenic process (Thompson et al. 2013). On the other hand, some studies have shown that high levels and/or activity of H2Bub1 and its E3 ligases may have an oncogenic effect (reviewed in (Wright et al. 2011)). USP22 is the best-characterized DUB of H2BK120ub1. USP22 overexpression was reported to be associated with more aggressive tumors and poor prognosis in breast cancer (Zhang et al. 2011).

Lysine 119 in H2As is the most frequently observed ubiquitination site. Really interesting new gene 1A (*RING1A*) and *RING1B*, and B-lymphoma Moloney murine leukemia virus insertion region 1 homolog (BMI1) are ubiquitin ligases responsible for the monoubiquitination of H2AK119 that plays a central role in transcriptional repression by coordinating with H3K27 trimethylation. USP16 and breast cancer type 1 susceptibility protein (BRCA1)-associated protein 1 (BAP1) are DUBs for H2AK119ub1. Mammals have two primary Polycomb group complexes, PRC1 and PRC2. H2A monoubiquitination is also involved in X inactivation. RING1B and H2Aub affect the initiation of imprinted and random X-chromosome inactivation. Loss of ubiquitylation of histone H2A in BRCA1-deficient mice resulted in disrupting structural heterochromatin and gene silencing integrity in the repeat regions (Zhu et al. 2011).

3.2.3.4 Histone Phosphorylation

Histone phosphorylation is a reversible PMT that usually occurs at serine (S), threonine(T), and tyrosine (Y) residues of histone tails and is controlled by various kinases and phosphatases. Histones H1, H2A, H2B, H3, and H4 are phosphorylated at multiple sites. It has been implicated in DNA repair, regulation of transcription, apoptosis, and chromatin remodeling (Shanmugam et al. 2018).

Phosphorylation of the histone H2A subtype, H2AX, at the Ser139 position occurs in response to DNA damage and is mediated by ATM and ATR (Podhorecka et al. 2010). Histone phosphorylation has been found to be associated with transcriptional regulation and gene expression, particularly genes that regulate cell cycle and proliferation. For example, H3S10 and 28, H2BS32 phosphorylations have been related to activation of epidermal growth factor (EGF)-mediated gene transcription. Aurora B, responsible for H3S10 phosphorylation, has been identified as being overexpressed in various solid tumors, including breast and colorectal cancers (Hosseini and Minucci 2018). An increase in H3S10 phosphorylation has been observed in breast cancer, esophageal squamous cell carcinoma, gastric cancer, glioblastoma, melanoma, and nasopharyngeal carcinoma (Komar and Juszczynski 2020). Small-cell lung cancers (SCLC) with c-MYC amplification/high expression have been shown to respond to Aurora B inhibitors (Helfrich et al. 2016).

H3Y41 phosphorylation and displacement of HP1α can lead to oncogene activation, inducing tumorigenesis. H3Y41 phosphorylation of Janus kinase 2 (JAK2) has been observed to cause disrupting chromatin binding by heterochromatin protein 1α (HP1α). Inhibition of JAK2 activity reduces the phosphorylation of H3Y41 in the promoter of the hematopoietic oncogene Imo2 and expression and also increases HP1 α binding at the same site in human leukemic cells (Dawson et al. 2009). Gene amplification, mutation, and/or rearrangement of JAK2 have been shown in several hematological malignancies.

3.2.3.5 Other Modifications

SUMOylation is a negative regulator and is known to reduce transcriptional activity. Small ubiquitin-like modifier (SUMO) pathway is involved in carcinogenesis, the regulation of DNA damage repair, immune responses, carcinogenesis, cell cycle progression, and apoptosis. Blocking sumoylation results in decreased proliferative capacity and induction of antitumor immune response in cancer cells. Key pathways related to cancer, such as PI3K/AKT/mTOR, JAK-STAT, MAPK/ERK cascade, TGF signaling, and EMT pathway, are subjected to SUMO control. Some tumor suppressor genes and proto-oncogenes are also SUMO targets (Shanmugam et al. 2018; Lara-Ureña et al. 2022).

O-GlcNAcylation is catalyzed by O-Linked N-acetylglucosamine (O-GlcNAc) transferase (OGT) and O-GlcNAcase (OGA). Alteration of these processes may lead to tumorigenesis (Forma et al. 2014). Low expression of OGA in hepatocellular carcinoma tissues has been suggested to be a prognostic marker for tumor recurrence (Zhu et al. 2012). It has been shown that global GlcNAcylation levels are significantly elevated in tumor tissues, and there is a significant increase in metastatic lymph nodes compared to the corresponding primary tumor tissues (Gu et al. 2010). Overexpression of OGT has been reported to alter mitotic histone post-translational modifications of histone H3 in Lys-9, Ser-10, Arg-17, and Lys-27 (Sakabe and Hart 2010).

3.2.4 Chromatin Remodelers

Chromatin remodeling complexes are regulators that remodel nucleosomes in an ATP-dependent manner and have essential roles in DNA damage repair, recombination, replication, and transcriptional control, and aberrations in this process can induce carcinogenesis. SWI/SNF, ISWI, INO80, and NuRD/Mi-2 are the best-characterized remodelers (Nair and Kumar 2012).

SWItch/Sucrose Non-Fermentable (SWI/SNF) chromatin remodeling complex uses energy from ATP dephosphorylation to alter chromatin accessibility by chromatin repositioning, exchanging specific or all nucleosome cores, and histone dimer eviction (Tsuda et al. 2021). The SWI/SNF complex is known to control transcription by regulating acetylated histone H3K27. Alterations in genes encoding SWI/SNF remodeling factors such as ARID1A have been identified in about 8% of human cancers. ARID1A has a role in the ability of the SWI/SNF complex to inhibit cell growth and prevent genomic instability (Krishnamurthy et al. 2022;

Tsuda et al. 2021). ARID1A mutations were observed in 13% of hepatocellular carcinoma, 9.6% of gastrointestinal adenocarcinoma, 2.5% of malignant melanoma, and 57% of ovarian clear cell carcinoma (Okawa et al. 2017).

Imitation switch (ISWI) family, which is included in the ATPase family, is involved in many cellular processes, such as transcriptional regulation, DNA damage response, repair, and recombination. ISWI subunits are thought to be involved in tumorigenesis by regulating oncogenic gene transcription. Somatic mutations, copy number variations, and gene fusions have been identified in various tumor types for ISWI subunits (Li et al. 2021b).

Ino80 ATPase is a member of the SNF2 family of ATPases and a component of the INO80 ATP-dependent chromatin remodeling complex (INO80). Ino80 overexpression has been shown to promote proliferation in the immortalized cervical epithelial cell line and non-small-cell lung cancer cells. It is thought that INO80 binds to enhancer regions near cancer-associated genes, promoting their expression (Hu et al. 2016; Zhang et al. 2017). Ino80 silencing also inhibited melanoma cell proliferation, anchorage-independent growth, and tumorigenesis (Zhou et al. 2016).

Nucleosome remodeling and deacetylase complex (Mi-2/NuRD) that function in gene repression contain histone deacetylases (HDAC1/2), metastasis-associated (MTA1/2) proteins, and methyl CpG-binding domain (MBD) proteins. Overexpression of MTA1 has been observed in gastrointestinal and esophageal carcinomas and breast adenocarcinomas (Fu et al. 2011; Toh and Nicolson 2009). It was shown that PML-RARa binds and recruits NuRD to target genes, including the tumor suppressor gene RARβ2. Knockdown of the NuRD complex in leukemic cells prevented histone deacetylation and chromatin compaction and promoted cellular differentiation by disrupting stable silencing and DNA and histone methylation (Morey et al. 2008).

3.2.5 miRNAs in Cancer

Understanding how cancer begins and progresses is essential for cancer prevention, early detection, and treatment. Since changes in gene expression also have important effects on cancer, microRNA (miRNA) research has also been a focus in recent years.

microRNAs are a type of non-coding RNA, 19–25 nucleotides in length, that regulate gene expression post-transcriptionally. A microRNA can target hundreds of genes and affect their expression (Lu and Rothenberg 2018). miRNA sequences can be located within introns, exons of non-coding RNAs and a intron of pre-mRNA (pre-messenger RNA). Most miRNAs are expressed by RNA polymerase II (RNA pol II), but some are transcribed by RNA polymerase III (Borchert et al. 2006; Lee et al. 2004).

3.2.5.1 miRNA Biogenesis and Functions

miRNA biogenesis occurs by two different pathways; canonical and non-canonical pathways.

3.2.5.1.1 Canonical miRNA Biogenesis

Most intergenic miRNAs use their own promoter region. miRNA sequences are located in exons or introns of non-coding RNAs. Polymerase II synthesizes pi-miRNAs containing at least 1 hairpin structure. pi-miRNAs are divided into structures called precursor miRNAs (pre-miRNAs). Each pre-miRNA is about 70 nt long, and this process takes place in the nucleus. Then, pre-miRNAs are exported to the cytoplasm. Drosha, DiGeorge syndrome critical region gene 8 (DGCR8), XPO5, and Ras-related nuclear protein (RAN) are involved in this process (Saliminejad et al. 2019). The microprocessor complex, consisting of Nuclear RNAase III DROSHA and its cofactor DGCR8, serves in the cleavage of pi-miRNA to form pre-miRNA (Nguyen et al. 2018). The Ran/GTP/Exportin 5 complex is involved in the transport of pre-miRNA into the cytoplasm. In the cytoplasm, the pre-miRNA is cleaved into a double strand, one of which is the passenger strand and the other is the guide strand. This process is catalyzed by Dicer, an RNAase III enzyme (Peng and Croce 2016). The mature miRNA gets its name from the 5′ or 3′ directionality of the strands. Both strands can be loaded into the Argonaute (AGO) protein family. Which strand will bind to AGO depends on the cell type and cell environment. The unloaded strand is identified as the passenger strand and is degraded by AGO2 (O'Brien et al. 2018).

Repression of transcription by miRNA is classically mediated by miRNA-induced silencing complex (miRISC). The miRISC allows to recognize 3'UTR region of the target mRNA. However, it is stated that mRNA can be recognized in the 5'-UTR and even in protein-coding sequences. The target mRNA is recognized by the sequences on it called miRNA response elements (Saliminejad et al. 2019). The degree of complementarity of miRNA with mRNA determines whether it is repressed by AGO2 or miRISC. Full complementarity between miRNA and mRNA activates AGO2 endonuclease activity and mRNA is cleaved (Fig. 3.8) (O'Brien et al. 2018).

It has been stated that miRNA can suppress translation in three different ways: (i) Ago2 interacts with TNRC6, which recruits the CCR4-NOT deadenylase complex. So, the mRNA is deadenylated and degraded. (ii) TNRC6 interacts with the Dcp 1/2 cap complex, which cleaves the 5′ capped mRNA and destabilizes the mRNA. (iii) With the binding of Ago 2, mRNA is rendered inaccessible for ribosome attachment and function, which inhibits the translation process. When Ago 2 binds, the mRNA cannot interact with the ribosome and the translation process is suppressed. Transfer of Ago 2 with mature miRNA to the nucleus is via

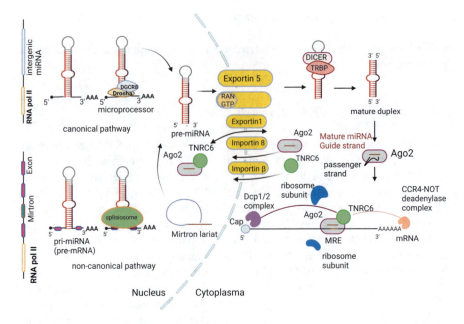

Fig. 3.8 Biogenesis and functions of miRNA

Importin 8, while TNRC6 is transported via Importin β. In the nucleus, RISC is assembled and RISC can be transported via Exportin1 (Fig. 3.8) (Liu et al. 2018).

It has long been known that miRNAs play a role in gene regulation post-transcriptionally. However, it has also been discovered to interact with long non-coding RNA (lncRNA), circular RNA (circRNA), and pseudogenes. This information indicates that while investigating the effects of miRNAs on diseases, the processes in question are much more complex. These functions of miRNAs will be explained in the sections on the effects of miRNAs in cancer.

3.2.5.1.2 Non-canonical miRNA Biogenesis

Although many different pathways have been described for non-canonical miRNA biogenesis, the well-recognized are Drosha- and Dicer-independent pathways. As an example of the Drosha/DGCR8-independent pathway, mirtrons produced from mRNA introns, have the property of being dicer substrates. Then it is included in the canonical pathway (O'Brien et al. 2018; Saliminejad et al. 2019). In the Dicer-independent pathway, the endogenous hairpin transcripts are short to become Dicer substrates and therefore require AGO2 (Fig. 3.8) (O'Brien et al. 2018).

3.2.6 The Role of miRNA in Cancer

Among the non-coding RNAs that play critical roles in gene regulation, microRNAs (miRNAs) are the most studied type of non-coding RNA in different types of cancer. The association between miRNAs and cancer was first discovered in CLL patients with 13q deletion. Two microRNAs (*miR-15a* and *miR-16-1*), deleted or downregulated, were discovered in the majority of CLL patients (Calin et al. 2002). After a while, it was determined that these microRNAs role as tumor suppressors by suppressing the *BCL2* (*B-cell lymphoma 2*) gene (Cimmino et al. 2005).

miRNAs act as tumor suppressors (oncosupressor-miR) or oncogenes (onko-miR) depending on the functions of the genes they target. One of the well-known oncosuppressor-miRs is let-7. Expression of let-7 has been shown to be decreased in various cancers and associated with poor prognosis (Boyerinas et al. 2010). OncomiRs generally contribute to tumor development by targeting genes that control cell division, differentiation, and apoptosis (Lujambio and Lowe 2012). *miR-21* is the first miRNA discovered in humans. As a result of transcript profiling studies conducted after many miRNA discoveries, *miR-21* was shown to be upregulated in various cancers such as breast cancer, chronic lymphoblastic leukemia, lung cancer, prostate cancer, colon cancer, and glioblastoma. Subsequent function studies have shown that miR-21 has oncogenic activity (Selcuklu et al. 2009).

3.2.6.1 Proliferation and miRNAs in Cancer

Suppression of cell differentiation and maintenance of proliferation is one of the very important mechanisms in tumorigenesis. The role of miRNAs in cell cycle progression was first proven by Hatfield et al. They showed that G1/S transition was suppressed when DICER-1 knockout in Drosophila germline stem cells. This proved that miRNAs have a role in the normal G1/S transition (Hatfield et al. 2005).

The E2F family of transcription factors controls cell proliferation. E2F1 acts as a tumor suppressor and induces transcription of the target gene in the transition from G1 to S stage. After *c-MYC* is activated, miR-17-92 inhibits the translation of E2F1. Since *C-MYC* also directly induces mir-17-92, this mechanism is evidence of a normal cell cycle process under normal conditions (Coller et al. 2007). The overexpression of miR-17-92 cluster has been demonstrated to have oncogenic functions in many cancer types (Kalkan and Atli 2016; Fang et al. 2017; Gruszka and Zakrzewska 2018).

3.2.6.2 Apoptosis and miRNAs in Cancer

Evasion of apoptosis is an important mechanism for tumor cells, and the cells can choose many different pathways for this. Although the most common mechanism is the loss of TP53 function, upregulation of anti-apoptotic regulators and suppression of pro-apoptotic regulators can also occur (Peng and Croce 2016).

Activation of miR-192, miR-194, and miR-215 by *TP53* and suppression of *MDM2* by targeting mRNA transcribed from the *MDM2* gene has been demonstrated in multiple myeloma. Because the *MDM2* gene is the negative regulator of *TP53* (Nag et al. 2013), downregulation of these miRNAs is an important mechanism in the development of multiple myelomas (Pichiorri et al. 2010). In a recent study, it was shown that the expression of miRNA-331-3p is downregulated in nasopharyngeal carcinoma patients and that overexpression of this miRNA leads to inhibition of phosphorylation of Phosphoinositide 3-kinase (PI3K) and Serine/threonine kinase (AKT). miRNA-331-3p has been shown to suppress proliferation and induce apoptosis (Xuefang et al. 2020).

3.2.6.3 Invasion, Metastasis, and miRNAs in Cancer

Epithelial-mesenchymal transition (EMT) is a very important mechanism for invasion and metastasis. Activation of EMT is required for cell migration and invasion, while mesenchymal-to-epithelial transition (MET) is required for metastasis outgrowth (Tan et al. 2018). EMT is characterized by loss of adhesion, decreased expression of E-cadherin, acquisition of mesenchymal markers, and mobilization of the cell.

Many transcription factors such as Snail, Slug, Twist, ZEB1, and ZEB2 are involved in the EMT process. The miR-205 and miR-200 family have been shown to be epithelial markers and suppressors. The miR-200 family target ZEB1/2 and act to suppress EMT. In contrast, ZEB1 directly binds to the promoter regions of miR-200 genes and represses its transcription. That is, there is a double negative feedback loop. While expression of the MiR-200 family is absent in metaplastic breast cancer cells, ZEB1 and ZEB2 are highly present in invasive mesenchymal cells (Zhang and Ma 2012). miR-99a inhibits the expression of E2F and adhesion G protein-coupled receptor E2 (ADGRE2), thereby suppressing the EMT process. miR-5188 targets the *Fork-headed Box Protein O1 (FOXO1)* gene and can activate the Wnt signaling pathway via β-catenin, thereby EMT is induced (Pan et al. 2021).

Thanks to the studies on the effects of miRNA on EMT, a lot of information has been obtained about cancer development and metastasis, and it is even among the subjects of drug resistance studies.

3.2.6.4 Angiogenesis and miRNAs in Cancer

One of the necessary mechanisms for tumor growth and metastasis is angiogenesis. It has been determined that miRNAs are effective in the mechanism of angiogenesis.

miR-34a is one of the most studied miRNAs in cancer and is known to have a suppressive effect on angiogenesis. miR-34a achieves this effect through the interactions of Silent Information Regulator 1 (Sirt1), Foxo1, Notch1 and Tp53. The mir-29 family also inhibits angiogenesis and tumorigenesis and has been shown to be downregulated in many varieties of cancers. miR-29b targets AKT3 and inhibits Akt3-mediated vascular endothelial growth factor (VEGF) and C-myc activations (Lahooti et al. 2021).

Considering that miRNAs are highly effective in angiogenesis and tumorigenesis, their potential to be a treatment target is quite high.

3.2.6.5 Non-canonical Function of miRNA in Cancer

For a long time, miRNAs were considered to suppress expression by targeting only mRNAs. however, in recent years, evidence has been presented that it both suppresses and increases expression. In recent studies, it has been shown that miRNAs also target the 5'UTR regions of mRNAs and have an effect on increasing transcription (Semina et al. 2021). It has been found that miR-1254 together with Ago/2 and iRISC, interacts with the 5'UTR region of mRNA of cell cycle and apoptosis regulator (CCAR1) and causes its upregulation, thus re-sensitizing mammary cancer cells resistant to tamoxifen (Li et al. 2016). Human miR-369-3 can activate the translation of tumor necrosis factor-α (TNF-α) mRNA when the cell cycle is stopped but suppresses it when cell division occurs. These data support that miRNAs have many functions in the cytoplasm, apart from targeting and suppressing mRNAs (Semina et al. 2021).

Evidence that miRNAs regulate expression in the nucleus has recently been found. It also performs the function of repressing transcription through traditional RISC in the nucleus. They also bind to promoter regions, alter the epigenetic profile, and regulate gene expression (Liu et al. 2018).

In the nucleus, the RNA-Ago complex can directly target non-coding transcripts and modify epigenetic modifications to serve as a scaffold on which epigenetic factors will be recruited. In a study, it was shown that three signaling molecules were activated in response to endoplasmic reticulum stress, and PERK, which is among these molecules, induced miR-211. It was determined that miR-211 increased methylation in the promoter of the proapoptotic transcription factor C/EBP homologous protein (CHOP), which resulted in decreased CHOP expression (Chitnis et al. 2012).

In addition to all these, it has been observed that miRNAs also connect with non-AGO proteins in tumor cells. Downregulation of miR-328 expression has been observed in the blast crisis of chronic myeloid leukemia (CML). It was found that miR-128 directly binds to hnRNP E2 and rescues the translation of the differentiation-inducing transcription factor CEBPA mRNA (Dragomir et al. 2022; Eiring et al. 2010).

The encoding of mRNA-encoded peptides (miPEP) by pri-miRNAs is one of the non-canonical actions of miRNAs. It has been determined that pri-miRNAs transcribed from MIR200A and MIR200B in prostate cancer encode miPEP200a and miPEP200b and these miPEPs show antioncogenic effect by inhibiting migration (Dragomir et al. 2022).

3.2.6.6 Deregulation of miRNA Expression in Cancer

After realizing that the expression of miRNAs was deregulated in tumor cells, many studies were conducted. Understanding the mechanisms that cause the dysregulation of cancer miRNA expression is very important for tumorigenesis, development, metastasis, and treatment.

One of the most common causes of miRNA expression changes in cancer cells is numerical and structural anomalies in the genome (such as amplification, deletion, and translocation). 13q deletions in CLL, which led to the establishment of the first association between miRNAs and cancer, are an example of decreased expression of miR-16-1 and miR-15a due to copy number loss (Calin et al. 2002). The miR-17-92 cluster has been amplified in lung and B-cell lymphoma, and it has been found to undergo a translocation that will lead to overexpression in T-cell acute lymphoblastic leukemia (Peng and Croce 2016). The relationship between chromosome breaks and miRNA localization was first discovered in the sample with t(8;17) anomaly. The miR-142 gene was determined to be located at a distance of 50 nt from the break point of chromosome 17, where it was included in t(8;17), and it was likely that the regulatory elements of miR-142 increased the expression of *C-MYC* (Calin and Croce 2006).

The expression of miRNAs is controlled by many different transcription factors. Two of these transcription factors are Tp53 and C-Myc, which are known to have important effects on tumorigenesis. C-Myc binds to the promoter of miR-17-92, which has oncogenic properties and activates its transcription. In addition, it suppresses the transcriptional activity of tumor suppressor miRNAs such as mir-15a, miR-26, miR-29, mir-30, and let-7 families (Chang et al. 2008). Expression of the miR-34 family is controlled by Tp53. When cell stress increases, miR-34 activates *TP53*. Expression of miR-145 is also induced by upregulated *TP53*. However, the miR-143/145 cluster is suppressed by the RAS signal. *RAS-responsive element-binding protein 1 (RREB1)* transcriptionally represses the miR-143/145 cluster, and then miR-143/145 represses the expression of *RREB1* (Ali Syeda et al. 2020).

One of the factors affecting miRNA expression is epigenetic changes. It has been determined that, like the hypermethylation of CpG islands in the promoters of tumor suppressor genes, the expression of miR-124 is also suppressed due to hypermethylation in their promoters in leukemia, lymphoma, breast, colon, and liver cancers (Lujambio et al. 2007; Ali Syeda et al. 2020).

Another mechanism that causes miRNA deregulation is mutations. The first discovered germline mutation in miRNA was detected in miR-16-1 (Calin et al. 2005). The most mutated miRNAs in the analysis of all cancers were MIR1324, MIR1303, and MIR4686, whereas MIR142, which has driver mutations in DLBCL, CLL, acute myeloid leukemia (ALL), and other kinds of lymphoma, was the most mutated miRNA in a particular cancer (Dragomir et al. 2022). Mutations or expression changes can be observed in DNA sequences encoding all proteins involved in miRNA biogenesis as well as in miRNA genes. Various mutations or change of expression have been detected in *DROSHA, DICER, DGCR8, AGO,* and *EXPORTIN*

5 genes, which are involved in miRNA biogenesis, in different cancer types (Ali Syeda et al. 2020; Peng and Croce 2016).

3.2.7 Circulating miRNA in Cancer

Extracellular miRNAs are highly durable and stable. Extracellular miRNAs exist as part of vesicles or as a soluble form of protein-containing complexes. HnRNPA2B1 and HnRNPA1 proteins regulate the loading of miRNAs into exosomes by identifying particular sequence patterns. As the suppression of neutral sphingomyelinase 2 (nSMase2), an enzyme involved in ceramide production, downregulates exosome secretion and releases exosomal miRNAs into the extracellular environment, exosomal miRNAs can be exported outside the cells through a ceramide-dependent mechanism. Although various distinct routes for miRNA entry into cells have been postulated, the mechanisms for exosomal miRNAs uptake by cells are currently poorly understood. Exosomes can enter cells through a variety of methods, including endocytosis, phagocytosis, and micropinocytosis. Another is a direct fusion of exosomes with the plasma membrane. Exosome-free miRNAs can also enter cells by way of certain receptors. Exosomes that contain miRNAs that are produced by tumor cells can be taken up by the recipient cells. MiRNAs can affect the development of tumors by promoting or inhibiting cell invasion, metastasis, and tumor neoangiogenesis. Exosomal miRNAs can potentially modify the extracellular matrix or attract and activate immune cells, which can both have an impact on the tumor microenvironment (Semina et al. 2021).

The first circulating miRNAs were discovered in patients with diffuse large B-cell lymphoma. As a result of subsequent studies, it was shown that miRNAs could be used to determine tumor grades or to evaluate treatment responses. Unlike mRNAs, their ability to stay for a long time without degradation also provides an advantage in using miRNAs as biomarkers (Smolarz et al. 2022).

3.2.8 miRNA-based Biomarkers in Cancer

After the discovery of the roles of miRNAs in cancer, it was inevitable to investigate the relationships between miRNAs and cancer types and disease prognosis. There is a large amount of data proving that many miRNAs can be diagnostic and prognostic markers. In addition to all these, miRNAs have become a treatment target in cancer.

There are many studies proving that miRNAs will show clinical benefits as diagnostic and prognostic markers (He et al. 2020). In a study investigating the role of miRNAs in triple-negative breast cancer (TNBC), databases such as PubMed, ScienceDirect, Springer, Web of Science, and Scopus were searched and 197/1233 articles were extensively reviewed. Many miRNAs have been reported that have the potential to be of prognostic and diagnostic importance, e.g., miR-9, miR-21, miR-93, miR-181a/b, miR-182, miR-221, miR-321, miR-155, miR-10b, miR-29, miR-222, miR-373, miR-145, miR-199a-5p, miR-200 family, miR-203, and miR-205 (Sabit et al. 2021).

MiR-155-5p, an oncogenic miRNA, regulates important transcription factors such as E2F2, hypoxia-inducible factor 1 (HIF1), and FOXO3. One study showed that the upregulation of miR-155-5p is associated with short overall survival in cases of chronic lymphocytic leukemia (CLL) (Papageorgiou et al. 2017).

Although hematuria is the most common symptom of bladder cancer (BC), hematuria is not a definitive diagnostic marker. In a study conducted, urinary cell-free microRNA expression differences were investigated to distinguish patients with BC from patients with hematuria, and the ratio of miR-612–miR-4511 was found to be significantly higher in BC (Piao et al. 2019).

One of the biggest problems in cancer treatment is the late detection of cancer. Plasma/serum circular miRNA can be used in the diagnosis of breast, colorectal, stomach, lung, pancreatic, and hepatocellular cancer. Circular miRNAs may contribute to the discovery of the primary origin of metastatic tumors of unknown primary tissue. In addition, circular miRNAs can be used as a marker in disease follow-up (Cui et al. 2019).

For example, it has been shown that miR-125b suppresses cell proliferation in ovarian, thyroid, and oral cancers, but induces proliferation in prostate cancers (Cui et al. 2019). Although hematuria is the most common symptom of bladder cancer (BC), hematuria is not a definitive diagnostic marker. In a study conducted, urinary cell-free microRNA expression differences were investigated to distinguish patients with BC from patients with hematuria, and the ratio of miR-6124 to miR-4511 was found to be significantly higher in BC (Piao et al. 2019). As another example, elevated levels of circulating miR-122 were found to correlate with metastatic recurrence in stage II-III breast cancer patients (Wu et al. 2012). In another study, it was determined that miR-375 and miR-200b in serum were expressed higher in patients with metastatic prostate cancer than in patients with localized cancer (Bryant et al. 2012).

3.2.9 miRNA-Based Therapies in Cancer

The regulatory role of miRNAs in many cancer types has made them a therapeutic target. The miRNA-based therapy methods in cancer have two approaches: increasing the activities of miRNAs that act as tumor suppressors and suppressing the functions of oncoMIRs.

Tumor suppressor miRNAs are downregulated in tumor cells and miRNA mimics are used to function as before. miRNA mimics are chemically modified (2'--O'methoxy) double-stranded RNA molecules (Menon et al. 2022). The size of miRNA is smaller than the protein, which gives it an advantage in terms of penetration into the cell. The first study to show the tumor suppressor function of Let-7 and its potential for treatment was conducted in 2008. In mouse models, it has been demonstrated that tumor growth can be inhibited by restoring let-7 (Esquela-Kerscher et al. 2008). Another study with mouse models of lung cancer demonstrated that metastasis and tumor growth could be suppressed through chemically synthesized miR-34a and a lipid-based delivery vehicle (Wiggins et al. 2010).

For the suppression of oncomiRs, small molecule inhibitors and complementary oligonucleotides such as anti-miRNA oligonucleotide (AMOs) (Amodeo et al. 2013), locked-nucleic acid antisense oligonucleotides (LNAs), antagomirs, and miRNA sponges have been developed. AMO is a DNA sequence complementary to the target miRNA and prevents the miRNA from binding to the target mRNA. LNA-AMOs are more stable and more sensitive than just AMOs. It was created as a result of the modification of AMOs. Antagomirs and miRNA sponges are longer nucleic acids that prevent miRNAs from binding to their targets (Mollaei et al. 2019; Fu et al. 2021). For example, in a study by Chen et al. (2014), it was shown that miRNA sponges successfully suppressed miR-23b expression both in vitro and in vivo, and reduced glioma angiogenesis, invasion, and migration (Chen et al. 2014).

3.2.9.1 Approaches for miRNA Therapeutic Delivery

Although the direct injection of miRNA mimics or inhibitors into tumor tissue is limited due to their application to localized and easily accessible solid tumors, it is an advantage that the probability of rejection by healthy organs is minimal. The development of a systemic delivery approach is needed to treat other types of cancer and metastatic tumors. For this, miRNAs must not deteriorate in the bloodstream in a short time, be able to be transported to target cells, and not cause an immunological response. Some chemical modifications are performed on miRNA oligonucleotides to increase miRNA stability and protect it from nucleases. LNAs are examples of modified nucleotides. LNA-anti-mir-122 has been shown to regulate the expression of mRNA in the liver of mice, depending on the level of miR-122 (Forterre et al. 2020).

Although viral and non-viral vectors are generally used for miRNA delivery, adverse immune responses occur against viral vectors. Tumor suppressor pri-miRNAs are inserted into a plasmid. A viral promoter, a restriction enzyme gene and an antibiotic resistance gene are contained in this plasmid. The plasmid is delivered to tumor cells in a viral vector and the mature miRNA suppresses translation or induces degradation of the target mRNA. The low cost of DNA plasmids is an advantage. Furthermore, the untranslated miRNA is transferred to the nucleus and its continuous expression is ensured. In addition, because it is translated in tumor cells, less off-target effects occur compared to synthetic miRNA sequences (Hosseinahli et al. 2018).

For the non-viral delivery system to be successful, it must prevent nuclease-mediated degradation and carry endogenous miRNA or miRNA-expressing vectors. Delivery can be accomplished using techniques such as gene gun, electroporation, or ultrasound, or using organic-based, inorganic-based, or polymer-based carriers.

Although non-viral systems have less toxicity and immunological effects, low transfection efficiency is considered a disadvantage of this method (Menon et al. 2022). Radiotherapy is used in the treatment of head and neck cancer, but its clinical effect is inhibited by both the side effects of radiation and radioresistance. RNA therapeutics therefore have great potential as radiosensitizers as they can target radioresistance-specific pathways. High-density lipoprotein nanoparticle (HDL NPs) was used in a head and neck cancer cell line in a 2022 study to deliver miR-34a. As a result of the study, it was observed that proliferation decreased and apoptosis increased (Dehghankelishadi et al. 2022). Besides biomaterials, polymeric vectors (PEIs, polylactic-co-glycolic acid/PLGA, chitosans, and dendrimers) and inorganic materials (gold, diamond, silica, and iron oxide) are also used in the non-viral vectors delivery system. Among these polymers, PLGA is an FDA-approved biodegradable polymer (Forterre et al. 2020; Menon et al. 2022).

Another miRNA delivery system is the use of outer membrane vesicles (OMVs) of *Escherichia coli* as nanoscale spherical vesicles (Menon et al. 2022). In a 2022 study, An inexpensive and potentially mass-produced method was found for the preparation of engineered OMV with overexpressed pre-miRNA. In this study, it was discovered that OMV can be discharged from parent *E. coli* and inherit an overexpressed tRNA$^{Lys\text{-}pre\text{-}miRNA}$ that is used directly for the treatment of tumors. It was suggested that the OMV-based platform is a flexible and effective method to directly and specifically target individualized tumor therapy (Cui et al. 2022).

Many studies have shown that the use of miRNA-based therapies together with other treatment options such as chemotherapy and radiotherapy induces the therapeutic effect and prevents drug resistance (He et al. 2020; Menon et al. 2022).

Understanding the molecular mechanism of cancer increases the chances of treatment success. Understanding the role of miRNAs in cancer also shows the potential to be used as a treatment target or tool in the future. However, one of the most important problems is that a miRNA has more than one target. Another problem is choosing the right miRNA delivery system. Today, pre-clinical and clinical studies continue. In the future, personalized treatment options based on miRNA are expected to be developed.

3.3 Conclusion

In this chapter, we have reviewed the known epigenetic mechanisms in normal cells and their roles in the carcinogenesis process. The molecular processes that lead to promoter hypermethylation, genome-wide DNA demethylation, histone modifications, and non-coding RNAs were highlighted in cancer cells. The long-held conventional belief that the genetic code is the primary determinant of cellular gene function and that its change is the primary cause of human diseases has been called into question by the epigenetic revolution that has occurred in the area of biology during the last decades. The packaging of the genome may be just as important as the genome itself in regulating the vital cellular activities necessary for maintaining a cellular identity as well as in the development of disease states like cancer, according to recent developments in the field of cancer epigenetics. All cells of an individual have the same genome, but they might have different epigenotypes depending on their epigenetic markings, which are suitable for different tissues, stages of development, or environmental conditions. The partially improved treatment approaches have been made possible by a deeper understanding of the worldwide patterns of these epigenetic modifications and their related changes in cancer. Several genetic and epigenetic abnormalities, including structural variants, copy number variations, single nucleotide polymorphisms, mutations, and epigenetic dysregulations, are addressed to cancer hallmarks. To advance personalized and precision medicine and improve cancer treatment, it is crucial to comprehend the intricate interplay of genetic and epigenetic modifications. Combinatorial promising approaches that combine several epigenetic therapeutic modalities with conventional chemotherapy have a strong chance of treating cancer successfully in the future. These methods may also enable cancer cells, particularly cancer stem cells, which are resistant to conventional chemotherapy, to become more sensitive. We may be able to successfully reset the altered cancer epigenome with increased knowledge of cancer stem cells and the development of more targeted epigenetic medicine.

Compliance with Ethical Standards Funding: The authors declare no competing financial interest.

Conflict of Interest: The authors declare no conflicts of interest.

Ethical Approval: This chapter does not contain any studies with human participants performed by the authors.

References

Albert M, Helin K (2010) Histone methyltransferases in cancer. Seminars in cell & developmental biology. Elsevier, pp 209–220

Ali Syeda Z, Langden SSS, Munkhzul C, Lee M, Song SJ (2020) Regulatory mechanism of microRNA expression in cancer. Int J Mol Sci 21

Allfrey VG, Faulkner R, Mirsky A (1964) Acetylation and methylation of histones and their possible role in the regulation of RNA synthesis. Proc Natl Acad Sci 51:786–794

Amodeo V, Bazan V, Fanale D, Insalaco L, Caruso S, Cicero G, Bronte G, Rolfo C, Santini S, Russo A (2013) Effects of anti-miR-182 on TSP-1 expression in human colon cancer cells: there is a sense in antisense? Expert Opin Ther Targets 17(11):1249–1261. https://doi.org/10.1517/14728222.2013.832206

Anvar Z, Chakchouk I, Demond H, Sharif M, Kelsey G, Van Den Veyver IB (2021) DNA methylation dynamics in the female germline and maternal-effect mutations that disrupt genomic imprinting. Genes 12:1214

Audia JE, Campbell RM (2016) Histone modifications and cancer. Cold Spring Harb Perspect Biol 8:A019521

Baba Y, Yagi T, Sawayama H, Hiyoshi Y, Ishimoto T, Iwatsuki M, Miyamoto Y, Yoshida N, Baba H (2018) Long interspersed Element-1 methylation level as A prognostic biomarker in gastrointestinal cancers. Digestion 97:26–30

Basu S, Dong Y, Kumar R, Jeter C, Tang DG (2021) Slow-cycling (dormant) cancer cells in therapy resistance, cancer relapse and metastasis. Seminars in cancer biology. Elsevier

Baylin SB, Jones PA (2016) Epigenetic determinants of cancer. Cold Spring Harb Perspect Biol 8: A019505

Begam N, Jamil K, Raju SG (2017) Promoter Hypermethylation of the ATM gene as A novel biomarker for breast cancer. Asian Pacific J Cancer Prevent 18:3003

Benedetti R, Conte M, Altucci L (2015) Targeting histone deacetylases in diseases: where are we? Antioxid Redox Signal 23:99–126

Berdasco M, Esteller M (2010) Aberrant epigenetic landscape in cancer: how cellular identity goes awry. Dev Cell 19:698–711

Berman BP, Weisenberger DJ, Aman JF, Hinoue T, Ramjan Z, Liu Y, Noushmehr H, Lange CPE, van Dijk JM, Tollenaar RAEM, Van Den Berg D, Laird PW (2012) Regions of focal DNA hypermethylation and long-range hypomethylation in colorectal cancer coincide with nuclear lamina–associated domains. Nat Genet 44(1):40–46. https://doi.org/10.1038/ng.969

Berkyurek AC, Suetake I, Arita K, Takeshita K, Nakagawa A, Shirakawa M, Tajima S (2014) The DNA methyltransferase Dnmt1 directly interacts with the SET and RING finger-associated (SRA) domain of the multifunctional protein Uhrf1 to facilitate accession of the catalytic center to hemi-methylated DNA. J Biol Chem 289:379–386

Bildik G, Liang X, Sutton MN, Bast RC Jr, Lu Z (2022) DIRAS3: an imprinted tumor suppressor gene that regulates RAS and PI3K-driven cancer growth motility autophagy and tumor dormancy. Mol Cancer Ther 21(1):25–37. https://doi.org/10.1158/1535-7163.MCT-21-0331

Bond DR, Uddipto K, Enjeti AK, Lee HJ (2020) Single-cell epigenomics in cancer: charting a course to clinical impact. Epigenomics 12:1139–1151

Borchert GM, Lanier W, Davidson BL (2006) RNA polymerase III transcribes human MicroRNAs. Nat Struct Mol Biol 13:1097–1101

Boyerinas B, Park SM, Hau A, Murmann AE, Peter ME (2010) The role of Let-7 in cell differentiation and cancer. Endocr Relat Cancer 17:F19–F36

Brinkman AB, Nik-Zainal S, Simmer F, Rodriguez-Gonzalez F, Smid M, Alexandrov LB, Butler A, Martin S, Davies H, Glodzik D (2019) Partially methylated domains are hypervariable in breast cancer and fuel widespread CpG island hypermethylation. Nat Commun 10:1–10

Bryant RJ, Pawlowski T, Catto JW, Marsden G, Vessella RL, Rhees B, Kuslich C, Visakorpi T, Hamdy FC (2012) Changes in circulating microRNA levels associated with prostate cancer. Br J Cancer 106:768–774

Calin GA, Croce CM (2006) MicroRNAs and chromosomal abnormalities in cancer cells. Oncogene 25:6202–6210

Calin GA, Dumitru CD, Shimizu M, Bichi R, Zupo S, Noch E, Aldler H, Rattan S, Keating M, Rai K, Rassenti L, Kipps T, Negrini M, Bullrich F, Croce CM (2002) Frequent deletions and down-regulation of micro-RNA genes Mir15 and Mir16 at 13q14 in chronic lymphocytic leukemia. Proc Natl Acad Sci U S A 99:15524–15529

Calin GA, Ferracin M, Cimmino A, Di Leva G, Shimizu M, Wojcik SE, Iorio MV, Visone R, Sever NI, Fabbri M, Iuliano R, Palumbo T, Pichiorri F, Roldo C, Garzon R, Sevignani C, Rassenti L, Alder H, Volinia S, Liu CG, Kipps TJ, Negrini M, Croce CM (2005) A microRNA signature associated with prognosis and progression in chronic lymphocytic leukemia. N Engl J Med 353: 1793–1801

Cavalli G, Heard E (2019) Advances in epigenetics link genetics to the environment and disease. Nature 571:489–499

Chang TC, Yu D, Lee YS, Wentzel EA, Arking DE, West KM, Dang CV, Thomas-Tikhonenko A, Mendell JT (2008) Widespread microRNA repression by Myc contributes to tumorigenesis. Nat Genet 40:43–50

Chase A, Cross NC (2011) Aberrations of EZH2 in cancer aberrations of EZH2 in cancer. Clin Cancer Res 17:2613–2618

Chauhan S, Sen S, Chauhan SS, Pushker N, Tandon R, Kashyap S, Vanathi M, Bajaj MS (2021) Stratifin in ocular surface squamous neoplasia and its association with P53. Acta Ophthalmol 99:E1483–E1491

Chen Y-F, Luo R-Z, Li Y, Cui B-K, Song M, Yang A-K, Chen W-K (2013) High expression levels of COX-2 and P300 are associated with unfavorable survival in laryngeal squamous cell carcinoma. Eur Arch Otorhinolaryngol 270:1009–1017

Chen L, Zhang K, Shi Z, Zhang A, Jia Z, Wang G, Pu P, Kang C, Han L (2014) A lentivirus-mediated Mir-23b sponge diminishes the malignant phenotype of glioma cells in vitro and in vivo. Oncol Rep 31:1573–1580

Cheng W, Li M, Cai J, Wang K, Zhang C, Bao Z, Liu Y, Wu A (2015) HDAC4, A prognostic and chromosomal instability marker, refines the predictive value of MGMT promoter methylation. J Neuro-Oncol 122:303–312

Cheng Y, He C, Wang M, Ma X, Mo F, Yang S, Han J, Wei X (2019) Targeting epigenetic regulators for cancer therapy: mechanisms and advances in clinical trials. Signal Transduct Target Ther 4:1–39

Cheng W-L, Feng P-H, Lee K-Y, Chen K-Y, Sun W-L, Van Hiep N, Luo C-S, Wu S-M (2021) The role of EREG/EGFR pathway in tumor progression. Int J Mol Sci 22:12828

Chitnis NS, Pytel D, Bobrovnikova-Marjon E, Pant D, Zheng H, Maas NL, Frederick B, Kushner JA, Chodosh LA, Koumenis C, Fuchs SY, Diehl JA (2012) Mir-211 is a prosurvival MicroRNA that regulates chop expression in a PERK-dependent manner. Mol Cell 48:353–364

Chuikov S, Kurash JK, Wilson JR, Xiao B, Justin N, Ivanov GS, McKinney K, Tempst P, Prives C, Gamblin SJ (2004) Regulation of P53 activity through lysine methylation. Nature 432:353–360

Cimmino A, Calin GA, Fabbri M, Iorio MV, Ferracin M, Shimizu M, Wojcik SE, Aqeilan RI, Zupo S, Dono M, Rassenti L, Alder H, Volinia S, Liu CG, Kipps TJ, Negrini M, Croce CM (2005) Mir-15 and Mir-16 induce apoptosis by targeting BCL2. Proc Natl Acad Sci U S A 102: 13944–13949

Clouaire T, Stancheva I (2008) Methyl-Cpg binding proteins: specialized transcriptional repressors or structural components of chromatin? Cell Mol Life Sci 65:1509–1522

Cohen I, Poręba E, Kamieniarz K, Schneider R (2011) Histone modifiers in cancer: friends or foes? Genes Cancer 2:631–647

Cole AJ, Clifton-Bligh R, Marsh DJ (2015) Histone H2B monoubiquitination: roles to play in human malignancy. Endocr Relat Cancer 22:T19–T33

Coller HA, Forman JJ, Legesse-Miller A (2007) "Myc'ed messages": Myc induces transcription of E2F1 while inhibiting its translation via a microRNA polycistron. PLoS Genet 3:E146

Costello JF, Frühwald MC, Smiraglia DJ, Rush LJ, Robertson GP, Gao X, Wright FA, Feramisco JD, Peltomäki P, Lang JC (2000) Aberrant Cpg-island methylation has non-random and tumour-type–specific patterns. Nat Genet 24:132–138

Cui M, Wang H, Yao X, Zhang D, Xie Y, Cui R, Zhang X (2019) Circulating microRNAs in cancer: potential and challenge. Front Genet 10:626

Cui C, Guo T, Zhang S, Yang M, Cheng J, Wang J, Kang J, Ma W, Nian Y, Sun Z, Weng H (2022) Bacteria-derived outer membrane vesicles engineered with over-expressed pre-Mirna as delivery nanocarriers for cancer therapy. Nanomedicine 45:102585

D'oto A, Tian Q-W, Davidoff AM, Yang J (2016) Histone demethylases and their roles in cancer epigenetics. J Med Oncol Ther 1:34

Dawson MA, Bannister AJ, Göttgens B, Foster SD, Bartke T, Green AR, Kouzarides T (2009) JAK2 phosphorylates histone H3Y41 and excludes HP1α from chromatin. Nature 461:819–822

De Souza CRT, Leal MF, Calcagno DQ, Costa Sozinho EK, Borges BDN, Montenegro RC, Dos Santos AKCR, Dos Santos SEB, Ribeiro HF, Assumpção PP (2013) MYC deregulation in gastric cancer and its clinicopathological implications. PLoS One 8:E64420

Deaton AM, Bird A (2011) Cpg Islands and the regulation of transcription. Genes Dev 25:1010–1022

Decato BE, Qu J, Ji X, Wagenblast E, Knott SR, Hannon GJ, Smith AD (2020) Characterization of universal features of partially methylated domains across tissues and species. Epigenetics Chromatin 13:1–14

Dehghankelishadi P, Maritz MF, Badiee P, Thierry B (2022) High density lipoprotein nanoparticle as delivery system for radio-sensitising Mirna: an investigation in 2D/3D head and neck cancer models. Int J Pharm 617:121585

Dell'Aversana C, Lepore I, Altucci L (2012) HDAC modulation and cell death in the clinic. Exp Cell Res 318:1229–1244

Deng L, Meng T, Chen L, Wei W, Wang P (2020) The role of ubiquitination in tumorigenesis and targeted drug discovery. Signal Transduct Target Ther 5:1–28

Di Cerbo V, Schneider R (2013) Cancers with wrong Hats: the impact of acetylation. Brief Funct Genomics 12:231–243

Dragomir MP, Knutsen E, Calin GA (2022) Classical and noncanonical functions of MiRNAs in cancers. Trends Genet 38:379–394

Dwane L, Gallagher WM, Chonghaile TN, O'connor DP (2017) The emerging role of non-traditional ubiquitination in oncogenic pathways. J Biol Chem 292:3543–3551

Eden A, Gaudet F, Waghmare A, Jaenisch R (2003) Chromosomal instability and tumors promoted by DNA hypomethylation. Science 300:455–455

Eiring AM, Harb JG, Neviani P, Garton C, Oaks JJ, Spizzo R, Liu S, Schwind S, Santhanam R, Hickey CJ, Becker H, Chandler JC, Andino R, Cortes J, Hokland P, Huettner CS, Bhatia R, Roy DC, Liebhaber SA, Caligiuri MA, Marcucci G, Garzon R, Croce CM, Calin GA, Perrotti D (2010) Mir-328 functions as an RNA decoy to modulate Hnrnp E2 regulation of Mrna translation in leukemic blasts. Cell 140:652–665

Esquela-Kerscher A, Trang P, Wiggins JF, Patrawala L, Cheng A, Ford L, Weidhaas JB, Brown D, Bader AG, Slack FJ (2008) The Let-7 Microrna reduces tumor growth in mouse models of lung cancer. Cell Cycle 7:759–764

Falkenberg KJ, Johnstone RW (2014) Histone deacetylases and their inhibitors in cancer, neurological diseases and immune disorders. Nat Rev Drug Discov 13:673–691

Fang LL, Wang XH, Sun BF, Zhang XD, Zhu XH, Yu ZJ, Luo H (2017) Expression, regulation and mechanism of action of the Mir-17-92 cluster in tumor cells (review). Int J Mol Med 40:1624–1630

Fatma H, Maurya SK, Siddique HR (2020) Epigenetic modifications of C-MYC: role in cancer cell reprogramming, progression and chemoresistance. Seminars in cancer biology. Elsevier

Felsenfeld G (2014) A brief history of epigenetics. Cold Spring Harb Perspect Biol 6:A018200

Ferguson-Smith AC (2011) Genomic imprinting: the emergence of an epigenetic paradigm. Nat Rev Genet 12:565–575

Forma E, Jóźwiak P, Bryś M, Krześlak A (2014) The potential role of O-Glcnac modification in cancer epigenetics. Cell Mol Biol Lett 19:438–460

Forterre A, Komuro H, Aminova S, Harada M (2020) A comprehensive review of cancer microrna therapeutic delivery strategies. Cancers (Basel) 12

Fraga MF, Ballestar E, Villar-Garea A, Boix-Chornet M, Espada J, Schotta G, Bonaldi T, Haydon C, Ropero S, Petrie K (2005) Loss of acetylation at Lys16 and trimethylation at Lys20 of histone H4 is A common hallmark of human cancer. Nat Genet 37:391–400

Fu J, Qin L, He T, Qin J, Hong J, Wong J, Liao L, Xu J (2011) The TWIST/Mi2/Nurd protein complex and its essential role in cancer metastasis. Cell Res 21:275–289

Fu Z, Wang L, Li S, Chen F, Au-Yeung KK, Shi C (2021) Microrna as an important target for anticancer drug development. Front Pharmacol 12:736323

Gao Y, Geng J, Hong X, Qi J, Teng Y, Yang Y, Qu D, Chen G (2014) Expression of P300 and CBP is associated with poor prognosis in small cell lung cancer. Int J Clin Exp Pathol 7:760

Garraway LA, Lander ES (2013) Lessons from the cancer genome. Cell 153:17–37

Gayther SA, Batley SJ, Linger L, Bannister A, Thorpe K, Chin S-F, Daigo Y, Russell P, Wilson A, Sowter HM (2000) Mutations truncating the EP300 acetylase in human cancers. Nat Genet 24:300–303

Goldberg AD, Allis CD, Bernstein E (2007) Epigenetics: a landscape takes shape. Cell 128:635–638

Greger V, Passarge E, Höpping W, Messmer E, Horsthemke B (1989) Epigenetic changes may contribute to the formation and spontaneous regression of retinoblastoma. Hum Genet 83:155–158

Gruszka R, Zakrzewska M (2018) The oncogenic relevance of Mir-17-92 cluster and its paralogous Mir-106b-25 and Mir-106a-363 clusters in brain tumors. Int J Mol Sci 19

Gu Y, Mi W, Ge Y, Liu H, Fan Q, Han C, Yang J, Han F, Lu X, Yu W (2010) Glcnacylation plays an essential role in breast cancer metastasis. Cancer Res 70:6344–6351

Guppy BJ, Jeusset LM, McManus KJ (2017) The relationship between DOT1L, histone H3 methylation, and genome stability in cancer. Curr Mol Biol Reports 3:18–27

Hájková H, Marková J, Haškovec C, Šárová I, Fuchs O, Kostečka A, Cetkovský P, Michalová K, Schwarz J (2012) Decreased DNA methylation in acute myeloid leukemia patients with DNMT3A mutations and prognostic implications of DNA methylation. Leuk Res 36:1128–1133

Han M, Jia L, Lv W, Wang L, Cui W (2019) Epigenetic enzyme mutations: role in tumorigenesis and molecular inhibitors. Front Oncol 9:194

Hanahan D (2022) Hallmarks of cancer: new dimensions. Cancer Discov 12:31–46

Handy DE, Castro R, Loscalzo J (2011) Epigenetic modifications: basic mechanisms and role in cardiovascular disease. Circulation 123:2145–2156

Hansen KD, Sabunciyan S, Langmead B, Nagy N, Curley R, Klein G, Klein E, Salamon D, Feinberg AP (2014) Large-scale hypomethylated blocks associated with Epstein-Barr virus-induced B-cell immortalization. Genome Res 24:177–184

Hassler MR, Egger G (2012) Epigenomics of cancer–emerging new concepts. Biochimie 94:2219–2230

Hatfield SD, Shcherbata HR, Fischer KA, Nakahara K, Carthew RW, Ruohola-Baker H (2005) Stem cell division is regulated by the microRNA pathway. Nature 435:974–978

He J, Nguyen AT, Zhang Y (2011a) KDM2b/JHDM1b, an H3k36me2-specific demethylase, is required for initiation and maintenance of acute myeloid Leukemia. Blood J Am Soc Hematol 117:3869–3880

He Y-F, Li B-Z, Li Z, Liu P, Wang Y, Tang Q, Ding J, Jia Y, Chen Z, Li L (2011b) Tet-mediated formation of 5-carboxylcytosine and its excision by TDG in mammalian DNA. Science 333: 1303–1307

He B, Zhao Z, Cai Q, Zhang Y, Zhang P, Shi S, Xie H, Peng X, Yin W, Tao Y, Wang X (2020) Mirna-based biomarkers, therapies, and resistance in cancer. Int J Biol Sci 16:2628–2647

Helfrich BA, Kim J, Gao D, Chan DC, Zhang Z, Tan A-C, Bunn PA (2016) Barasertib (AZD1152), A small molecule Aurora B inhibitor, inhibits the growth of SCLC cell lines in vitro and in Vivobarasertib inhibits small-cell lung cancer cell line growth. Mol Cancer Ther 15:2314–2322

Hon GC, Hawkins RD, Caballero OL, Lo C, Lister L, Pelizzola M, Valsesia A, Ye Z, Kuan S, Edsall LE, Camargo AA, Stevenson BJ, Ecker JR, Bafna V, Strausberg RL, Simpson AJ, Ren B (2012) Global DNA hypomethylation coupled to repressive chromatin domain formation and gene silencing in breast cancer. Genome Res 22(2):246–258. https://doi.org/10.1101/gr. 125872.111

Hosseinahli N, Aghapour M, Duijf PHG, Baradaran B (2018) Treating cancer with Microrna replacement therapy: a literature review. J Cell Physiol 233:5574–5588

Hosseini A, Minucci S (2018) Alterations of histone modifications in cancer. Epigenetics in human disease. Elsevier

Hu J, Liu J, Chen A, Lyu J, Ai G, Zeng Q, Sun Y, Chen C, Wang J, Qiu J (2016) Ino80 promotes cervical cancer tumorigenesis by activating Nanog expression. Oncotarget 7:72250

Huang B, Laban M, Leung CH, Lee L, Lee C, Salto-Tellez M, Raju G, Hooi S (2005) Inhibition of histone deacetylase 2 increases apoptosis and p21Cip1/WAF1 expression, independent of histone deacetylase 1. Cell Death Differ 12:395–404

Huang J, Sengupta R, Espejo AB, Lee MG, Dorsey JA, Richter M, Opravil S, Shiekhattar R, Bedford MT, Jenuwein T (2007) P53 is regulated by the lysine demethylase LSD1. Nature 449: 105–108

Igarashi S, Suzuki H, Niinuma T, Shimizu H, Nojima M, Iwaki H, Nobuoka T, Nishida T, Miyazaki Y, Takamaru H (2010) A novel correlation between LINE-1 hypomethylation and the malignancy of gastrointestinal stromal tumorsline-1 hypomethylation and malignancy of gists. Clin Cancer Res 16:5114–5123

Ikeda K, Shiraishi K, Eguchi A, Shibata H, Yoshimoto K, Mori T, Baba Y, Baba H, Suzuki M (2013) Long interspersed nucleotide element 1 hypomethylation is associated with poor prognosis of lung adenocarcinoma. Ann Thorac Surg 96:1790–1794

Ito S, Shen L, Dai Q, Wu SC, Collins LB, Swenberg JA, He C, Zhang Y (2011) Tet proteins can convert 5-methylcytosine to 5-formylcytosine and 5-carboxylcytosine. Science 333:1300–1303

Izzo A, Schneider R (2010) Chatting histone modifications in mammals. Brief Funct Genomics 9: 429–443

Jelinic P, Shaw P (2007) Loss of imprinting and cancer. J Pathol 211(3):261–268. 10.1002/(ISSN) 1096-9896. https://doi.org/10.1002/path.v211:3 10.1002/path.2116

Jeusset LM, McManus KJ (2019) Developing targeted therapies that exploit aberrant histone ubiquitination in cancer. Cell 8:165

Jones PA, Baylin SB (2007) The epigenomics of cancer. Cell 128:683–692

Kalkan R, Atli EI (2016) The impacts of miRNAs in glioblastoma progression. Crit Rev Eukaryot Gene Expr 26(2):137–142. 10.1615/CritRevEukaryotGeneExpr.v26.i2. https://doi.org/10.1615/CritRevEukaryotGeneExpr.2016015964

Kao S-H, Wu K-J, Lee W-H (2016) Hypoxia, epithelial-mesenchymal transition, and TET-mediated epigenetic changes. J Clin Med 5:24

Karagiannis D, Rampias T (2021) HDAC inhibitors: dissecting mechanisms of action to counter tumor heterogeneity. Cancers 13:3575

Karpf AR, Matsui S-I (2005) Genetic disruption of cytosine DNA methyltransferase enzymes induces chromosomal instability in human cancer cells. Cancer Res 65:8635–8639

Keum N, Giovannucci E (2019) Global burden of colorectal cancer: emerging trends, risk factors and prevention strategies. Nat Rev Gastroenterol Hepatol 16:713–732

Khan S, Jankowska A, Mahfouz R, Dunbar A, Sugimoto Y, Hosono N, Hu Z, Cheriyath V, Vatolin S, Przychodzen B (2013) Multiple mechanisms deregulate EZH2 and histone H3 lysine 27 epigenetic changes in myeloid malignancies. Leukemia 27:1301–1309

Khor GH, Froemming GRA, Zain RB, Abraham MT, Omar E, Tan SK, Tan AC, Vincent-Chong VK, Thong KL (2013) DNA methylation profiling revealed promoter hypermethylation-induced silencing of P16, DDAH2 and DUSP1 in primary oral squamous cell carcinoma. Int J Med Sci 10:1727

Kim J, Bretz CL, Lee S (2015) Epigenetic instability of imprinted genes in human cancers. Nucleic Acids Res 43(22):10689–10699. https://doi.org/10.1093/nar/gkv867

Klemm SL, Shipony Z, Greenleaf WJ (2019) Chromatin accessibility and the regulatory epigenome. Nat Rev Genet 20:207–220

Kohli RM, Zhang Y (2013) TET enzymes, TDG and the dynamics of DNA demethylation. Nature 502:472–479

Komar D, Juszczynski P (2020) Rebelled epigenome: histone H3S10 phosphorylation and H3S10 kinases in cancer biology and therapy. Clin Epigenetics 12:1–14

Krishnamurthy N, Kato S, Lippman S, Kurzrock R (2022) Chromatin remodeling (SWI/SNF) complexes, cancer, and response to immunotherapy. J Immunother Cancer 10:E004669

Kukreja L, Li CJ, Ezhilan S, Iyer VR, Kuo JS (2021) Emerging epigenetic therapies for brain tumors. NeuroMolecular Med:1–9

Kulis M, Esteller M (2010) DNA methylation and cancer. Adv Genet 70:27–56

Lachner M, O'carroll D, Rea S, Mechtler K, Jenuwein T (2001) Methylation of histone H3 lysine 9 creates A binding site for HP1 proteins. Nature 410:116–120

Laczmanska I, Karpinski P, Bebenek M, Sedziak T, Ramsey D, Szmida E, Sasiadek MM (2013) Protein tyrosine phosphatase receptor-like genes are frequently hypermethylated in sporadic colorectal cancer. J Hum Genet 58:11–15

Ladelfa MF, Peche LY, Toledo MF, Laiseca JE, Schneider C, Monte M (2012) Tumor-specific MAGE proteins as regulators of P53 function. Cancer Lett 325:11–17

Lafave LM, Savage RE, Buenrostro JD (2022) Single-cell epigenomics reveals mechanisms of cancer progression. Ann Rev Cancer Biol 6:167–185

Lahooti B, Poudel S, Mikelis CM, Mattheolabakis G (2021) Mirnas as anti-angiogenic adjuvant therapy in cancer: synopsis and potential. Front Oncol 11:705634

Lara-Ureña N, Jafari V, García-Domínguez M (2022) Cancer-associated dysregulation of Sumo regulators: proteases and ligases. Int J Mol Sci 23:8012

Lee JE, Kim M-Y (2021) Cancer epigenetics: past, present and future. Seminars in cancer biology. Elsevier

Lee Y, Kim M, Han J, Yeom KH, Lee S, Baek SH, Kim VN (2004) Microrna genes are transcribed by RNA polymerase II. EMBO J 23:4051–4060

Li Y, Seto E (2016) Hdacs and HDAC inhibitors in cancer development and therapy. Cold Spring Harb Perspect Med 6:A026831

Li G, Wu X, Qian W, Cai H, Sun X, Zhang W, Tan S, Wu Z, Qian P, Ding K, Lu X, Zhang X, Yan H, Song H, Guang S, Wu Q, Lobie PE, Shan G, Zhu T (2016) CCAR1 5' UTR as a natural miRancer of Mir-1254 overrides tamoxifen resistance. Cell Res 26:655–673

Li Y, Li Z, Zhu W-G (2019) Molecular mechanisms of epigenetic regulators as activatable targets in cancer theranostics. Curr Med Chem 26(8):1328–1350. https://doi.org/10.2174/0929867324666170921101947

Li Y, Chen X, Lu C (2021a) The interplay between DNA and histone methylation: molecular mechanisms and disease implications. EMBO Rep 22:E51803

Li Y, Gong H, Wang P, Zhu Y, Peng H, Cui Y, Li H, Liu J, Wang Z (2021b) The emerging role of ISWI chromatin remodeling complexes in cancer. J Exp Clin Cancer Res 40:1–27

Li Y, Fan Z, Meng Y, Liu S, Zhan H (2023) Blood-based DNA methylation signatures in cancer: a systematic review. Biochim Biophys Acta (BBA) - Mol Basis Dis 1869(1):166583. https://doi.org/10.1016/j.bbadis.2022.166583

Liang G, Weisenberger DJ (2017) DNA methylation aberrancies as a guide for surveillance and treatment of human cancers. Epigenetics 12:416–432

Liang T-J, Wang H-X, Zheng Y-Y, Cao Y-Q, Wu X, Zhou X, Dong S-X (2017) APC hypermethylation for early diagnosis of colorectal cancer: a meta-analysis and literature review. Oncotarget 8:46468

Liang Y, Xu P, Zou Q, Luo H, Yu W (2019) An epigenetic perspective on tumorigenesis: loss of cell identity, enhancer switching, and namirna network. seminars in cancer biology. Elsevier, pp 1–9

Liu H, Lei C, He Q, Pan Z, Xiao D, Tao Y (2018) Nuclear functions of mammalian micrornas in gene regulation, immunity and cancer. Mol Cancer 17:64

Liu F, Lu X, Zhou X, Huang H (2021a) APC gene promoter methylation as a potential biomarker for lung cancer diagnosis: a meta-analysis. Thoracic Cancer 12:2907–2913

Liu X-Y, Guo C-H, Xi Z-Y, Xu X-Q, Zhao Q-Y, Li L-S, Wang Y (2021b) Histone methylation in pancreatic cancer and its clinical implications. World J Gastroenterol 27:6004

Ljungman M, Parks L, Hulbatte R, Bedi K (2019) The role of H3K79 methylation in transcription and the DNA damage response. Mutat Res/Rev Mutat Res 780:48–54

López-Moyado IF, Tsagaratou A, Yuita H, Seo H, Delatte B, Heinz S, Benner C, Rao A (2019) Paradoxical association of TET loss of function with genome-wide DNA hypomethylation. Proc Natl Acad Sci 116:16933–16942

Losman J-A, Kaelin WG (2013) What a difference a hydroxyl makes: mutant IDH,(R)-2-hydroxyglutarate, and cancer. Genes Dev 27:836–852

Lozano-Urena A, Jimenez-Villalba E, Pinedo-Serrano A, Jordan-Pla A, Kirstein M, Ferron SR (2021) Aberrations of genomic imprinting in glioblastoma formation. Front Oncol 11:630482

Lu TX, Rothenberg ME (2018) Microrna. J Allergy Clin Immunol 141:1202–1207
Lujambio A, Lowe SW (2012) The microcosmos of cancer. Nature 482:347–355
Lujambio A, Ropero S, Ballestar E, Fraga MF, Cerrato C, Setién F, Casado S, Suarez-Gauthier A, Sanchez-Cespedes M, Git A, Spiteri I, Das PP, Caldas C, Miska E, Esteller M (2007) Genetic unmasking of an epigenetically silenced microrna in human cancer cells. Cancer Res 67:1424–1429
Mallm JP, Windisch P, Biran A, Gal Z, Schumacher S, Glass R, Herold-Mende C, Meshorer E, Barbus M, Rippe K (2020) Glioblastoma initiating cells are sensitive to histone demethylase inhibition due to epigenetic deregulation. Int J Cancer 146:1281–1292
Malpeli G, Innamorati G, Decimo I, Bencivenga M, Nwabo Kamdje AH, Perris R, Bassi C (2019) Methylation dynamics of RASSF1A and its impact on cancer. Cancers 11:959
Marei HE, Althani A, Afifi N, Hasan A, Caceci T, Pozzoli G, Morrione A, Giordano A, Cenciarelli C (2021) P53 Signaling in cancer progression and therapy. Cancer Cell Int 21:1–15
Markouli M, Strepkos D, Basdra EK, Papavassiliou AG, Piperi C (2021) Prominent role of histone modifications in the regulation of tumor metastasis. Int J Mol Sci 22:2778
Marsh DJ, Dickson K-A (2019) Writing histone Monoubiquitination in human malignancy—the role of RING finger E3 ubiquitin ligases. Genes 10:67
Marusyk A, Janiszewska M, Polyak K (2020) Intratumor heterogeneity: the Rosetta stone of therapy resistance. Cancer Cell 37:471–484
Mazloumi Z, Farahzadi R, Rafat A, Asl KD, Karimipour M, Montazer M, Movassaghpour AA, Dehnad A, Charoudeh HN (2022) Effect of aberrant DNA methylation on cancer stem cell properties. Exp Mol Pathol 104757
McGranahan N, Swanton C (2017) Clonal heterogeneity and tumor evolution: past, present, and the future. Cell 168:613–628
McGuire A, Casey M, Shalaby A, Kalinina O, Curran C, Webber M, Callagy G, Holian E, Bourke E, Kerin M (2019) Quantifying Tip60 (Kat5) stratifies breast cancer. Sci Rep 9:1–14
Melling N, Grimm N, Simon R, Stahl P, Bokemeyer C, Terracciano L, Sauter G, Izbicki JR, Marx AH (2016) Loss of H2Bub1 expression is linked to poor prognosis in nodal negative colorectal cancers. Pathol Oncol Res 22:95–102
Menon A, Abd-Aziz N, Khalid K, Poh CL, Naidu R (2022) Mirna: a promising therapeutic target in cancer. Int J Mol Sci 23
Mollaei H, Safaralizadeh R, Rostami Z (2019) MicroRNA replacement therapy in cancer. J Cell Physiol 234:12369–12384
Morey L, Brenner C, Fazi F, Villa R, Gutierrez A, Buschbeck M, Nervi C, Minucci S, Fuks F, Di Croce L (2008) MBD3, a component of the Nurd complex, facilitates chromatin alteration and deposition of epigenetic marks. Mol Cell Biol 28:5912–5923
Morgan A, Davies TJ, McAuley MT (2018) The role of DNA methylation in ageing and cancer. Proc Nutr Soc 77:412–422
Motavalli R, Marofi F, Nasimi M, Yousefi M, Khiavi FM (2021) Association of hippo signalling pathway with epigenetic changes in cancer cells and therapeutic approaches: a review. Anti-Cancer Agents Med Chem 21:1520–1528
Moyal L, Lerenthal Y, Gana-Weisz M, Mass G, So S, Wang S-Y, Eppink B, Chung YM, Shalev G, Shema E (2011) Requirement of ATM-dependent monoubiquitylation of histone H2B for timely repair of DNA double-strand breaks. Mol Cell 41:529–542
Mudbhary R, Hoshida Y, Chernyavskaya Y, Jacob V, Villanueva A, Fiel MI, Chen X, Kojima K, Thung S, Bronson RT (2014) UHRF1 overexpression drives DNA hypomethylation and hepatocellular carcinoma. Cancer Cell 25:196–209
Muller S, Filippakopoulos P, Knapp S (2011) Bromodomains as therapeutic targets. Expert Rev Mol Med 13
Nag S, Qin J, Srivenugopal KS, Wang M, Zhang R (2013) The MDM2-P53 pathway revisited. J Biomed Res 27:254–271
Nair SS, Kumar R (2012) Chromatin remodeling in cancer: a gateway to regulate gene transcription. Mol Oncol 6:611–619

Neganova ME, Klochkov SG, Aleksandrova YR, Aliev G (2022) Histone modifications in epigenetic regulation of cancer: perspectives and achieved progress. Seminars in cancer biology. Elsevier, pp 452–471

Neidhart M (2015) DNA methylation and complex human disease. Academic Press

Neophytou CM, Trougakos IP, Erin N, Papageorgis P (2021) Apoptosis deregulation and the development of cancer multi-drug resistance. Cancers 13:4363

Neri F, Rapelli S, Krepelova A, Incarnato D, Parlato C, Basile G, Maldotti M, Anselmi F, Oliviero S (2017) Intragenic DNA methylation prevents spurious transcription initiation. Nature 543:72–77

Nguyen TA, Park J, Dang TL, Choi YG, Kim VN (2018) Microprocessor depends on hemin to recognize the apical loop of primary microrna. Nucleic Acids Res 46:5726–5736

Nishiyama A, Nakanishi M (2021) Navigating the DNA methylation landscape of cancer. Trends Genet 37:1012–1027

O'Brien J, Hayder H, Zayed Y, Peng C (2018) Overview of microrna biogenesis, mechanisms of actions, and circulation. Front Endocrinol (Lausanne) 9:402

Okawa R, Banno K, Iida M, Yanokura M, Takeda T, Iijima M, Kunitomi-Irie H, Nakamura K, Adachi M, Umene K (2017) Aberrant chromatin remodeling in gynecological cancer. Oncol Lett 14:5107–5113

Ortiz-Barahona, V., Joshi, R. S. & Esteller, M. Use of DNA methylation profiling in translational oncology. Seminars in cancer biology. 2020 Elsevier

Pan G, Liu Y, Shang L, Zhou F, Yang S (2021) EMT-associated micrornas and their roles in cancer stemness and drug resistance. Cancer Commun (Lond) 41:199–217

Papageorgiou SG, Kontos CK, Diamantopoulos MA, Bouchla A, Glezou E, Bazani E, Pappa V, Scorilas A (2017) Microrna-155-5p overexpression in peripheral blood mononuclear cells of chronic lymphocytic leukemia patients is a novel, independent molecular biomarker of poor prognosis. Dis Markers 2017:2046545

Pappalardo XG, Barra V (2021) Losing DNA methylation at repetitive elements and breaking bad. Epigenetics Chromatin 14:1–21

Pasqualucci L, Dominguez-Sola D, Chiarenza A, Fabbri G, Grunn A, Trifonov V, Kasper LH, Lerach S, Tang H, Ma J (2011) Inactivating mutations of acetyltransferase genes in B-cell lymphoma. Nature 471:189–195

Pastor WA, Aravind L, Rao A (2013) Tetonic shift: biological roles of TET proteins in DNA demethylation and transcription. Nat Rev Mol Cell Biol 14:341–356

Patani N, Jiang WG, Newbold RF, Mokbel K (2011) Histone-modifier gene expression profiles are associated with pathological and clinical outcomes in human breast cancer. Anticancer Res 31:4115–4125

Peifer M, Fernández-Cuesta L, Sos ML, George J, Seidel D, Kasper LH, Plenker D, Leenders F, Sun R, Zander T (2012) Integrative genome analyses identify key somatic driver mutations of small-cell lung cancer. Nat Genet 44:1104–1110

Peixoto P, Cartron P-F, Serandour AA, Hervouet E (2020) From 1957 to nowadays: a brief history of epigenetics. Int J Mol Sci 21:7571

Peng Y, Croce CM (2016) The role of micrornas in human cancer. Signal Transduct Target Ther 1:15004

Pfeifer GP (2018) Defining driver DNA methylation changes in human cancer. Int J Mol Sci 19:1166

Piao XM, Jeong P, Kim YH, Byun YJ, Xu Y, Kang HW, Ha YS, Kim WT, Lee JY, Woo SH, Kwon TG, Kim IY, Moon SK, Choi YH, Cha EJ, Yun SJ, Kim WJ (2019) Urinary cell-free microrna biomarker could discriminate bladder cancer from benign hematuria. Int J Cancer 144:380–388

Pichiorri F, Suh SS, Rocci A, De Luca L, Taccioli C, Santhanam R, Zhou W, Benson DM Jr, Hofmainster C, Alder H, Garofalo M, Di Leva G, Volinia S, Lin HJ, Perrotti D, Kuehl M, Aqeilan RI, Palumbo A, Croce CM (2010) Downregulation of P53-inducible micrornas 192, 194, and 215 impairs the P53/MDM2 autoregulatory loop in multiple myeloma development. Cancer Cell 18:367–381

Podhorecka M, Skladanowski A, Bozko P (2010) H2AX phosphorylation: its role in DNA damage response and cancer therapy. J Nucl Acids 2010

Prenzel T, Begus-Nahrmann Y, Kramer F, Hennion M, Hsu C, Gorsler T, Hintermair C, Eick D, Kremmer E, Simons M (2011) Estrogen-dependent gene transcription in human breast cancer cells relies upon proteasome-dependent monoubiquitination of histone H2B. Cancer Res 71: 5739–5753

Qin W, Wolf P, Liu N, Link S, Smets M, Mastra FL, Forné I, Pichler G, Hörl D, Fellinger K (2015) DNA methylation requires a DNMT1 ubiquitin interacting motif (UIM) and histone ubiquitination. Cell Res 25:911–929

Rebel VI, Kung AL, Tanner EA, Yang H, Bronson RT, Livingston DM (2002) Distinct roles for CREB-binding protein and P300 in hematopoietic stem cell self-renewal. Proc Natl Acad Sci 99:14789–14794

Recasens A, Munoz L (2019) Targeting cancer cell dormancy. Trends Pharmacol Sci 40:128–141

Renaud S, Loukinov D, Alberti L, Vostrov A, Kwon Y-W, Bosman FT, Lobanenkov V, Benhattar J (2011) BORIS/CTCFL-mediated transcriptional regulation of the Htert telomerase gene in testicular and ovarian tumor cells. Nucleic Acids Res 39:862–873

Roberti A, Valdes AF, Torrecillas R, Fraga MF, Fernandez AF (2019) Epigenetics in cancer therapy and nanomedicine. Clin Epigenetics 11:1–18

Robinson NJ, Parker KA, Schiemann WP (2020) Epigenetic plasticity in metastatic dormancy: mechanisms and therapeutic implications. Ann Transl Med 8

Rodríguez-Paredes M, Bormann F, Raddatz G, Gutekunst J, Lucena-Porcel C, Köhler F, Wurzer E, Schmidt K, Gallinat S, Wenck H (2018) Methylation profiling identifies two subclasses of squamous cell carcinoma related to distinct cells of origin. Nat Commun 9:1–9

Rondinelli B, Rosano D, Antonini E, Frenquelli M, Montanini L, Huang D, Segalla S, Yoshihara K, Amin SB, Lazarevic D (2015) Histone demethylase JARID1C inactivation triggers genomic instability in sporadic renal cancer. J Clin Invest 125:4625–4637

Ropero S, Fraga MF, Ballestar E, Hamelin R, Yamamoto H, Boix-Chornet M, Caballero R, Alaminos M, Setien F, Paz MF (2006) A truncating mutation of HDAC2 in human cancers confers resistance to histone deacetylase inhibition. Nat Genet 38:566–569

Ross SE and Bogdanovic O (2019) TET enzymes, DNA demethylation and pluripotency. Biochem Soc Trans 47:875–885

Sabit H, Cevik E, Tombuloglu H, Abdel-Ghany S, Tombuloglu G, Esteller M (2021) Triple negative breast cancer in the era of Mirna. Crit Rev Oncol Hematol 157:103196

Sakabe K, Hart GW (2010) O-Glcnac transferase regulates mitotic chromatin dynamics. J Biol Chem 285:34460–34468

Salhab A, Nordström K, Gasparoni G, Kattler K, Ebert P, Ramirez F, Arrigoni L, Müller F, Polansky JK, Cadenas C, Hengstler JG, Lengauer T, Manke T, DEEP Consortium, Walter J (2018) A comprehensive analysis of 195 DNA methylomes reveals shared and cell-specific features of partially methylated domains. Genome Biol 19(1). https://doi.org/10.1186/s13059-018-1510-5

Saliminejad K, Khorram Khorshid HR, Soleymani Fard S, Ghaffari SH (2019) An overview of micrornas: biology, functions, therapeutics, and analysis methods. J Cell Physiol 234:5451–5465

Schroeder DI, Lott P, Korf I, LaSalle JM (2011) Large-scale methylation domains mark a functional subset of neuronally expressed genes. Genome Res 21(10):1583–1591

Schübeler D (2015) Function and information content of DNA methylation. Nature 517:321–326

Selcuklu SD, Donoghue MT, Spillane C (2009) Mir-21 as a key regulator of oncogenic processes. Biochem Soc Trans 37:918–925

Seligson DB, Horvath S, McBrian MA, Mah V, Yu H, Tze S, Wang Q, Chia D, Goodglick L, Kurdistani SK (2009) Global levels of histone modifications predict prognosis in different cancers. Am J Pathol 174:1619–1628

Semina EV, Rysenkova KD, Troyanovskiy KE, Shmakova AA, Rubina KA (2021) Micrornas in cancer: from gene expression regulation to the metastatic niche reprogramming. Biochemistry (Mosc) 86:785–799

Shanmugam MK, Arfuso F, Arumugam S, Chinnathambi A, Jinsong B, Warrier S, Wang LZ, Kumar AP, Ahn KS, Sethi G (2018) Role of novel histone modifications in cancer. Oncotarget 9:11414

Sharif J, Muto M, Takebayashi S-I, Suetake I, Iwamatsu A, Endo TA, Shinga J, Mizutani-Koseki Y, Toyoda T, Okamura K (2007) The SRA protein Np95 mediates epigenetic inheritance by recruiting Dnmt1 to methylated DNA. Nature 450:908–912

Sharma S, Kelly TK, Jones PA (2010) Epigenetics in cancer. Carcinogenesis 31:27–36

Shekhawat J, Gauba K, Gupta S, Choudhury B, Purohit P, Sharma P, Banerjee M (2021) Ten–eleven translocase: key regulator of the methylation landscape in cancer. J Cancer Res Clin Oncol 147:1869–1879

Shema E, Tirosh I, Aylon Y, Huang J, Ye C, Moskovits N, Raver-Shapira N, Minsky N, Pirngruber J, Tarcic G (2008) The histone H2B-specific ubiquitin ligase RNF20/Hbre1 acts as a putative tumor suppressor through selective regulation of gene expression. Genes Dev 22:2664–2676

Śledzińska P, Bebyn MG, Furtak J, Kowalewski J, Lewandowska MA (2021) Prognostic and predictive biomarkers in gliomas. Int J Mol Sci 22:10373

Smiley JA, Kundracik M, Landfried DA, Barnes SR, Axhemi AA (2005) Genes of the thymidine salvage pathway: thymine-7-hydroxylase from a rhodotorula glutinis cDNA library and iso-orotate decarboxylase from Neurospora crassa. Biochimica Et Biophysica Acta (BBA)-General Subjects 1723:256–264

Smith ZD, Meissner A (2013) DNA methylation: roles in mammalian development. Nat Rev Genet 14:204–220

Smolarz B, Durczyński A, Romanowicz H, Szyłło K, Hogendorf P (2022) Mirnas in cancer (review of literature). Int J Mol Sci 23

Song Z, Wei Z, Wang Q, Zhang X, Tao X, Wu N, Liu X, Qian J (2020) The role of DOT1L in the proliferation and prognosis of gastric cancer. Biosci Reports 40

Sterling J, Menezes SV, Abbassi RH, Munoz L (2021) Histone lysine demethylases and their functions in cancer. Int J Cancer 148:2375–2388

Tahiliani M, Koh KP, Shen Y, Pastor WA, Bandukwala H, Brudno Y, Agarwal S, Iyer LM, Liu DR, Aravind L (2009) Conversion of 5-methylcytosine to 5-hydroxymethylcytosine in mammalian DNA by MLL partner TET1. Science 324:930–935

Tan W, Liu B, Qu S, Liang G, Luo W, Gong C (2018) Micrornas and cancer: key paradigms in molecular therapy. Oncol Lett 15:2735–2742

Thompson LL, Guppy BJ, Sawchuk L, Davie JR, McManus KJ (2013) Regulation of chromatin structure via histone post-translational modification and the link to carcinogenesis. Cancer Metastasis Rev 32:363–376

Toh Y, Nicolson GL (2009) The role of the MTA family and their encoded proteins in human cancers: molecular functions and clinical implications. Clin Exp Metastasis 26:215–227

Tsuda M, Fukuda A, Kawai M, Araki O, Seno H (2021) The role of the SWI/SNF chromatin remodeling complex in pancreatic ductal adenocarcinoma. Cancer Sci 112:490–497

Tucci V, Isles AR, Kelsey G, Ferguson-Smith AC, Bartolomei MS, Benvenisty N, Bourc'his D, Charalambous M, Dulac C, Feil R (2019) Genomic imprinting and physiological processes in mammals. Cell 176:952–965

Turpin M, Salbert G (2022) 5-Methylcytosine turnover: mechanisms and therapeutic implications in cancer. Front Mol Biosci 9

Uddin MS, Al Mamun A, Alghamdi BS, Tewari D, Jeandet P, Sarwar MS, Ashraf GM (2020) Epigenetics of glioblastoma multiforme: from molecular mechanisms to therapeutic approaches. Seminars in cancer biology. Elsevier

Van Tongelen A, Loriot A, De Smet C (2017) Oncogenic roles of DNA hypomethylation through the activation of cancer-germline genes. Cancer Lett 396:130–137

Varambally S, Cao Q, Mani R-S, Shankar S, Wang X, Ateeq B, Laxman B, Cao X, Jing X, Ramnarayanan K (2008) Genomic loss of Microrna-101 leads to overexpression of histone methyltransferase EZH2 in cancer. Science 322:1695–1699

Vaughan RM, Dickson BM, Cornett EM, Harrison JS, Kuhlman B, Rothbart SB (2018) Comparative biochemical analysis of UHRF proteins reveals molecular mechanisms that uncouple UHRF2 from DNA methylation maintenance. Nucleic Acids Res 46:4405–4416

Vogelstein B, Kinzler KW (2004) Cancer genes and the pathways they control. Nat Med 10:789–799

Vogelstein B, Papadopoulos N, Velculescu VE, Zhou S, Diaz LA Jr, Kinzler KW (2013) Cancer genome landscapes. Science 339:1546–1558

Vrba L, Futscher BW (2019) DNA methylation changes in biomarker loci occur early in cancer progression. F1000Research:8

Waddington CH (2012) The epigenotype. Int J Epidemiol 41:10–13

Wagner KW, Alam H, Dhar SS, Giri U, Li N, Wei Y, Giri D, Cascone T, Kim J-H, Ye Y (2013) KDM2A promotes lung tumorigenesis by epigenetically enhancing ERK1/2 signaling. J Clin Invest 123:5231–5246

Waitkus MS, Diplas BH, Yan H (2015) Isocitrate dehydrogenase mutations in gliomas. Neuro-Oncology 18:16–26

Wang H, Wang S, Shen L, Chen Y, Zhang X, Zhou J, Wang Z, Hu C, Yue W, Wang H (2010) Chk2 down-regulation by promoter hypermethylation in human bulk gliomas. Life Sci 86:185–191

Wang L, Gural A, Sun X-J, Zhao X, Perna F, Huang G, Hatlen MA, Vu L, Liu F, Xu H (2011) The leukemogenicity of AML1-ETO is dependent on site-specific lysine acetylation. Science 333:765–769

Wang E, Kawaoka S, Yu M, Shi J, Ni T, Yang W, Zhu J, Roeder RG, Vakoc CR (2013) Histone H2B ubiquitin ligase RNF20 is required for MLL-rearranged leukemia. Proc Natl Acad Sci 110:3901–3906

Wiggins JF, Ruffino L, Kelnar K, Omotola M, Patrawala L, Brown D, Bader AG (2010) Development of a lung cancer therapeutic based on the tumor suppressor microrna-34. Cancer Res 70:5923–5930

Wong SH, Goode DL, Iwasaki M, Wei MC, Kuo H-P, Zhu L, Schneidawind D, Duque-Afonso J, Weng Z, Cleary ML (2015) The H3K4-methyl epigenome regulates leukemia stem cell oncogenic potential. Cancer Cell 28:198–209

Wright DE, Wang C-Y, Kao C-F (2011) Flickin'the ubiquitin switch: the role of H2B ubiquitylation in development. Epigenetics 6:1165–1175

Wu X, Zhang Y (2017) TET-mediated active DNA demethylation: mechanism, function and beyond. Nat Rev Genet 18:517–534

Wu X, Somlo G, Yu Y, Palomares MR, Li AX, Zhou W, Chow A, Yen Y, Rossi JJ, Gao H, Wang J, Yuan YC, Frankel P, Li S, Ashing-Giwa KT, Sun G, Wang Y, Smith R, Robinson K, Ren X, Wang SE (2012) De novo sequencing of circulating mirnas identifies novel markers predicting clinical outcome of locally advanced breast cancer. J Transl Med 10:42

Wu Y, Chen P, Jing Y, Wang C, Men Y-L, Zhan W, Wang Q, Gan Z, Huang J, Xie K (2015) Microarray analysis reveals potential biological functions of histone H2B monoubiquitination. PLoS One 10:E0133444

Xiang Q, He X, Mu J, Mu H, Zhou D, Tang J, Xiao Q, Jiang Y, Ren G, Xiang T (2019) The phosphoinositide hydrolase phospholipase C delta1 inhibits epithelial-mesenchymal transition and is silenced in colorectal cancer. J Cell Physiol 234:13906–13916

Xu W, Yang H, Liu Y, Yang Y, Wang P, Kim S-H, Ito S, Yang C, Wang P, Xiao M-T (2011) Oncometabolite 2-hydroxyglutarate is a competitive inhibitor of α-ketoglutarate-dependent dioxygenases. Cancer Cell 19:17–30

Xuefang Z, Ruinian Z, Liji J, Chun Z, Qiaolan Z, Jun J, Yuming C, Junrong H (2020) Mir-331-3p inhibits proliferation and promotes apoptosis of nasopharyngeal carcinoma cells by targeting Elf4b-PI3K-AKT pathway. Technol Cancer Res Treat 19:1533033819892251

Yang J, Altahan AM, Hu D, Wang Y, Cheng P-H, Morton CL, Qu C, Nathwani AC, Shohet JM, Fotsis T (2015) The role of histone demethylase KDM4B in Myc signaling in neuroblastoma. JNCI:107

Yang Y, Sun J, Chen T, Tao Z, Zhang X, Tian F, Zhou X, Lu D (2017) Tat-interactive protein-60KDA (TIP60) regulates the tumorigenesis of lung cancer in vitro. J Cancer 8:2277

Yang G-J, Ko C-N, Zhong H-J, Leung C-H, Ma D-L (2019) Structure-based discovery of a selective KDM5A inhibitor that exhibits anti-cancer activity via inducing cell cycle arrest and senescence in breast cancer cell lines. Cancers 11:92

Yang M-Y, Lin P-M, Yang C-H, Hu M-L, Chen I-Y, Lin S-F, Hsu C-M (2021) Loss of ZNF215 imprinting is associated with poor five-year survival in patients with cytogenetically abnormal-acute myeloid leukemia. Blood Cells Mol Dis 90:102577. https://doi.org/10.1016/j.bcmd.2021.102577

Yap DB, Chu J, Berg T, Schapira M, Cheng S-WG, Moradian A, Morin RD, Mungall AJ, Meissner B, Boyle M (2011) Somatic mutations at EZH2 Y641 act dominantly through a mechanism of selectively altered PRC2 catalytic activity, to increase H3K27 trimethylation. Blood J Am Soc Hematol 117:2451–2459

Yildirim O, Li R, Hung J-H, Chen PB, Dong X, Ee L-S, Weng Z, Rando OJ, Fazzio TG (2011) Mbd3/NURD complex regulates expression of 5-Hydroxymethylcytosine marked genes in embryonic stem cells. Cell 147:1498–1510

Yin Y-W, Jin H-J, Zhao W, Gao B, Fang J, Wei J, Zhang DD, Zhang J, Fang D (2015) The histone acetyltransferase GCN5 expression is elevated and regulated by C-Myc and E2F1 transcription factors in human colon cancer. Gene Expr 16:187

You JS, Jones PA (2012) Cancer genetics and epigenetics: two sides of the same coin? Cancer Cell 22:9–20

Zhang J, Ma L (2012) Microrna control of epithelial-mesenchymal transition and metastasis. Cancer Metastasis Rev 31:653–662

Zhang Y, Yao L, Zhang X, Ji H, Wang L, Sun S, Pang D (2011) Elevated expression of USP22 in correlation with poor prognosis in patients with invasive breast cancer. J Cancer Res Clin Oncol 137:1245–1253

Zhang C, Zhong JF, Stucky A, Chen X-L, Press MF, Zhang X (2015) Histone acetylation: novel target for the treatment of acute lymphoblastic Leukemia. Clin Epigenetics 7:1–10

Zhang S, Zhou B, Wang L, Li P, Bennett B, Snyder R, Garantziotis S, Fargo D, Cox AD, Chen L (2017) INO80 is required for oncogenic transcription and tumor growth in non-small cell lung cancer. Oncogene 36:1430–1439

Zhang T-J, Zhou J-D, Zhang W, Lin J, Ma J-C, Wen X-M, Yuan Q, Li X-X, Xu Z-J, Qian J (2018) H19 overexpression promotes leukemogenesis and predicts unfavorable prognosis in acute myeloid leukemia. Clin Epigenetics 10(1):1–12. https://doi.org/10.1186/s13148-018-0486-z

Zhang W, Klinkebiel D, Barger CJ, Pandey S, Guda C, Miller A, Akers SN, Odunsi K, Karpf AR (2020) Global DNA Hypomethylation in epithelial ovarian cancer: passive demethylation and association with genomic instability. Cancers 12:764

Zhao Z, Shilatifard A (2019) Epigenetic modifications of histones in cancer. Genome Biol 20:1–16

Zhao S, Allis CD, Wang GG (2021) The language of chromatin modification in human cancers. Nat Rev Cancer 21:413–430

Zheng Y, Hlady RA, Joyce BT, Robertson KD, He C, Nannini DR, Kibbe WA, Achenbach CJ, Murphy RL, Roberts LR (2019) DNA methylation of individual repetitive elements in hepatitis C virus infection-induced hepatocellular carcinoma. Clin Epigenetics 11:1–13

Zhou B, Wang L, Zhang S, Bennett BD, He F, Zhang Y, Xiong C, Han L, Diao L, Li P (2016) INO80 governs superenhancer-mediated oncogenic transcription and tumor growth in melanoma. Genes Dev 30:1440–1453

Zhou X, Jiao D, Dou M, Zhang W, Hua H, Chen J, Li Z, Li L, Han X (2020) Association of APC gene promoter methylation and the risk of gastric cancer: a meta-analysis and bioinformatics study. Medicine 99

Zhu Q, Pao GM, Huynh AM, Suh H, Tonnu N, Nederlof PM, Gage FH, Verma IM (2011) BRCA1 tumour suppression occurs via heterochromatin-mediated silencing. Nature 477:179–184

Zhu Q, Zhou L, Yang Z, Lai M, Xie H, Wu L, Xing C, Zhang F, Zheng S (2012) O-Glcnacylation plays a role in tumor recurrence of hepatocellular carcinoma following liver transplantation. Med Oncol 29:985–993

Part II
Cancer Specific Epigenetic Alterations

Chapter 4
Epigenetics in the Diagnosis, Prognosis, and Therapy of Cancer

Leilei Fu and Bo Liu

Abstract As cancer is the leading cause of death worldwide, there is an urgent necessity to discover novel diagnostic and prognostic biomarkers as well as therapeutic targets for this dreadful disease. The progression of various malignancies can be the result of abnormalities in multiple epigenetic regulations. Over the past decades, major epigenetics categories, including histone modification, DNA methylation, noncoding RNAs, and chromatin remodeling, have been reported to be involved in tumor genesis and development. Therefore, epigenetic changes can be used as clinical biomarkers for the diagnosis of cancer and to predict the prognosis. Moreover, epigenetic regulators have emerged as promising drug targets for cancer therapy. This review delineates the latest evidence in epigenetics alterations in cancer and discusses their potential contribution to the diagnosis, prognosis, and therapy of cancer. These diagnostic, prognostic, and therapeutic strategies based on epigenetics might bring hope to reducing the high fatality rate of malignancies.

Keywords Epigenetics · DNA methylation · Histone modification · Noncoding RNAs · Chromatin remodeling · Diagnosis and prognosis of cancer · Therapeutic strategy

Abbreviations

ALC1	Amplification in liver cancer 1
AML	Acute myeloid leukemia
ARID2	AT-rich interactive domain 2
AUC	Area under the curve

L. Fu (✉)
Sichuan Engineering Research Center for Biomimetic Synthesis of Natural Drugs, School of Life Science and Engineering, Southwest Jiaotong University, Chengdu, China

B. Liu (✉)
State Key Laboratory of Biotherapy and Cancer Center, West China Hospital, Sichuan University, Collaborative Innovation Center of Biotherapy, Chengdu, China

© The Author(s), under exclusive license to Springer Nature Switzerland AG 2023
R. Kalkan (ed.), *Cancer Epigenetics*, Epigenetics and Human Health 11,
https://doi.org/10.1007/978-3-031-42365-9_4

B-CLL	B-cell chronic lymphocytic leukemia
BET	Bromodomain and extra-terminal
BMI1	B-cell-specific Moloney murine leukemia virus insertion site1
BRDs	Bromodomain-containing proteins
CIMP	CpG island methylator phenotype
circRNAs	Circular RNA
CRC	Colorectal cancer
DNMT1	DNA methyltransferase 1
EHMT1/2	Euchromatin histone-lysine-n-methyltransferases I and II
EOC	Epithelial ovarian cancer
ESCC	Esophageal squamous cell carcinoma
EZH2	Enhancer of zeste homolog 2
FAL1	Focally amplified lncRNA on chromosome 1
GBM	Glioblastoma
GEO	Gene Expression Omnibus
GNAT	GCN5-related n-acetyltransferase
HAT	Histone acetyltransferase
HCC	Hepatocellular carcinoma
HDAC	histone deacetylase
hnRNPL	Heterogeneous nuclear ribonucleoprotein L
KDMs	Histone lysine demethylases
lncRNAs	Long noncoding RNAs
m1A	N1-methyladenosine
m3C	N3-methylcytosine
m5C	5-methylcytosine
m6A	N6-methyl-adenosine
m7G	N7-methylguanosine
METTL3	Methyltransferase-like 3
MGMT	O6 methylguanine DNA methyltransferase
miRNA	MicroRNA
MYC	Myelocytomatosis oncogene
NAD+	Nicotinamide adenine dinucleotide
ncRNAs	Noncoding RNAs
NSCLC	Non-small-cell lung cancer
PARP1	Poly (ADP-ribose) polymerase 1
PCAF	p300/CBP-related factor
piRNAs	PIWI-interacting RNAs
PR C1	Polycomb repressive complex 1
PR C2	Polycomb repressive complex 2
SIRTs	Sirtuins
SND1	Staphylococcal nuclease domain-containing protein 1
TCGA	The Cancer Genome Atlas
TET	Ten-eleven translocation
TKI	Tyrosine-kinase inhibitor

TNBC	Triple-negative breast cancer
WGBS	Whole-genome bisulfite sequencing
YBX1	Y-box binding protein 1

4.1 Epigenetics and Cancer

Classical genetics assumes that the molecular basis of heredity is nucleic acid, and the genetic information of life is stored in the base sequences of nucleic acids. Changes in the base sequences will cause changes in the phenotype of an organism, which can be transmitted from one generation to the next. However, with the development of genetics, it has been found that modifications at the DNA, histone, and chromosome levels can also cause changes in gene expression patterns that can be inherited. Modifications that alter the genome without affecting the DNA sequence can not only affect the development of the individual but also be passed on to future generations, and epigenetics refers to such changes in gene expression levels based on nongenetic sequence changes (Skvortsova et al. 2018). Epigenetics is currently divided into several categories, including histone modification, DNA methylation, noncoding RNAs (ncRNAs), and chromatin remodeling (Akone et al. 2020; Zhang et al. 2020a) (Fig. 4.1).

4.1.1 Histone Modification

Histone modifications mainly include acetylation, methylation, phosphorylation, and ubiquitination (Zhu et al. 2021). These modifications can recruit recognition proteins that recognize the modification sites (Zhang and Pradhan 2014), which in turn recruit other transcription factors or can form complexes with numerous physiological functions for transcriptional regulation (Lambert et al. 2018). Acetylation and methylation are among the most widely affected genetic pathways in tumors (Taby and Issa 2010), and many proteins that modify specific histones or bind specific histone modification sites have dysregulated activity in tumors (Rice et al. 2007). Histone acetyltransferase (HAT) and histone deacetylase (HDAC) are capable of acetylating or deacetylating a variety of nonhistone proteins, including p53, retinoblastoma (Rb), and myelocytomatosis oncogene (MYC) (Dang and Wei 2021; Lafon-Hughes et al. 2008; Wagner et al. 2014). HDAC is overexpressed in a variety of tumors, resulting in loss of histone acetylation and silencing of tumor suppressor gene expression (Yoon and Eom 2016). HDAC is divided into four classes, Among them, 11 subtypes, including class I, II, and IV, are all Zn^{2+}-dependent proteins; seven subtypes of class III, Sir1–7, use nicotinamide adenine dinucleotide (NAD+) as the catalytic active site (Li and Seto 2016; Porter et al. 2017; Sixto-López et al. 2020). HDAC inhibitors are potential antitumor compounds, and many studies have

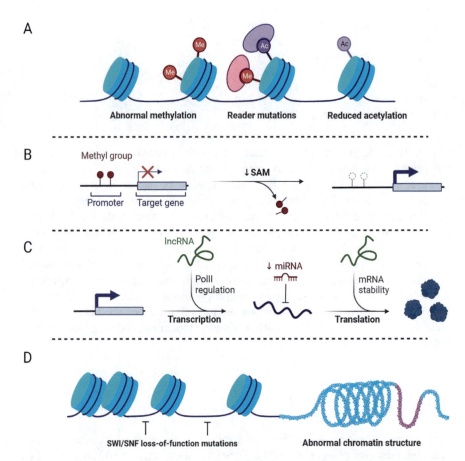

Fig. 4.1 Applications of epigenetics in cancer include histone modifications, DNA methylation, noncoding RNAs, and chromatin remodeling. (**a**) Histone modifications. Histone modifications such as acetylation and methylation are among the most widely affected epigenetic pathways in tumors. (**b**) DNA methylation. Loss of DNA methylation results in abnormal transcription of target genes. (**c**) Noncoding RNA. Noncoding RNA regulation can alter the transcription and translation of oncogene targets. (**d**) Chromatin remodeling. Regulation of chromatin remodeling factor SWI/SNF

shown that abnormal expression of HDACs is associated with a variety of tumors. By analyzing the expression of HDACs in 13 tumors (chronic lymphocytic leukemia, gastric cancer, breast cancer, colon cancer, liver cancer, medulloblastoma, non-small-cell lung cancer (NSCLC), lymphoma, neuroblastoma, ovarian cancer, pancreatic cancer, prostate cancer, and kidney cancer), the expression of class I HDACs was found in 11 types of tumors, indicating that class I HDACs might play a key role in tumorigenesis and invasion, and might be a promising antitumor target (Chun 2015). Histone methylation occurs at the N-terminal lysine or arginine residues of H3 and H4 histones (Yi et al. 2017). Mutations or altered expression of histone methyl modifications and methyl-binding proteins are associated with increased incidence of a variety of different cancers. For example, H3K27me3 methyltransferase is upregulated in some cancers, including prostate cancer, breast cancer, and lymphoma (Duan et al. 2020). Importantly, activating point mutations in Enhancer of zeste homolog 2 (EZH2) were recently found to be associated with B-cell lymphomas, which is consistent with the notion that EZH2 is oncogenic (Duan et al. 2020). Therefore, epigenetic drugs targeting acetyl and methyl groups may have clinical implications in cancer therapy (Fig. 4.1a).

4.1.2 DNA Methylation

In tumors, genome-wide and individual gene methylation patterns are often altered. DNA methylation can add methyl groups to DNA molecules without changing the DNA sequence, thereby regulating the effect of genetic expression. Many recent studies have shown that DNA aberrant methylation is closely related to the occurrence, development, and carcinogenesis of tumors. Aberrant DNA methylation in malignancies is mainly caused by DNA hypermethylation or hypomethylation (Nishiyama and Nakanishi 2021). Genome-wide hypomethylation is frequently detected in tumor genomes, which is often considered a hallmark of cancer cells (Dong et al. 2014). In contrast, DNA hypermethylation often leads to transcriptional repression and reduced gene expression, often occurring in specific CpG-enriched regions that result in the silencing of tumor suppressor genes (Pfeifer 2018). Therefore, changes in DNA methylation levels and changes in specific gene methylation levels can be used as tumor diagnostic indicators. DNA hydroxymethylation, another type of DNA modification, is produced in mammals mainly by sequential oxidative catalytic reactions of the ten-eleven translocation (TET) gene family, and the expression of TET family member TET2 is reduced in various hematopoietic malignancies, including acute myeloid leukemia and myeloproliferative disorders (Heiblig et al. 2015). Similar to DNA methylation, methylation modification packages also appear on RNA, including N6-methyladenosine (m6A), 5-methylcytosine (m5C), N1-methyladenosine (m1A), N3-methylcytosine (m3C), and N7-methylguanosine (m7G) (Yang et al. 2021). Among them, m6A is the most common RNA modification in mammals, which is related to a variety of malignancies, such as acute myeloid leukemia (AML) (Kumar et al. 2021), glioblastoma (Cui et al. 2017), breast cancer (Shi et al. 2020b), and hepatoblastoma (Liu et al. 2019; Ma et al. 2019; Sun et al. 2019b). In the process of m6A methylation, methyltransferase-like 3(METTL3) is the key methyltransferase, which can affect tumor formation by regulating the m6A modification in mRNA through key oncogenes or tumor suppressor genes (Wang et al. 2020b) (Fig. 4.1b).

4.1.3 Noncoding RNAs

With the development of genomics and bioinformatics, especially the massive application of high-throughput sequencing technologies, scientists have discovered an increasing number of nonprotein-coding transcription units like ncRNAs. Long noncoding RNAs (lncRNAs) play a vital role in diverse important biological processes, and lncRNAs can regulate gene expression in a variety of cells during early mammalian development (Fatica and Bozzoni 2014). Alterations of lncRNAs in cancer cells have also been found to be closely associated with tumor formation, progression, and metastasis. It has also been found that ncRNAs, especially microRNA (miRNA), are involved in the development of inflammatory responses and they are important for stabilizing and maintaining the genotypic characteristics of some cell types (Li et al. 2016). miRNA can affect oncogene expression, induce apoptosis, and participate in downstream regulation of oncogenes in tumor cells. In B-cell chronic lymphocytic leukemia (B-CLL), the cluster consisting of miR-15a and miR-16-1 is frequently absent or down-expressed, and this change is associated with the development of B-CLL (Bottoni et al. 2005). P53 is a well-known tumor suppressor gene, and its tumor suppressor effect partly comes from the transcriptional activation of tumor suppressor miRNA-miR-34a. In tumorigenesis, p53 often shows low expression, resulting in transcriptional repression of miR-34a (Shi et al. 2020a). Epigenetic mechanisms are also important causes of altered miRNA expression in cancer. During cancer development, lncRNAs are involved in the regulation of multiple epigenetic complexes that repress or activate gene expression. For example, lncRNA can bind to multiprotein complexes to regulate carcinogenesis. Polycomb repressive complex 1 (PRC1) and polycomb repressive complex 2 (PRC2) are known oncogenes that can cause many malignancies. The lncRNA, named Focally amplified lncRNA on chromosome 1 (FAL1), can bind to B-cell-specific Moloney murine leukemia virus insertion site1 (BMI1), a subunit of PRC1. In ovarian cancer, FAL1 has been reported to accelerate cancer progression and shorten patient survival time. The binding of FAL1 to BMI1 prevents BMI1 degradation to stabilize the PRC1 complex, which allows PRC1 to occupy and repress the promoters of target genes such as p21, leading to cell cycle dysregulation and increased chances of tumorigenesis (Hu et al. 2014) (Fig. 4.1c).

4.1.4 Chromatin Remodeling

During DNA transcription, chromatin changes from a tight superhelical structure to an open sparse structure, the structural change that does not alter the DNA base sequence is called chromatin remodeling (Goldberg et al. 2007). Chromatin remodeling is an important mechanism in epigenetic modification patterns, and chromatin remodeling regulates processes such as gene transcription, DNA repair, and programmed cell death. The chromatin remodeling enzyme amplification in liver cancer1 (ALC1), a potential oncogene, is activated in the presence of both poly (ADP-ribose) polymerase 1 (PARP1) and NAD+, driving nucleosomes to restructure chromatin (Ooi et al. 2021). Interestingly, chromatin remodeling complex ISWI complexes remodel in nucleosome arrays and nucleosome free zones, thereby regulating gene expression (Kwon et al. 2016), heterochromatin establishment and replication (Culver-Cochran and Chadwick 2013), DNA repair (Atsumi et al. 2015), as well as the coordination of rRNA gene expression (Erdel and Rippe 2011). Studies have shown that SWI/SNF subunits are highly mutated in a variety of cancers, including ovarian cancer, pancreatic cancer, kidney cancer, liver cancer, and bladder cancers (Kadoch et al. 2013) (Fig. 4.1d).

4.2 Epigenetics in Cancer Diagnosis

Notably, aberrant epigenetic modifications in organisms are usually closely associated with the occurrence and development of multiple cancers. For instance, DNA methylation, one of the first discovered epigenetic forms, is involved in a variety of cellular physiological functions and plays an essential role in the occurrence of diseases, especially cancer. Some studies have reported that the occurrence of specific types of cancer can be detected earlier by detecting changes in DNA methylation; thus, DNA methylation has a very high potential to be used as a biomarker for cancer diagnosis (Michalak et al. 2019). Importantly, a big database DNA methylation analysis showed that DNA methylation is tumor-specific and allows better detection of the primary tumor site, serving as a powerful diagnostic marker for primary cancers and leading to more precise and personalized therapies (Moran et al. 2016). Currently, lung cancer and colorectal cancer (CRC) are the top two causes of cancer death worldwide, which are often diagnosed at an advanced stage, missing the best time for therapy. DNA methylation, one of the most intensively studied epigenetic forms, is expected to significantly contribute to the early diagnosis of lung cancer and CRC. For example, a series of novel epigenetic regulatory molecules have been reported to help improve the diagnostic efficacy of standard clinical markers for diagnosis of lung cancer (Diaz-Lagares et al. 2016). Moreover, genome-wide hypomethylation is frequently captured in the early stages of CRC (Jung et al. 2020). In summary, in-depth studies of epigenetics hold the promise of greatly improving the early clinical diagnosis of multiple cancers and reducing the persistently high cancer mortality rate.

4.2.1 DNA Methylation as a Biomarker of Cancer Diagnosis

As mentioned above, DNA methylation has exciting potential to be a biomarker in cancer diagnosis. A genome-wide methylation analysis based on whole-genome bisulfite sequencing (WGBS) data and validated with methylation data from the

Cancer Genome Atlas (TCGA) lung cancer cohort showed that some well-known methylation biomarkers for lung cancer, namely, *SHOX2*, *POU4F2*, *BCAT1*, *HOXA9*, and *PTGDR*, were all captured significantly. In addition, two novel hypermethylated genes, *HIST1H4F* and *HIST1H4I*, were significantly observed in both gene sets, with the area under the curve (AUC) of 0.89 and 0.90, respectively, reflecting the potential of both *HIST1H4F* and *HIST1H4I* as biomarkers for lung cancer diagnosis. Interestingly, the predictive potency of the combination was greater than that of the individual genes with an AUC of 0.95. Interestingly, TCGA pan-cancer methylation analysis showed that hypermethylation of *HIST1H4F* has the potential as a diagnostic biomarker for multiple cancers with AUCs of 0.9–1(Dong et al. 2019a). According to a new integrated epigenomic-transcriptomic analysis of lung cancer, eight novel hypermethylated driver genes, namely, *PCDH17*, *IRX1*, *ITGA5*, *HSPB6*, *TBX5*, *ADCY8*, *GALNT13*, and *TCTEX1D1*, were identified and validated with for predicting an AUC of 0.965 in lung cancer patients, demonstrating reliable clinical diagnostic value (Sun et al. 2021). Moreover, *ITPKA* gene body methylation could also be regarded as a novel diagnostic biomarker for lung cancer with an AUC of 0.93 in the TCGA-lung cancer cohort (Wang et al. 2016). Hitherto, there are two FDA-approved methylation-related diagnostic biomarkers in CRC, of which *SEPT9* is a single-gene methylation biomarker and *NDRG4* and *BMP3* are multigene methylation biomarkers, both of which currently demonstrate compelling clinical value. In addition, *SDC2*, *VIM*, *APC*, *MGMT*, *SFRP1*, *SFRP2*, and *NDRG4* are the most frequently reported methylation biomarkers with promising applications in the early clinical diagnosis of CRC (Müller and Győrffy 2022). In breast cancer, hypomethylation of *SEPTIN7*, *TRIM27*, *LIMD2*, and *LDHA* is often associated with malignant phenotype, while *APC*, *RARB*, *GSTP1*, *DAPK*, and *SFN* are frequently methylated in patients, and all these dysregulated methylation genes are of great diagnostic value (Sher et al. 2022). In a recent genome-wide methylation analysis of 91 esophageal squamous cell carcinoma (ESCC) cases in China, aberrant methylation of six genes was found to be associated with ESCC progression, namely, *PAX9*, *THSD4*, *TWIST1*, *EPB41L3*, *GPX3*, and *COL14A1*, and was similarly validated in TCGA data (Xi et al. 2022). In a meta-analysis of malignant mesothelioma, *APC*, *miR-34b/c*, and *WIF1* were found to be promising diagnostic biomarkers, and further exploration of their diagnostic capabilities is necessary (Vandenhoeck et al. 2021). In cervical cancer, a modeling analysis based on TCGA methylation data yielded four potential diagnostic markers, namely, *RAB3C*, *GABRA2*, *ZNF257*, and *SLC5A8*, with AUCs of 94.2, 100, 100, and 100% in the Gene Expression Omnibus (GEO) dataset to distinguish between cancer and paracancerous tissue, which showed exciting diagnostic results (Xu et al. 2019) (Fig. 4.2).

4 Epigenetics in the Diagnosis, Prognosis, and Therapy of Cancer

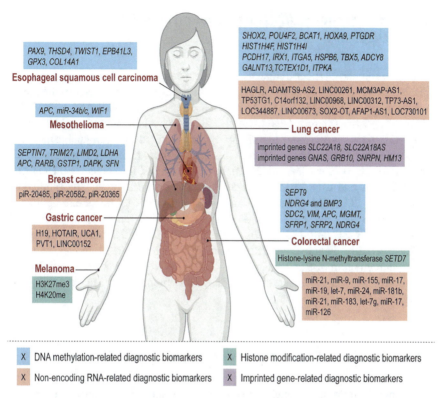

Fig. 4.2 The application of epigenetics in early cancer diagnosis. Currently, the potential of many epigenetic forms as early diagnostic markers has been widely reported in multiple cancers. Various cancer diagnostic markers have emerged in the four major epigenetic forms, DNA methylation, histone modification, non-encoding RNA, and imprinted genes

4.2.2 Histone Modification as a Biomarker of Cancer Diagnosis

Recently, abnormalities in histone modifications have been frequently captured in malignancies, reflecting the powerful potential of histone modifications in cancer diagnosis (Riedel et al. 2015). The results of a recent study suggest that upregulation of Histone-lysine N-methyltransferase *SETD7* implies the occurrence of CRC and could serve as a potential biomarker for CRC (Duan et al. 2018). In addition, by detecting abnormalities in histone H3K27me3 and histone H4K20me, melanoma can be definitively diagnosed in a population with benign nevi (Davis et al. 2020) (Fig. 4.2).

4.2.3 Non-encoding RNAs as Biomarkers of Cancer Diagnosis

Notably, as an early cancer diagnosis strategy, the detection of emerging molecular ncRNAs has gradually made good progress. In NSCLC patients, a set of diagnostic lncRNAs HAGLR, ADAMTS9-AS2, LINC00261, MCM3AP-AS1, TP53TG1, C14orf132, LINC00968, LINC00312, TP73-AS1, LOC344887, LINC00673, SOX2-OT, AFAP1-AS1, and LOC730101 were reported to distinguish cancer and paraneoplastic tissues with a high AUC as 0.98 ± 0.01, reflecting a strong clinical application value (Sulewska et al. 2022). In gastric cancer, lncRNAs H19, HOTAIR, UCA1, PVT1, and LINC00152 were identified as potential diagnostic biomarkers (Fattahi et al. 2020). In addition, abnormalities in miRNAs are often captured in carcinogenesis. In CRC, miR-21, miR-9, miR-155, miR-17, miR-19, let-7, miR-24, miR-181b, miR-21, miR-183, let-7 g, miR-17, and miR-126 are all promising biomarkers for early diagnosis of CRC (Moridikia et al. 2018). Moreover, in breast cancer, upregulation of PIWI-interacting RNAs (piRNAs) piR-20,485, piR-20,582, and piR-20,365 could be used as early diagnostic biomarkers (Maleki Dana et al. 2020) (Fig. 4.2).

4.2.4 Imprinted Genes as Biomarkers of Cancer Diagnosis

Importantly, imprinted genes are also contributors to cancer diagnosis. Overexpression of imprinted *SLC22A18* and *SLC22A18AS* gene has been reported to promote the occurrence and progression of NSCLC, and both genes can be used for early diagnosis of NSCLC (Noguera-Uclés et al. 2020). Another study showed that the imprinted genes *GNAS*, *GRB10*, *SNRPN*, and *HM13* are also diagnostic markers for early-stage lung cancer, which are expected to translate to the clinic (Zhou et al. 2021). Similarly, the imprinted genes *GNAS*, *GRB10*, and *SNRPZ* also showed convincing diagnostic performance in another study (Shen et al. 2020) (Fig. 4.2).

4.3 Epigenetic in Cancer Prognosis

The epigenetic variation of cancer patients may enable the tumor to obtain the ability to adapt to the treatment, which will lead to a poor prognosis. In recent years, the research on epigenetics in cancer prognosis mainly focuses on the prognostic grading of cancer patients and the development of specific prognostic markers (Wong et al. 2019). Using epigenetic differences such as DNA methylation to classify patients and provide different targeted treatment methods is a research hotspot at present. The change in DNA methylation level exists in many cancers.

Therefore, the identification and development of prognostic markers can be used to indicate tumor metastasis, recurrence, and 5-year survival rate. Now, the latest genome-wide epigenomics method makes it feasible to construct a comprehensive map of cancer methylation groups and may bring a standardized method for epigenetic prediction of cancer prognosis (Grady et al. 2021). The research content of epigenetics in cancer prognosis can be roughly divided into four aspects: DNA methylation, histone modification, chromatin remodeling, and functional noncoding RNA. This chapter will discuss the above four parts.

4.3.1 DNA Methylation in Cancer Prognosis

DNA methylation is an important part of epigenetic research, and differences in DNA methylation enable altered transcription of gene expression. Studies have shown that most of cancer patients, mainly gastric cancer, CRC, and liver cancer, tend to have some degree of changes in DNA methylation status (Mehdipour et al. 2020). Therefore, identifying abnormally methylated genes may provide new ideas for developing prognostic markers for cancer. Through studies of different liver cancer patients, it has been found that DNA methylation changes might contribute to a poorer prognosis. For instance, treatment with α-Interferon for liver cancer patients usually cannot achieve better curative effect when the patient's miR-26a (an epigenetic marker) expression is high. Markers of tissue DNA methylation are equally important in the prediction of liver cancer prognosis and can serve as potential prognostic biomarkers for the staging of hepatocellular carcinoma mainly: keratin 19,5-hydroxymethylcytosine. In hepatocellular carcinoma, an analysis targeting the promoter methylation status of 105 possible tumor suppressor genes found that low methylation frequency might increase the risk of the poor prognosis (Tricarico et al. 2020). Meanwhile, the detection of DNA methylation abnormalities has also achieved certain results in the diagnosis and prognosis of gastrointestinal tumor. For example, the CpG island methylator phenotype (CIMP), first discovered in CRC, might be a potential prognostic marker for CRC as well as gastric cancer. One study, which analyzed more than 600 CRC patients, suggested that CIMP was potentially associated with poor prognosis in microsatellite stable CRC patients (Liu et al. 2020). It may become a new basis for patient prognostic stratification through methylation profiling of cancer patient samples as well as existing therapeutic targets. Evidence has been presented that there is epigenetic differential regulation of Lag3 by DNA methylation in tumor cells and normal immune cells, and Niklas et al. found that methylation of the promoter resulted in lower amounts of Lag3 expression, which negatively correlated with poor prognosis (Klumper et al. 2020). O6 methylguanine DNA methyltransferase (MGMT) gene promoter methylation levels and can be used as a basis for histologic stratification of patients with glioblastoma (GBM), as well as for prediction of posttreatment survival (Mansouri et al. 2019). An important reason that can lead to a poor prognosis of cancer is that tumors may acquire drug resistance in treatment, and an increasing number of studies have shown that epigenetic modification of mRNA has a certain link with tumor drug resistance. mRNA modifications include m6A, and Fukumoto T. et al. found that m6A modification in fzd10 mRNA is positively correlated with its stability and may be associated with resistance to PARP inhibitors in BRCA mutated epithelial ovarian cancer (EOC) (Fukumoto et al. 2019).

4.3.2 Histone Modifications in Cancer Prognosis

Histone modification refers to the process by which histones undergo methylation, acetylation, phosphorylation, and other modifications under the action of related enzymes. An increasing number of studies have shown that the modification of histones is inseparable from the occurrence and development of cancer. Similar to methylation of DNA, in which methylation of histones is more widespread in histone modification studies. Currently, some scholars focus their eyes on the synergistic roles of DNA and histone modifications in cancer progression, and G9a histone methyltransferase and DNA methyltransferase I were found to be significantly overexpressed in hepatocellular carcinoma. Both are synergistically associated with poor prognosis in hepatocellular carcinoma, and thus, intervention in hepatocellular carcinoma at the epigenetic level may be achieved through inhibition of G9a and DNMT1 (Barcena-Varela et al. 2019). Studies targeting epigenetic alterations have found that histone methylation and modifying enzymes may play a role in prognosis prediction in various cancers. Methylation of histone H3K27me3 has been associated with breast cancer migration and may serve as a prognostic marker (Hsieh et al. 2020). Studies have shown that inhibition of euchromatin histone-lysine-n-methyltransferases I and II (EHMT1/2) may reverse partial ovarian cancer resistance to PARP inhibitors (Watson et al. 2019). Histone modifications are closely associated with epigenetic alterations in gene expression and have significant research potential in both the identification of cancer subtypes and the development of predictive markers for patient survival. CDX2 is a prognostic biomarker for colorectal, and the histone deacetylases HDAC4 and HDAC5 can repress CDX2 expression (Graule et al. 2018). Nowadays, new concepts reveal that the tumor microenvironment may be associated with histone modifications, and a persistent hypoxic microenvironment has been found in pancreatic cancer capable of altering histone methylation (Li et al. 2021). The hypoxic tumor microenvironment serves as a potential judgment for cancer malignancy, which also provides a new direction for predicting the prognosis of tumors through histone modification status. The role of epigenetics in multiple myeloma has similarly received attention from investigators, and some modifying enzymes that alter the acetylation status of histones may be involved in the progression of multiple myeloma. The development of inhibitors targeting histone deacetylases may offer new therapeutic possibilities for multiple myeloma (Ohguchi et al. 2018).

4.3.3 Histone Variants and Chromatin Remodeling in Cancer Prognosis

Chromatin remodeling is a process in which chromatin unfolds during gene-initiated expression and depends on three dynamic properties of nucleosomes, including remodeling, enzyme induced covalent modification, and repositioning. In which histone variants can be incorporated into nucleosomes and displace existing nucleosomal subunits, a process that is tightly regulated by chromatin remodeling factors, such as the SWR1 complex. At the same time, these variants can also affect chromatin remodeling and thus transcriptional regulation. In addition to mutations in genes involved in chromatin remodeling that are frequently observed in many types of cancer, for example, studies of chromatin remodeling genes suggest that SMARCA4 may be involved in neuroblastoma tumorigenesis, numerous studies have also shown that histone variants can predict prognosis in various cancers. John Blenis and Ana P. Gomes et al. collaborate to discover that the histone H3 variant H3.3, under the action of histone chaperones that regulate metastasis, deposits at metastasis to induce transcription factor promoters and promote tumor metastasis. Standard histone intercalation is reduced in chromatin, leading to the deposition of poor prognosis genes in tumors (Gomes et al. 2019).

The family of chromatin regulators plays an important role in chromatin remodeling. Among these, cBAF is the most abundant of the SWI/SNF complexes. ARID1A is the largest subunit homologous to cBAF, and its gene is mutated at a frequency of up to 50–60% during carcinogenesis. There are studies showing partial or complete inactivation of arid2 expressed protein in liver cancer. Researchers have found that the C2H2 domain of ARID2 can recruit DNMT1 to the promoter of the transcription factor Snail, which elevates DNA methylation and inhibits the transcription of snail, thereby inhibiting epithelial mesenchymal transition in liver cancer cells. Therefore, mutations in the C2H2 domain of ARID2 promote liver cancer metastasis and reduce the 5-year survival rate of patients with liver cancer (Jiang et al. 2020).

Additional studies identified the chromatin structure regulators SND1 and RHOA as independent predictors of poor prognosis in glioma patients. SND1 can remodel chromatin conformation, allowing transcriptional upregulation of RHOA. Thus, it activates the Cyclin/CDK signaling pathway, which enables the G1/S phase transition of the glioma cell cycle and promotes glioma cell proliferation, migration, and invasion (Yu et al. 2019). In the development of CRC, the chromatin remodeling genes PRMT1 and SMARCA4 have higher expression; thus, inhibition of PRMT1/SMARCA 4 may be used as an intervention strategy to prolong the overall survival of patients with intestinal cancer (Yao et al. 2021). Whereas in hepatic cell carcinoma (HCC) cells, overexpression of the chromatin remodeler Hells epigenetically silences multiple tumor suppressor genes, thereby promoting HCC cell proliferation and migration, thus manifesting as more aggressive and worse patient outcomes in clinicopathological features (Law et al. 2019).

4.3.4 Noncoding RNAs in Cancer Prognosis

ncRNAs refer to RNAs that do not encode proteins. In the development of tumors, ncRNAs are involved in the process of their proliferation, differentiation, and metastasis, which play an extremely important role in the prognosis of tumors. Currently, a variety of ncRNAs are found to serve as tumor markers, becoming a hotspot of tumor research in recent years. miRNAs contain only 22–24 nucleotides, but by imperfect base complementarity, miRNAs can match and silence multiple mRNAs. In terms of tumor therapy, mir-302a affects tumor migration with the acquisition of drug resistance. Studies have suggested its role as a candidate prognostic predictor in CRC by targeting NFIB and CD44 (Sun et al. 2019a). Circular RNAs (circRNAs) do not have covalently closed loops at the 5 'and 3' ends, but studies have found their ability to participate in transcription, regulation of translation, and localization of proteins. In the occurrence of multiple cancers, circRNAs can act on miRNAs, for example, CIRS-7 promotes CRC progression by blocking the tumor suppressive effect of miR-7. While in bladder cancer, circACVR2A acted as a miRNA sponge to regulate miR-626 to inhibit cancer cell proliferation and metastasis. This provides a new idea for the prognosis and treatment of bladder cancer (Dong et al. 2019b). In addition, it was demonstrated through autophagy-related experiments that the level of circCDYL increased in the tissues of breast cancer, thus promoting the level of autophagy in breast cancer cells and reducing the survival of breast cancer patients with curative effect, and this process was associated with miR-1275-ATG7/ULK1-AUTOP (Liang et al. 2020b). lncRNAs are a class of RNAs >200 bp in length that lack an effective open reading frame, sequestering little or no protein coding sequence. Analysis through multiple data means in recent years has shown that the expression of lncRNAs is associated with prognosis in multiple tumors (Zhang et al. 2020b). Acting with YBX1 in cells by specifically expressing lncRNA DSCAM-AS1 in tumors, adversely affecting the prognosis of tumors. In vitro and in vivo experiments have demonstrated the important regulatory roles of lncRNAs in tumorigenesis and development. In recent years, lncRNA research has become a current research hotspot. Multiple understudied lncRNAs were found to potentially play a role in tumor prognosis. LINC02273 contributed to the metastasis of breast cancer and increased the metastasis associated protein hnRNPL, which in turn activated the AGR2 axis, providing new protein markers for the cure and prognosis of breast cancer (Xiu et al. 2019). Whereas in CRC, lncRNA LINRIS may serve as an independent biomarker for its prognosis, which plays an important role in CRC by inhibiting aerobic glycolysis (Wang et al. 2019). It is believed that with the development of biotechnology, more novel ncRNAs will be found to function in cancer treatment and play more important roles in the prognosis of patients (Fig. 4.3).

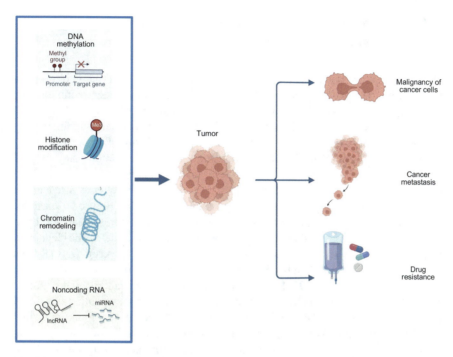

Fig. 4.3 Impact of epigenetic alterations on cancer prognosis. In the absence of alterations in the nuclear DNA sequence, tumor tissues exhibit diverse biological features through epigenetic modification pathways such as DNA methylation, histone modification, chromatin remodeling, and functional ncRNAs. Different degrees of epigenetic modification can alter cancer cell proliferation speed, migration ability and even lead to drug resistance. Therefore, DNA methylation, histone modification status, and so on can be used to predict the prognostic performance of cancer patients

4.4 Epigenetics in Cancer Therapy

As previously mentioned, epigenetic dysregulation has a far-reaching impact on gene expression, DNA replication, and DNA repair, which is closely associated with tumorigenesis and tumor progression. Unlike genetic mutations, epigenetic changes in cancer epigenome are mainly enzyme-catalyzed and probably reversible, which provides ideal targets for cancer treatment (Bianco and Gevry 2012; Lopez-Camarillo et al. 2019). So far, most epigenetic drugs are designed to modulate DNA methylation- and histone acetylation-related enzymes, including DNA methyltransferases, histone acetyltransferases, histone deacetylases, and histone methyltransferases, which are known as of importance targets of cancer treatment (Miranda Furtado et al. 2019).

4.4.1 Targeting DNA Methylation-Related Enzymes in Cancer Treatment

DNA methylation is the main epigenetic mechanism and a well-publicized epigenetic marker, in which the cytosine bases in CpG island are covalently modified by methyl groups. The DNMT family of enzymes, including DNMT1, DNMT3a, and DNMT3b, plays a pivotal role in the methylation process by catalyzing the transfer of a methyl group of s-adenosyl-l-methionine to DNA (Bestor 2000; Li et al. 1992; Okano et al. 1999; Schapira and Arrowsmith 2016). Hereunto, many DNMT inhibitors were discovered. For example, azacytidine and decitabine have been originally used to treat myelodysplastic syndrome and subsequently used in the treatment of chronic myelomonocytic (Derissen et al. 2013). Both can inactivate DNMT by forming an irreversible covalent complex with it (Stresemann and Lyko 2008). In addition, many other types of inhibitors, such as non-nucleoside chemicals, were developed in recent years. Procaine has been found to keep DNMT from interacting with DNA, whereby the promoter regions of CDKN2A and RAR β are less methylated (Li et al. 2018). In gastric cancer cells, procaine was detected to promote apoptosis and inhibit proliferation, suggesting the therapeutic potential of procaine on cancer as a DNMT inhibitor. MC3343, however, was found to exhibit proliferative activity on osteosarcoma cells by causing cell cycle arrest at the G0-G1 or G2-M phases, which can be attributed to its inhibitory activity against DNMT1, DNMT3a, and DNMT3b expression and biological activity (Manara et al. 2018). As an analog of MC3343, MC3353 acts as a DNMT inhibitor, displaying strong demethylation ability and providing reactivation of the silenced gene (Zwergel et al. 2019). By modulating the genes involved in osteoblast differentiation, MC3353 showed its activity in primary osteosarcoma cell models. In addition, compounds 3b and 4a can inhibit DNMT and therefore impair acute myelogenous leukemia cells KG1 and CRC cells HCT116 proliferation (Pechalrieu et al. 2020).

4.4.2 Targeting Histone Acetylation-Related Enzymes in Cancer Treatment

Being a crucial epigenetic regulation, the acetylation of histone lysine affects cell differentiation and proliferation by intervening the interaction between transcription factors and the regulatory sequence of oncogenes (Kulka et al. 2020). This process is generally controlled by related enzymes including KATs and KDACs. Among them, potential epigenetic targets HATs and HDAC are responsible for adding and deleting acetyl group to lysine residues (Tapadar et al. 2020). They are also known as "Writers" and "Erasers" of epigenetic modifications. Besides, the bromodomain-containing proteins (BRDs) functions as "Readers" to decode those acetylated lysine and consequently recruit chromatin regulators to control gene expression (Hillyar et al. 2020; Wang et al. 2021).

HATs include several enzymes. Epigenetic therapies targeting HATs mainly focus on the GCN5-related n-acetyltransferase (GNAT) family, including GCN5 and p300/CBP-related factor (PCAF) (Trisciuoglio et al. 2018; Wang et al. 2021). As a selective and effective catalytic inhibitor of p300/CBP, A-485 competed with acetyl-CoA and thus selectively inhibited proliferation across lineage-specific tumor types, including androgen receptor-positive prostate cancer and several hematologic malignancies (Lasko et al. 2017). P300 and CBP highly express in five gastric cancer cell lines, in which compound C646 affects cell cycle and promotes cell apoptosis to exert antitumor effects by selectively inhibiting P300 and CBP (Wang et al. 2017). Garcinol was discovered to exhibit inhibitory activity against human esophageal cancer cell lines KYSE150 and KYSE450 for migration and invasion in a dose-dependent manner via blocking p300 and TGF-β1 signaling pathway, which influences the cell cycle and induces apoptosis (Wang et al. 2020a).

HDACs take charge of removing acetyl groups on lysine residues of histone proteins. Considering that HDACs are high expression in cancer cells, accumulating explorations have focused on HADCs. Up to now, 18 subtypes of HDACs have been discovered, and their inhibitors have been identified and applied in clinical trials, including vorinostat, romidepsin, belinostat, and panobinostat (Fan et al. 2021). Besides, several compounds are being studied. VS13, a quinoline derivative, can potently inhibit HDAC6 in nanomolar concentration, showing its anti-proliferation activity against uveal melanoma cell line (Nencetti et al. 2021). Unlike VS13, compound 12a selectively inhibited subtype HDAC2, and consequently restrained A549 cells from migration and colony formation, and induced apoptosis and G2/M cell cycle arrest (Wang et al. 2021; Xie et al. 2017). AES-135 possesses potent pancreatic cancer cells cytotoxicity in vitro and prolonging the survival time of the pancreatic cancer mouse model, which can be attributed to its inhibitory effect on HDAC3, HDAC6, and HDAC11 (Shouksmith et al. 2019). Generally, HDACs inhibitors can restrain tumor proliferation by affecting apoptosis, differentiation, cell migration, and cell cycle arrest (Wang et al. 2021). As the class III HDAC family, the sirtuins (SIRTs) family is also considered as a therapeutic target of cancer treatment, whose pharmacological inhibition remodels the chromatin state and results in a blurring of the boundaries between transcriptional activity and static chromatin (Manzotti et al. 2019; Wang et al. 2021). SIRTs are a series of nicotinamide adenine dinucleotide dependent enzymes, including intranuclear SIRT1, SIRT6, and SIRT7, intramitochondrial SIRT3–5, and cytoplasmic SIRT2 (Houtkooper et al. 2012). Recently, the high level of SIRT1 has been found to be associated with recurrence and poor prognosis in patients with lung adenocarcinoma, which is highly sensitive to the treatment of gefitinib, a tyrosine kinase inhibitor (TKI). It has been reported that SIRT1 inhibitor TV-6 can enhance TKI therapeutic effect. The combined administration of TV-6 and TKI results in tumor regression in xenograft mouse models and improved sensitivity of tumor cells to TKI (Sun et al. 2020). Besides, gastric carcinoma cell lines are also sensitive to TV-6 with activating p53 or inhibiting autophagic flux (Ke et al. 2020). TM, a thiomyristoyl lysine compound, has been discovered as antiproliferative agent against many human cancer cells and breast cancer mouse models. TM inhibits SIRT2, promoting the ubiquitination and degradation of c-Myc and therefore exhibiting anticancer activity (Jing et al. 2016). Selectively targeting SIRT2, γ-mangostin, a natural product, was found to inhibit MDA-MD-231 and MCF-7 breast cancer cells proliferation by improving the acetylation level of α-tubulin (Yeong et al. 2020). As a SIRT1 activator, compound F0911-7667 induces autophagic cell death via the AMPK-mTOR-ULK complex and induces mitochondrial phagocytosis through the SIRT1-PINK1-Parkin pathway in glioblastoma cells (Yao et al. 2018). SIRT6 agonists UBCS039 and MDL-800 can induce autophagic cell death in several human tumors by activating the deacetylation of SIRT6 (Huang et al. 2018; Iachettini et al. 2018).

BRDs is a highly conserved protein superfamily containing eight families, in which the bromodomain and extra-terminal (BET) families are the most widely studied, containing BRD2, BDR3, BRD4, and BRDT (Wang et al. 2021). BRDs serve on readers of histone acetylation, causing epigenetic regulation of gene expression (Liang et al. 2020a). Many small molecules targeting BRDs have been discovered to develop cancer therapies. In phase I clinical trials, small-molecule pan-BET inhibitor ABBV-075 was discovered to be promising in tolerance and therapeutic activity in highly pretreated patients with refractory solid tumors (Piha-Paul et al. 2019). Further chemical modifications of ABBV-075 obtained compound 38, which exhibited higher inhibitory efficiency than clinical candidate OTX-015 (Li et al. 2020). In addition to pan-inhibitors, selective BRD inhibitors have also been identified. For instance, small-molecule FL-411 has been reported to be involved in autophagy-related cell death via targeting BRD4, which blocked BRD4-AMPK interaction and thereby activating the autophagic pathway in breast cancer (Ouyang et al. 2017). Bioavailable chemicals GSK452, GSK737, and GSK217 were screened to have ideal solubility, cell efficacy, as well as pharmacokinetics (Aylott et al. 2021). It has been reported that dBET6, a BET protein degrading agent, possesses antitumor activity against hematology and solid cancer and presents higher efficiency than first generation drugs, such as dBET1 and JQ1 (Bauer et al. 2021). Moreover, bBET6 also lowered the resistance of cancer to immuno- and chemotherapy.

4.4.3 Targeting Other Epigenetic Biomarkers in Cancer Treatment

Other important elements of epigenetic regulation are histone methylation enzymes, including histone H3K79 methyltransferase DOT1L, H3K27 methyltransferase enhancer of zeste homolog 2 EZH2, and histone lysine demethylases (KDMs) (Wang et al. 2021). It has been reported that DOT1L promoted triple-negative breast cancer (TNBC) progression via the binding between c-Myc and p300 acetyltransferase (Cho et al. 2015; Wang et al. 2021). DOT1L inhibitor PsA-309 was identified to have an inhibitory effect on the proliferation of human breast cancer cell lines MDA-MB-231 and Hs578T via selectively suppressing H3K79

4 Epigenetics in the Diagnosis, Prognosis, and Therapy of Cancer   155

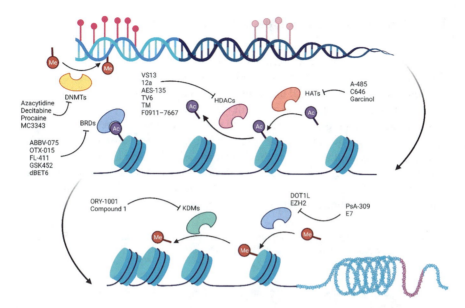

Fig. 4.4 Epigenetic interventions in cancer therapy. Currently, most epigenetic drugs are designed to inhibit DNA methylation and histone acetylation-related enzymes, including DNMTs, HATs, HDACs, KDMs, as well as histone acetylation readers BRDs, histone methylases DOT1L, and EZH2, suppressing tumorigenesis and malignant progression by inhibiting epigenetic dysregulation

methylation (Byun et al. 2019). PROTAC E7 was discovered to inhibit EZH2. E7-mediated degradation of EZH2 resulted in decreased proliferation rates in diffuse large B-cell lymphoma cell WSU-DLCL-2, human lung cancer cell A549, and NSCLC cell NCI-H1299 (Wang et al. 2021). Besides, inhibiting histone demethylation-related KDMs can also be a therapeutic strategy in cancer treatment. By inhibiting KDM1A, ORY-1001 can compromise leukemic stem cell activity in acute leukemia (Maes et al. 2018). ORY-1001 impaired cancer cell proliferation and increased survival in rodent xenograft models of acute leukemia. In addition, KDM5A inhibitor compound 1 has been reported to suppress the demethylation of H3K4me3, leading to p16 and p27 accumulation, which consequently resulted in cell cycle arrest and inhibited proliferation of various KDM5A-overexpressing breast cancer cells (Yang et al. 2019). Additionally, it has been proposed that miRNAs associated with cancer are therapeutic targets of cancer. Several miRNA inhibition therapies have been developed, for example, the combination treatment of anti-miRNA oligonucleotides with a low dose of sunitinib exhibited a significant synergistic antitumor effect in pancreatic ductal adenocarcinoma cells (Song et al. 2017) (Fig. 4.4).

4.5 Conclusions and Perspectives

To date, cancer incidence and mortality rates remain high worldwide; therefore, the discovery of novel cancer diagnostic and prognostic biomarkers and the search for emerging and effective targeted therapies are major attention worldwide. Epigenetics is closely related to the occurrence and development of many types of cancers, and abnormalities in multiple epigenetic regulations often lead to the progression of malignancies; therefore, an in-depth analysis of the key mechanisms of epigenetics in cancer development will help in the early diagnosis of cancer and greatly improve cancer treatment. At present, several epigenetic biomarkers have made great progress in the research of cancer diagnosis, prognosis, and treatment, but how to translate them into clinical practice is still a challenge for us.

Acknowledgments This work was supported by grants from National Natural Science Foundation of China (Grant No. 82172649) and Fundamental Research Funds for the Central Universities (Grant No. 2682021CX088).

Compliance with Ethical Standards Conflict of Interest: The authors declare that they have no conflict of interest.

Research involving human participants and/or animals: This article does not contain any studies with human participants performed by any of the authors.

References

Akone SH, Ntie-Kang F, Stuhldreier F, Ewonkem MB, Noah AM, Mouelle SEM, Müller R (2020) Natural products impacting DNA methyltransferases and histone deacetylases. Front Pharmacol 11:992

Atsumi Y, Minakawa Y, Ono M, Dobashi S, Shinohe K, Shinohara A, Takeda S, Takagi M, Takamatsu N, Nakagama H et al (2015) ATM and SIRT6/SNF2H mediate transient H2AX stabilization when DSBs form by blocking HUWE1 to allow efficient γH2AX foci formation. Cell Rep 13:2728–2740

Aylott HE, Atkinson SJ, Bamborough P, Bassil A, Chung CW, Gordon L, Grandi P, Gray JRJ, Harrison LA, Hayhow TG et al (2021) Template-hopping approach leads to potent, selective, and highly soluble Bromo and Extraterminal domain (BET) second Bromodomain (BD2) inhibitors. J Med Chem 64:3249–3281

Barcena-Varela M, Caruso S, Llerena S, Alvarez-Sola G, Uriarte I, Latasa MU, Urtasun R, Rebouissou S, Alvarez L, Jimenez M et al (2019) Dual targeting of histone methyltransferase G9a and DNA-methyltransferase 1 for the treatment of experimental hepatocellular carcinoma. Hepatology 69:587–603

Bauer K, Berghoff AS, Preusser M, Heller G, Zielinski CC, Valent P, Grunt TW (2021) Degradation of BRD4–a promising treatment approach not only for hematologic but also for solid cancer. Am J Cancer Res 11:530–545

Bestor TH (2000) The DNA methyltransferases of mammals. Hum Mol Genet 9:2395–2402

Bianco S, Gevry N (2012) Endocrine resistance in breast cancer: from cellular signaling pathways to epigenetic mechanisms. Transcription 3:165–170

Bottoni A, Piccin D, Tagliati F, Luchin A, Zatelli MC, degli Uberti EC (2005) miR-15a and miR-16-1 down-regulation in pituitary adenomas. J Cell Physiol 204:280–285

Byun WS, Kim WK, Han HJ, Chung HJ, Jang K, Kim HS, Kim S, Kim D, Bae ES, Park S et al (2019) Targeting histone methyltransferase DOT1L by a novel Psammaplin a analog inhibits growth and metastasis of triple-negative breast cancer. Mol Ther Oncolytics 15:140–152

Cho MH, Park JH, Choi HJ, Park MK, Won HY, Park YJ, Lee CH, Oh SH, Song YS, Kim HS et al (2015) DOT1L cooperates with the c-Myc-p300 complex to epigenetically derepress CDH1 transcription factors in breast cancer progression. Nat Commun 6:7821

Chun P (2015) Histone deacetylase inhibitors in hematological malignancies and solid tumors. Arch Pharm Res 38:933–949

Cui Q, Shi H, Ye P, Li L, Qu Q, Sun G, Sun G, Lu Z, Huang Y, Yang CG et al (2017) M(6)a RNA methylation regulates the self-renewal and tumorigenesis of glioblastoma stem cells. Cell Rep 18:2622–2634

Culver-Cochran AE, Chadwick BP (2013) Loss of WSTF results in spontaneous fluctuations of heterochromatin formation and resolution, combined with substantial changes to gene expression. BMC Genomics 14:740

Dang F, Wei W (2021) Targeting the acetylation signaling pathway in cancer therapy. Paper presented at: seminars in cancer biology (Elsevier)

Davis LE, Shalin SC, Tackett AJ (2020) Utility of histone H3K27me3 and H4K20me as diagnostic indicators of melanoma. Melanoma Res 30:159–165

Derissen EJ, Beijnen JH, Schellens JH (2013) Concise drug review: azacitidine and decitabine. Oncologist 18:619–624

Diaz-Lagares A, Mendez-Gonzalez J, Hervas D, Saigi M, Pajares MJ, Garcia D, Crujerias AB, Pio R, Montuenga LM, Zulueta J et al (2016) A novel epigenetic signature for early diagnosis in lung cancer. Clin Cancer Res 22:3361–3371

Dong S, Li W, Wang L, Hu J, Song Y, Zhang B, Ren X, Ji S, Li J, Xu P et al (2019a) Histone-related genes are Hypermethylated in lung cancer and Hypermethylated HIST1H4F could serve as a Pan-cancer biomarker. Cancer Res 79:6101–6112

Dong W, Bi J, Liu H, Yan D, He Q, Zhou Q, Wang Q, Xie R, Su Y, Yang M et al (2019b) Circular RNA ACVR2A suppresses bladder cancer cells proliferation and metastasis through miR-626/EYA4 axis. Mol Cancer 18:95

Dong Y, Zhao H, Li H, Li X, Yang S (2014) DNA methylation as an early diagnostic marker of cancer. Biomed Rep 2:326–330

Duan B, Bai J, Qiu J, Wang J, Tong C, Wang X, Miao J, Li Z, Li W, Yang J, Huang C (2018) Histone-lysine N-methyltransferase SETD7 is a potential serum biomarker for colorectal cancer patients. EBioMedicine 37:134–143

Duan R, Du W, Guo W (2020) EZH2: a novel target for cancer treatment. J Hematol Oncol 13:104

Erdel F, Rippe K (2011) Chromatin remodelling in mammalian cells by ISWI-type complexes–where, when and why? FEBS J 278:3608–3618

Fan W, Zhang L, Wang X, Jia H, Zhang L (2021) Discovery of potent histone deacetylase inhibitors with modified phenanthridine caps. J Enzyme Inhib Med Chem 36:707–718

Fatica A, Bozzoni I (2014) Long non-coding RNAs: new players in cell differentiation and development. Nat Rev Genet 15:7–21

Fattahi S, Kosari-Monfared M, Golpour M, Emami Z, Ghasemiyan M, Nouri M, Akhavan-Niaki H (2020) LncRNAs as potential diagnostic and prognostic biomarkers in gastric cancer: a novel approach to personalized medicine. J Cell Physiol 235:3189–3206

Fukumoto T, Zhu H, Nacarelli T, Karakashev S, Fatkhutdinov N, Wu S, Liu P, Kossenkov AV, Showe LC, Jean S et al (2019) N(6)-methylation of adenosine of FZD10 mRNA contributes to PARP inhibitor resistance. Cancer Res 79:2812–2820

Goldberg AD, Allis CD, Bernstein E (2007) Epigenetics: a landscape takes shape. Cell 128:635–638

Gomes AP, Ilter D, Low V, Rosenzweig A, Shen ZJ, Schild T, Rivas MA, Er EE, McNally DR, Mutvei AP et al (2019) Dynamic incorporation of histone H3 variants into chromatin is essential for acquisition of aggressive traits and metastatic colonization. Cancer Cell 36:402–417 e413

Grady WM, Yu M, Markowitz SD (2021) Epigenetic alterations in the gastrointestinal tract: current and emerging use for biomarkers of cancer. Gastroenterology 160:690–709

Graule J, Uth K, Fischer E, Centeno I, Galvan JA, Eichmann M, Rau TT, Langer R, Dawson H, Nitsche U et al (2018) CDX2 in colorectal cancer is an independent prognostic factor and regulated by promoter methylation and histone deacetylation in tumors of the serrated pathway. Clin Epigenetics 10:120

Heiblig M, El Hamri M, Salles G, Thomas X (2015) Epigenetics and adult acute myeloid leukemia. In: Current Pharmacogenomics and Personalized Medicine (Formerly Current Pharmacogenomics), vol 13, pp 117–133

Hillyar C, Rallis KS, Varghese J (2020) Advances in epigenetic cancer therapeutics. Cureus 12: e11725

Houtkooper RH, Pirinen E, Auwerx J (2012) Sirtuins as regulators of metabolism and healthspan. Nat Rev Mol Cell Bio 13:225–238

Hsieh IY, He J, Wang L, Lin B, Liang Z, Lu B, Chen W, Lu G, Li F, Lv W et al (2020) H3K27me3 loss plays a vital role in CEMIP mediated carcinogenesis and progression of breast cancer with poor prognosis. Biomed Pharmacother 123:109728

Hu X, Feng Y, Zhang D, Zhao SD, Hu Z, Greshock J, Zhang Y, Yang L, Zhong X, Wang LP et al (2014) A functional genomic approach identifies FAL1 as an oncogenic long noncoding RNA that associates with BMI1 and represses p21 expression in cancer. Cancer Cell 26:344–357

Huang ZM, Zhao JX, Deng W, Chen YY, Shang JL, Song K, Zhang L, Wang CX, Lu SY, Yang XY et al (2018) Identification of a cellularly active SIRT6 allosteric activator. Nat Chem Biol 14: 1118-+

Iachettini S, Trisciuoglio D, Rotili D, Lucidi A, Salvati E, Zizza P, Di Leo L, Del Bufalo D, Ciriolo MR, Leonetti C et al (2018) Pharmacological activation of SIRT6 triggers lethal autophagy in human cancer cells. Cell Death Dis 9:996

Jiang H, Cao HJ, Ma N, Bao WD, Wang JJ, Chen TW, Zhang EB, Yuan YM, Ni QZ, Zhang FK et al (2020) Chromatin remodeling factor ARID2 suppresses hepatocellular carcinoma metastasis via DNMT1-snail axis. Proc Natl Acad Sci U S A 117:4770–4780

Jing H, Hu J, He B, Abril YLN, Stupinski J, Weiser K, Carbonaro M, Chiang YL, Southard T, Giannakakou P et al (2016) A SIRT2-selective inhibitor promotes c-Myc Oncoprotein degradation and exhibits broad anticancer activity. Cancer Cell 29:297–310

Jung G, Hernández-Illán E, Moreira L, Balaguer F, Goel A (2020) Epigenetics of colorectal cancer: biomarker and therapeutic potential. Nat Rev Gastroenterol Hepatol 17:111–130

Kadoch C, Hargreaves DC, Hodges C, Elias L, Ho L, Ranish J, Crabtree GR (2013) Proteomic and bioinformatic analysis of mammalian SWI/SNF complexes identifies extensive roles in human malignancy. Nat Genet 45:592–601

Ke XY, Qin QS, Deng TY, Liao YY, Gao SJ (2020) Heterogeneous responses of gastric cancer cell lines to Tenovin-6 and synergistic effect with chloroquine. Cancers 12

Klumper N, Ralser M, Bawden EG, Landsberg J, Zarbl R, Kristiansen G, Toma M, Ritter M, Holzel M, Ellinger J, Dietrich D (2020) LAG3 (LAG-3, CD223) DNA methylation correlates with LAG3 expression by tumor and immune cells, immune cell infiltration, and overall survival in clear cell renal cell carcinoma. J Immunother Cancer 8:e000552

Kulka LAM, Fangmann PV, Panfilova D, Olzscha H (2020) Impact of HDAC inhibitors on protein quality control systems: consequences for precision medicine in malignant disease. Front Cell Dev Biol 8:425

Kumar S, Nagpal R, Kumar A, Ashraf MU, Bae YS (2021) Immunotherapeutic potential of m6A-modifiers and MicroRNAs in controlling acute myeloid Leukaemia. Biomedicine 9

Kwon SY, Grisan V, Jang B, Herbert J, Badenhorst P (2016) Genome-wide mapping targets of the metazoan chromatin remodeling factor NURF reveals nucleosome remodeling at enhancers, Core promoters and gene insulators. PLoS Genet 12:e1005969

Lafon-Hughes L, Di Tomaso MV, Méndez-Acuña L, Martínez-López W (2008) Chromatin-remodelling mechanisms in cancer. Mutation Res/Rev Mutation Res 658:191–214

Lambert SA, Jolma A, Campitelli LF, Das PK, Yin Y, Albu M, Chen X, Taipale J, Hughes TR, Weirauch MT (2018) The human transcription factors. Cell 172:650–665

Lasko LM, Jakob CG, Edalji RP, Qiu W, Montgomery D, Digiammarino EL, Hansen TM, Risi RM, Frey R, Manaves V et al (2017) Discovery of a selective catalytic p300/CBP inhibitor that targets lineage-specific tumours. Nature 550:128–132

Law CT, Wei L, Tsang FH, Chan CY, Xu IM, Lai RK, Ho DW, Lee JM, Wong CC, Ng IO, Wong CM (2019) HELLS regulates chromatin remodeling and epigenetic silencing of multiple tumor suppressor genes in human hepatocellular carcinoma. Hepatology 69:2013–2030

Li E, Bestor TH, Jaenisch R (1992) Targeted mutation of the DNA methyltransferase gene results in embryonic lethality. Cell 69:915–926

Li H, Peng C, Zhu C, Nie S, Qian X, Shi Z, Shi M, Liang Y, Ding X, Zhang S et al (2021) Hypoxia promotes the metastasis of pancreatic cancer through regulating NOX4/KDM5A-mediated histone methylation modification changes in a HIF1A-independent manner. Clin Epigenetics 13:18

Li J, Meng H, Bai Y, Wang K (2016) Regulation of lncRNA and its role in cancer metastasis. Oncol Res 23:205–217

Li Y, Seto E (2016) HDACs and HDAC inhibitors in cancer development and therapy. Cold Spring Harb Perspect Med 6

Li YC, Wang Y, Li DD, Zhang Y, Zhao TC, Li CF (2018) Procaine is a specific DNA methylation inhibitor with anti-tumor effect for human gastric cancer. J Cell Biochem 119:2440–2449

Li Z, Xiao S, Yang Y, Chen C, Lu T, Chen Z, Jiang H, Chen S, Luo C, Zhou B (2020) Discovery of 8-methyl-pyrrolo[1,2-a]pyrazin-1(2H)-one derivatives as highly potent and selective Bromodomain and extra-terminal (BET) Bromodomain inhibitors. J Med Chem 63:3956–3975

Liang DL, Yu YF, Ma ZH (2020a) Novel strategies targeting bromodomain-containing protein 4 (BRD4) for cancer drug discovery. Eur J Med Chem 200:112426

Liang G, Ling Y, Mehrpour M, Saw PE, Liu Z, Tan W, Tian Z, Zhong W, Lin W, Luo Q et al (2020b) Autophagy-associated circRNA circCDYL augments autophagy and promotes breast cancer progression. Mol Cancer 19:65

Liu A, Wu Q, Peng D, Ares I, Anadon A, Lopez-Torres B, Martinez-Larranaga MR, Wang X, Martinez MA (2020) A novel strategy for the diagnosis, prognosis, treatment, and chemoresistance of hepatocellular carcinoma: DNA methylation. Med Res Rev 40:1973–2018

Liu L, Wang J, Sun G, Wu Q, Ma J, Zhang X, Huang N, Bian Z, Gu S, Xu M et al (2019) M(6)a mRNA methylation regulates CTNNB1 to promote the proliferation of hepatoblastoma. Mol Cancer 18:188

Lopez-Camarillo C, Gallardo-Rincon D, Alvarez-Sanchez ME, Marchat LA (2019) Pharmaco-epigenomics: on the road of translation medicine. Adv Exp Med Biol 1168:31–42

Ma S, Chen C, Ji X, Liu J, Zhou Q, Wang G, Yuan W, Kan Q, Sun Z (2019) The interplay between m6A RNA methylation and noncoding RNA in cancer. J Hematol Oncol 12:121

Maes T, Mascaro C, Tirapu I, Estiarte A, Ciceri F, Lunardi S, Guibourt N, Perdones A, Lufino MMP, Somervaille TCP et al (2018) ORY-1001, a potent and selective covalent KDM1A inhibitor, for the treatment of acute leukemia. Cancer Cell 33:495-+

Maleki Dana P, Mansournia MA, Mirhashemi SM (2020) PIWI-interacting RNAs: new biomarkers for diagnosis and treatment of breast cancer. Cell Biosci 10:44

Manara MC, Valente S, Cristalli C, Nicoletti G, Landuzzi L, Zwergel C, Mazzone R, Stazi G, Arimondo PB, Pasello M et al (2018) A Quinoline-based DNA methyltransferase inhibitor as a possible adjuvant in osteosarcoma therapy. Mol Cancer Ther 17:1881–1892

Mansouri A, Hachem LD, Mansouri S, Nassiri F, Laperriere NJ, Xia D, Lindeman NI, Wen PY, Chakravarti A, Mehta MP et al (2019) MGMT promoter methylation status testing to guide therapy for glioblastoma: refining the approach based on emerging evidence and current challenges. Neuro-Oncology 21:167–178

Manzotti G, Ciarrocchi A, Sancisi V (2019) Inhibition of BET proteins and histone deacetylase (HDACs): crossing roads in cancer therapy. Cancers 11

Mehdipour P, Murphy T, De Carvalho DD (2020) The role of DNA-demethylating agents in cancer therapy. Pharmacol Ther 205:107416

Michalak EM, Burr ML, Bannister AJ, Dawson MA (2019) The roles of DNA, RNA and histone methylation in ageing and cancer. Nat Rev Mol Cell Biol 20:573–589

Miranda Furtado CL, Dos Santos Luciano MC, Silva Santos RD, Furtado GP, Moraes MO, Pessoa C (2019) Epidrugs: targeting epigenetic marks in cancer treatment. Epigenetics 14:1164–1176

Moran S, Martínez-Cardús A, Sayols S, Musulén E, Balañá C, Estival-Gonzalez A, Moutinho C, Heyn H, Diaz-Lagares A, de Moura MC et al (2016) Epigenetic profiling to classify cancer of unknown primary: a multicentre, retrospective analysis. Lancet Oncol 17:1386–1395

Moridikia A, Mirzaei H, Sahebkar A, Salimian J (2018) MicroRNAs: potential candidates for diagnosis and treatment of colorectal cancer. J Cell Physiol 233:901–913

Müller D, Győrffy B (2022) DNA methylation-based diagnostic, prognostic, and predictive biomarkers in colorectal cancer. Biochim Biophys Acta Rev Cancer 1877:188722

Nencetti S, Cuffaro D, Nuti E, Ciccone L, Rossello A, Fabbi M, Ballante F, Ortore G, Carbotti G, Campelli F et al (2021) Identification of histone deacetylase inhibitors with (arylidene)aminoxy scaffold active in uveal melanoma cell lines. J Enzyme Inhib Med Chem 36:34–47

Nishiyama A, Nakanishi M (2021) Navigating the DNA methylation landscape of cancer. Trends Genet 37:1012–1027

Noguera-Uclés JF, Boyero L, Salinas A, Cordero Varela JA, Benedetti JC, Bernabé-Caro R, Sánchez-Gastaldo A, Alonso M, Paz-Ares L, Molina-Pinelo S (2020) The roles of imprinted SLC22A18 and SLC22A18AS gene overexpression caused by promoter CpG Island Hypomethylation AS diagnostic and prognostic biomarkers for non-small cell lung cancer patients. Cancers 12

Ohguchi H, Hideshima T, Anderson KC (2018) The biological significance of histone modifiers in multiple myeloma: clinical applications. Blood Cancer J 8:83

Okano M, Bell DW, Haber DA, Li E (1999) DNA methyltransferases Dnmt3a and Dnmt3b are essential for de novo methylation and mammalian development. Cell 99:247–257

Ooi SK, Sato S, Tomomori-Sato C, Zhang Y, Wen Z, Banks CAS, Washburn MP, Unruh JR, Florens L, Conaway RC, Conaway JW (2021) Multiple roles for PARP1 in ALC1-dependent nucleosome remodeling. Proc Natl Acad Sci U S A 118

Ouyang L, Zhang L, Liu J, Fu LL, Yao DH, Zhao YQ, Zhang SY, Wang G, He G, Liu B (2017) Discovery of a small-molecule Bromodomain-containing protein 4 (BRD4) inhibitor that induces AMP-activated protein kinase-modulated autophagy-associated cell death in breast cancer. J Med Chem 60:9990–10012

Pechalrieu D, Dauzonne D, Arimondo PB, Lopez M (2020) Synthesis of novel 3-halo-3-nitroflavanones and their activities as DNA methyltransferase inhibitors in cancer cells. Eur J Med Chem 186:111829

Pfeifer GP (2018) Defining driver DNA methylation changes in human cancer. Int J Mol Sci 19: 1166

Piha-Paul SA, Sachdev JC, Barve M, LoRusso P, Szmulewitz R, Patel SP, Lara PN, Chen XT, Hu BB, Freise KJ et al (2019) First-in-human study of Mivebresib (ABBV-075), an Oral Pan-inhibitor of Bromodomain and extra terminal proteins, in patients with relapsed/refractory solid tumors. Clin Cancer Res 25:6309–6319

Porter NJ, Mahendran A, Breslow R, Christianson DW (2017) Unusual zinc-binding mode of HDAC6-selective hydroxamate inhibitors. Proc Natl Acad Sci 114:13459–13464

Rice K, Hormaeche I, Licht J (2007) Epigenetic regulation of normal and malignant hematopoiesis. Oncogene 26:6697–6714

Riedel SS, Neff T, Bernt KM (2015) Histone profiles in cancer. Pharmacol Ther 154:87–109

Schapira M, Arrowsmith CH (2016) Methyltransferase inhibitors for modulation of the epigenome and beyond. Curr Opin Chem Biol 33:81–87

Shen R, Cheng T, Xu C, Yung RC, Bao J, Li X, Yu H, Lu S, Xu H, Wu H et al (2020) Novel visualized quantitative epigenetic imprinted gene biomarkers diagnose the malignancy of ten cancer types. Clin Epigenetics 12:71

Sher G, Salman NA, Khan AQ, Prabhu KS, Raza A, Kulinski M, Dermime S, Haris M, Junejo K, Uddin S (2022) Epigenetic and breast cancer therapy: promising diagnostic and therapeutic applications. Semin Cancer Biol 83:152–165

Shi X, Kaller M, Rokavec M, Kirchner T, Horst D, Hermeking H (2020a) Characterization of a p53/miR-34a/CSF1R/STAT3 feedback loop in colorectal cancer. Cell Mol Gastroenterol Hepatol 10:391–418

Shi Y, Zheng C, Jin Y, Bao B, Wang D, Hou K, Feng J, Tang S, Qu X, Liu Y et al (2020b) Reduced expression of METTL3 promotes metastasis of triple-negative breast cancer by m6A methylation-mediated COL3A1 up-regulation. Front Oncol 10:1126

Shouksmith AE, Shah F, Grimard ML, Gawel JM, Raouf YS, Geletu M, Berger-Becvar A, de Araujo ED, Luchman HA, Heaton WL et al (2019) Identification and characterization of AES-135, a Hydroxamic acid-based HDAC inhibitor that prolongs survival in an Orthotopic mouse model of pancreatic cancer. J Med Chem 62:2651–2665

Sixto-López Y, Bello M, Correa-Basurto J (2020) Exploring the inhibitory activity of valproic acid against the HDAC family using an MMGBSA approach. J Comput Aided Mol Des 34:857–878

Skvortsova K, Iovino N, Bogdanović O (2018) Functions and mechanisms of epigenetic inheritance in animals. Nat Rev Mol Cell Biol 19:774–790

Song J, Ouyang Y, Che J, Li X, Zhao Y, Yang K, Zhao X, Chen Y, Fan C, Yuan W (2017) Potential value of miR-221/222 as diagnostic, prognostic, and therapeutic biomarkers for diseases. Front Immunol 8:56

Stresemann C, Lyko F (2008) Modes of action of the DNA methyltransferase inhibitors azacytidine and decitabine. Int J Cancer 123:8–13

Sulewska A, Niklinski J, Charkiewicz R, Karabowicz P, Biecek P, Baniecki H, Kowalczuk O, Kozlowski M, Modzelewska P, Majewski P et al (2022) A signature of 14 long non-coding RNAs (lncRNAs) as a step towards precision diagnosis for NSCLC. Cancers 14

Sun JT, Li GF, Liu YW, Ma MY, Song KF, Li HX, Zhu DX, Tang XJ, Kong JY, Yuan X (2020) Targeting histone deacetylase SIRT1 selectively eradicates EGFR TKI-resistant cancer stem cells via regulation of mitochondrial oxidative phosphorylation in lung adenocarcinoma. Neoplasia 22:33–46

Sun L, Fang Y, Wang X, Han Y, Du F, Li C, Hu H, Liu H, Liu Q, Wang J et al (2019a) miR-302a inhibits metastasis and Cetuximab resistance in colorectal cancer by targeting NFIB and CD44. Theranostics 9:8409–8425

Sun T, Wu R, Ming L (2019b) The role of m6A RNA methylation in cancer. Biomed Pharmacother 112:108613

Sun X, Yi J, Yang J, Han Y, Qian X, Liu Y, Li J, Lu B, Zhang J, Pan X et al (2021) An integrated epigenomic-transcriptomic landscape of lung cancer reveals novel methylation driver genes of diagnostic and therapeutic relevance. Theranostics 11:5346–5364

Taby R, Issa JPJ (2010) Cancer epigenetics. CA Cancer J Clin 60:376–392

Tapadar S, Fathi S, Wu B, Sun CQ, Raji I, Moore SG, Arnold RS, Gaul DA, Petros JA, Oyelere AK (2020) Liver-targeting class I selective histone deacetylase inhibitors potently suppress hepatocellular tumor growth as standalone agents. Cancers (Basel) 12

Tricarico R, Nicolas E, Hall MJ, Golemis EA (2020) X- and Y-linked chromatin-modifying genes as regulators of sex-specific cancer incidence and prognosis. Clin Cancer Res 26:5567–5578

Trisciuoglio D, Di Martile M, Del Bufalo D (2018) Emerging role of histone acetyltransferase in stem cells and cancer. Stem Cells Int 2018:8908751

Vandenhoeck J, van Meerbeeck JP, Fransen E, Raskin J, Van Camp G, Op de Beeck K, Lamote K (2021) DNA methylation as a diagnostic biomarker for malignant mesothelioma: a systematic review and meta-analysis. J Thorac Oncol 16:1461–1478

Wagner T, Brand P, Heinzel T, Krämer OH (2014) Histone deacetylase 2 controls p53 and is a critical factor in tumorigenesis. Biochim Biophys Acta *1846*:524–538

Wang J, Wu M, Zheng D, Zhang H, Lv Y, Zhang L, Tan HS, Zhou H, Lao YZ, Xu HX (2020a) Garcinol inhibits esophageal cancer metastasis by suppressing the p300 and TGF-beta1 signaling pathways. Acta Pharmacol Sin 41:82–92

Wang T, Kong S, Tao M, Ju S (2020b) The potential role of RNA N6-methyladenosine in cancer progression. Mol Cancer 19:88

Wang Y, Lu JH, Wu QN, Jin Y, Wang DS, Chen YX, Liu J, Luo XJ, Meng Q, Pu HY et al (2019) LncRNA LINRIS stabilizes IGF2BP2 and promotes the aerobic glycolysis in colorectal cancer. Mol Cancer 18:174

Wang Y, Xie Q, Tan H, Liao M, Zhu S, Zheng LL, Huang H, Liu B (2021) Targeting cancer epigenetic pathways with small-molecule compounds: therapeutic efficacy and combination therapies. Pharmacol Res 173:105702

Wang YM, Gu ML, Meng FS, Jiao WR, Zhou XX, Yao HP, Ji F (2017) Histone acetyltransferase p300/CBP inhibitor C646 blocks the survival and invasion pathways of gastric cancer cell lines. Int J Oncol 51:1860–1868

Wang YW, Ma X, Zhang YA, Wang MJ, Yatabe Y, Lam S, Girard L, Chen JY, Gazdar AF (2016) ITPKA gene body methylation regulates gene expression and serves as an early diagnostic marker in lung and other cancers. J Thorac Oncol 11:1469–1481

Watson ZL, Yamamoto TM, McMellen A, Kim H, Hughes CJ, Wheeler LJ, Post MD, Behbakht K, Bitler BG (2019) Histone methyltransferases EHMT1 and EHMT2 (GLP/G9A) maintain PARP inhibitor resistance in high-grade serous ovarian carcinoma. Clin Epigenetics 11:165

Wong CC, Li W, Chan B, Yu J (2019) Epigenomic biomarkers for prognostication and diagnosis of gastrointestinal cancers. Semin Cancer Biol 55:90–105

Xi Y, Lin Y, Guo W, Wang X, Zhao H, Miao C, Liu W, Liu Y, Liu T, Luo Y et al (2022) Multiomic characterization of genome-wide abnormal DNA methylation reveals diagnostic and prognostic markers for esophageal squamous-cell carcinoma. Signal Transduct Target Ther 7:53

Xie R, Yao Y, Tang P, Chen G, Liu X, Yun F, Cheng C, Wu X, Yuan Q (2017) Design, synthesis and biological evaluation of novel hydroxamates and 2-aminobenzamides as potent histone deacetylase inhibitors and antitumor agents. Eur J Med Chem 134:1–12

Xiu B, Chi Y, Liu L, Chi W, Zhang Q, Chen J, Guo R, Si J, Li L, Xue J et al (2019) LINC02273 drives breast cancer metastasis by epigenetically increasing AGR2 transcription. Mol Cancer 18:187

Xu W, Xu M, Wang L, Zhou W, Xiang R, Shi Y, Zhang Y, Piao Y (2019) Integrative analysis of DNA methylation and gene expression identified cervical cancer-specific diagnostic biomarkers. Signal Transduct Target Ther 4:55

Yang B, Wang JQ, Tan Y, Yuan R, Chen ZS, Zou C (2021) RNA methylation and cancer treatment. Pharmacol Res 174:105937

Yang GJ, Ko CN, Zhong HJ, Leung CH, Ma DL (2019) Structure-based discovery of a selective KDM5A inhibitor that exhibits anti-cancer activity via inducing cell cycle arrest and senescence in breast cancer cell lines. Cancers 11

Yao B, Gui T, Zeng X, Deng Y, Wang Z, Wang Y, Yang D, Li Q, Xu P, Hu R et al (2021) PRMT1-mediated H4R3me2a recruits SMARCA4 to promote colorectal cancer progression by enhancing EGFR signaling. Genome Med 13:58

Yao ZQ, Zhang X, Zhen YQ, He XY, Zhao SM, Li XF, Yang B, Gao F, Guo FY, Fu LL et al (2018) A novel small-molecule activator of Sirtuin-1 induces autophagic cell death/mitophagy as a potential therapeutic strategy in glioblastoma. Cell Death Dis 9:767

Yeong KY, Khaw KY, Takahashi Y, Itoh Y, Murugaiyah V, Suzuki T (2020) Discovery of gamma-mangostin from Garcinia mangostana as a potent and selective natural SIRT2 inhibitor. Bioorg Chem 94:103403

Yi X, Jiang X, Li X, Jiang DS (2017) Histone lysine methylation and congenital heart disease: from bench to bedside (review). Int J Mol Med 40:953–964

Yoon S, Eom GH (2016) HDAC and HDAC inhibitor: from cancer to cardiovascular diseases. Chonnam Med J 52:1–11

Yu L, Xu J, Liu J, Zhang H, Sun C, Wang Q, Shi C, Zhou X, Hua D, Luo W et al (2019) The novel chromatin architectural regulator SND1 promotes glioma proliferation and invasion and predicts the prognosis of patients. Neuro-Oncology 21:742–754

Zhang G, Pradhan S (2014) Mammalian epigenetic mechanisms. IUBMB life 66:240–256

Zhang W, Qu J, Liu GH, Belmonte JCI (2020a) The ageing epigenome and its rejuvenation. Nat Rev Mol Cell Biol 21:137–150

Zhang Y, Huang YX, Wang DL, Yang B, Yan HY, Lin LH, Li Y, Chen J, Xie LM, Huang YS et al (2020b) LncRNA DSCAM-AS1 interacts with YBX1 to promote cancer progression by forming a positive feedback loop that activates FOXA1 transcription network. Theranostics 10:10823–10837

Zhou J, Cheng T, Li X, Hu J, Li E, Ding M, Shen R, Pineda JP, Li C, Lu S et al (2021) Epigenetic imprinting alterations as effective diagnostic biomarkers for early-stage lung cancer and small pulmonary nodules. Clin Epigenetics 13:220

Zhu D, Zhang Y, Wang S (2021) Histone citrullination: a new target for tumors. Mol Cancer 20:90

Zwergel C, Schnekenburger M, Sarno F, Battistelli C, Manara MC, Stazi G, Mazzone R, Fioravanti R, Gros C, Ausseil F et al (2019) Identification of a novel quinoline-based DNA demethylating compound highly potent in cancer cells. Clin Epigenetics 11:68

Chapter 5
Clinical Studies and Epi-Drugs in Various Cancer Types

Taha Bahsi, Ezgi Cevik, Zeynep Ozdemir, and Haktan Bagis Erdem

Abstract Cancer is characterized by modifications in the epigenetic mechanisms and chromatin functions. Alterations in gene expression, as well as the development and progression of cancer, have been related to the disruption of epigenetic processes. It is worth noting that epigenetic changes are not distinct or mutually exclusive events but rather interact with one another to create subsequent changes. Recent efforts to sequence the cancer genome have shown that many epigenetic regulators are frequently altered in a variety of malignancies. Since early diagnosis has a significant role in the effective management of cancer, the development of novel epigenetic biomarkers is extremely promising in this field. Moreover, epigenomics are not only being used as diagnostic signatures but also prognostic and predictive indicators to direct personalized treatment options. This information has also become more relevant for understanding the molecular mechanisms of epigenetic control in both healthy and pathological circumstances. Additionally, the reversible nature of epigenetic aberrations has sparked the development of the promising field of epigenetic treatment, which has already given patients with malignancies characterized by epigenetic modifications new therapeutic options. Some of these therapeutics, focusing on the epigenome, have already been approved by the FDA. From a clinical standpoint, we covered epigenetics in numerous common and relevant tumors in this chapter.

Keywords Epigenetics · Cancer · Early diagnosis · Prognosis prediction · Targeted treatment

Abbreviations

TCGA	The Cancer Genome Atlas
HEP	The Human Epigenome Project

T. Bahsi (✉) · E. Cevik · Z. Ozdemir · H. B. Erdem
Department of Medical Genetics, Ankara Etlik City Hospital, Ankara, Turkey
e-mail: taha.bahsi@saglik.gov.tr

SNPs	Single nucleotide polymorphisms
NSCLC	Non-small cell lung cancer
MSP	Methylation-specific PCR
qMSP	Quantitative methylation-specific PCR
CMI	Cumulative methylation index
KMTs	Lysine methyltransferases
KDMs	Lysine demethylases
AR	Androgen receptor
NEPC	Neuroendocrine prostate cancer
lncRNAs	Long noncoding RNAs
BL	Burkitt lymphoma
DLBCL	Diffuse large B-cell lymphoma
SS	Sezary syndrome
SMAD1	Small mothers against decapentaplegic homolog1
TET	Ten-eleven translocation
TME	Tumor microenvironment

5.1 Introduction

Over 100 separate and distinct diseases fall under the umbrella term "cancer," in which abnormal cells divide out of control and have the potential to infiltrate surrounding tissues. Because cancer is the second leading cause of mortality worldwide, significant efforts have been made to understand and treat this disease in recent years (Siegel et al. 2022). Studies on gene expression and DNA methylation provided early evidence of an epigenetic connection to cancer. The validation of the concept that cancer is not only caused by genetic changes but also epigenetic alterations has sped up during the last decade. Whole-genome sequencing data demonstrate that several epigenetic regulators are the target of mutations and epimutations in cancer cells, with an intriguing interplay between the two. Almost all aspects of epigenetic regulation are impacted by the prevalence of cancer mutations involving these genes, including critical actors in DNA methylation, histone modifications, and chromatin architecture. The discovery of mutations in writers, readers, and erasers, as well as changes in the epigenetic landscape in malignancies, not only suggests a causative role for these elements in the development and evolution of cancer but also offers prospective therapeutic targets. The reversible nature of epigenetic aberrations has given rise to the exciting field of epigenetic therapy, which has already introduced novel therapeutic options for patients with epigenetic malignancies, laying the groundwork for new and individualized medicine.

The first epigenetic abnormality found in human malignancies was aberrant DNA methylation (Feinberg and Vogelstein 1983). Together with histone modifications and other chromatin-related proteins, DNA methylation offers a stable gene silencing mechanism that plays a crucial role in controlling gene expression and chromatin architecture. Cancer cells may exhibit abnormalities such as promoter hypermethylation, which silences tumor suppressor genes; global hypomethylation, which has been linked to genomic instability; and changes in DNA methylation at imprinting regulatory regions, which results in the loss of imprinting (Dawson and Kouzarides 2012). The so-called shores, which are low-density methylation regions close to CpG islands, show significant diversity in DNA methylation across various cancer types, including hypomethylation and hypermethylation (Doi et al. 2009). De novo methylation occurs in several genes related to CpG islands in cancer. Multiple CpG islands are frequently methylated, which is known as the "CpG island methylator phenotype (CIMP)" and have been observed in a variety of malignancies. The CIMP status relates to distinct clinicopathological characteristics in individual cancer types, and it can provide information for cancer diagnosis and patient stratification for various therapeutic regimens (Issa 2004). One of the first found alterations of the epigenome in human cancer, the loss of 5-methylcytosine (5mC), occurs mostly in repetitive DNA sequences, coding regions, and introns. It has been proposed that the substantial DNA demethylation seen during tumor growth is the cause of the ongoing cellular variety seen in malignancies. The degree of hypomethylation of genomic DNA increases during the progression of a neoplasm from hyperplasia to invasive and potentially metastatic cancer (Feinberg and Vogelstein 1983). Combinations of different chemical modifications on nucleosomal histones can also influence transcriptional repression or activation, referring to the "histone code" (Jenuwein and Allis 2001). Histone changes like H3K4/K20/K36/K79 methylation, H3/H4 lysine acetylation, asymmetric H3R17me2, and H4R3me2 are examples of those connected to active transcription. Specialized "reader" proteins frequently detect these alterations and then enlist additional components to produce an open chromatin structure (Black et al. 2012). Other histone changes, such as H3K27 methylation, H3/H4 deacetylation, asymmetric H3R2me2, and symmetric H4R3me2, inhibit transcription (Otani et al. 2009). Along with histone modifications, ATP-dependent chromatin remodeling mechanisms play important roles in transcription by facilitating access of DNA-binding proteins and the transcriptional machinery to DNA to promote expression. In eukaryotes, four families of chromatin remodeling complexes have been identified: the switching defective/sucrose non-fermenting (SWI/SNF) family, the imitation-switch (ISWI) family, the nucleosome remodeling and histone deacetylase complex (NuRD), and the inositol 80 (INO80) family (Audia and Campbell 2016). Finally, noncoding RNAs (ncRNAs) mediating cellular activities, such as signal transduction, chromatin remodeling, transcription, and post-transcriptional alterations, have been found to act as both tumor suppressors and oncogenic drivers in many cancer types. Designing more effective therapies may be made possible by gaining a deeper comprehension of the complicated network interactions that ncRNAs control (Anastasiadou et al. 2018).

The Cancer Genome Atlas (TCGA), a breakthrough cancer genomics program, molecularly described approximately 20,000 primary cancer and matched normal samples from 33 cancer types. This collaborative project between the National

Cancer Institute and the National Human Genome Research Institute started in 2006 and involved researchers from various institutes and disciplinary backgrounds. More than 2,5 petabytes of genomic, epigenomic, transcriptomic, and proteomic data were produced by TCGA over the following 12 years (https://portal.gdc.cancer.gov/). The information will continue to be accessible to the research community for use by everyone and has already improved our capacity to identify, treat, and prevent cancer. On the other hand, The Human Epigenome Project (HEP) intends to detect, document, and analyze genome-wide DNA methylation patterns of all human genes in all major tissues. Depending on the tissue type and illness state, differently methylated cytosines produce distinctive patterns. These MVPs (methylation variable positions) are typical epigenetic biomarkers, promising to improve our capacity to comprehend and diagnose human disease, much like single nucleotide polymorphisms (SNPs). The Human Epigenome Project (HEP) is a public/private collaboration led by Human Epigenome Consortium members. MVPs recognized as part of the HEP will be public under the HEP data release policy (https://www.epigenome.org/).

As it is known, early cancer diagnosis could detect malignancies when outcomes are better, and treatment is less morbid. Effective screening paradigms are limited to a restricted number of neoplasms and are type specific. In this regard, there are few research examining the simultaneous detection and localization of several cancer types, using cfDNA or other analytes. The Circulating Cell-free Genome Atlas (CCGA; NCT02889978) study sought to determine whether genome-wide cfDNA sequencing combined with machine learning could detect and localize a variety of cancer types with sufficient accuracy to be considered for a general population-based cancer screening program. Using informative methylation patterns and cfDNA sequencing, over 50 different cancer types and stages were identified in this study. Further analysis of this test is warranted in prospective population-level investigations given the potential benefit of early diagnosis in fatal cancers (Liu et al. 2020b).

In this chapter, we outline the most well-studied chromatin and epigenetic abnormalities discovered in common malignancies, their mechanisms of occurrence, and how they influence the development and spread of cancer. We also give instances of how epigenetic biomarkers could be used as early diagnostic tools, as well as how epigenetic inhibitors are created and describe how to use them therapeutically.

5.2 Epigenetic Studies Regarding the Most Common Solid Tumor Types

5.2.1 Lung Cancer

Lung cancer is the leading cause of cancer-related death globally (Siegel et al. 2022). The basic hypothesis explaining the pathophysiology of tumorigenesis in lung cancer is based on the sequential occurrence of genetic and epigenetic changes (Lantuejoul et al. 2009). Because epigenetic disruption plays such a crucial role in this process, epigenomic biomarkers found in tissue or body fluids may be extremely helpful in enhancing the effectiveness of screening or diagnostic techniques or providing alternative approaches (Duruisseaux and Esteller 2018). A key element in the development of lung cancer has been identified as abnormal DNA methylation. For example, promoter hypermethylation frequently occurs in non-small cell lung cancer (NSCLC) and inhibits the production of tumor suppressors. To predict prognosis and responsiveness to conventional therapy, DNA methylation-based biomarkers have thus been thoroughly researched.

Existing commercial in vitro diagnostic tests are based on either DNA methylation of specific genes or microRNA measurement. The Epi proLung (Epigenomics AG, Berlin, Germany) is a commercially available IVD test that uses DNA methylation biomarkers to diagnose lung cancer. Using bisulfite converted free-circulating DNA from the blood and a PCR assay, Epi proLung assesses *SHOX2* and *PTGER4* gene methylation. The ability of the test to distinguish between benign diseases and lung cancer has been well established (https://www.epigenomics.com).

The *CDKN2A* gene, which is hypermethylated in NSCLC, is one of the most studied genes in terms of methylation status. Numerous studies have investigated the possibility of using CDKN2A hypermethylation as a biomarker to predict lung cancer. The findings suggest that methylation of this gene could be used as a biomarker for NSCLC diagnosis, but it was not suitable for screening. Researchers discovered that while bronchoalveolar fluid/sputum samples had higher sensitivity, serum samples had higher specificity (Tuo et al. 2018). In comparison to other research, the study by Bing et al. demonstrated that the combined detection of *SHOX2*, *RASSF1A*, and *PTGER4* gene methylation in plasma increased detection sensitivity in lung cancer diagnosis and attained high sensitivity across all histological subtypes of lung cancer. The overall diagnosis of lung cancer may be improved by using this new noninvasive test of three biomarkers, which has potential clinical utility. It may be used alone or in conjunction with existing imaging detection techniques (Wei et al. 2021). In the study by Liu et al., methylation in the *CDO1, TAC1, HOXA7, HOXA9, SOX17,* and *ZFP42* genes was examined in preoperative plasma and urine samples of individuals with suspicious nodules on CT imaging. The findings of this study demonstrate that all six genes in plasma and *CDO1, TAC1, HOXA9,* and *SOX17* in urine can be found to be methylated more frequently in cancer patients compared to controls (Liu et al. 2020a). A 2019 study reported an ultrasensitive high-throughput targeted DNA methylation sequencing method for ctDNA identification. With this model, tumor-specific ctDNA from the plasma of patients with pulmonary nodules was identified, and satisfactory susceptibility and specificity to early-stage lung cancer were demonstrated. This approach, complemented by CT scanning, holds great promise for a revolutionary screening or diagnostic test to identify lung cancer in a noninvasive way at its early and treatable stage (Liang et al. 2019).

Epigenetic alterations are excellent prospective therapeutic targets for anticancer therapies due to their adaptability and reversible nature. The research of epigenetic

Table 5.1 Ongoing clinical trials targeting epigenetic regulators in non-small cell lung cancer (NSCLC) (as of September 17, 2022)

NCT number	Drugs	Epigenetic target
NCT01928576	Azacitidine + entinostat + nivolumab	DNMT, HDAC
NCT03233724	Decitabine + tetrahydrouridine + pembrolizumab	DNMT, HDAC
NCT03220477	Guadecitabine + mocetinostat + pembrolizumab	DNMT, HDAC
NCT02664181	Nivolumab + decitabine + tetrahydrouridine	DNMT
NCT02250326	Nab-paclitaxel + azacitidine	DNMT
NCT02546986	Azacitidine + pembrolizumab	DNMT
NCT02437136	Entinostat + pembro	HDAC
NCT02638090	Pembrolizumab + vorinostat	HDAC

therapeutics is becoming increasingly important as a potential therapeutic approach in lung cancer, considering the significant role of epigenetic mechanisms in the biology of lung cancer and the degree of treatment resistance. In research by Szejniuk et al. for neoadjuvant chemotherapy and monitoring response, *RASSF1A/RARB2* methylation demonstrated promising predictive power (Szejniuk et al. 2019). Also, a 2010 study showed that loss of *IGFBP-3* expression mediated by promoter-hypermethylation results in a reduction of tumor cell sensitivity to cisplatin in NSCLC (Ibanez de Caceres et al. 2010). In terms of treatment, patients with recurrent metastatic NSCLC were given azacitidine (DNMTi) and entinostat (HDAC inhibitor or HDACi) in a phase I/II research, and the results showed improved clinical outcomes. The scientists discovered a biomarker associated with better outcomes by observing a decrease in methylation levels in genes (such as *APC, RASSF1A, CDH13,* and *CDKN2A*) evaluated from serial blood samples of patients with poor prognosis. In addition, four out of 19 patients responded better to more advanced anticancer therapies, indicating that treatment with these two epigenetic medicines helped these patients maintain their beneficial effects over time (Juergens et al. 2011). Additionally, the use of DNMTi resensitizes cancer cells when they become resistant to EGFR-TKIs in addition to chemoresistance. One putative acquired resistance mechanism to gefitinib is methylation of the EGFR promoter, and gefitinib plus azacitidine treatment together inhibited cancer cell proliferation and caused apoptosis (Li et al. 2013). Furthermore, because immune checkpoint blockade (ICB) therapy is effective in treating non-small cell lung cancer (NSCLC), the number of patients receiving it has rapidly expanded over the past few years (Doroshow et al. 2019). Numerous genetic and transcriptome indicators have been developed to identify patients who are most likely to respond to this medication. These biomarkers, however, are insufficient to precisely predict the response to ICB therapy. In addition to genomic and transcriptomic indicators, epigenetic alterations have also been linked to ICB therapeutic responsiveness (Loo Yau et al. 2019). In the study of Kim et al., it was demonstrated that methylation patterns can provide insight into molecular determinants underlying the clinical benefit of ICB therapy (Kim et al. 2020). Certain therapeutic trials are studying DNMTi and ICB combinations, like azacitidine with pembrolizumab and decitabine with nivolumab (Table 5.1).

Intratumoral heterogeneity is significantly influenced by numerous aspects of the immune microenvironment and lung tumor that affect epigenetic changes in cancer cells. The epigenome is a promising therapeutic target for the treatment of cancer and immunological compartments. To achieve the clinical benefit, more investigation of the precise epigenetic alterations that are both inherent and acquired throughout different treatments is needed. Despite being a fascinating area, the epigenome is incredibly complex, and effective lung cancer treatment will require precise, multidimensional reprogramming (Chao and Pecot 2021).

5.2.2 Breast Cancer

Among women, breast cancer is the most commonly diagnosed cancer with an estimated 2.3 million new cases worldwide as well as the major cause of cancer mortality (Sung et al. 2021). Despite the apparent increase in breast cancer incidence, the mortality rate from 2009 to 2018 reduced by about 27%. This reduction was owing to aggressive screening, early diagnosis, and novel therapeutic advances (Ahmad 2019). The term epigenetics, which has been well-known in carcinogenesis, can be a promising component of the mortality rate reduction steps for breast cancer.

Early diagnosis of breast cancer is associated with higher survival rates. Numerous studies have investigated the DNA methylation status particularly to identify tests for early diagnosis, primarily in blood-based samples (Brown et al. 2022). Moreover, an article reported that both CpG hypermethylation and hypomethylation may be significant events in breast carcinogenesis (de Almeida et al. 2019). In another study, the predictive utility of a novel risk score constructed using blood DNA methylation array data, methylation-based breast cancer risk score (mBCRS) was evaluated. The accuracy of breast cancer prediction was significantly improved by the inclusion of the mBCRS in existing genetic and questionnaire-based information (Kresovich et al. 2022). However, due to the inter/intratumoral heterogeneity of breast cancer, the use of one methylated gene in ccfDNA has limited accuracy for cancer detection, since it might be specific for one subtype and presumably will not serve for another, which leads to misdiagnosis. Consequently, various studies have been carried out for developing several gene panels, applying different assays to improve the test sensitivity (Constancio et al. 2020; Sher et al. 2022). The sensitivity and specificity of these promising assays for detecting early breast cancer have been reported above 80%, comparable to mammography screening, and moreover was higher in stages II and III breast cancer compared with stage I (Brown et al. 2022). In a review that reports methylation biomarkers for breast cancer detection, a panel using methylation-specific PCR (MSP) including $DAPK1_{me}$ and $RASSF1A_{me}$ showed the highest sensitivity with 96%, as well as other studies attempted to assemble panels for detection using quantitative methylation-specific PCR (qMSP), achieving sensitivities above 80%. Additionally, some panels reached

100% specificity for detection of breast cancer (Constancio et al. 2020). Furthermore, another study mentioned in this review reported a six-gene panel with higher sensitivity than mammography for detecting breast cancer with tumor sizes under 1 cm (Shan et al. 2016). Currently, Therascreen® PITX2 RQG and IvyGene are the only DNA methylation-based assays that were developed and validated as diagnostic and prognostic CE-IVD in the EU and USA, respectively (Sher et al. 2022). Besides, a blood-based multi-cancer screening approach called the Galleri™ test may be able to detect cancer earlier and overcome the limitations of organ-specific screening tests. It can detect methylation patterns of cfDNA and identify the tumor's tissue of origin accurately. It has a reported sensitivity and specificity of >95% for stage III and IV breast cancer, yet <50% for stage I and II, emphasizing the limitations of using this test for early diagnosis of breast cancer (Brown et al. 2022). There have been several attempts to develop epigenetic risk classifiers for breast cancer to date, but they only had limited success, which could be for several reasons. First, most studies have only used blood samples for DNA-methylation analyses; however, breast cancer is an epithelial disease by definition, and thus immune cells in the blood might not be an appropriate tissue. Second, the timing of the sample collection for epigenetic analysis is critical. For instance, samples obtained during cancer treatment will display the effects of the treatment. Third, in contrast to polygenic risk scores, epigenetic risk signatures reflect cell programs; as a result, techniques that a priori choose a sizable number of CpGs for inclusion in the epigenetic signatures are more likely to be suitable. Fourth, the epigenome of a specific tissue can be modified by the presence of cancer. For example, in patients with ovarian cancer, a higher granulocyte to lymphocyte ratio is seen in the blood, which alters the DNA-methylation signature observed when peripheral blood mononuclear cells are assessed (Pashayan et al. 2020).

In addition to early diagnosis, several biomarkers can be used to predict the prognosis of breast cancer patients (Constancio et al. 2020). A study described a cumulative methylation index (CMI), which includes the methylation of 6 genes ($AKR1B1_{me}$, $HOXB4_{me}$, $RASGRF2_{me}$, $RASSF1_{me}$, $HIT1H3C_{me}$, and $TM6SF1_{me}$), and progression-free survival (PFS) and overall survival (OS) were considerably shorter in metastatic breast cancer patients with high levels of the CMI (Visvanathan et al. 2017). Another study investigating methylation patterns of several genes in ccfDNA showed that breast cancer patients with positive $SOX17_{me}$ and $WNT5A_{me}$ demonstrated shorter OS, although $KLK10_{me}$ correlated with more relapses and shorter disease-free interval (Panagopoulou et al. 2019). In another research, triple-negative breast cancer (TNBC) cases with *BRCA1* hypermethylated and *BRCA1* mutated were compared. It is shown that *BRCA1* promoter hypermethylation is twice as common as *BRCA1* pathogenic variants in early-stage TNBC and that hypermethylated and mutated cases have similar beneficial outcomes after adjuvant chemotherapy, indicating that BRCA1 hypermethylation could serve as a prognostic biomarker that is distinguishable even in tumor tissue samples with low-cellularity (Glodzik et al. 2020). Not only DNA methylation but also histone modifications could be used for prognosis prediction. In a study, it was found that inactivation of

HDAC7, directly or via inhibition of *HDAC1* and *HDAC3*, can lead to the inhibition of the cancer stem cell phenotype (Caslini et al. 2019).

Since DNA methylation, histone modifications, and chromatin remodeling are some of the key features that are frequently altered during the breast cancer process, the epigenome is also a potential therapeutic target (Brown et al. 2022). However, the implementation of epigenetic therapies for breast cancer in clinical practice has often been limited to hematological tumors, because epigenetic alterations are cell-type specific, and solid tumors are quite heterogeneous, making them difficult to target (Navada, 2021). To date, only two types of epigenetic drugs(epi-drugs) are US Food and Drug Administration (FDA)-approved for clinical use in hematological malignancies: DNMT inhibitors (DNMTi) and histone deacetylase small molecule inhibitors (HDACi). Both DNMTi and HDCAi have displayed antitumor effects in preclinical studies for breast cancer, yet the clinical benefits of these drugs are still being investigated (Pasculli et al. 2018). Besides, due to the high correlation between their expression and breast cancer progression, ncRNAs can be used as a promising biomarkers for early diagnosis and prognosis detection (Wang et al. 2022a). There are some ncRNA-based drugs in preclinical and clinical trials as well, yet none of them are approved for clinical utilization. The most promising approach for breast cancer treatment is the use of epigenetic medications to overcome de novo or acquired drug-resistance mechanisms (Garcia-Martinez et al. 2021). Therefore, the use of epi-drugs is likely to evolve to include combinations with cytotoxic chemotherapies, endocrine therapy, targeted medications, radiotherapy, and immunotherapy (Morel et al. 2020). The published clinical trials that have been conducted for investigating epigenetic therapies alone or combined with other systemic therapies for the management of breast cancer are summarized (Table 5.2). Nevertheless, none of these epigenetic therapies are currently FDA approved for the treatment of breast cancer.

DNMTi, azacitidine and decitabine, are widely researched in the field of cancer. They have shown antitumor activity in both preclinical ER-positive and TNBC breast cancer models (Brown et al. 2022). There is a clinical trial underway investigating the use of azacitidine alone (NCT04891068), but to date, there are no published trials in the use of DNMT inhibitors alone for the management of breast cancer. However, trials are being conducted to investigate the use of azacitidine and decitabine in combination with other systemic therapies (Clinicaltrials.gov identifiers NCT01349959, NCT05381038, and NCT02957968 accessed on September 17, 2022). Furthermore, it is remarkable that the hypermethylation of *BRCA1* is being investigated to predict the response to poly(ADP-ribose) polymerase (PARP) inhibitors in patients with breast and/or ovarian cancer due to the growing number of clinical trials based on it (Berdasco and Esteller 2019).

HDAC enzymes also have an important role in transcriptional regulation in ER-positive breast cancer. HDACi, such as entinostat, vorinostat (suberanilohydroxamic acid/SAHA), dacinostat, valproate, etc., have been demonstrated to induce antitumor activity in preclinical ER-positive and TNBC models. Similar to DNMTi, the activity of single-agent HDACi in breast cancer management has also failed to translate into clinical practice (Brown et al. 2022). On the contrary,

Table 5.2 Published trials of epigenetic therapies in breast cancer (*BC* breast cancer. *ORR* overall response rate. *OS* overall survival. *PFS* progression-free survival. Ref: reference. *NR* not reached. *NA* not applicable. *ER* estrogen receptor. *HR* hormone receptor. *HER2* human epidermal growth factor receptor 2)

Trial number	Epigenetic target	Treatment	Patient population	Phase	ORR (%)	OS (months)	PFS (months)	Ref
Epigenetic therapies alone								
NCT00132002	HDAC	Vorinostat	Advanced BC	2	0	NR	NR	(Luu et al. 2008)
NCT02391480	BET	Mivebresib	Solid organ tumors	1	0	NR	1.8	(Piha-Paul et al. 2019)
Epigenetic therapies + endocrine therapy								
NCT02833155	HDAC	Entinostat + exemestane	Advanced HR(+) BC	1	14.3	28.1	4.3	(Wang et al. 2021)
NCT00676663	HDAC	Entinostat +/- exemestane	Advanced ER (+) BC	2	NA	28.1 vs. 19.8 (EE vs. EP)	4.3 vs. 2.3 (EE vs. EP)	(Yardley et al. 2013)
NCT02115282	HDAC	Entinostat +/- exemestane	Advanced ER (+) BC	3	5.8 vs. 5.6 (EE vs. EP)	23.4 vs. 21.7 (EE vs. EP)	3.3 vs. 3.1 (EE vs. EP)	(Connolly et al. 2021)
NCT02623751	HDAC	KHK2375 + exemestane	Advanced HR(+), HER2(-)BC	1	0	NR	13.9 (median)	(Masuda et al. 2021)
NCT02482753	HDAC	Tucidinostat +/- exemestane	Advanced ER (+) BC	3	18 vs. 9 (ET vs. EP)	NR	7.4 vs. 3.8 (ET vs. EP)	(Jiang et al. 2019)
NCT00365599	HDAC	Vorinostat + tamoxifen	Advanced ER (+) BC	2	19	29	10.3	(Munster et al. 2011)
Epigenetic therapies + cytotoxic agents								
NCT00368875	HDAC	Vorinostat + paclitaxel + bevacizumab	Advanced BC	1/2	55	29.4	11.9	(Ramaswamy et al. 2012)

NCT00404508	HDAC	Magnesium valproate + hydralazine + chemotherapy	Solid organ tumors	2	24	NR	NR	(Candelaria et al. 2007)
NCT00331955	HDAC	Vorinostat + doxorubicin	Solid organ tumors	1	8	NR	NR	(Munster et al. 2009)
Epigenetic therapies + targeted therapy								
NCT02811497	DNMT	Azacitidine + durvalumab	ER(+)/HER2(-) BC and solid tumors	2	NR	5	1.9	(Taylor et al. 2020)
NCT00996515	DNMT	Azacitidine + erlotinib	Solid organ tumors	1	7	7.5	2	(Bauman et al. 2012)
NCT01087554	HDAC	Vorinostat + sirolimus	Solid organ tumors	1	3	10.3	2	(Park et al. 2016)
NCT00258349	HDAC	Vorinostat + trastuzumab	Advanced HER2(+) BC	1/2	7	9.3	1.5	(Goldstein et al. 2017)

several studies investigating the use of HDACi in combination with endocrine therapy demonstrated activity in drug-resistant ER-positive models and reported that HDACi possibly acts by re-sensitizing the cells to endocrine therapy (Sabnis et al. 2011; Thomas et al. 2011).

Bromo- and extra-terminal domain (BET) proteins are a subgroup of the bromodomain (BRD) family proteins that have been linked to transcriptional upregulation of multiple genes involved in cell cycle control, with important oncogenic potential. BET inhibitors are currently being researched for use in cancer therapy. There have been numerous preclinical studies evaluating BET inhibitors in TNBC, and they show promise in growth inhibition both in vitro and in vivo (Shu et al. 2016; da Motta et al. 2017). Limited efficacy has been shown in a phase 1 trial of the BET inhibitor mivebresib including in patients with breast cancer (Piha-Paul et al. 2019).

Epi-drugs combined with targeted therapies are also promising. Clinical trials ongoing for investigating epigenetic therapies alone or combined with other therapies in breast cancer are summarized (Table 5.3). Although CDK4/6 inhibitors are currently the gold standard treatment of metastatic ER-positive cancer in the first-line setting, investigations on epigenetics in the context of CDK4/6 inhibitors are lacking. As mentioned before, in the treatment of patients who have developed resistance to conventional endocrine therapy, HDACi have shown promise. These investigations, however, were conducted in populations that had never received CDK4/6 inhibitor therapy. There is no consensus on the best treatment for CDK4/6 inhibitor-resistant disease at the moment. Therefore, further research into the use of HDACi in CDK4/6 resistant populations is essential, in order to expand therapy options for this population and to evaluate the efficacy of HDACi in comparison to current therapeutic regimens (Brown et al. 2022). Moreover, it has been demonstrated that PARP inhibitors are beneficial for patients with *BRCA1* methylation (Berdasco and Esteller 2019). Particularly, preclinical models of TNBC have shown efficacy for PARP inhibitors combined with HDACi (Marijon et al. 2018). Olaparib has also been demonstrated to increase the sensitivity of BRCA wild-type TNBC to Olaparib when combined with BET inhibitors (Yang et al. 2017).

The epigenome can be a target for therapy as well as predictive of response to therapy. A study has reported that EZH2 inhibition may enhance antitumor immunity in ErbB2+ breast cancer, suggesting the role of targeting the polycomb repressor complex 2 (PRC2) combined with anti-ERBB2 agents as a strategy to improve responses and overcoming resistance in aggressive ERBB2+ disease (Hirukawa et al. 2019). One other study, researching the role of H3K4 methyltransferase *KMT2C* in ER+ breast cancer, showed that deletion or loss-of-function mutations are associated with a decreased response to aromatase inhibitors (Gala et al. 2018). A study, investigating the association between the transcription factor *SOX9* and tamoxifen resistance, indicated that *HDAC5*, whose transcription is stimulated by *C-MYC*, is crucial for *SOX9* deacetylation and nuclear localization in tamoxifen-resistant breast cancer, hence targeting *C-MYC/HDAC5/SOX9* axis may be beneficial (Xue et al. 2019). Another study investigating the histone H3 lysine 4 (H3K4) demethylase *KDM5B* demonstrated that higher KDM5 activity increases tumor cell

5 Clinical Studies and Epi-Drugs in Various Cancer Types 177

Table 5.3 Ongoing trials of epigenetic therapies in breast cancer (*BC* breast cancer. *ER* estrogen receptor. *HR* hormone receptor. *HER2* human epidermal growth factor receptor 2. *DNMTi* DNA methyltransferase inhibitor. *HDACi* histone deacetylase inhibitor. *BETi* bromodomain and extra-terminal inhibitor)

Epigenetic target	Treatment	Trial number	Phase	Conditions	Status
DNMT	Azacitidine	NCT04891068	2	High risk, early stage BC	Recruiting
	Azacitidine + nab-paclitaxel	NCT00748553	1/2	Advanced/metastatic solid tumors/BC	Completed
	Azacitadine + fulvestrant	NCT02374099	2	Advanced ER(+), HER2(-) BC, progressed on an aromatase inhibitor	Terminated
	Decitabine	NCT00030615	1	Advanced solid tumors	Completed
	Decitabine + carboplatin	NCT03295552	2	Metastatic TNBC	Terminated
	Decitabine + paclitaxel	NCT03282825	1	Advanced HER2(-) BC	Unknown
	Decitabine + pembrolizumab followed by neoadjuvant chemotherapy	NCT02957968	2	Locally advanced, HER2(-) BC	Active, not recruiting
DNMT + HDAC	Azacitadine + entinostat	NCT01349959	2	Advanced TNBC or ER(+), HER2(-) BC	Active, not recruiting
	Decitabine + panobinostat prior to tamoxifen	NCT01194908	1/2	Metastatic or locally advanced TNBC	Terminated
HDAC	Entinostat + capecitabine	NCT03473639	1	Advanced HER2(-) and stage 1–3 HER2(-), with residual disease following neoadjuvant chemotherapy	Recruiting
	Entinostat + exemestane	NCT02820961	1	Locally recurrent/metastatic, ER(+) BC	Completed
	Entinostat + atezolizumab	NCT02708680	1/2	Advanced TNBC	Completed
	Entinostat + aromatase inhibitor (AI)	NCT00828854	2	ER(+) BC	Completed
	Entinostat +/- exemestane	NCT02115282	3	Advanced HR(+) BC	Active, not recruiting
	Entinostat +/- exemestane	NCT03538171	3	Advanced HR(+) BC	Active, not recruiting
	Entinostat +/- exemestane	NCT03291886	2	Advanced HR(+) BC	Completed
	Entinostat + lapatinib + trastuzumab	NCT01434303	1	Advanced HER2(+) BC	Completed
	Entinostat + neoadjuvant anastrozole	NCT01234532	2	Early stage TNBC	Terminated
	Entinostat + nivolumab + ipilimumab	NCT02453620	1	Advanced HR(+) BC	

(continued)

Table 5.3 (continued)

Epigenetic target	Treatment	Trial number	Phase	Conditions	Status
	Atezolizumab + entinostat vs. fulvestrant	NCT03280563	1/2	Advanced HR(+), HER2(−) BC	Active, not recruiting
	Vorinostat + olaparib	NCT03742245	1	Advanced HER2(+) BC	Recruiting
	Vorinostat + paclitaxel +/− trastuzumab	NCT00574587	1/2	Advanced HER2(+) BC	Completed
	Vorinostat + capecitabine	NCT00719875	1	Advanced BC	Completed
	Vorinostat + tamoxifen	NCT00365599	2	Advanced HR(+) BC	Completed
	Vorinostat + ixabepilone	NCT01084057	1	Advanced BC	Completed
	Tamoxifen + pembrolizumab +/− vorinostat	NCT04190056	2	Advanced HR(+), PD1 > 10%	Recruiting
	Vorinostat + trastuzumab	NCT00258349	1/2	Advanced HER2(+) BC	Completed
	Tucidinostat + chemotherapy	NCT04582955	NA	TNBC	Recruiting
	Tucidinostat + cisplatin	NCT04192903	2	TNBC	Recruiting
	Tucidinostat + surufatinib + fulvestrant	NCT05186545	2	Advanced HR(+)/HER2(−) BC	Recruiting
	Tucidinostat + fulvestrant	NCT05047848	NA	Advanced HR(+)/HER2(−) BC	Recruiting
	Fluzoparib + tucidinostat/camrelizumab	NCT05085626	2	Advanced HER2(−)/HRD(+) BC	Recruiting
	Tucidinostat + abemaciclib + fulvestrant	NCT05464173	1/2	Previously treated with palbociclib in HR(+)/HER2(−) advanced BC	Recruiting
	Tucidinostat + neoadjuvant treatment	NCT05400993	2	Stage II–III HR(+)/HER2(−) BC	Recruiting
	Belinostat + ribociclib	NCT04315233	1	Advanced TNBC	Recruiting
	Romidepsin + cisplatin + nivolumab	NCT02393794	1/2	Advanced TNBC	Active, not recruiting
	Panobinostat	NCT00777049	2	HER2(−) locally recurrent or metastatic BC	Completed
	Panobinostat + trastuzumab + paclitaxel	NCT00788931	1	Advanced HER2(+) BC	Completed
BET	Molibresib + fulvestrant	NCT02964507	1	Advanced HR(+) BC	Completed
	Alobresib + fulvestrant/exemestane	NCT02392611	1	Advanced HR(+) BC	Completed
	ZEN003694 + talazoparib	NCT03901469	2	Advanced TNBC, BRCA wild type	Recruiting

transcriptomic heterogeneity and risk of resistance to endocrine therapy (Hinohara et al. 2018).

5.2.3 Colorectal Cancer

Colorectal cancer (CRC) is a common disease and continues to be the second highest cause of death from cancer in developed nations (Siegel et al. 2022). As with gene mutations, it appears that many of the epigenetic changes present in a typical cancer cell contribute to the development of CRC by sequentially altering important tumor suppressor and oncogenes. This is how epigenetic changes are thought to play a role in the normal-polyp-cancer sequence. The two main molecular pathways, through which CRCs primarily originate, are chromosomal instability and microsatellite instability (MSI). The third class of CRCs referred to as having a "CpG island methylator phenotype (or CIMP)" has recently been discovered and is distinguished by a high prevalence of DNA hypermethylation. As a result, we now understand a lot more about the molecular mechanisms that control the growth of colorectal cancers.

The DNA mismatch repair (MMR) mechanism is typically inactivated in MSI due to hypermethylation or mutations in the genes *MLH1, MSH2, MSH6,* and *PMS2*. Owing to the inactivation of these genes, repetitive microsatellite sequences, some of which are found in the exons of putative tumor suppressor genes, begin to accumulate DNA replication mistakes (Toh et al. 2021). While DNA hypermethylation can silence tumor suppressor genes, global DNA hypomethylation is believed to affect CRC development by causing loss of chromosomal instability and global imprinting (Tse et al. 2017). In many malignancies, including CRC, repetitive transposable DNA elements like the LINE-1 or short interspersed nucleotide elements (SINE, or Alu) sequences frequently exhibit genome-wide hypomethylation (Goel et al. 2010).

The best approach for minimizing CRC-related mortality still seems to be early diagnosis and removal of premalignant lesions. However, the limitations of today's screening techniques include their cost, invasiveness, and poor patient compliance, which results in tumors being discovered later and having a worse prognosis. There is a need for additional reliable noninvasive biomarker assays for the early diagnosis of CRC, as the two most used fecal screening tests, fecal occult blood test (FOBT) and fecal immunochemical test (FIT), have inadequate diagnostic accuracy. Because epigenetic modifications tend to occur more frequently than genetic mutations during the early stages of colorectal carcinogenesis, they may be more useful as the next generation of diagnostic biomarkers for the identification of colonic polyps and malignancies.

The *SEPT9* gene, which encodes septin 9, a GTP-binding protein involved in actin dynamics, is one of the most extensively researched noninvasive DNA methylation biomarkers for CRC diagnosis. This biomarker is marketed as the Epi proColon test (Epigenomics), which was authorized by the FDA in 2016 as the first molecular blood-based CRC screening assay. According to the data that are available today, Epi proColon® 2.0 CE has the potential to be a sensitive and feasible screening option for individuals who reject colonoscopy screening (Song et al. 2017). Methylation of the VIM gene, which encodes for the intermediate filament protein vimentin, is another noninvasive methylation biomarker commonly encountered for CRC diagnosis. Using fecal samples rather than blood samples may increase the diagnostic accuracy of this biomarker (Muller and Gyorffy 2022). Given VIM methylation's satisfactory performance in fecal samples, the ColoSure test has also been developed using this biomarker (LabCorp). Nevertheless, ColoSure has not yet been approved by the FDA to be used as a CRC screening test (Jung et al. 2020). To enhance diagnostic accuracy, various potential combinations of methylation biomarkers have been presented. Cologuard (Exact Sciences), the first FDA-approved stool-based multi-target panel for CRC screening, combines a molecular assay for three biomarkers (seven *KRAS* mutant sites and the methylation status of *NDRG4* and *BMP3*) with immunohistochemical testing for hemoglobin (Imperiale et al. 2014). Finally, the most widely used epigenetic biomarker in current clinical practice is the examination of somatic MLH1 promoter methylation in CRCs, demonstrating loss of *MLH1* and/or *PMS2* protein expression. The currently proposed method for identifying patients with Lynch syndrome is universal testing for MMR proteins and/or MSI analysis in CRC patients (Cerretelli et al. 2020). However, the most common cause of MLH1 inactivation is somatic inactivation caused by biallelic promoter hypermethylation. As a result, in patients with CRC without *MLH1* expression, *MLH1* hypermethylation analysis is widely used in clinical practice to differentiate between Lynch syndrome and sporadic CRCs with MMR deficiency (Anghel et al. 2021). In a 2022 study conducted in Israel, researchers devised a single-molecule-based liquid biopsy technique to evaluate several epigenetic characteristics from a 1 ml plasma sample. They have shown its value for the diagnosis of CRC, along with the highly sensitive detection of protein biomarkers. Their research identifies EPINUC as a liquid biopsy technique that can evaluate a variety of histone, DNA, and protein biomarkers with single-molecule accuracy. EPINUC distinguishes between CRC patients and healthy people with a high degree of specificity and sensitivity. The ability of this multiparametric method to identify people with early-stage cancer has been demonstrated (Fedyuk et al. 2022).

The most accurate method of determining CRC patient prognosis currently requires pathological staging of the tumor as well as evaluation of specific histological features of the tumor. However, CIMP status has emerged as the most promising biomarker candidate for predicting the prognosis of CRC patients. Cancers that test positive for CIMP have a poor prognosis overall (Zlobec et al. 2012). Aside from hypermethylation of various genes/loci, there is rising evidence that DNA hypomethylation status is associated with the prognosis of CRC patients. For example, hypomethylation of LINE-1 sequences in tumors has been extensively researched and is linked to poor survival results in CRC patients (Akimoto et al. 2021). Furthermore, hypermethylation of several recognized tumor suppressor genes has been linked to poor results. For example, in patients with CRC, hypermethylation of *CDKN2A* (particularly at the p16INK4A promoter) in tissue and blood was demonstrated to be related to poor prognosis as well as a higher risk

5 Clinical Studies and Epi-Drugs in Various Cancer Types

Table 5.4 The combination of epigenetic drugs and immunotherapy in CRC

NCT number	Drug combinations	Epigenetic mechanism
NCT03576963	Guadecitabine + nivolumab	DNMT
NCT02260440	Azacitidine + pembrolizumab	DNMT
NCT02811497	CC-486 (oral azacitidine) + durvalumab	DNMT
NCT02805660	Mocetinostat + durvalumab	HDAC
NCT02419417	BMS-986158 + nivolumab	BET
NCT02959437	INCB057643/INCB059872 + pembrolizumab + epacadostat	BET/LSD

of recurrence and distant metastasis (Zou et al. 2002). Further, HPP1 and HLTF have emerged in recent years as two of the most promising noninvasive methylation biomarkers for disease monitoring in CRC patients. The methylation status of *HPP1* and *HLTF* has been linked to advanced disease stages, tumor aggressiveness, poor survival, and tumor recurrence. Furthermore, HPP1 methylation levels in cfDNA could be used to identify patients with metastatic CRC who might respond to chemotherapy with bevacizumab early in treatment (Herbst et al. 2017). In summary, DNA methylation biomarkers have been generally related to prognosis and survival, but data on their utility in specific clinical scenarios, which could influence existing treatment regimens, is still relatively sparse.

Inhibitors of the enzymes responsible for DNA methylation (DNMTs and HDACs) and histone modification (HMTs and HDMs), as well as medications that therapeutically modulate miRNA expression, are examples of epigenetic modifying drugs. Some of these medications have been tested in preclinical or early-phase clinical trials in CRC. In preclinical research, epigenetic modifiers have shown to have positive synergistic effects when combined with other drugs. For example, the combination of pembrolizumab plus azacitidine is safe and tolerated but linked with relatively limited clinical activity in patients with chemotherapy-refractory CRC (Kuang et al. 2022). While the number of available epigenetic modifiers is growing, proof of a definite survival benefit in CRC patients receiving these medications is still lacking. Furthermore, none of the existing epigenetic modifiers have progressed past phase II clinical studies, owing to safety concerns. Ongoing clinical trials can be seen in Table 5.4.

5.2.4 Prostate Cancer

Prostate cancer (PCa) is the second most common cancer among men worldwide, with the estimated sixth leading cause of cancer death (Siegel et al. 2022). The most well-studied epigenetic modifications in prostate cancer are DNA methylation alterations, which have been researched throughout carcinogenesis and disease development. There are various instances of PCa-related genes that get hypermethylated and silenced. Indeed, data suggest that de novo promoter hypermethylation is a more prevalent way of gene inactivation in this illness than traditional genetic processes such as mutation, deletion, or translocation. Glutathione S-transferase Pi (GSTP1) is a gene that encodes an intracellular detoxification enzyme that plays a crucial role in oxidative repair. GSTP1 promoter hypermethylation is the most common somatic change in PCa, occurring in more than 85% of tumors. GSTP1 hypermethylation appears to occur early in prostate carcinogenesis, as seen in a subset of preneoplastic high-grade intraepithelial neoplasia, indicating that its loss plays a formative role in the change to a neoplastic phenotype (Henrique and Jeronimo 2004). Other genes that are often hypermethylated in PCa are the tumor suppressors *APC, RAR*, and *RASSF1A,* the cellular adhesion gene *CDH1*, the cell cycle control gene *CCND2*, and the DNA repair gene *MGMT* (Saghafinia et al. 2018). Furthermore, prostate cancers have lower TET2 expression, which stimulates androgen receptor signaling and increases invasion (Takayama et al. 2015). Several lysine methyltransferases (KMTs) and lysine demethylases (KDMs) are also implicated in prostate cancer growth and progression. For example, the H3K36 KMT NSD2 is a cofactor of the androgen receptor (AR), which is overexpressed in metastatic prostate cancer (Ezponda et al. 2013). Additionally, NSD2 increases prostate cancer metastasis and epithelial-mesenchymal transition (EMT) (Aytes et al. 2018). EZH2 has also been linked to the development of NEPC (neuroendocrine prostate cancer) and antiandrogen resistance (Dardenne et al. 2016).

Many DNA methylation abnormalities are found in 70% or more of prostate cancer cases but not in normal prostate. Therefore, these modifications might significantly improve present clinical decision-making as standalone biomarkers or in conjunction with genetic changes (Savio and Bapat 2015). Numerous DNA methylation changes, such as CpG island promoter methylation of *GSTP1, APC, RARB, RASSF1*, and *PTGS2*, among dozens to hundreds of others, are extremely specific and frequent in prostate tumors (Aryee et al. 2013). If these changes could be found in the blood or urine of asymptomatic individuals, they may be a valuable biomarker for prostate cancer screening. In this regard, numerous novel ways to identify cancer DNA alterations, including mutations and DNA methylation changes in circulating tumor DNA and/or urine, give proof of principle that such a DNA methylation-guided screening tool for prostate cancer may be achievable. Additionally, immunohistochemical or immunofluorescent techniques to identify global changes in histone marks or hydroxymethylcytosine content levels, as well as DNA methylation modifications, may be effective in helping in the tissue-based diagnosis of prostate cancer, particularly if standard morphological and IHC features are insufficient (Seligson et al. 2005). DNA hypermethylation modifications are also particularly promising for tracking tumor burden and therapy response, since they are stable over time, are highly disease-specific, and can occur quite often in prostate cancer (Yegnasubramanian et al. 2019).

Novel treatment approaches that can target these epigenetic processes are being developed because of our growing understanding of the epigenetic mechanisms behind the onset and progression of prostate cancer. Decitabine, azacitidine, and guadecitabine, a novel prodrug, are currently available DNA methyltransferase

5 Clinical Studies and Epi-Drugs in Various Cancer Types

Table 5.5 The combination of epigenetic drugs in PCa

NCT number	Drug combination	Epigenetic target
NCT00503984	Azacitidine + docetaxel/prednisone (100)	DNMT
NCT02998567	Guadecitabine + pembrolizumab	DNMT
NCT00331955	Vorinostat + doxorubicin	HDAC
NCT00663832	Panobinostat + docetaxel + prednisone	HDAC
NCT00878436	Panobinostat + bicalutamide	HDAC
NCT04179864	Tazemetostat + abiraterone + prednisone or enzalutamide	EZH2
NCT03480646	CPI-1205 + enzalutamide or abiraterone	EZH2
NCT02711956	ZEN003694 + enzalutamide (101)	BET

inhibitors. These drugs have not been successful when used alone to treat prostate cancer or many other solid organ tumors. As previously established, histone changes are common in prostate cancer. HDAC inhibitors, like DNMT inhibitors, have not been successful as single treatments in most solid organ tumors. Several studies have demonstrated that these epigenetic drugs might modify the immunogenicity and immune response of cancer cells, increasing their proclivity to respond to immunotherapies. These hypotheses are now being investigated in a variety of cancers, including prostate cancer. Considering the frequent upregulation of EZH2 in aggressive and neuroendocrine prostate cancer, there is significant interest in evaluating new EZH2 inhibitors (e.g., tazemetostat, CPI-1205) in comprehensive preclinical and clinical investigations (Morel et al. 2021). Another intriguing new class of medicines is those that target BET-bromodomain readers of histone acetylation marks, such as BRD4. Because BRD4 has been demonstrated to be crucial in the control of MYC and AR, both of which play critical roles in prostate cancer start and progression, BRD4 inhibitors, including the first-in-class JQ-1 and numerous other medicines in this class, are being vigorously studied for prostate cancer treatment in preclinical and clinical settings (Asangani et al. 2014). Examples of ongoing/completed clinical trials of epigenetic drug combinations can be seen in Table 5.5.

Prostate cancer is a complicated disease that is influenced by both hereditary and epigenetic factors. New strategies to cure or prevent prostate cancer are anticipated to emerge as a result of a better knowledge of the role of epigenetic factors in the disease's onset or progression. Furthermore, measuring the expression of epigenetic markers or variables may help us diagnose prostate cancer earlier, give more accurate prognostic information, or treat patients more effectively. More research using larger datasets is required to develop more clinically useful epigenetic biomarkers (Kumaraswamy et al. 2021).

5.2.5 Cutaneous Melanoma

Melanoma is an extremely aggressive tumor that is responsible for less than 5% of all skin cancers yet 80% of skin cancer-related deaths (Bertolotto 2013). Epigenetics is becoming a significant regulatory concept in melanoma biology, in addition to genetic and transcriptional control (Guo et al. 2021). The main epigenetic modifications strongly correlated with melanoma are DNA methylation, histone modification, noncoding RNA, and the recently identified N6-methyladenosine (m6A) RNA methylation (Moran et al. 2018).

DNA methylation is the most extensively investigated epigenetic alteration in melanoma as in other cancers. A review demonstrated the importance of searching for differentially methylated CpG sites in diagnosis, emphasizing that one focus can be on melanoma cell-specific differential methylation patterns in cfDNA that depend on a single, significant CpG-rich gene promoter regions such as *RASSF1A*, *LINE-1*, or *ODC1* (Santourlidis et al. 2022). Additionally, focal DNA hypermethylation of the promoters of some particular tumor suppressors, such as *PTEN*, *P16INK4A*, *P14ARF*, *RASSF1A*, and *MGMT,* has been well demonstrated in melanoma, and functional deficit of these genes is associated with melanoma progression (Guo et al. 2021). A study found that loss of global DNA methylation has been related to continuous overexpression of *PD-L1*, which has been associated with inhibited host antitumor response (Chatterjee et al. 2018). DNA methylation patterns have also been investigated as potential clinical indicators for melanoma diagnosis and prognosis. In a study, methylation of *PTEN* has been identified as a key prognostic indicator of poor prognosis in melanoma (Micevic et al. 2017). The same study demonstrated that *RASSF1A* methylation in serum cfDNA has been linked to a noticeably worse prognosis in patients with stage IV melanoma before biochemotherapy administration, hence which suggests circulating methylated DNA in serum may serve as a potential biomarker for disease outcome and therapeutic response (Mori et al. 2005). A meta-analysis showed that hypermethylation of claudin 11, *MGMT*, p16, retinoic acid receptor β, and *RASSF1A* was significantly higher not only in melanoma patients but also in patients with metastasis (Guo et al. 2019a). Another study indicated that the four-DNA methylation signature (KLHL121, GBP5, OCA2, and RAB37) was strongly related to overall survival in melanoma patients; besides, the prediction accuracy of this signature was noticeably higher than that of known biomarkers (Guo et al. 2019b).

In the context of treatment, DNA methylation is mostly being investigated in the form of combined therapies, especially with immunotherapy. Current studies exploring combination therapy of DNMTi and immune-checkpoint inhibitors demonstrate synergism (Micevic et al. 2017). A study showed that DNMT inhibition increased the expression of viral defense genes, and in melanoma patients, this gene signature was associated with a better response to anti-CTLA4 therapy (Chiappinelli et al. 2015). A phase I clinical trial, researching the use of guadecitabine in combination with ipilimumab in patients with unresectable stage III/IV melanoma (NCT02608437), indicated antitumor activity (Di Giacomo et al. 2019). In another clinical trial, investigating the combination of a DNMT inhibitor and temozolomide chemotherapy (NCT00715793), an overall 1-year survival rate of 56% was shown (Tawbi et al. 2013).

Histone marker profiling has also been studied as a potential diagnostic and prognostic tool in melanoma. A study indicated that upregulation of the *EZH2* and its resultant histone modification H3K27me3 was associated with increased tumor thickness, nodal involvement, and resistance to immune checkpoint blockade (Hoffmann et al. 2020). A study reported that amplification of *SETDB1*, by causing increased methylation of histone H3 on lysine 9 (H3K9), exhibited oncogenic activity in an established tumor (Shi et al. 2017). In contrast, several studies demonstrated that H3K9 demethylation by two different H3K9 demethylases, *LSD1* and *JMJD2C*, inhibits melanomagenesis (Guo et al. 2021). Furthermore, *LSD1* was found to be associated with resistance to anti-PD-1 agents in melanoma, suggesting *LSD1* as a possible target for immunotherapy (Sheng et al. 2018). Another histone demethylase, *PHF8*, was also found to play a role in TGFβ signal regulation and melanoma invasion (Moubarak et al. 2022). Aside from histone methylation, acetylation modifications also have an impact on melanoma biology. A study showed that a melanoma-driven transcription factor, *MITF* (microphthalmia-associated transcription factor), is regulated by a histone acetyltransferase, p300; hence, *MITF* can be a promising target, whose expression indicates melanoma response to p300 HAT inhibition (Kim et al. 2019).

As with DNA methylation, histone modifications have been investigated in therapy, mostly as combination therapies due to the early evidence suggesting that epi-drugs may prepare the microenvironment for subsequent immunotherapy and targeted therapies. Clinical trials that are ongoing to evaluate the use of epi-drugs in melanoma are summarized (Table 5.6). A review reported that, through the control of DNA damage repair, intracellular ROS production, and *PD-L1* expression, respectively, HDAC inhibitors can also synergistically increase the effectiveness of radiotherapy, MAPK pathway-targeted therapy, and immunotherapy (Guo et al. 2021). In preclinical trials, HDACi overcome acquired resistance to *BRAF* or *MEK* inhibitors by controlling PI3K signaling (Gallagher et al. 2018). A study emphasizes the value of targeting P300 as a synergistic treatment strategy to sensitize melanoma cells to inhibition of the MAPK pathway (Zhang et al. 2021). In recent years, new epigenetic targets have been discovered. One of them, Corin, is a synthetic drug that targets both HDAC and *LSD1* within the CoREST complex, a complex that has been implicated in carcinogenesis. According to a recent study, Corin inhibited melanoma growth in a mouse xenograft model and exhibited significant efficacy against several melanoma cell lines compared to HDAC inhibitors or LSD inhibitors (Kalin et al. 2018).

In addition to DNA methylation and histone modification, noncoding RNA and m6A RNA methylation are two more crucial epigenetic modification paradigms that are found to be linked to several cancer features, such as migration and invasion, metastasis, antitumor immunity (Guo et al. 2021). The early focus of ncRNA research in melanoma was on their contributions to their functions in cancer biology hallmarks, but then, their functions in the tumor microenvironment, including as angiogenesis, metastatic niche formation, and T cell dysfunction, began to receive increasing attention (Guo et al. 2021). What is more, N6-methyladenosine (m6A) RNA methylation is another major modification that has been identified (Zhao et al. 2017). The first RNA N6-methyladenosine (m6A) demethylase in eukaryotic cells has been identified as the fat mass and obesity-associated protein (FTO) (Azzam

Table 5.6 Ongoing trials of epigenetic therapies in cutaneous melanoma (*DNMT* DNA methyltransferase. *HDAC* histone deacetylase inhibitor. *NSCLC* non-small cell lung cancer. *CRC* colorectal cancer)

Epigenetic target	Trial number	Approach	Phase	Status
DNMT	NCT02608437	Guadecitabine + ipilimumab for unresectable disease	1	Unknown
	NCT00398450	Azacitidine + interferon alfa for metastatic melanoma	1	Completed, no results reported
	NCT02816021	Azacitidine + pembrolizumab for metastatic melanoma	2	Active, not recruiting
	NCT00217542	Azacitidine + r-interferon alfa2b for stage III/IV unresectable disease	1	Completed, no results reported
	NCT02223052	Oral azacitidine bioequivalence study	1	Completed, no results reported
	NCT00715793	Decitabine and temozolomide for metastatic melanoma	1/2	Completed
	NCT00030615	Decitabine for advanced solid tumors	1	Completed, no results reported
HDAC	NCT03765229	Entinostat + pembrolizumab for non-inflamed stage III/IV melanoma	2	Recruiting
	NCT02437136	Entinostat + pembrolizumab for NSCLC, melanoma, CRC	1/2	Active, not recruiting
	NCT02836548	Vorinostat for advanced melanoma	1/2	Unknown
	NCT03590054	Abexinostat + pembrolizumab for advanced solid tumors	1	Active, not recruiting
	NCT01065467	Panobinostat for metastatic melanoma	1	Completed, no results reported
	NCT02032810	Panobinostat + ipilimumab for unresectable III/IV melanoma	1	Active, not recruiting
	NCT00185302	MS-275 for metastatic melanoma	2	Completed, no results reported
	NCT03008018	KA2507(HDAC6 inhibitor) for solid tumors	1	Completed, no results reported

et al. 2022). A study reported that the expression of m6A demethylase FTO is markedly increased in melanoma which contributes to oncogenesis as well as greater responsiveness to anti-PD-1 agents by coordinating *PD-1*, *CXCR4*, and *SOX10* expression (Yang et al. 2019b). Other studies demonstrate that the methyltransferases METTL3/14 and ALKBH5 (m6A demethylase alkylation repair homolog 5) control the response to anti-PD-1 blockade by altering the tumor microenvironment (Li et al. 2020; Wang et al. 2020). Further research in this field may yield new insights that result in groundbreaking developments in the treatment of melanoma.

5.2.6 Epigenetic Studies Regarding the Most Common Hematological Malignancies

5.2.6.1 Myeloid Leukemias

While acute myeloid leukemia (AML) affects tens of thousands of patients each year, several groups have survival rates as low as 25% due to the broad mutational profile of the disease (Heimbruch et al. 2021). DNA methylation heterogeneity, which is pathophysiologically important for the neoplastic process, is a key feature of AML (Toyota et al. 2001). As *DNMT3A* and *TET2* mutations might lead HSCs to a preleukemic state (Sato et al. 2016), evaluation of aberrant DNA methylation patterns can be useful for the early detection of AML with no clinical manifestations (Yang et al. 2019a). Although identification of diagnostic epigenetic biomarkers remains a challenge, global initiatives like the International Cancer Genome Consortium (ICGC) and the European Community initiative BLUEPRINT Consortium enable the analysis of epigenomic changes in AML patients (Yang et al. 2019a).

Fifteen to twenty-five percent of AML patients have *DNMT3A* enzyme mutations, that appear early in clonal evolution and remain detectable after malignant transformation, enabling for more rapid progression. The most frequent mutation is located in codon R882 (DNMT3AR882mut) (Park et al. 2020). During oxidative phosphorylation, the wild-type IDH enzyme catalyzes the conversion of isocitrate to α-ketoglutarate (α-KG); however, when *IDH* is mutated, α-KG is transformed to an oncometabolite, 2-HG, which causes inhibition of α-KG-dependent enzymes like histone/DNA demethylases and 5-methlycytosine hydroxylase, epigenetic dysregulation occurs and an oncogenic process starts (Govindarajan et al. 2022). Several studies revealed that *FLT3*, *NPM1*, and *IDH1/2* gene mutations frequently co-occur with *DNMT3A* mutations. Moreover, patients who have this combination of mutations tend to exhibit a larger proportion of blast cells as well as poorer outcomes (Loghavi et al. 2014; Bezerra et al. 2020). Likewise, *TET2* and *IDH1/2* mutations alter the epigenome via modulating hydroxymethylation-like *DNMT3A*, and these mutations were found to be persistent in AML patients from the time of diagnosis till recurrence, associated with poor prognosis as well (Wang et al. 2019). It's important to note that *IDH1/2* and *TET2* mutations in myeloid neoplasms have been found to be mutually exclusive (Inoue et al. 2016). However, studies on *IDH1/2* mutations for prognosis prediction are controversial; further research is required to accurately figure out the impact of these mutations (Ok et al. 2019).

Histone modifications have been implicated in the pathogenesis of many diseases, including cancer as well. Interestingly, a hallmark of AML is the aberrant recruitment of HDACs by mutant or fusion proteins, such as *PML-RARα*, *PLZF-RARα*, and *AML1-ETO* causing leukemogenesis via aberrant gene silencing. For example, the chimeric protein *AML1-ETO* recruits *HDAC1*, *HDAC2*, and *HDAC3*, which silences *AML1* target genes, enabling differentiation arrest and transformation (Jose-Eneriz et al. 2019). A study revealed a mechanism by which *HDAC3* expression is associated with chemotherapy resistance by regulating AKT activation, suggesting that this may be a strategy to overcome chemoresistance (Long et al. 2017). Another study identified, in *FLT3-ITD+* AML cells, *FLT3* inhibition causes *HDAC8* upregulation, which is leading to TKI resistance by p53 inactivation and promotes leukemia maintenance (Long et al. 2020). Likewise, in another study, *SIRT3*, a class III HDAC, was found to be contributed to chemoresistance by influencing mitochondrial metabolism and decreasing ROS production (Ma et al. 2019). Similarly, *GCN5*, a HAT, was shown to be responsible for resistance to all-trans retinoic acid (ATRA) treatment in non-APL AML cells (Kahl et al. 2019). As well as acetylation, histone methylation profiles also have importance in AML. A study showed that loss-of-function mutations in *SETD2*, a member of the KMT3 family, cause chemoresistance to DNA-damaging treatment by altering cell cycle checkpoints (Dong et al. 2019). A study about *KMT2A*, another histone lysine methyltransferase, demonstrated that AML with partial tandem duplications (PTD) in *KMT2A* has a unique gene expression profile and that concurrent *DNMT3A* and *NRAS* mutations were linked to a poor clinical outcome in this subset of AML (Hinai et al. 2019). Similarly, a study investigating *EZH2* reported that low levels of EZH2 protein are associated with disease relapse and multidrug resistance (Gollner et al. 2017). According to another study, high levels of *EZH1* expression in *AML1-ETO*-positive patients are related to worse overall survival (Dou et al. 2019). As like in AML, *EZH2* mutations have also been observed in chronic myeloid leukemia (CML), and *EZH2* overexpression has been shown to be regulated by *BCR-ABL1* indicating that TKI and EZH inhibitor combination may be beneficial to eliminate the residual disease burden (Rinke et al. 2020). Furthermore, mutations in the *ASXL1* gene, which participates in epigenetic regulation, are found to be frequently seen in malignant myeloid diseases and correlated with worse prognosis (Asada et al. 2018). HDMs are also linked to the leukemic phenotype. In a study, *KDM6A* inactivation is shown to be linked to disease progression, treatment resistance, and poor survival (Stief et al. 2020).

Alterations in ncRNAs, particularly in miRNAs, have also been reported in AML. Furthermore, miRNAs can either serve as oncogenes or tumor suppressors. For instance, it was shown that miR-9 functioning as an oncogene in *MLL*-rearranged AML yet acts as a tumor suppressor in pediatric AML with t(8;21) (Wallace and O'Connell 2017). The same miR-9 was also found to play a role in daunorubicin resistance (Liu et al. 2019). Likewise, inhibition of the miRNA let-7 (MIRLET7) family, which functions as tumor suppressors via targeting a number of oncogenes including *NRAS*, *KRAS*, and *MYC*, has been shown to be associated with chemoresistance (Chirshev et al. 2019).

In the context of CML epigenetics, several studies have also been conducted to predict disease progression. A study investigating mutations at diagnosis and blast crisis (BC) in CML found numerous variants, such as *ASXL1*, *IKZF1*, *RUNX1*, and *SETD1B*, in patients with poor outcomes and in all patients at BC, suggesting that this could be a promising biomarker for offering prognostic data to support a risk-adapted therapy strategy (Branford et al. 2018). Moreover, studies searching for gene expression signatures in risk assessment of CML revealed that epigenetic alterations, which are present at diagnosis, can lead to the emergence of *BCR-*

ABL1-independent clones that result in both TKI resistance and BC transformation (Krishnan et al. 2022).

The optimal treatment of AML in clinical practice is still difficult at the moment. Age used to be the only criterion to decide whether a patient was a candidate for intensive chemo-regimens or other treatments like epi-drugs. But today, comorbidities can determine suitability for epigenetic therapy based on patient fragility. Completed trials and active trials for investigating epigenetic therapies alone or combined with other therapies in AML and CML are summarized (Table 5.7).

Since traditional chemotherapy has high morbidity and relatively low efficacy and due to the stronger relation between DNMTs and pathogenesis and prognosis of AML, hypomethylating drugs have generated a lot of interest (Wong et al. 2019). To date, there are more than 200 clinical trials ongoing that are investigating the utility of DNMTs in AML either as a single agent or in combination (accessed on September 23, 2022). A phase 1/2 trial searching treatment with the single-agent guadecitabine (NCT01261312) reported that 80% of patients benefited from guadecitabine with a 2-year survival rate of 21% (Chung et al. 2019). DNMTs are being further tested in combination with other drugs, such as chemotherapy agents, ATRA, *BET* inhibitors, *DOT1L* inhibitors, immune checkpoint inhibitors, *BCL-2* inhibitors, and especially HDAC inhibitors. In another phase 2 study, pracinostat, which has limited single-agent activity in AML, was shown to achieve a 52% response rate in patients aged ≥ 65 years with newly diagnosed AML, when combined with azacitidine (Garcia-Manero et al. 2019). Similarly, another phase I study (NCT02203773) evaluating decitabine or azacitidine in combination with the antiapoptotic B-cell lymphoma 2 protein inhibitor venetoclax was conducted on previously untreated AML patients aged ≥65 years and who were unsuitable for standard induction therapy. Despite 30% of patients having infections as a side effect, the outcomes were optimistic, with 61% of patients achieving complete remission or complete remission with incomplete bone marrow repair (DiNardo et al. 2018). The correlation between *TET2* and *IDH1/2* is an important hallmark for the therapy of hematological malignancies. Initially, enasidenib was approved by the FDA for the treatment of adult patients with relapsed or refractory acute myeloid leukemia with an isocitrate dehydrogenase-2 (*IDH2*) mutation on August 1, 2017 (Kim 2017). Then, ivosidenib received FDA approval for the treatment of people with relapsed or refractory acute myeloid leukemia who have susceptible *IDH1* mutation on July 20, 2018 (161). Recently, on May 25, 2022, FDA approved ivosidenib in combination with azacitidine (in injection form) for newly diagnosed AML with a susceptible *IDH1* mutation, in adults 75 years of age or older, or who cannot receive intensive induction chemotherapy due to comorbidities (Norsworthy et al. 2019). The approval was based on a randomized, multicenter, double-blind, placebo-controlled trial (NCT03173248) that included 146 patients (Montesinos et al. 2022).

Histone modifications are also another target for hematological malignancies, either as single-agent or in combination with other therapies. HDAC inhibitors have been investigated in more than 70 clinical trials, and vorinostat is the most tested one

Table 5.7 Completed and active clinical trials of epigenetic therapies in myeloid leukemias (*DNMT* DNA methyltransferase. *HDAC* histone deacetylase. *LSD1* lysine-specific demethylase. *DOT1L* DOT1-like histone H3K79 methyltransferase. *ATRA* all-trans retinoic acid. *AML* acute myeloid leukemia. *MDS* myelodysplastic syndrome. *CML* chronic myeloid leukemia. *CMML* chronic myelomonocytic leukemia. *R/R AML* relapsed/refractory acute myeloid leukemia)

Epigenetic target	Trial number	Approach	Phase	Status
DNMT	NCT00887068	Azacitidine for posttransplant prevention of AML and MDS relapse	3	Completed
	NCT01074047	Azacitidine for AML	3	Completed
	NCT01350947	Azacitidine for CML	2	Completed
	NCT00416598	Decitabine as maintenance therapy for AML	2	Completed
	NCT00042003	Decitabine for refractory CML	2	Completed
	NCT01261312	Guadecitabine for AML, MDS	1/2	Completed
	NCT02348489	Guadecitabine for AML	3	Completed
	NCT03701295	Azacitidine + pinometostat for AML	1/2	Completed
	NCT04022785	Azacitidine + PLX51107 for AML, MDS	1	Recruiting
	NCT03466294	Azacitidine + venetoclax for AML	2	Active, not recruiting
	NCT03769532	Azacitidine + pembrolizumab for NPM1 mutated AML	2	Recruiting
	NCT02845297	Azacitidine + pembrolizumab AML	2	Active, not recruiting
	NCT02397720	Azacitidine + nivolumab +/- ipilimumab for R/R or newly diagnosed AML	2	Recruiting
	NCT03825367	Azacitidine + nivolumab for childhood R/R AML	1/2	Recruiting
	NCT02775903	Azacitidine + durvalumab for MDS, AML	2	Completed
	NCT03173248	Azacitidine + ivosidenib for IDH1 mutant AML	3	Active, not recruiting
	NCT02085408	Decitabine + clofarabine for AML	3	Active, not recruiting
	NCT01303796	Decitabine + sapacitabine for elderly AML	3	Completed
	NCT03404193	Decitabine + venetoclax for AML, MDS	2	Recruiting
	NCT03844815	Decitabine + venetoclax for AML	1	Recruiting
	NCT03941964	Decitabine or azacitidine + Venetoclax for AML	3	Completed
	NCT02996474	Decitabine + pembrolizumab for R/R AML	1/2	Completed
	NCT03969446	Decitabine + pembrolizumab +/- Venetoclax for AML, MDS	1	Recruiting
	NCT02890329	Decitabine + ipilimumab for R/R MDS or AML	1	Active, not recruiting
	NCT02096055	Guadecitabine + idarubicin + cladribine for AML	2	Completed
	NCT02124174	Azacitidine + valproate for AML, MDS	2	Recruiting

(continued)

Table 5.7 (continued)

Epigenetic target	Trial number	Approach	Phase	Status
DNMT + HDAC	NCT01617226	Azacitidine + vorinostat for AML, MDS	2	Completed
	NCT01522976	Azacitidine +/- lenalidomide/vorinostat for MDS, CML	2	Active, not recruiting
	NCT00392353	Azacitidine + vorinostat for AML, MDS	1/2	Active, not recruiting
	NCT00313586	Azacitidine + entinostat for AML, CMML, MDS	2	Completed
	NCT00946647	Azacitidine + panobinostat for AML, CMML, MDS	1/2	Completed
	NCT00414310	Decitabine +/- valproate for MDS, AML	2	Completed
	NCT00867672	Decitabine +/- valproate and ATRA for AML	2	Completed
	NCT00479232	Decitabine + vorinostat for AML, MDS	1	Completed
	NCT00691938	Decitabine + panobinostat for AML, MDS	1/2	Completed
HDAC	NCT00305773	Vorinostat for AML	2	Completed
	NCT01451268	Panobinostat for MDS, AML	1/2	Unknown
	NCT00062075	Romidepsin for R/R AML	2	Completed
	NCT01550224	Vorinostat + temozolomide for R/R AML	2	Completed
	NCT00656617	Vorinostat + idarubicin + cytarabine for MDS/AML	2	Completed
	NCT04326764	Vorinostat + HSC + cytarabine + daunorubicin hydrochloride + idarubicin for AML with younger patients	3	Completed
	NCT01802333	Vorinostat + chemotherapy for younger AML	3	Completed
	NCT00840346	Panobinostat + idarubicin + cytarabine for elderly AML	1/2	Completed
	NCT00006240	Phenylbutyrate + dexamethasone + sargramostim for R/R AML	2	Completed
	NCT00462605	Entinostat + sargramostim for MDS or R/R AML, ALL	2	Completed
LSD1	NCT02273102	Tranylcypromine + ATRA for R/R AML, MDS	1	Completed
	NCT02717884	Tranylcypromine + ATRA + cytarabine for AML	1/2	Unknown
	NCT02842827	IMG-7289 + ATRA for AML, MDS	1	Completed
DOT1L	NCT03701295	Azacitidine + pinometostat for AML	1/2	Completed
	NCT03724084	Pinometostat + cytarabine + daunorubicin in AML with MLL rearrangement	1/2	Active, not recruiting

with more than 30. Other HDACi are being studied for the treatment of myeloid leukemias are panobinostat, valproate, romidepsin, entinostat, phenylbutyrate, and pracinostat. A phase 2 trial (NCT00656617) reported that vorinostat combined with idarubicin and cytarabine was safe and effective in AML (Garcia-Manero et al. 2012). Following the promising results from a phase 2 trial searching the efficacy of pracinostat plus azacytidine therapy in older patients, a multicenter, double-linked, randomized phase 3 trial was conducted; however, the trial was terminated due to lack of efficacy (ClinicalTrials.gov identifier NCT03151408 accessed on September 24, 2022). Moreover, another study (NCT00313586) demonstrated that entinostat in combination with azacitidine for the treatment of therapy-related myeloid neoplasms is not recommended for treatment and is also linked to increased toxicity (Prebet et al. 2016). Furthermore, in a phase 2 trial (NCT00867672) investigating decitabine alone or combined with valproate and ATRA, decitabine plus ATRA resulted in a higher remission rate and a clinically significant extension of survival in AML, without additional toxicity; however, no difference in survival rate was observed with valproate (Lubbert et al. 2020). Romidepsin was approved by the FDA for the treatment of cutaneous or peripheral T-cell lymphoma, and phase 1/2 trials on romidepsin for AML treatment are ongoing as a single agent or in combination. Not only HDACs, but also HMTs and HDMs can be targeted by epi-drugs. Currently, a trial (NCT03724084) is evaluating the treatment with pinometostat, a DOT1-like histone H3K79 methyltransferase (*DOT1L*) inhibitor, combined with standard chemotherapy agents like daunorubicin and cytarabine in AML with *MLL* rearrangement (Yi and Ge 2022). A well-known HDM, lysine-specific demethylase (*LSD1*), is also under investigation, especially in combination with ATRA. A study (NCT02273102) searching tranylcypromine, an *LSD1* inhibitor, plus ATRA treatment in AML/MDS, indicated that *LSD1* inhibition sensitizes AML cells to ATRA and may overcome ATRA resistance (Tayari et al. 2021). Ongoing studies in this area may produce fresh insights that lead to revolutionary discoveries in the treatment of myeloid leukemias.

5.2.6.2 Lymphoid Leukemias

Eighty percent of all children leukemia cases are caused by acute lymphoblastic leukemia (ALL), and despite having a 90% cure rate in children, it is still a major cause of morbidity and mortality in both children and adults (Iacobucci and Mullighan 2017; Gebarowska et al. 2021). ALL arises from hematopoietic cells of either the B-cell precursor lineage (BCP-ALL) or, less commonly, T-cell lineage (T-ALL). Both groups include a variety of subtypes that are often characterized by chromosomal alterations that are believed to be leukemia-initiating lesions, with secondary somatic DNA copy number alterations and sequence mutations promoting leukemogenesis. As with other cancers, epigenomics is a promising field for ALL relapse prediction at diagnosis, subtype categorization, disease progression, and treatment.

Commonly, the important alterations for the prognosis of ALL are high hyperdiploidy, *ETV6-RUNX1*, *BCR-ABL1*, iAMP21, and *KMT2A* (also known as *MLL1*) rearrangements. A study revealed 300 highly variable methylated CpG sites among ALL samples (Milani et al. 2010), suggesting that variable methylation of these sites was able to predict the relapse risk in *ETV6/RUNX1* and high hyperdiploid ALLs, two subtypes that are often associated with a favorable prognosis (Meyer and Hermiston 2019). The Philadelphia chromosome is detected in 3–5%

of pediatric cases, moreover is an indicator of poor prognosis and targeted therapies with imatinib or dasatinib in ALL (Nordlund and Syvanen 2018). Patients with *BCR-ABL1* can also be distinguished by subtype-specific DNA methylation profiles, yet DNA methylation in *BCR-ABL1* ALL likely has an indirect impact; thus, further investigation is needed (de Barrios and Parra 2021). Complex intrachromosomal amplification of chromosome 21 (iAMP21) is most prevalent in older children and also was related to poor prognosis initially, which is now improved with intensive treatment regimens. *KMT2A* (lysine methyltransferase 2A; also known as *MLL*, mixed lineage leukemia) rearrangements (*KMT2A*-r) are detected in approximately 5–6% of whole ALL cases, 75% of infants with B-ALL, and especially in those under 6 months of age (Stahl et al. 2016; Tasian and Hunger 2017). Furthermore, *KMT2A*-r is associated with an unfavorable prognosis, elevated relapse frequency, and intrinsic drug resistance (Sanjuan-Pla et al. 2015, Tasian et al. 2015). It is worth noting that the most frequent *KMT2A*-r in ALL causes H3K4 methyltransferase domain deletion. However, these rearrangements also lead to fusions with partners that also serve to covalently modify histones, and *KMT2A* has over 80 different gene-fusion partners (Rao and Dou 2015). For instance, in *KMT2A-AF4* or *KMT-AF10* ALLs, *AF4* and *AF10* interact with the *DOT1L* methyltransferase, which increases the expression of *HOX* gene via inducing H3 lysine 79 methylation (H3K79). Not only the methylation pattern but also histone acetylation is modified by *KMT2A* rearrangements; the regulation of antiapoptotic genes like *RUNX1*, *MCL1*, and *BCL2* is an example of this. In addition to these alterations, loss-of-function mutations in *CREBBP*, a histone acetyltransferase, was shown to be associated with poor response to glucocorticoids (Gao et al. 2017). Similarly, a study revealed that increased *HDAC4* expression is related to high leukocyte levels and an impaired response to glucocorticoids, as well as overexpression of *HDAC1*, *HDAC2*, *HDAC4*, and *HDAC11* was significantly correlated with unfavorable prognosis (Gruhn et al. 2013). Interestingly, mutations in *KDM6A* (also known as *UTX*), a H3K27me3 histone demethylase that serves as a tumor suppressor gene, have been exclusively found in male T-ALL patients. Since, , *KDM6A* escapes X-inactivation in females, single copy loss of the gene does not affect females yet leads to tumor development in males (Van der Meulen et al. 2015). Additionally, loss-of-function mutations in *DNMT3A* (Mackowska et al. 2021) and *EZH2* (Dawson and Kouzarides 2012) genes are linked to poor prognosis in T-ALL. miRNAs are also thought to play a role in oncogenesis and are currently being studied by researchers as prospective therapeutic targets and prognostic indicators (Drobna et al. 2018). For instance, a study demonstrated that patients with low miR-128b expression have markedly poor prognosis and poor response to glucocorticoid treatment (Nemes et al. 2015). Another study has shown that miR-1246, miR-1248, and miR-429 may be key factors in T-ALL relapse (Luo et al. 2018). Clinical use of miRNA-based profiling is still rather uncommon due to the vagueness of data, and more studies on miRNA expression levels in various T-ALL subtypes are needed.

The understanding of de novo and recurrent ALL has improved, owing to significant progress in genetic and epigenetic profiling of ALL, which has improved the patient risk classification (Tasian and Hunger 2017). There is currently no known credible biomarker other than the minimal residual disease (MRD) for identifying patients at a higher risk of relapse, particularly in T-ALL (Karrman and Johansson 2017). Several studies have attempted to predict ALL relapse using DNA methylation signatures that can be detected at diagnosis. In this context, CpG island methylator phenotype (CIMP) classification seems to foresee relapse regardless of MRD (Borssen et al. 2016). However, assessing the results of these studies is challenging due to the minimal overlap of the DNA methylation profiles detected or CpG sites studied and the small number of ALL patients in any research, subtype, or treatment group. One other explanation for the inability to accurately predict relapse using DNA methylation profiles at diagnosis is that the small clones that evolve to relapse are present at diagnosis in varied numbers among patients and consequently do not expose a significant DNA methylation signature (Nordlund and Syvanen 2018). For instance, a past study found that CIMP+ (hypermethylation phenotype with a higher number of methylated loci) T-ALL patients had a higher relapse rate and mortality rate compared to CIMP- (hypomethylation phenotype with a lower number of methylated loci) (Roman-Gomez et al. 2005). These findings were confirmed by the same group in a study published one year later, and probably used a patient population that overlapped partially (Roman-Gomez et al. 2006). In contrast, a recent study searching the clinical meaning of genome-wide promoter methylation profiles in T-ALL demonstrated that hypomethylation was associated with a higher incidence of relapse (Touzart et al. 2020). Therefore, the clinical significance of DNA methylation in this malignancy has fundamentally changed in light of recent technical developments, and thus more research is needed. In addition to methylation, histone modifications might be helpful for relapse prediction. For instance, in BCP-ALL patients with t(4;11) translocation, high levels of *IRX1* expression were shown to be related to an elevated risk of relapse via *HOXB4* activation (Kühn et al. 2016). Alterations in *KMT2A* (also known as MLL) and *CREBBP* were also found to be associated with increased relapse frequency and poor prognosis in the context of BCP-ALL (de Barrios and Parra 2021).

Inhibitors of DNA methyltransferases have been used to treat hematologic malignancies such as AML and MDS. A study (NCT01861002) revealed that children with relapsed or refractory AML could be treated with azacitidine combined with chemotherapy; however, neither of the patients with ALL responded to azacytidine (Sun et al. 2018). To better understand if azacitidine or decitabine might be candidate therapies in ALL, more research is required. Several studies have also examined HDACi as potential therapeutics in ALL, although they are less efficient and more toxic in vivo than they seemed to be in vitro (Xu et al. 2021). A trial (NCT01483690) reported that decitabine combined with vorinostat was not feasible in pediatric B-ALL patients due to the high toxicity, despite encouraging response rates (Burke et al. 2020). Studies are underway investigating the utility of both DNMTi and HDACi in the treatment of ALL, especially in combination with other agents, such as BCL-2 inhibitors, tyrosine kinase inhibitors, proteasome inhibitors, and mTOR inhibitors. Clinical trials involving epigenetic therapies for the treatment of ALL are summarized (Table 5.8).

Table 5.8 Clinical trials of epigenetic therapies in acute lymphoblastic leukemia (ALL) (*DNMT* DNA methyltransferase. *HDAC* histone deacetylase. *DOT1L* DOT1-like histone H3K79 methyltransferase. *AML* acute myeloid leukemia. *CML* chronic myeloid leukemia. *R/R AML* relapsed/refractory)

Epigenetic target	Trial number	Approach	Phase	Status
DNMT	NCT02828358	Azacitidine + combination chemotherapy for ALL and KMT2A gene rearrangement	2	Active, not recruiting
	NCT01861002	5-Azacytidine + combination chemotherapy for R/R ALL	1	Completed
	NCT02458235	Azacitidine + donor lymphocyte infusion to prevent relapse after stem cell transplantation	2	Completed
	NCT05149378	Azacitidine + venetoclax for R/R T-ALL	2	Recruiting
	NCT05376111	Azacitidine + venetoclax for newly diagnosed T-ALL	2	Recruiting
	NCT00349596	Low-dose decitabine for R/R ALL	1	Completed
	NCT00042796	Decitabine for R/R AML or ALL	1	Terminated
	NCT02264873	Decitabine, dose escalation study for ALL, AML	1	Completed
	NCT03132454	Palbociclib and sorafenib, decitabine, or dexamethasone for R/R leukemia	1	Recruiting
DNMT + HDAC	NCT01483690	Decitabine + vorinostat with chemotherapy for relapsed ALL	1/2	Terminated (toxicity)
	NCT00882206	Decitabine + vorinostat for relapsed lymphoblastic lymphoma or ALL	2	Terminated
	NCT00275080	Decitabine + vorinostat for advanced solid tumors or R/R non-Hodgkin's lymphoma, AML, ALL, or CML	1	Completed
	NCT00075010	Decitabine + valproate in R/R leukemia or myelodysplastic syndromes	1/2	Completed
HDAC	NCT01422499	Vorinostat for children with relapsed solid tumor, lymphoma, or leukemia	1/2	Completed
	NCT00816283	Vorinostat + dasatinib for accelerated phase or blastic phase CML or ALL	1	Completed
	NCT00217412	Vorinostat +/- isotretinoin for young patients with R/R solid tumors, lymphoma, or leukemia	1	Completed
	NCT02083250	Vorinostat + fludarabine, clofarabine, and busulfan for acute leukemia in R/R undergoing donor stem cell transplant	1	Completed
	NCT02419755	Vorinostat + bortezomib for younger patients with R/R MLL rearranged hematologic malignancies	2	Terminated
	NCT01312818	Vorinostat + bortezomib + dexamethasone for R/R ALL	2	Terminated
	NCT02553460	Vorinostat + bortezomib (total therapy for infants with ALL)	1/2	Active, not recruiting

(continued)

Table 5.8 (continued)

Epigenetic target	Trial number	Approach	Phase	Status
	NCT03117751	Vorinostat + (total therapy XVII for newly diagnosed patients with ALL and lymphoma)	2/3	Recruiting
	NCT01321346	Panobinostat for children with refractory hematologic malignancies	1	Completed
	NCT02518750	Panobinostat + bortezomib + liposomal vincristine (re-induction therapy for relapsed pediatric T-ALL or lymphoma)	2	Terminated
DOT1L	NCT02141828	Pinometostat for pediatric R/R leukemias bearing a rearrangement of the MLL gene	1	Completed

Novel therapeutic approaches targeting *KMT2A*-r ALL are under development. A vital element of *KMT2A*-r oncogenesis is *DOT1L*. A phase 1 study (NCT02141828) reported that pinometostat in children with R/R *KMT2A*-r leukemia has a tolerable safety profile yet poor clinical efficacy when used as monotherapy (Shukla et al. 2015). Epigenetic silencing of tumor suppressor genes through hypermethylation of the promoter region CpG island is another trait of *KMT2A*-r ALL; thus, numerous studies have shown that demethylating drugs like azacytidine and decitabine selectively reverse aberrant DNA methylation and induce apoptosis in *KMT2A*-r ALL cells (Xu et al. 2021). Inhibition of *BRD4*, which promotes transcription of *MYC* and other oncogenes, was found to be associated with antileukemic activity via downregulation of *KMT2A*-r and *MYC* target genes (Xu et al. 2021). Although preclinical evidence points to their potential value, the function of bromodomain inhibitors in lymphoid malignancies is still poorly understood (Meyer and Hermiston 2019).

Moreover, chimeric antigen receptor (CAR) T cells directed against CD19 (CART19) are effective in B-cell malignancies, yet molecular parameters to predict the clinical outcome of CART19 therapy are not well understood. For this purpose, a study demonstrated that DNA methylation profiles of pre-infusion CART19 cells may predict which patients with a B cell malignancy benefit from CAR T-cell therapy the most. Therefore, larger and prospective clinical trials are required in this regard (Garcia-Prieto et al. 2022).

In recent years, genomic and epigenomic research has significantly increased our understanding of the pathogenesis of chronic lymphocytic leukemia (CLL) by revealing a huge number of novel alterations that may be responsible for the progression of the disease (Landau et al. 2015; Beekman et al. 2018). Two main molecular categories have been identified from an immunogenetic perspective: those with mutated IG heavy-chain variable region (IGHV) genes (M-CLL) and those with unmutated IGHV genes (U-CLL). In addition to these subtypes, recent epigenetic studies also identified a third subtype with an intermediate profile with moderate levels of IGHV mutation. Nevertheless, further investigation is necessary to comprehend the biological significance of this subtype (Delgado et al. 2020). Epigenomics of CLL can also be a target in the context of drug resistance. A study demonstrated that the main resistance to *BTK* inhibitors in diffuse large B-cell lymphoma results from epigenetic rather than genetic alterations; moreover, the same mechanism in CLL was observed (Shaffer 3rd et al. 2021).

5.2.6.3 Lymphomas

Lymphoma is the most frequent lymphoid malignancy and one of the top ten malignancies globally (Siegel et al. 2022). It is divided into two types: Hodgkin's lymphoma (HL) and non-lymphoma Hodgkin's (NHL). NHL accounts for approximately 90% of all lymphomas, with the other 10% referred to as HL (Armitage et al. 2017). Lymphomas are a diverse group of malignancies defined by clonal lymphoproliferation. Epigenetic dysregulations, which are widespread in hematological malignancies such as lymphomas, have been found in recent years. It is now clear that epigenetic dysregulations in lymphoid neoplasms are mostly induced by genetic changes in genes encoding enzymes involved in histone or chromatin modifications (Chebly et al. 2021). Distinct patterns of DNA methylation, histone modifications, miRNA expression, and, more recently, long noncoding RNAs (lncRNAs) have been found to be critical in the preservation of T-cell identity and the growth of B cells, implying that epigenetic alterations are a major mechanism in numerous forms of lymphomas (Zhang et al. 2012).

Mutations in genes involved in DNA methylation and demethylation have been found in lymphomas, resulting in incorrect methylation patterns. These mutations may be shared by several lymphoma subtypes or exclusive to a single subtype. Mutations that result in abnormal focalized DNA hypermethylation can mute tumor suppressor genes, while mutations that result in widespread genomic DNA hypomethylation can induce genomic instability. For instance, in MYC-induced T-cell lymphomas, *DNMT1* was found to be significantly involved in the de novo methylation during carcinogenesis, preventing and maintaining the tumor phenotype (Peters et al. 2013). *DNTM1* is overexpressed in Burkitt lymphoma (BL) patients, indicating its critical involvement in BL pathogenesis beyond its role in normal B-cell development (Robaina et al. 2015). Also, mutations in *DNMT3A* have been reported in a variety of hematological disorders, including T-cell lymphomas. A significant prevalence of biallelic mutations is reported in T-cell neoplasms, suggesting that total loss of *DNMT3A* is a key event during the formation of these neoplasms (Yang et al. 2015). *DNMT3A* mutations co-occur with *TET2* (Ten-Eleven Translocation 2) mutations at a high frequency (73%), particularly in angioimmunoblastic T-cell lymphoma (AITL) and peripheral T-cell lymphoma (PTCL), implying carcinogenic collaboration involving cytosine methylation and demethylation processes (Couronne et al. 2012). Mutations in *DNMT3A* have also been found in cutaneous T-cell lymphoma (CTCL), with a decreased or total loss of expression in Sezary syndrome (SS), an aggressive subtype of CTCL (da Silva Almeida et al. 2015). Additionally, DNMT3B overexpression is predominantly reported in BL patients, contributing to the BL's DNA methylation pattern with *DNMT1* (Robaina et al. 2015). In addition, *DNMT3B* has been identified in diffuse large B-cell lymphoma (DLBCL) as a prognostic marker linked with treatment resistance and poor survival (Poole et al. 2017).

The methylation heterogeneity can be used to identify effective therapeutic responses. Relapsed DLBCL patients had more methylation heterogeneity at diagnosis than non-relapsed patients, implying a relapse-associated methylation signature in this malignancy (Pan et al. 2015). At relapse, a decrease in intra-tumor methylation heterogeneity induces clonal tumor cell selection, reinforcing the concept that methylation heterogeneity is dynamic and can be used as a predictive pre- and posttreatment biomarker. Furthermore, chemoresistant DLBCL patients have hypermethylation of SMAD1 (small mothers against decapentaplegic homolog 1) gene of which silencing was caused by hypermethylation. After exposure to a modest dosage of DNMT inhibitors, SMAD1 reactivation and chemosensitization were restored, confirming the notion that in certain lymphomas, more especially in DLBCL, DNA methylation might predict the response to therapy (Clozel et al. 2013).

DNA demethylation is a dynamic process involving *TET* (ten-eleven translocation) and *IDH* (isocitrate dehydrogenase) proteins that plays crucial roles in the transcriptional activation of silenced genes. Numerous studies have shown that *TET* or *IDH* genes may play a role in carcinogenesis, since their inactivation might cause aberrant histone/DNA methylation patterns. *TET2* mutations have been found in a variety of hematopoietic neoplasms, both in myeloid and lymphoid malignancies (Nakajima and Kunimoto 2014). AITL, one of the most common T-cell lymphomas, frequently has *TET2* mutations (50%) or *IDH2* mutations (20–30%). Surprisingly, these two mutations are not mutually exclusive and coexist in the same tumor in 60–100% of IDH2-mutated AITL patients (Inoue et al. 2016).

The most researched histone modifications in lymphomas are histone methylation and histone acetylation. Mutations in histone modification genes were identified, as well as abnormal specific methylation or acetylation profiles. Some mutations' ability to cause cancer has been clearly established; however, for other mutations, just correlative information is available, while functional research investigations are still being conducted (Chebly et al. 2021). A family of methyltransferases called KMT2 promotes transcription by inducing an open chromatin conformation, enabling H3K4 methylation. *KMT2D (MLL2)* is one of the most altered genes in follicular lymphoma (70%–90%) and DLBCL (30%) (Morin et al. 2011). Additionally, in 2022, scientists discovered *MYCN* to be a new oncogenic driver in PTCL. They also demonstrated that *MYCN* directly collaborates with *EZH2* as a transcriptional coactivator of the MYCN-driven gene expression program, which may be successfully targeted by *EZH2* degradation or dephosphorylation in conjunction with HDAC inhibition (Vanden Bempt et al. 2022). Furthermore, multicenter research focused on miRNAs as biomarkers in the treatment of lymphomas may provide interesting results, as several miRNAs have shown significant associations with medication resistance or sensitivity (Chebly et al. 2021).

Numerous novel medications targeting epigenetic alterations are now being studied in clinical trials for patients with hematological neoplasms such as lymphomas. The FDA approved the oral histone deacetylase inhibitor (HDI), vorinostat, in 2006 and the intravenous HDI, romidepsin, in 2009 for CTCL; for PTCL, romidepsin was approved in 2011, followed by belinostat in 2014. All three medicines block the three HDAC classes (I, II, and IV) that are important in lymphoma etiology (Booth and Collins 2021). Recent studies have focused on the use of HDIs in conjunction with other therapies. For example, in PTCL, phase I/II trials have been published investigating romidepsin in combination with CHOP (cyclophosphamide, doxorubicin, vincristine, and prednisolone) as first-line therapy, as well as combinations with gemcitabine, ICE (ifosfamide, cisplatin, and etoposide), or the antimetabolite pralatrexate (Chihara et al. 2015; Dupuis et al. 2015; Pellegrini et al. 2016; Amengual et al. 2018). While romidepsin and vorinostat are being studied in over 50 and 70 clinical trials, respectively, as monotherapy or in combination with other medications, other HDACi, including entinostat (MS-275), panobinostat (LBH589), resminostat (4SC-201), abexinostat (PCI24781), mocetinostat (MGCD0103), and others, are being studied in phase 1/2 trials for the treatment of B-cell or T-cell lymphomas (Chebly et al. 2021). Additionally, researchers in China conducted a multicenter phase II clinical study combining chidamide, a class I selective oral HDI, with a prednisone, etoposide, and thalidomide (CPET) regimen. The current investigation indicated that the oral CPET regimen was efficient for untreated AITL patients with tolerable effects in the Chinese population (Wang et al. 2022b). Tazemetostat, an oral EZH2 inhibitor, showed clinically meaningful, durable responses and was well tolerated in heavily pretreated patients with relapsed or refractory follicular lymphoma (Morschhauser et al. 2020). Several new EZH2 inhibitor agents have recently been discovered, and several of these are now being evaluated in clinical trials. The FDA has approved DNA demethylating drugs such as decitabine and azacitidine for the treatment of myelodysplastic syndrome (MDS) and acute myeloid leukemia (AML). Decitabine is now being studied in a phase 4 study as a monotherapy in relapsed and refractory DLBCL (NCT03579082) and relapsed or refractory T-lymphoblastic lymphoma (NCT03558412). Decitabine is being investigated in a few different clinical studies, either alone or in combination with other medications (Hu et al. 2021). While 5-azacytidine is currently not licensed for the treatment of AITL patients, a clinical trial demonstrated significant responses to 5-azacytidine not only in AITL patients with an associated myeloid neoplasm but also in 4 out of 7 AITL patients without a myeloid association indicating the effect of 5-azacytidine on AITL is not restricted to patients with associated myeloid neoplasm (Lemonnier et al. 2018) (Table 5.9).

In conclusion, new epigenetic monotherapies or combinations are currently being tested in clinical trials as phases 1–4. Increased knowledge of the epigenetic alterations driving lymphoid malignancy has aided in developing new potential therapeutics. Several medications targeting epigenetic modifiers demonstrated significant efficacy, and studies are currently underway to investigate new combinations of epigenetic pharmaceuticals and chemotherapies or even immunotherapies. Despite the amazing progress made in cancer epigenetic treatments, further understanding of

Table 5.9 Epigenetic therapies in lymphomas (a) primary mediastinal B-cell lymphoma (PMBL), (b) anaplastic large-cell lymphoma, (c) mantle cell lymphoma, (d) mycosis fungoides

Drug	Epigenetic mechanism	Indication	Status
Vorinostat	HDACi	CTCL	Approved
Romidepsin	HDACi	CTCL	Approved
Belinostat	HDACi	Relapsed/refractory PTCL	Approved
Decitabine	DNMTi	Relapsed/refractory DLBCL Relapsed/refractory T-lymphoblastic lymphoma	Clinical trials
Tazemetostat	EZH2i	DLBCL, FL, and PMBCL[a]	Clinical trials
CPI-1205	EZH2i	B-cell lymphomas	Clinical trials
SHR2554	EZH2i	Relapsed/refractory mature lymphoid neoplasms	Clinical trials
PF-06821497	EZH2i	FL	Clinical trials
Entinostat	HDACi	ALCL[b], AITL	Clinical trials
Panobinostat	HDACi	DLBCL, CTCL, MCL[c], AITL, PTCL	Clinical trials
Resminostat	HDACi	CTCL, MF[d], SS	Clinical trials
Abexinostat	HDACi	DLBCL, FL, MCL	Clinical trials
Mocetionostat	HDACi	DLBCL, FL	Clinical trials
Cobomarsen	miR-155 inhibitor	MF, DLBCL, ATLL	Clinical trials

the biological effects of epigenetic therapies, as well as the discovery of response mechanisms, is required. This will help us learn how to restore the abnormal epigenome and design tailored therapeutics.

5.3 Conclusion

In this chapter, we reviewed epigenetics in several prevalent and important cancers from a clinical perspective. Even though we now have a tremendous amount of knowledge about the role of epigenetic modifications in cancer development and progression, numerous unsolved questions still remain.

One of these question marks is epigenetic heterogeneity in cancer. Heterogeneity is a well-known trait of cancer and is also one of the obstacles to the development of effective treatments. To support this observation, sequencing of different regions of cancer mass, especially with single-cell sequencing (sc-seq) provides valuable information on epigenetic heterogeneity in cancer; however, most sc-seq performed so far has been limited in exploring genetic heterogeneity. Furthermore, epigenetic heterogeneity is not only a feature of the tumor cells but also the cells comprising the tumor microenvironment (TME), and this is also another unsolved question. A study exemplifying this showed that the activity of TET demethylase is inhibited by hypoxia (Thienpont et al. 2016). In another study, *ZEB1*, a key regulator of EMT, has been shown to keep the invasive EMT status via promoting the expression of a histone methyltransferase, *SETD1B*, which maintains *ZEB1* expression in a positive feedback loop (Lindner et al. 2020). Research is ongoing to better understand the relationship between tumor cells and TME and to develop new targeted agents.

Contrasting with increasing knowledge about the role of the epigenome in cancer, only nine drugs that target epigenome have been approved. To date, the approved epigenetic antitumor agents are DNMT inhibitors (azacitidine, decitabine), HDAC inhibitors (vorinostat, romidepsin, belinostat, panobinostat), IDH mutation inhibitors (enasidenib, ivosidenib), and an *EZH2* inhibitor (tazemetostat). Due to the limitations of these drugs and as cancer is arising from numerous genetic and epigenetic molecular processes, the combination of agents targeting both genetic and epigenetic alterations is needed to use for overcoming not only cancer itself but also the therapeutic resistance, in several cases.

The limitations of profiling epigenetic modifications will be solved by the development of next-generation sequencing technology, which enables high-resolution sequencing of huge numbers of cells, as well as standardization of techniques for collecting, processing, and evaluating data. Thus, cancer epigenetics will be better elucidated and the clinical use of epigenomics will become widespread.

Disclosure of Potential Conflicts of Interest
The authors have no conflicts of interest to declare.

References

Ahmad A (2019) Breast cancer statistics: recent trends. Adv Exp Med Biol 1152:1–7

Akimoto N, Zhao M, Ugai T, Zhong R, Lau MC, Fujiyoshi K et al (2021) Tumor Long interspersed nucleotide Element-1 (LINE-1) hypomethylation in relation to age of colorectal cancer diagnosis and prognosis. Cancers (Basel). 13(9)

Amengual JE, Lichtenstein R, Lue J, Sawas A, Deng C, Lichtenstein E et al (2018) A phase 1 study of romidepsin and pralatrexate reveals marked activity in relapsed and refractory T-cell lymphoma. Blood 131(4):397–407

Anastasiadou E, Jacob LS, Slack FJ (2018) Non-coding RNA networks in cancer. Nat Rev Cancer 18(1):5–18

Anghel SA, Ionita-Mindrican CB, Luca I, Pop AL (2021) Promising epigenetic biomarkers for the early detection of colorectal cancer: a systematic review. Cancers (Basel). 13(19)

Armitage JO, Gascoyne RD, Lunning MA, Cavalli F (2017) Non-Hodgkin lymphoma. Lancet 390(10091):298–310

Aryee MJ, Liu W, Engelmann JC, Nuhn P, Gurel M, Haffner MC et al (2013) DNA methylation alterations exhibit intraindividual stability and interindividual heterogeneity in prostate cancer metastases. Sci Transl Med 5(169):169ra10

Asada S, Goyama S, Inoue D, Shikata S, Takeda R, Fukushima T et al (2018) Mutant ASXL1 cooperates with BAP1 to promote myeloid leukaemogenesis. Nat Commun 9(1):2733

Asangani IA, Dommeti VL, Wang X, Malik R, Cieslik M, Yang R et al (2014) Therapeutic targeting of BET bromodomain proteins in castration-resistant prostate cancer. Nature 510(7504):278–282

Audia JE, Campbell RM (2016) Histone modifications and cancer. Cold Spring Harb Perspect Biol 8(4):a019521

Aytes A, Giacobbe A, Mitrofanova A, Ruggero K, Cyrta J, Arriaga J et al (2018) NSD2 is a conserved driver of metastatic prostate cancer progression. Nat Commun 9(1):5201

Azzam SK, Alsafar H, Sajini AA (2022) FTO m6A demethylase in obesity and cancer: implications and underlying molecular mechanisms. Int J Mol Sci 23(7):3800

Bauman J, Verschraegen C, Belinsky S, Muller C, Rutledge T, Fekrazad M et al (2012) A phase I study of 5-azacytidine and erlotinib in advanced solid tumor malignancies. Cancer Chemother Pharmacol 69(2):547–554

Beekman R, Chapaprieta V, Russinol N, Vilarrasa-Blasi R, Verdaguer-Dot N, Martens JHA et al (2018) The reference epigenome and regulatory chromatin landscape of chronic lymphocytic leukemia. Nat Med 24(6):868–880

Berdasco M, Esteller M (2019) Clinical epigenetics: seizing opportunities for translation. Nat Rev Genet 20(2):109–127

Bertolotto C (2013) Melanoma: from melanocyte to genetic alterations and clinical options. Scientifica (Cairo) 2013:635203

Bezerra MF, Lima AS, Pique-Borras MR, Silveira DR, Coelho-Silva JL, Pereira-Martins DA et al (2020) Co-occurrence of DNMT3A, NPM1, FLT3 mutations identifies a subset of acute myeloid leukemia with adverse prognosis. Blood 135(11):870–875

Black JC, Van Rechem C, Whetstine JR (2012) Histone lysine methylation dynamics: establishment, regulation, and biological impact. Mol Cell 48(4):491–507

Booth S, Collins G (2021) Epigenetic targeting in lymphoma. Br J Haematol 192(1):50–61

Borssen M, Haider Z, Landfors M, Noren-Nystrom U, Schmiegelow K, Asberg AE et al (2016) DNA methylation adds prognostic value to minimal residual disease status in Pediatric T-cell acute lymphoblastic Leukemia. Pediatr Blood Cancer 63(7):1185–1192

Branford S, Wang P, Yeung DT, Thomson D, Purins A, Wadham C et al (2018) Integrative genomic analysis reveals cancer-associated mutations at diagnosis of CML in patients with high-risk disease. Blood 132(9):948–961

Brown LJ, Achinger-Kawecka J, Portman N, Clark S, Stirzaker C, Lim E (2022) Epigenetic therapies and biomarkers in breast cancer. Cancers (Basel) 14(3)

Burke MJ, Kostadinov R, Sposto R, Gore L, Kelley SM, Rabik C et al (2020) Decitabine and Vorinostat with chemotherapy in relapsed Pediatric acute lymphoblastic Leukemia: a TACL pilot study. Clin Cancer Res 26(10):2297–2307

Candelaria M, Gallardo-Rincon D, Arce C, Cetina L, Aguilar-Ponce JL, Arrieta O et al (2007) A phase II study of epigenetic therapy with hydralazine and magnesium valproate to overcome chemotherapy resistance in refractory solid tumors. Ann Oncol 18(9):1529–1538

Caslini C, Hong S, Ban YJ, Chen XS, Ince TA (2019) HDAC7 regulates histone 3 lysine 27 acetylation and transcriptional activity at super-enhancer-associated genes in breast cancer stem cells. Oncogene 38(39):6599–6614

Cerretelli G, Ager A, Arends MJ, Frayling IM (2020) Molecular pathology of lynch syndrome. J Pathol 250(5):518–531

Chao YL, Pecot CV. Targeting epigenetics in lung cancer. Cold Spring Harb Perspect Med. 2021;11(6)

Chatterjee A, Rodger EJ, Ahn A, Stockwell PA, Parry M, Motwani J et al (2018) Marked global DNA hypomethylation is associated with constitutive PD-L1 expression in melanoma. iScience 4:312–325

Chebly A, Chouery E, Ropio J, Kourie HR, Beylot-Barry M, Merlio JP et al (2021) Diagnosis and treatment of lymphomas in the era of epigenetics. Blood Rev 48:100782

Chiappinelli KB, Strissel PL, Desrichard A, Li H, Henke C, Akman B et al (2015) Inhibiting DNA methylation causes an interferon response in cancer via dsRNA including endogenous retroviruses. Cell 162(5):974–986

Chihara D, Oki Y, Westin JR, Nastoupil L, Fayad LE, Samaniego F et al (2015) High response rate of Romidepsin in combination with ICE (Ifosfamide, carboplatin and etoposide) in patients with relapsed or refractory peripheral T-cell lymphoma: updates of phase I trial. Blood 126(23):3987

Chirshev E, Oberg KC, Ioffe YJ, Unternaehrer JJ (2019) Let-7 as biomarker, prognostic indicator, and therapy for precision medicine in cancer. Clin Transl Med 8(1):24

Chung W, Kelly AD, Kropf P, Fung H, Jelinek J, Su XY et al (2019) Genomic and epigenomic predictors of response to guadecitabine in relapsed/refractory acute myelogenous leukemia. Clin Epigenetics 11(1):106

Clozel T, Yang S, Elstrom RL, Tam W, Martin P, Kormaksson M et al (2013) Mechanism-based epigenetic chemosensitization therapy of diffuse large B-cell lymphoma. Cancer Discov 3(9):1002–1019

Connolly RM, Zhao F, Miller KD, Lee MJ, Piekarz RL, Smith KL et al (2021) E2112: randomized phase III trial of endocrine therapy plus Entinostat or placebo in hormone receptor-positive advanced breast cancer. A trial of the ECOG-ACRIN cancer research group. J Clin Oncol 39(28):3171–3181

Constancio V, Nunes SP, Henrique R, Jeronimo C (2020) DNA methylation-based testing in liquid biopsies as detection and prognostic biomarkers for the four major cancer types. Cell 9(3)

Couronne L, Bastard C, Bernard OA (2012) TET2 and DNMT3A mutations in human T-cell lymphoma. N Engl J Med 366(1):95–96

da Motta LL, Ledaki I, Purshouse K, Haider S, De Bastiani MA, Baban D et al (2017) The BET inhibitor JQ1 selectively impairs tumour response to hypoxia and downregulates CA9 and angiogenesis in triple negative breast cancer. Oncogene 36(1):122–132

da Silva Almeida AC, Abate F, Khiabanian H, Martinez-Escala E, Guitart J, Tensen CP et al (2015) The mutational landscape of cutaneous T cell lymphoma and Sezary syndrome. Nat Genet 47(12):1465–1470

Dardenne E, Beltran H, Benelli M, Gayvert K, Berger A, Puca L et al (2016) N-Myc induces an EZH2-mediated transcriptional program driving neuroendocrine prostate cancer. Cancer Cell 30(4):563–577

Dawson MA, Kouzarides T (2012) Cancer epigenetics: from mechanism to therapy. Cell 150(1):12–27

de Almeida BP, Apolonio JD, Binnie A, Castelo-Branco P (2019) Roadmap of DNA methylation in breast cancer identifies novel prognostic biomarkers. BMC Cancer 19(1):219

de Barrios O, Parra M (2021) Epigenetic control of infant B cell precursor acute lymphoblastic Leukemia. Int J Mol Sci 22(6):3127

Delgado J, Nadeu F, Colomer D, Campo E (2020) Chronic lymphocytic leukemia: from molecular pathogenesis to novel therapeutic strategies. Haematologica 105(9):2205–2217

Di Giacomo AM, Covre A, Finotello F, Rieder D, Danielli R, Sigalotti L et al (2019) Guadecitabine plus ipilimumab in unresectable melanoma: the NIBIT-M4 clinical trial. Clin Cancer Res 25(24):7351–7362

DiNardo CD, Pratz KW, Letai A, Jonas BA, Wei AH, Thirman M et al (2018) Safety and preliminary efficacy of venetoclax with decitabine or azacitidine in elderly patients with previously untreated acute myeloid leukaemia: a non-randomised, open-label, phase 1b study. Lancet Oncol 19(2):216–228

Doi A, Park IH, Wen B, Murakami P, Aryee MJ, Irizarry R et al (2009) Differential methylation of tissue- and cancer-specific CpG Island shores distinguishes human induced pluripotent stem cells, embryonic stem cells and fibroblasts. Nat Genet 41(12):1350-3

Dong Y, Zhao X, Feng X, Zhou Y, Yan X, Zhang Y et al (2019) SETD2 mutations confer chemoresistance in acute myeloid leukemia partly through altered cell cycle checkpoints. Leukemia 33(11):2585–2598

Doroshow DB, Sanmamed MF, Hastings K, Politi K, Rimm DL, Chen L et al (2019) Immunotherapy in non-small cell lung cancer: facts and hopes. Clin Cancer Res 25(15):4592–4602

Dou L, Yan F, Pang J, Zheng D, Li D, Gao L et al (2019) Protein lysine 43 methylation by EZH1 promotes AML1-ETO transcriptional repression in leukemia. Nat Commun 10(1):5051

Drobna M, Szarzynska-Zawadzka B, Daca-Roszak P, Kosmalska M, Jaksik R, Witt M et al (2018) Identification of endogenous control miRNAs for RT-qPCR in T-cell acute lymphoblastic Leukemia. Int J Mol Sci 19(10)

Dupuis J, Morschhauser F, Ghesquieres H, Tilly H, Casasnovas O, Thieblemont C et al (2015) Combination of romidepsin with cyclophosphamide, doxorubicin, vincristine, and prednisone in previously untreated patients with peripheral T-cell lymphoma: a non-randomised, phase 1b/2 study. Lancet Haematol 2(4):e160–e165

Duruisseaux M, Esteller M (2018) Lung cancer epigenetics: from knowledge to applications. Semin Cancer Biol 51:116–128

Ezponda T, Popovic R, Shah MY, Martinez-Garcia E, Zheng Y, Min DJ et al (2013) The histone methyltransferase MMSET/WHSC1 activates TWIST1 to promote an epithelial-mesenchymal transition and invasive properties of prostate cancer. Oncogene 32(23):2882–2890

Fedyuk V, Erez N, Furth N, Beresh O, Andreishcheva E, Shinde A et al (2022) Multiplexed, singlemolecule, epigenetic analysis of plasma-isolated nucleosomes for cancer diagnostics. Nat Biotechnol

Feinberg AP, Vogelstein B (1983) Hypomethylation distinguishes genes of some human cancers from their normal counterparts. Nature 301(5895):89–92

Gala K, Li Q, Sinha A, Razavi P, Dorso M, Sanchez-Vega F et al (2018) KMT2C mediates the estrogen dependence of breast cancer through regulation of ERalpha enhancer function. Oncogene 37(34):4692–4710

Gallagher SJ, Gunatilake D, Beaumont KA, Sharp DM, Tiffen JC, Heinemann A et al (2018) HDAC inhibitors restore BRAF-inhibitor sensitivity by altering PI3K and survival signalling in a subset of melanoma. Int J Cancer 142(9):1926–1937

Gao C, Zhang RD, Liu SG, Zhao XX, Cui L, Yue ZX et al (2017) Low CREBBP expression is associated with adverse long-term outcomes in paediatric acute lymphoblastic leukaemia. Eur J Haematol 99(2):150–159

Garcia-Manero G, Abaza Y, Takahashi K, Medeiros BC, Arellano M, Khaled SK et al (2019) Pracinostat plus azacitidine in older patients with newly diagnosed acute myeloid leukemia: results of a phase 2 study. Blood Adv 3(4):508–518

Garcia-Manero G, Tambaro FP, Bekele NB, Yang H, Ravandi F, Jabbour E et al (2012) Phase II trial of vorinostat with idarubicin and cytarabine for patients with newly diagnosed acute myelogenous leukemia or myelodysplastic syndrome. J Clin Oncol 30(18):2204–2210

Garcia-Prieto CA, Villanueva L, Bueno-Costa A, Davalos V, Gonzalez-Navarro EA, Juan M et al (2022) Epigenetic profiling and response to CD19 chimeric antigen receptor T-cell therapy in B-cell malignancies. J Natl Cancer Inst 114(3):436–445

Gebarowska K, Mroczek A, Kowalczyk JR, Lejman M (2021) MicroRNA as a prognostic and diagnostic marker in T-cell acute lymphoblastic Leukemia. Int J Mol Sci 22(10)

Glodzik D, Bosch A, Hartman J, Aine M, Vallon-Christersson J, Reutersward C et al (2020) Comprehensive molecular comparison of BRCA1 hypermethylated and BRCA1 mutated triple negative breast cancers. Nat Commun 11(1):3747

Goel A, Xicola RM, Nguyen TP, Doyle BJ, Sohn VR, Bandipalliam P et al (2010) Aberrant DNA methylation in hereditary nonpolyposis colorectal cancer without mismatch repair deficiency. Gastroenterology 138(5):1854–1862

Goldstein LJ, Zhao F, Wang M, Swaby RF, Sparano JA, Meropol NJ et al (2017) A phase I/II study of suberoylanilide hydroxamic acid (SAHA) in combination with trastuzumab (Herceptin) in patients with advanced metastatic and/or local chest wall recurrent HER2-amplified breast cancer: a trial of the ECOG-ACRIN cancer research group (E1104). Breast Cancer Res Treat 165(2):375–382

Gollner S, Oellerich T, Agrawal-Singh S, Schenk T, Klein HU, Rohde C et al (2017) Loss of the histone methyltransferase EZH2 induces resistance to multiple drugs in acute myeloid leukemia. Nat Med 23(1):69–78

Govindarajan V, Shah AH, Di L, Rivas S, Suter RK, Eichberg DG et al (2022) Systematic review of epigenetic therapies for treatment of IDH-mutant glioma. World Neurosurg 162:47–56

Gruhn B, Naumann T, Gruner D, Walther M, Wittig S, Becker S et al (2013) The expression of histone deacetylase 4 is associated with prednisone poor-response in childhood acute lymphoblastic leukemia. Leuk Res 37(10):1200–1207

Guo Y, Long J, Lei S (2019a) Promoter methylation as biomarkers for diagnosis of melanoma: a systematic review and meta-analysis. J Cell Physiol 234(5):7356–7367

Guo W, Wang H, Li C (2021) Signal pathways of melanoma and targeted therapy. Signal Transduct Target Ther 6(1):424

Guo W, Zhu L, Zhu R, Chen Q, Wang Q, Chen JQ (2019b) A four-DNA methylation biomarker is a superior predictor of survival of patients with cutaneous melanoma. elife:8

Heimbruch KE, Meyer AE, Agrawal P, Viny AD, Rao S (2021) A cohesive look at leukemogenesis: the cohesin complex and other driving mutations in AML. Neoplasia 23(3):337–347

Henrique R, Jeronimo C. Molecular detection of prostate cancer: a role for GSTP1 hypermethylation. Eur Urol 2004;46(5):660–9; discussion 9

Herbst A, Vdovin N, Gacesa S, Ofner A, Philipp A, Nagel D et al (2017) Methylated free-circulating HPP1 DNA is an early response marker in patients with metastatic colorectal cancer. Int J Cancer 140(9):2134–2144

Hinai A, Pratcorona M, Grob T, Kavelaars FG, Bussaglia E, Sanders MA et al (2019) The landscape of KMT2A-PTD AML: concurrent mutations, gene expression signatures, and clinical outcome. Hema 3(2):e181

Hinohara K, Wu HJ, Vigneau S, McDonald TO, Igarashi KJ, Yamamoto KN et al (2018) KDM5 histone demethylase activity links cellular transcriptomic heterogeneity to therapeutic resistance. Cancer Cell 34(6):939–53 e9

Hirukawa A, Singh S, Wang J, Rennhack JP, Swiatnicki M, Sanguin-Gendreau V et al (2019) Reduction of global H3K27me(3) enhances HER2/ErbB2 targeted therapy. Cell Rep 29(2):249–57 e8

Hoffmann F, Niebel D, Aymans P, Ferring-Schmitt S, Dietrich D, Landsberg J (2020) H3K27me3 and EZH2 expression in melanoma: relevance for melanoma progression and response to immune checkpoint blockade. Clin Epigenetics 12(1):24

Hu J, Wang X, Chen F, Ding M, Dong M, Yang W et al (2021) Combination of decitabine and a modified regimen of cisplatin, cytarabine and dexamethasone: a potential salvage regimen for relapsed or refractory diffuse large B-cell lymphoma after second-Line treatment failure. Front Oncol 11:687374

Iacobucci I, Mullighan CG (2017) Genetic basis of acute lymphoblastic Leukemia. J Clin Oncol 35(9):975–983

Ibanez de Caceres I, Cortes-Sempere M, Moratilla C, Machado-Pinilla R, Rodriguez-Fanjul V, Manguan-Garcia C et al (2010) IGFBP-3 hypermethylation-derived deficiency mediates cisplatin resistance in non-small-cell lung cancer. Oncogene 29(11):1681–1690

Imperiale TF, Ransohoff DF, Itzkowitz SH, Levin TR, Lavin P, Lidgard GP et al (2014) Multitarget stool DNA testing for colorectal-cancer screening. N Engl J Med 370(14):1287–1297

Inoue S, Lemonnier F, Mak TW (2016) Roles of IDH1/2 and TET2 mutations in myeloid disorders. Int J Hematol 103(6):627–633

Issa JP (2004) CpG Island methylator phenotype in cancer. Nat Rev Cancer 4(12):988–993

Jenuwein T, Allis CD (2001) Translating the histone code. Science 293(5532):1074–1080

Jiang Z, Li W, Hu X, Zhang Q, Sun T, Cui S et al (2019) Tucidinostat plus exemestane for postmenopausal patients with advanced, hormone receptor-positive breast cancer (ACE): a randomised, double-blind, placebo-controlled, phase 3 trial. Lancet Oncol 20(6):806–815

Juergens RA, Wrangle J, Vendetti FP, Murphy SC, Zhao M, Coleman B et al (2011) Combination epigenetic therapy has efficacy in patients with refractory advanced non-small cell lung cancer. Cancer Discov 1(7):598–607

Jung G, Hernandez-Illan E, Moreira L, Balaguer F, Goel A (2020) Epigenetics of colorectal cancer: biomarker and therapeutic potential. Nat Rev Gastroenterol Hepatol 17(2):111–130

Kahl M, Brioli A, Bens M, Perner F, Kresinsky A, Schnetzke U et al (2019) The acetyltransferase GCN5 maintains ATRA-resistance in non-APL AML. Leukemia 33(11):2628–2639

Kalin JH, Wu M, Gomez AV, Song Y, Das J, Hayward D et al (2018) Targeting the CoREST complex with dual histone deacetylase and demethylase inhibitors. Nat Commun 9(1):53

Karrman K, Johansson B (2017) Pediatric T-cell acute lymphoblastic leukemia. Genes Chromosomes Cancer 56(2):89–116

Kim ES (2017) Enasidenib: First Global Approval. Drugs 77(15):1705–1711

Kim JY, Choi JK, Jung H (2020) Genome-wide methylation patterns predict clinical benefit of immunotherapy in lung cancer. Clin Epigenetics 12(1):119

Kim E, Zucconi BE, Wu M, Nocco SE, Meyers DJ, McGee JS et al (2019) MITF expression predicts therapeutic vulnerability to p300 inhibition in human melanoma. Cancer Res 79(10): 2649–2661

Kresovich JK, Xu Z, O'Brien KM, Shi M, Weinberg CR, Sandler DP et al (2022) Blood DNA methylation profiles improve breast cancer prediction. Mol Oncol 16(1):42–53

Krishnan V, Kim DDH, Hughes TP, Branford S, Ong ST (2022) Integrating genetic and epigenetic factors in chronic myeloid leukemia risk assessment: toward gene expression-based biomarkers. Haematologica 107(2):358–370

Kuang C, Park Y, Augustin RC, Lin Y, Hartman DJ, Seigh L et al (2022) Pembrolizumab plus azacitidine in patients with chemotherapy refractory metastatic colorectal cancer: a single-arm phase 2 trial and correlative biomarker analysis. Clin Epigenetics 14(1):3

Kühn A, Löscher D, Marschalek R (2016) The IRX1/HOXA connection: insights into a novel t (4;11)- specific cancer mechanism. Oncotarget 7(23)

Kumaraswamy A, Welker Leng KR, Westbrook TC, Yates JA, Zhao SG, Evans CP et al (2021) Recent advances in epigenetic biomarkers and epigenetic targeting in prostate cancer. Eur Urol 80(1):71–81

Landau DA, Tausch E, Taylor-Weiner AN, Stewart C, Reiter JG, Bahlo J et al (2015) Mutations driving CLL and their evolution in progression and relapse. Nature 526(7574):525–530

Lantuejoul S, Salameire D, Salon C, Brambilla E (2009) Pulmonary preneoplasia–sequential molecular carcinogenetic events. Histopathology 54(1):43–54

Lemonnier F, Dupuis J, Sujobert P, Tournillhac O, Cheminant M, Sarkozy C et al (2018) Treatment with 5-azacytidine induces a sustained response in patients with angioimmunoblastic T-cell lymphoma. Blood 132(21):2305–2309

Li N, Kang Y, Wang L, Huff S, Tang R, Hui H et al (2020) ALKBH5 regulates anti-PD-1 therapy response by modulating lactate and suppressive immune cell accumulation in tumor microenvironment. Proc Natl Acad Sci U S A 117(33):20159–20170

Li XY, Wu JZ, Cao HX, Ma R, Wu JQ, Zhong YJ et al (2013) Blockade of DNA methylation enhances the therapeutic effect of gefitinib in non-small cell lung cancer cells. Oncol Rep 29(5): 1975–1982

Liang W, Zhao Y, Huang W, Gao Y, Xu W, Tao J et al (2019) Non-invasive diagnosis of early-stage lung cancer using high-throughput targeted DNA methylation sequencing of circulating tumor DNA (ctDNA). Theranostics 9(7):2056–2070

Lindner P, Paul S, Eckstein M, Hampel C, Muenzner JK, Erlenbach-Wuensch K et al (2020) EMT transcription factor ZEB1 alters the epigenetic landscape of colorectal cancer cells. Cell Death Dis 11(2):147

Liu Y, Lei P, Qiao H, Sun K, Lu X, Bao F et al (2019) miR-9 enhances the chemosensitivity of AML cells to daunorubicin by targeting the EIF5A2/MCL-1 Axis. Int J Biol Sci 15(3):579–586

Liu MC, Oxnard GR, Klein EA, Swanton C, Seiden MV, Consortium C (2020b) Sensitive and specific multi-cancer detection and localization using methylation signatures in cell-free DNA. Ann Oncol 31(6):745–759

Liu B, Ricarte Filho J, Mallisetty A, Villani C, Kottorou A, Rodgers K et al (2020a) Detection of promoter DNA methylation in urine and plasma aids the detection of non-small cell lung cancer. Clin Cancer Res 26(16):4339–4348

Loghavi S, Zuo Z, Ravandi F, Kantarjian HM, Bueso-Ramos C, Zhang L et al (2014) Clinical features of de novo acute myeloid leukemia with concurrent DNMT3A, FLT3 and NPM1 mutations. J Hematol Oncol 7(1):74

Long J, Fang WY, Chang L, Gao WH, Shen Y, Jia MY et al (2017) Targeting HDAC3, a new partner protein of AKT in the reversal of chemoresistance in acute myeloid leukemia via DNA damage response. Leukemia 31(12):2761–2770

Long J, Jia MY, Fang WY, Chen XJ, Mu LL, Wang ZY et al (2020) FLT3 inhibition upregulates HDAC8 via FOXO to inactivate p53 and promote maintenance of FLT3-ITD+ acute myeloid leukemia. Blood 135(17):1472–1483

Loo Yau H, Ettayebi I, De Carvalho DD (2019) The cancer epigenome: exploiting its vulnerabilities for immunotherapy. Trends Cell Biol 29(1):31–43

Lubbert M, Grishina O, Schmoor C, Schlenk RF, Jost E, Crysandt M et al (2020) Valproate and retinoic acid in combination with decitabine in elderly nonfit patients with acute myeloid Leukemia: results of a Multicenter, randomized, 2 × 2. Phase II Trial J Clin Oncol 38(3): 257–270

Luo M, Zhang Q, Xia M, Hu F, Ma Z, Chen Z et al (2018) Differential co-expression and regulatory network analysis uncover the relapse factor and mechanism of T cell acute Leukemia. Mol Ther Nucleic Acids 12:184–194

Luu TH, Morgan RJ, Leong L, Lim D, McNamara M, Portnow J et al (2008) A phase II trial of vorinostat (suberoylanilide hydroxamic acid) in metastatic breast cancer: a California cancer Consortium study. Clin Cancer Res 14(21):7138–7142

Ma J, Liu B, Yu D, Zuo Y, Cai R, Yang J et al (2019) SIRT3 deacetylase activity confers chemoresistance in AML via regulation of mitochondrial oxidative phosphorylation. Br J Haematol 187(1):49–64

Mackowska N, Drobna-Sledzinska M, Witt M, Dawidowska M (2021) DNA methylation in t-cell acute Lymphoblastic Leukemia: in search for clinical and biological meaning. Int J Mol Sci 22(3):1388

Marijon H, Lee DH, Ding L, Sun H, Gery S, de Gramont A et al (2018) Co-targeting poly (ADP-ribose) polymerase (PARP) and histone deacetylase (HDAC) in triple-negative breast cancer: higher synergism in BRCA mutated cells. Biomed Pharmacother 99:543–551

Masuda N, Tamura K, Yasojima H, Shimomura A, Sawaki M, Lee MJ et al (2021) Phase 1 trial of entinostat as monotherapy and combined with exemestane in Japanese patients with hormone receptor-positive advanced breast cancer. BMC Cancer 21(1):1269

Meyer LK, Hermiston ML (2019) The epigenome in pediatric acute lymphoblastic leukemia: drug resistance and therapeutic opportunities. Cancer Drug Resist 2(2):313–325

Micevic G, Theodosakis N, Bosenberg M (2017) Aberrant DNA methylation in melanoma: biomarker and therapeutic opportunities. Clin Epigenetics 9:34

Milani L, Lundmark A, Kiiialainen A, Nordlund J, Flaegstad T, Forestier E et al (2010) DNA methylation for subtype classification and prediction of treatment outcome in patients with childhood acute lymphoblastic leukemia. Blood 115(6):1214–1225

Montesinos P, Recher C, Vives S, Zarzycka E, Wang J, Bertani G et al (2022) Ivosidenib and Azacitidine in IDH1-mutated acute myeloid Leukemia. N Engl J Med 386(16):1519–1531

Moran B, Silva R, Perry AS, Gallagher WM (2018) Epigenetics of malignant melanoma. Semin Cancer Biol 51:80–88

Morel D, Jeffery D, Aspeslagh S, Almouzni G, Postel-Vinay S (2020) Combining epigenetic drugs with other therapies for solid tumours–past lessons and future promise. Nat Rev Clin Oncol 17(2):91–107

Morel KL, Sheahan AV, Burkhart DL, Baca SC, Boufaied N, Liu Y et al (2021) EZH2 inhibition activates a dsRNA-STING-interferon stress axis that potentiates response to PD-1 checkpoint blockade in prostate cancer. Nat Cancer 2(4):444–456

Mori T, O'Day SJ, Umetani N, Martinez SR, Kitago M, Koyanagi K et al (2005) Predictive utility of circulating methylated DNA in serum of melanoma patients receiving biochemotherapy. J Clin Oncol 23(36):9351–9358

Morin RD, Mendez-Lago M, Mungall AJ, Goya R, Mungall KL, Corbett RD et al (2011) Frequent mutation of histone-modifying genes in non-Hodgkin lymphoma. Nature 476(7360):298–303

Morschhauser F, Tilly H, Chaidos A, McKay P, Phillips T, Assouline S et al (2020) Tazemetostat for patients with relapsed or refractory follicular lymphoma: an open-label, single-arm, multicentre, phase 2 trial. Lancet Oncol 21(11):1433–1442

Moubarak RS, de Pablos-Aragoneses A, Ortiz-Barahona V, Gong Y, Gowen M, Dolgalev I et al (2022) The histone demethylase PHF8 regulates TGFbeta signaling and promotes melanoma metastasis. Sci Adv 8(7):eabi7127

Muller D, Gyorffy B (2022) DNA methylation-based diagnostic, prognostic, and predictive biomarkers in colorectal cancer. Biochim Biophys Acta Rev Cancer 1877(3):188722

Munster PN, Marchion D, Thomas S, Egorin M, Minton S, Springett G et al (2009) Phase I trial of vorinostat and doxorubicin in solid tumours: histone deacetylase 2 expression as a predictive marker. Br J Cancer 101(7):1044–1050

Munster PN, Thurn KT, Thomas S, Raha P, Lacevic M, Miller A et al (2011) A phase II study of the histone deacetylase inhibitor vorinostat combined with tamoxifen for the treatment of patients with hormone therapy-resistant breast cancer. Br J Cancer 104(12):1828–1835

Nakajima H, Kunimoto H (2014) TET2 as an epigenetic master regulator for normal and malignant hematopoiesis. Cancer Sci 105(9):1093–1099

Nemes K, Csoka M, Nagy N, Mark A, Varadi Z, Danko T et al (2015) Expression of certain leukemia/lymphoma related microRNAs and its correlation with prognosis in childhood acute lymphoblastic leukemia. Pathol Oncol Res 21(3):597–604

Nordlund J, Syvanen AC (2018) Epigenetics in pediatric acute lymphoblastic leukemia. Semin Cancer Biol 51:129–138

Norsworthy KJ, Luo L, Hsu V, Gudi R, Dorff SE, Przepiorka D et al (2019) FDA approval summary: Ivosidenib for relapsed or refractory acute myeloid Leukemia with an isocitrate Dehydrogenase-1 mutation. Clin Cancer Res 25(11):3205–3209

Ok CY, Loghavi S, Sui D, Wei P, Kanagal-Shamanna R, Yin CC et al (2019) Persistent IDH1/2 mutations in remission can predict relapse in patients with acute myeloid leukemia. Haematologica 104(2):305–311

Otani J, Nankumo T, Arita K, Inamoto S, Ariyoshi M, Shirakawa M (2009) Structural basis for recognition of H3K4 methylation status by the DNA methyltransferase 3A ATRX-DNMT3-DNMT3L domain. EMBO Rep 10(11):1235–1241

Pan H, Jiang Y, Boi M, Tabbo F, Redmond D, Nie K et al (2015) Epigenomic evolution in diffuse large B-cell lymphomas. Nat Commun 6:6921

Panagopoulou M, Karaglani M, Balgkouranidou I, Biziota E, Koukaki T, Karamitrousis E et al (2019) Circulating cell-free DNA in breast cancer: size profiling, levels, and methylation patterns lead to prognostic and predictive classifiers. Oncogene 38(18):3387–3401

Park H, Garrido-Laguna I, Naing A, Fu S, Falchook GS, Piha-Paul SA et al (2016) Phase I dose-escalation study of the mTOR inhibitor sirolimus and the HDAC inhibitor vorinostat in patients with advanced malignancy. Oncotarget 7(41):67521–67531

Park DJ, Kwon A, Cho BS, Kim HJ, Hwang KA, Kim M et al (2020) Characteristics of DNMT3A mutations in acute myeloid leukemia. Blood Res 55(1):17–26

Pasculli B, Barbano R, Parrella P (2018) Epigenetics of breast cancer: biology and clinical implication in the era of precision medicine. Semin Cancer Biol 51:22–35

Pashayan N, Antoniou AC, Ivanus U, Esserman LJ, Easton DF, French D et al (2020) Personalized early detection and prevention of breast cancer: ENVISION consensus statement. Nat Rev Clin Oncol 17(11):687–705

Pellegrini C, Dodero A, Chiappella A, Monaco F, Degl'Innocenti D, Salvi F et al (2016) A phase II study on the role of gemcitabine plus romidepsin (GEMRO regimen) in the treatment of relapsed/refractory peripheral T-cell lymphoma patients. J Hematol Oncol 9:38

Peters SL, Hlady RA, Opavska J, Klinkebiel D, Novakova S, Smith LM et al (2013) Essential role for Dnmt1 in the prevention and maintenance of MYC-induced T-cell lymphomas. Mol Cell Biol 33(21):4321–4333

Piha-Paul SA, Sachdev JC, Barve M, LoRusso P, Szmulewitz R, Patel SP et al (2019) First-in-human study of Mivebresib (ABBV-075), an Oral Pan-inhibitor of bromodomain and extra terminal proteins, in patients with relapsed/refractory solid Tumors. Clin Cancer Res 25(21): 6309–6319

Poole CJ, Zheng W, Lodh A, Yevtodiyenko A, Liefwalker D, Li H et al (2017) DNMT3B overexpression contributes to aberrant DNA methylation and MYC-driven tumor maintenance in T-ALL and Burkitt's lymphoma. Oncotarget 8(44):76898–76920

Prebet T, Sun Z, Ketterling RP, Zeidan A, Greenberg P, Herman J et al (2016) Azacitidine with or without Entinostat for the treatment of therapy-related myeloid neoplasm: further results of the E1905 north American Leukemia intergroup study. Br J Haematol 172(3):384–391

Ramaswamy B, Fiskus W, Cohen B, Pellegrino C, Hershman DL, Chuang E et al (2012) Phase I-II study of vorinostat plus paclitaxel and bevacizumab in metastatic breast cancer: evidence for vorinostat-induced tubulin acetylation and Hsp90 inhibition in vivo. Breast Cancer Res Treat 132(3):1063–1072

Rao RC, Dou Y (2015) Hijacked in cancer: the KMT2 (MLL) family of methyltransferases. Nat Rev Cancer 15(6):334–346

Rinke J, Chase A, Cross NCP, Hochhaus A, Ernst T (2020) EZH2 in myeloid malignancies. Cell 9(7)

Robaina MC, Mazzoccoli L, Arruda VO, Reis FR, Apa AG, de Rezende LM et al (2015) Deregulation of DNMT1, DNMT3B and miR-29s in Burkitt lymphoma suggests novel contribution for disease pathogenesis. Exp Mol Pathol 98(2):200–207

Roman-Gomez J, Jimenez-Velasco A, Agirre X, Castillejo JA, Navarro G, Garate L et al (2006) Promoter hypermethylation and global hypomethylation are independent epigenetic events in lymphoid leukemogenesis with opposing effects on clinical outcome. Leukemia 20(8): 1445–1448

Roman-Gomez J, Jimenez-Velasco A, Agirre X, Prosper F, Heiniger A, Torres A (2005) Lack of CpG Island methylator phenotype defines a clinical subtype of T-cell acute lymphoblastic leukemia associated with good prognosis. J Clin Oncol 23(28):7043–7049

Sabnis GJ, Goloubeva O, Chumsri S, Nguyen N, Sukumar S, Brodie AM (2011) Functional activation of the estrogen receptor-alpha and aromatase by the HDAC inhibitor entinostat sensitizes ER-negative tumors to letrozole. Cancer Res 71(5):1893–1903

Saghafinia S, Mina M, Riggi N, Hanahan D, Ciriello G (2018) Pan-cancer landscape of aberrant DNA methylation across human Tumors. Cell Rep 25(4):1066–80 e8

San Jose-Eneriz E, Gimenez-Camino N, Agirre X, Prosper F (2019) HDAC inhibitors in acute myeloid Leukemia. Cancers (Basel). 11(11):1794

Sanjuan-Pla A, Bueno C, Prieto C, Acha P, Stam RW, Marschalek R et al (2015) Revisiting the biology of infant t(4;11)/MLL-AF4+ B-cell acute lymphoblastic leukemia. Blood 126(25): 2676–2685

Santourlidis S, Schulz WA, Arauzo-Bravo MJ, Gerovska D, Ott P, Bendhack ML et al (2022) Epigenetics in the diagnosis and therapy of malignant melanoma. Int J Mol Sci 23(3)

Sato H, Wheat JC, Steidl U, Ito K (2016) DNMT3A and TET2 in the pre-leukemic phase of hematopoietic disorders. Front Oncol 6:187

Savio AJ, Bapat B (2015) Beyond the Island: epigenetic biomarkers of colorectal and prostate cancer. Methods Mol Biol 1238:103–124

Seligson DB, Horvath S, Shi T, Yu H, Tze S, Grunstein M et al (2005) Global histone modification patterns predict risk of prostate cancer recurrence. Nature 435(7046):1262–1266

Shaffer AL 3rd, Phelan JD, Wang JQ, Huang D, Wright GW, Kasbekar M et al (2021) Overcoming acquired epigenetic resistance to BTK inhibitors. Blood Cancer Discov 2(6):630–647

Shan M, Yin H, Li J, Li X, Wang D, Su Y et al (2016) Detection of aberrant methylation of a six-gene panel in serum DNA for diagnosis of breast cancer. Oncotarget 7(14):18485–18494

Sheng W, LaFleur MW, Nguyen TH, Chen S, Chakravarthy A, Conway JR et al (2018) LSD1 ablation stimulates anti-tumor immunity and enables checkpoint blockade. Cell 174(3):549–63 e19

Sher G, Salman NA, Khan AQ, Prabhu KS, Raza A, Kulinski M et al (2022) Epigenetic and breast cancer therapy: promising diagnostic and therapeutic applications. Semin Cancer Biol 83:152–165

Shi X, Tasdogan A, Huang F, Hu Z, Morrison SJ, DeBerardinis RJ (2017) The abundance of metabolites related to protein methylation correlates with the metastatic capacity of human melanoma xenografts. Sci Adv 3(11):eaao5268

Shu S, Lin CY, He HH, Witwicki RM, Tabassum DP, Roberts JM et al (2016) Response and resistance to BET bromodomain inhibitors in triple-negative breast cancer. Nature 529(7586):413–417

Shukla N, O'Brien MM, Silverman LB, Pauly M, Wetmore C, Loh ML et al (2015) Preliminary report of the phase 1 study of the DOT1L inhibitor, Pinometostat, EPZ-5676, in children with relapsed or refractory MLL-r acute Leukemia: safety, exposure and target inhibition. Blood 126(23):3792

Siegel RL, Miller KD, Fuchs HE, Jemal A (2022) Cancer statistics, 2022. CA Cancer J Clin 72(1):7–33

Song L, Jia J, Peng X, Xiao W, Li Y (2017) The performance of the SEPT9 gene methylation assay and a comparison with other CRC screening tests: a meta-analysis. Sci Rep 7(1):3032

Stahl M, Kohrman N, Gore SD, Kim TK, Zeidan AM, Prebet T (2016) Epigenetics in cancer: a Hematological perspective. PLoS Genet 12(10):e1006193

Stief SM, Hanneforth AL, Weser S, Mattes R, Carlet M, Liu WH et al (2020) Loss of KDM6A confers drug resistance in acute myeloid leukemia. Leukemia 34(1):50–62

Sun W, Triche T Jr, Malvar J, Gaynon P, Sposto R, Yang X et al (2018) A phase 1 study of azacitidine combined with chemotherapy in childhood leukemia: a report from the TACL consortium. Blood 131(10):1145–1148

Sung H, Ferlay J, Siegel RL, Laversanne M, Soerjomataram I, Jemal A et al (2021) Global cancer statistics 2020: GLOBOCAN estimates of incidence and mortality worldwide for 36 cancers in 185 countries. CA Cancer J Clin 71(3):209–249

Szejniuk WM, Robles AI, McCulloch T, Falkmer UGI, Roe OD (2019) Epigenetic predictive biomarkers for response or outcome to platinum-based chemotherapy in non-small cell lung cancer, current state-of-art. Pharmacogenomics J 19(1):5–14

Takayama K, Misawa A, Suzuki T, Takagi K, Hayashizaki Y, Fujimura T et al (2015) TET2 repression by androgen hormone regulates global hydroxymethylation status and prostate cancer progression. Nat Commun 6:8219

Tasian SK, Hunger SP (2017) Genomic characterization of paediatric acute lymphoblastic leukaemia: an opportunity for precision medicine therapeutics. Br J Haematol 176(6):867–882

Tasian SK, Loh ML, Hunger SP (2015) Childhood acute lymphoblastic leukemia: integrating genomics into therapy. Cancer 121(20):3577–3590

Tawbi HA, Beumer JH, Tarhini AA, Moschos S, Buch SC, Egorin MJ et al (2013) Safety and efficacy of decitabine in combination with temozolomide in metastatic melanoma: a phase I/II study and pharmacokinetic analysis. Ann Oncol 24(4):1112–1119

Tayari MM, Santos HGD, Kwon D, Bradley TJ, Thomassen A, Chen C et al (2021) Clinical responsiveness to all-trans retinoic acid is potentiated by LSD1 inhibition and associated with a quiescent transcriptome in myeloid malignancies. Clin Cancer Res 27(7):1893–1903

Taylor K, Loo Yau H, Chakravarthy A, Wang B, Shen SY, Ettayebi I et al (2020) An open-label, phase II multicohort study of an oral hypomethylating agent CC-486 and durvalumab in advanced solid tumors. J Immunother Cancer 8(2)

Thienpont B, Van Dyck L, Lambrechts D (2016) Tumors smother their epigenome. Mol Cell Oncol 3(6):e1240549

Thomas S, Thurn KT, Bicaku E, Marchion DC, Munster PN (2011) Addition of a histone deacetylase inhibitor redirects tamoxifen-treated breast cancer cells into apoptosis, which is opposed by the induction of autophagy. Breast Cancer Res Treat 130(2):437–447

Toh JWT, Phan K, Reza F, Chapuis P, Spring KJ (2021) Rate of dissemination and prognosis in early and advanced stage colorectal cancer based on microsatellite instability status: systematic review and meta-analysis. Int J Color Dis 36(8):1573–1596

Navada SC (2021) Therapeutics and DNA methylation inhibitors, Ed. Tollefsbol T, Clinical Epipenetics.

Touzart A, Boissel N, Belhocine M, Smith C, Graux C, Latiri M et al (2020) Low level CpG Island promoter methylation predicts a poor outcome in adult T-cell acute lymphoblastic leukemia. Haematologica 105(6):1575–1581

Toyota M, Kopecky KJ, Toyota MO, Jair KW, Willman CL, Issa JP (2001) Methylation profiling in acute myeloid leukemia. Blood 97(9):2823–2829

Tse JWT, Jenkins LJ, Chionh F, Mariadason JM (2017) Aberrant DNA methylation in colorectal cancer: what should we target? Trends Cancer 3(10):698–712

Tuo L, Sha S, Huayu Z, Du K (2018) P16(INK4a) gene promoter methylation as a biomarker for the diagnosis of non-small cell lung cancer: an updated meta-analysis. Thorac Cancer 9(8): 1032–1040

Van der Meulen J, Sanghvi V, Mavrakis K, Durinck K, Fang F, Matthijssens F et al (2015) The H3K27me3 demethylase UTX is a gender-specific tumor suppressor in T-cell acute lymphoblastic leukemia. Blood 125(1):13–21

Vanden Bempt M, Debackere K, Demeyer S, Van Thillo Q, Meeuws N, Fernandez CP et al (2022) Aberrant MYCN expression drives oncogenic hijacking of EZH2 as a transcriptional activator in peripheral T cell lymphoma. Blood

Visvanathan K, Fackler MS, Zhang Z, Lopez-Bujanda ZA, Jeter SC, Sokoll LJ et al (2017) Monitoring of serum DNA methylation as an early independent marker of response and survival in metastatic breast cancer: TBCRC 005 prospective biomarker study. J Clin Oncol 35(7): 751–758

Wallace JA, O'Connell RM (2017) MicroRNAs and acute myeloid leukemia: therapeutic implications and emerging concepts. Blood 130(11):1290–1301

Wang R, Gao X, Yu L (2019) The prognostic impact of tet oncogene family member 2 mutations in patients with acute myeloid leukemia: a systematic-review and meta-analysis. BMC Cancer 19(1):389

Wang L, Hui H, Agrawal K, Kang Y, Li N, Tang R et al (2020) M(6) a RNA methyltransferases METTL3/14 regulate immune responses to anti-PD-1 therapy. EMBO J 39(20):e104514

Wang X, Tian L, Lu J, Ng IO-L (2022a) Exosomes and cancer–diagnostic and prognostic biomarkers and therapeutic vehicle. Oncogenesis 11(1):54

Wang J, Zhang Q, Li Q, Mu Y, Jing J, Li H et al (2021) Phase I study and pilot efficacy analysis of Entinostat, a novel histone deacetylase inhibitor, in Chinese postmenopausal women with hormone receptor-positive metastatic breast cancer. Target Oncol 16(5):591–599

Wang Y, Zhang M, Song W, Cai Q, Zhang L, Sun X et al (2022b) Chidamide plus prednisone, etoposide, and thalidomide for untreated angioimmunoblastic T-cell lymphoma in a Chinese population: a multicenter phase II trial. Am J Hematol 97(5):623–629

Wei B, Wu F, Xing W, Sun H, Yan C, Zhao C et al (2021) A panel of DNA methylation biomarkers for detection and improving diagnostic efficiency of lung cancer. Sci Rep 11(1):16782

Wong KK, Lawrie CH, Green TM (2019) Oncogenic roles and inhibitors of DNMT1, DNMT3A, and DNMT3B in acute myeloid leukaemia. Biomark Insights 14:1177271919846454

Xu H, Yu H, Jin R, Wu X, Chen H (2021) Genetic and epigenetic targeting therapy for Pediatric acute lymphoblastic Leukemia. Cell 10(12):3349

Xue Y, Lian W, Zhi J, Yang W, Li Q, Guo X et al (2019) HDAC5-mediated deacetylation and nuclear localisation of SOX9 is critical for tamoxifen resistance in breast cancer. Br J Cancer 121(12):1039–1049

Yang L, Rau R, Goodell MA (2015) DNMT3A in haematological malignancies. Nat Rev Cancer 15(3):152–165

Yang S, Wei J, Cui YH, Park G, Shah P, Deng Y et al (2019b) M(6)a mRNA demethylase FTO regulates melanoma tumorigenicity and response to anti-PD-1 blockade. Nat Commun 10(1):2782

Yang X, Wong MPM, Ng RK (2019a) Aberrant DNA methylation in acute myeloid Leukemia and its clinical implications. Int J Mol Sci 20(18)

Yang L, Zhang Y, Shan W, Hu Z, Yuan J, Pi J et al (2017) Repression of BET activity sensitizes homologous recombination-proficient cancers to PARP inhibition. Sci Transl Med 9(400)

Yardley DA, Ismail-Khan RR, Melichar B, Lichinitser M, Munster PN, Klein PM et al (2013) Randomized phase II, double-blind, placebo-controlled study of exemestane with or without entinostat in postmenopausal women with locally recurrent or metastatic estrogen receptor-positive breast cancer progressing on treatment with a nonsteroidal aromatase inhibitor. J Clin Oncol 31(17):2128–2135

Yegnasubramanian S, De Marzo AM, Nelson WG (2019) Prostate cancer epigenetics: from basic mechanisms to clinical implications. Cold Spring Harb Perspect Med 9(4)

Yi Y, Ge S (2022) Targeting the histone H3 lysine 79 methyltransferase DOT1L in MLL-rearranged leukemias. J Hematol Oncol 15(1):35

Zhang JA, Mortazavi A, Williams BA, Wold BJ, Rothenberg EV (2012) Dynamic transformations of genome-wide epigenetic marking and transcriptional control establish T cell identity. Cell 149(2):467–482

Zhang F, Tang X, Fan S, Liu X, Sun J, Ju C et al (2021) Targeting the p300/NONO axis sensitizes melanoma cells to BRAF inhibitors. Oncogene 40(24):4137–4150

Zhao BS, Roundtree IA, He C (2017) Post-transcriptional gene regulation by mRNA modifications. Nat Rev Mol Cell Biol 18(1):31–42

Zlobec I, Bihl MP, Foerster A, Rufle A, Terracciano L, Lugli A (2012) Stratification and prognostic relevance of Jass's molecular classification of colorectal cancer. Front Oncol 2:7

Zou HZ, Yu BM, Wang ZW, Sun JY, Cang H, Gao F et al (2002) Detection of aberrant p16 methylation in the serum of colorectal cancer patients. Clin Cancer Res 8(1):188–191

Chapter 6
Epigenetic Regulation in Breast Cancer Tumor Microenvironment

Bhavjot Kaur, Priya Mondal, and Syed Musthapa Meeran

Abstract Breast cancer is an exceedingly complex disease that is driven by multiple factors and aberrantly regulated pathways. Mounting evidence suggests that the aggressive nature of breast cancer is highly influenced by its microenvironment, called the tumor microenvironment (TME). Bidirectional cross talk between the cancer cells and the cells of the immune system helps in the reshaping of the TME into an immunosuppressive and pro-tumorigenic milieu through a process called tumor immunoediting. However, the molecular mechanisms underlying the plasticity of TME have not been thoroughly explored. Recent studies have shown the participation of epigenetic dysregulation, such as DNA methylation, histone modifications, and ncRNA-mediated gene silencing, in the regulation of the plastic nature of the TME. Thus, in order to obtain a better clinical response, altering epigenetics in conjunction with immunotherapy may be a potential therapeutic strategy. This chapter reviews the role of various innate and adaptive immune cells in breast cancer TME and how epigenetics modifications drive this immunosuppression. We also summarize the effects of epigenetic modulators on the TME and the potential of these epigenetic modulators to improve the prognosis of breast cancer patients.

Keywords Epigenetics · Breast cancer · Tumor microenvironment · Epigenetic modulators · Tumor immunology

---

B. Kaur
Department of Biochemistry, CSIR-Central Food Technological Research Institute, Mysuru, Karnataka, India

P. Mondal · S. M. Meeran (✉)
Department of Biochemistry, CSIR-Central Food Technological Research Institute, Mysuru, Karnataka, India

Academy of Scientific and Innovative Research (AcSIR), Ghaziabad, India
e-mail: s.musthapa@cftri.res.in

Abbreviations

ADCC	Antibody-dependent cell-mediated cytotoxicity
APCs	Antigen-presenting cells
CAFs	Cancer-associated fibroblasts
DCs	Dendritic cells
ECM	Extracellular matrix
EMT	Epithelial to mesenchymal transition
HCC	Hepatocellular carcinoma
MDSCs	Myeloid-derived suppressor cells
MHC-II	Major histocompatibility complex class II
snRNA	Small nuclear ribonucleic acid RNA
TAM	Tumor-associated macrophage
TGF-β	Transforming growth factor-β
TME	Tumor microenvironment
TNBC	Triple-negative breast cancer
VEGF	Vascular endothelial growth factor

6.1 Introduction

According to the GLOBOCAN 2020 reports, breast cancer has the highest incidence and mortality rates among women worldwide (Sung et al. 2021). Current treatment options available for breast cancer can be divided into three major therapeutic approaches: surgery, radiation, and anticancer drugs (classical chemotherapy, hormonal therapy, and targeted therapy). However, owing to the heterogeneity of the bulk tumor, which contains a high degree of diversity among the cancer cells, the treatment of breast cancer becomes a strenuous task. Thus, there is a need to explore and better understand the molecular differences between the bulk tumor, which is composed of multiple cell types, and the specific contribution of these cells in the prognosis of breast cancer.

As the breast tumor progresses, the tumor becomes infiltrated by immune cells in an attempt for an antitumor response. However, the tumor-secreted factors help in reshaping the antitumor response to a suppressed or pro-tumorigenic response. This immunosuppressive niche is called the tumor microenvironment (TME), which includes not only the immune cells but also stromal cells, cytokines and chemokines, blood vessels, and extracellular matrix (ECM). This further influences bidirectional cross talk between the tumor cells and the cells of the TME and plays a critical role in the progression of breast cancer (Baghban et al. 2020). However, the molecular mechanisms underlying the reshaping of the TME are not thoroughly explored.

Recent studies have revealed epigenomic signatures in the breast TME, which associate with pro-tumorigenic mechanisms, such as tumor-associated macrophage (TAM) polarization, T cell and natural killer cells exhaustion, dysfunctional dendritic cells, and activation of cancer-associated fibroblast. This epigenetic dysregulation of immune signatures hence helps in breast cancer progression and metastasis and also creates a hindrance for the immunotherapy and chemotherapy to work efficiently (Jeschke et al. 2015). Interestingly, unlike genetic alterations, epigenetic alterations such as DNA methylation, histone modifications, and ncRNA-mediated gene silencing are reversible in nature and hence act as attractive targets for disease therapy, including breast cancer therapy. Thus, given the novel role of epigenetics in the TME, it makes an attractive target for the use of epigenetic modulatory drugs, also called as epi-drugs, along with the use of immunomodulators for breast cancer therapy, the idea of which is being tested in many clinical trials (Lodewijk et al. 2021).

In this chapter, we have discussed the role of TME in breast cancer progression and how epigenetic modifications help in modulating the TME from an antitumorigenic to an immunosuppressive environment. Further, we have also discussed the application of epigenetic modulators in immune activation for the treatment and management of breast cancer.

6.2 Tumor Microenvironment in Breast Cancer

A favorable TME is marked by the immunosuppressive effect of the immune cells on tumor sculpting during cancer immunoediting (Mondal et al. 2021). Immunoediting is a process where the immune system establishes complex and dynamic bidirectional cross talk with the tumor cells, thus promoting tumor progression (Mittal et al. 2014; Vesely and Schreiber 2013). This process comprises three phases: (1) elimination phase, also called immunosurveillance, where the malignant cells are destroyed by the immune system in an attempt to fight against cancer; (2) equilibrium phase, where a balance is established between the immune cells and the tumor cells that survived the elimination phase; and (3) escape phase, where the immune cells fail to limit tumor growth causing clinically apparent disease. Therefore, given the various innate and adaptive immune cells involved in tumor immunoediting, we have discussed the functions of the major immune subpopulations in breast cancer TME, summarized in Table 6.1.

6.2.1 Tumor-Associated Macrophages (TAMs)

Macrophages, a part of the innate immune system, are derived from a type of white blood cells, called monocytes, which play crucial a role in the host defense, where they help in modulating the immune response and also in the phagocytosis and destruction of the pathogens. To carry on these functions, macrophages are polarized to different phenotypes, M1 and M2, depending upon the stimuli (Mills 2015). M1-polarized macrophages are known to have a pro-inflammatory phenotype,

Table 6.1 Cell populations contributing to the development of breast cancer tumor microenvironment

TME components	Pathway	Contribution	References
M2/TAMs	Breast cancer cells secrete M-CSF	Promotes M2 polarization, causing high cell proliferation of grade 3 ER-negative breast cancer	(Sousa et al. 2015)
	High hyaluronan accumulation in tumor	Causes high TAMs infiltration, high tumor volume, lymph node metastasis, increased relapse rate, and low overall survival among breast cancer patients	(Tiainen et al. 2015)
	Tumor cells secrete lactate	Promotes M2 polarization via activation of ERK/STAT3 pathway, promotes breast cancer cell proliferation, migration, and angiogenesis	(Mu et al. 2018)
	M2-polarized macrophages induce EMT in cancer cells	Promotes migration and invasion of breast cancer cells	(Chen et al. 2022)
	A high SOX2 expression in breast cancer cells increases ICAM-1 and CCL3 secretion	Recruitment of TAMs via regulation of NFAT, STAT3, and NF-κB signaling pathways promotes breast cancer metastasis	(Mou et al. 2015)
T cells	Activated Treg cells express high levels of CCR8	Increases Treg cell proliferation and suppressive activities causing low overall and disease-free survival of breast cancer patients	(Plitas et al. 2016)
	B cells induce the conversion of resting CD4+ T cells to FoxP3+ Treg cells	Suppressed T cell activity promotes lung metastasis in mouse 4T1 breast tumor model	(Olkhanud et al. 2011)
	IL-17-producing γδ T cells promote polarization and expansion of neutrophils	Suppressed cytotoxic T lymphocytes promote lymph node metastasis in mouse breast tumor model	(Coffelt et al. 2015)
	Naïve CD4+ T cells infiltrate breast tumors via CCL18-dependent chemotaxis	Differentiation into Tregs, causing reduced disease-free survival of breast cancer patients	(Su et al. 2017)
NK cells	Tumor-infiltrating immature NK cells express reduced cytotoxic granzyme signature	Activation of cancer stem cells via Wnt signaling, thus contributing to breast cancer progression	(Thacker et al. 2023)
	Reduced expression of CX3CL1 chemokine	Low infiltration of NK cells, T cells, and DC cells, contributing to low overall and disease-	(Park et al. 2012)

(continued)

Table 6.1 (continued)

TME components	Pathway	Contribution	References
		free survival of breast cancer patients	
Dendritic cells	CTLA-4-expressing breast cancer cells activate ERK and STAT3 signaling in DCs	Reduction in the maturation of DCs and impairment antigen presentation ability of DCs and also reduced IL-2, IL-6, TNF-α, and IFN-γ secretion, causing hindrance in the differentiation of naïve T cells to Th1 effector cells	(Chen et al. 2017)
	Dysfunctional DCs with reduced HLA-DR, CD40, and CD86 expression with reduced IL-12 secretion	Reduced MLR response and T cell proliferation in breast cancer patients	(Satthaporn et al. 2004)
	Reduced CD11c expression correlates with reduced CXCL12 and CXCL13 secretion	Decrease in tumor-infiltrating lymphocytes (TILs) and tertiary lymphoid structures (TLSs), causing low recurrence-free and overall survival in TNBC patients	(Lee et al. 2018)
MDSCs	Expansion of IL-10 and IL-8 secreting MDSCs	Decrease in T cell proliferation in metastatic breast cancer patients	(Bergenfelz et al. 2015)
	Expansion of IL-6 secreting CD11b+/CXCR2 MDSCs	Increased T cell exhaustion and induction of EMT in breast cancer cells, promoting breast cancer growth and metastasis	(Zhu et al. 2017)
	Enrichment of MDSCs in the peripheral blood of breast cancer patients	Association with de novo metastatic breast cancer with ER negativity and liver and bone metastasis	(Bergenfelz et al. 2020)
	Expansion of MDSCs via an increase in IDO expression	Reduced T cell proliferation and Th cell polarization and induction of apoptosis in T cells cause lymph node metastasis in breast cancer patients	(Yu et al. 2013)
CAFs	Hypoxia induces HIF-1α/HPER signaling in CAFs	Increase in expression of VEGF in breast cancer cells and CAFs, thus promoting hypoxia-dependent tumor angiogenesis	(De Francesco et al. 2013)
	Downregulation of caveolin-1	Regulation of RB pathway, causing poor clinical outcome in tamoxifen-treated breast cancer patient	(Mercier et al. 2008)
	Compression-induced glycolysis in CAFs	Induces EMT and angiogenesis in breast cancer	(Kim et al. 2019a)

(continued)

Table 6.1 (continued)

TME components	Pathway	Contribution	References
Endothelial cells	Overexpression of YAP signaling pathway in breast cancer cells	Induction of CTGF and ANG-2 in endothelial cells, thus promoting angiogenesis	(Yan et al. 2022)
	Overexpression of EGFL6 in cancer and endothelial cells	Promotes EMT and stemness of cancer cells and induces tumor angiogenesis	(An et al. 2019)

where they activate the immune response and thus have antitumorigenic properties. These macrophages are known to secrete pro-inflammatory cytokines, such as IL-1β, TNF, IL-6, IL-12, IL-23, and IL-18, and also express high levels of major histocompatibility complex class II (MHC-II) molecules, CD68, CD80, and CD86 cell surface markers (Chávez-Galán et al. 2015; Orecchioni et al. 2019). On the other hand, M2-polarized macrophages have an anti-inflammatory phenotype, which thereby suppresses the immune system response and thus have pro-tumorigenic properties, and are hence deemed as TAMs (Jayasingam et al. 2020). M2 macrophages secrete anti-inflammatory cytokines, such as, IL-10, and TGF-β, where TGF-β plays a pivotal role in the activation of epithelial to mesenchymal transition (EMT) pathway in cancer cells, thus contributing to cancer metastasis (Dong et al. 2019; Duque and Descoteaux 2014; Yadav and Shankar 2019). Therefore, a high infiltration of the M2 macrophages in breast cancer has been linked to a more aggressive tumor phenotype. A study by Linde et al. (2018) observed that in HER2+ breast cancer cells HER2 signaling activates the NF-κB signaling pathway. As a result, this signaling pathway transcriptionally activates CCL2, which is a chemokine and plays a role in recruiting macrophages. These macrophages further induce Wnt-1, downregulating E-cadherin junctions in cancer cells and promoting metastasis (Linde et al. 2018). Another chemokine CCL18, secreted by TAMs, bound to the PITPNM3 transmembrane receptor present on the breast cancer cells and was shown to induce ECM adherence and metastasis of breast cancer cells (Chen et al. 2011). Dysregulation of NF-κB signaling in M2 macrophages causes an upregulation of HSPG2, which is essential for the stiffness of ECM in breast cancer (De Paolis et al. 2022). Moreover, a meta-analysis study revealed that high infiltration of CD68 TAMs relates to worse overall and disease-free survival in breast cancer patients, which can be used as a biomarker for breast cancer progression (Zhao et al. 2017).

One of the crucial characteristics of solid tumors, such as breast cancer, is hypoxia due to abnormal cancer cell growth and vasculature. This results in regions within the tumor with low blood supply causing low oxygen concentrations. This hypoxic niche influences the cancer cells to secrete chemoattractants, further recruiting TAMs (Henze and Mazzone 2016). This hypoxic condition in breast tumors further upregulates VEGF expression in TAMs resulting in an increased tumor vascularity (Obeid et al. 2013). Further, hypoxia also promotes immune evasion via TAMs secreted immunosuppressive cytokines (Henze and Mazzone 2016).

6.2.2 Helper T Cells and Cytotoxic T Cells

T cells, a part of the adaptive immune system, is derived from a type of leukocyte called lymphocytes. These lymphocytes originate in the bone marrow and mature in the thymus, called T cells or T lymphocytes. Upon maturation, they are released into the bloodstream as naïve T cells, where they encounter a recognizable antigen-presenting cell (APC) with the help of T cell receptors (TCRs). CD4+ T cells, upon activation following antigen recognition, are known to differentiate into helper T (Th) cells (Th1, Th2, and Th17 cells), which play a key role in immune activation, and CTLA4$^+$CD4$^+$CD25$^+$ T regulatory (Treg) cells, which are known to act as immunosuppressors (Zhu and Paul 2008). Th cells are known to secrete cytokines, such as IFN-γ, TNF-α, IL-4, IL-5, IL-13, and IL-22, that help in the activation of B cells, macrophages and DCs and also increase T cell-mediated cytotoxicity. On the other hand, Treg cells, also known as suppressor T cells, play a crucial role in modulating the immune response by inhibiting autoimmune responses. These cells secrete IL-10 and IL-35, which have anti-inflammatory functions, thus limiting tissue damage and promoting wound healing (Luckheeram et al. 2012). Therefore, a high ratio of Treg/Th cells has been observed in stage IV breast cancer patients and has been associated with impaired immune function (Wang et al. 2012). TSLP, an epithelium-derived cytokine, induced CD4+ T cell immunity against high-grade breast tumors by causing terminal differentiation of breast cancer cells (Boieri et al. 2022). CD8+ T cells, recognized by MHC class I molecules, are also called killer T cells or cytotoxic cells, are activated upon recognition with APCs, mainly dendritic cells, with the help of TCRs by forming a complex with MHC I receptor, and further carry out cytotoxic functions against the target cell. CD8+ T cells viability was reduced due to dysregulated tryptophan catabolism, which reduced distant metastasis-free survival and overall survival in breast cancer patients (Greene et al. 2019). Moreover, a high infiltration of CD8+ T cells in ER-negative and ER-positive breast tumors associated with favorable clinical outcomes (Ali et al. 2014; Mahmoud et al. 2011).

T cells are also known to express two immune checkpoint proteins or receptors, called programmed cell death protein 1 (PD-1) and cytotoxic T-lymphocyte-associated protein 4 (CTLA-4), which help in the regulation of T cell cytotoxic immune response. PD-1 binds to its ligand programmed cell death ligand 1 (PD-L1), whereas, CTLA-4 binds to its ligands, the B7 family of proteins, CD80 and CD86. These receptors, under normal conditions, bind to their ligands and help keep the immune response in check. However, cancer cells, are known to overexpress these ligands, and when they bind to their respective receptors, they reduce T cell proliferation and inactivate their cytotoxic activity, which leads to an immunosuppressive microenvironment (Saleh et al. 2019). TGF-β present in the TME induces the secretion of exosomal PD-L1, which causes CD8+ T cell dysfunction via early phosphorylation of TCR signalome in breast cancer (Chatterjee et al. 2021).

6.2.3 Natural Killer Cells (NK Cells)

Natural killer cells are a type of white blood cells and a component of the innate immune system. These are large granular lymphocytes that have cytotoxic properties and play a key role in providing the first line of defense against physiologically stressed cells such as malignant cells and virus-infected cells. However, the modulation of NK cells in the TME causes low cytotoxicity of the NK cells, which leads to NK cell exhaustion. This causes suppression of the immune system, thus helping in cancer progression. In a study demonstrated by Krneta et al. (2016), NK cells in the breast TME were shown to have an immature phenotype, which upon treatment with IL-12 and anti-TGFβ, induced NK cells maturity and activation (Krneta et al. 2016). In another study, it was demonstrated that hypoxia helps breast cancer cells evade NK cell-mediated cytotoxicity via the activation of autophagy, which is achieved by the degradation of NK-derived granzyme B (Baginska et al. 2013). However, upon neoadjuvant chemotherapy, breast cancer patients were observed to have increased peripheral NK cell activity, which correlated with the disappearance of axillary nymph node metastasis (Kim et al. 2019b). Moreover, NK cells from the peripheral blood of metastatic breast cancer patients showed TGF-β-derived metabolic deficits, which caused reduced IFN-γ secretion and cytotoxicity (Slattery et al. 2021).

PD-1, an inhibitory receptor mainly found on the surface of T cells, which upon binding to the PD-L1 ligand on the tumor cells, leads to the inactivation of T cells. However, studies have shown that NK cells also harness PD-1 receptors, which cause NK cell exhaustion upon binding to PD-L1 on tumor cells. Triple-negative breast cancer (TNBC) cells secrete IL-18, increasing PD-1 expression on CD56dimCD16dim/− NK cells, further correlating with poor prognosis among breast cancer patients (Park et al. 2017). Further, in a study demonstrated by Juliá et al. (2018), NK cell-mediated cytotoxicity against TNBC cells was significantly increased upon treatment with avelumab (Juliá et al. 2018). Avelumab is an FDA-approved human IgG1 anti-PD-L1 monoclonal antibody medication that induces antibody-dependent cell-mediated cytotoxicity (ADCC) against cancer cells. Another receptor commonly found on the NK cell surface is the NKG2D receptor, which is known to recognize several ligands, such as NKG2DL, on the surface of malignant cells. NKG2D–NKG2DL interaction of the NK cells to the malignant cells leads to NK cell activation and cytotoxic activity. However, protein expression of NKG2DLs, mainly histocompatibility complex class I chain-related proteins A and B (MICA/B), is significantly lower in advanced-stage breast cancer patients with nodal metastasis (Shen et al. 2017).

6.2.4 Dendritic Cells (DCs)

Dendritic cells, also known as antigen-presenting cells (APCs), link innate and adaptive immunity. DCs are known to capture the antigen, which is further processed and expressed on the surface of the DCs in the form of a peptide via MHC molecules to be recognized by the T cells. This interaction causes the naïve CD4+ and CD8+ T cells to differentiate into antigen-specific memory T cells. Along with MHC/peptide complex expression, DCs secrete pro-inflammatory cytokines, such as TNFα and IL-12, essential for T cell activation (de Winde et al. 2020; Tang et al. 2017). However, in the TME, the DC functions are altered by two mechanisms: (1) tumor cells alter the potential of hematopoietic progenitor cells to differentiate into functional DC, thus suppressing DC function, and (2) where the tumor cells alter the DC maturation mechanism, by promoting an early but dysfunctional maturation of DCs. DCs isolated from breast cancer patients were characterized by a more mature phenotype and impaired IL-12 production, which is associated with the differentiation of naïve T cells into Th1 cells (Della Bella et al. 2003). DCs in breast cancer patients were found to have decreased antigen presentation, which led to a defective T cell immune response (Gabrilovich et al. 1997; Satthaporn et al. 2004). Further, plasmacytoid DCs from breast cancer patients show reduced potential to secrete IFN-γ, which favors FoxP3$^+$ Treg differentiation from naïve T cells, thus leading to an immunosuppressive microenvironment. Moreover, upon treatment with exogenous IFN-γ, a substantial reduction in Treg cell differentiation was observed, suggesting a defect in the IFN-γ production by DCs, a cause for immunosuppression (Sisirak et al. 2012). Patients with high infiltration of CD1a DCs showed longer disease-free survival, bone metastasis-free survival, and overall survival in breast cancer patients (Giorello et al. 2021). Another study showed that type 1 conventional DCs are essential for CD4+ and CD8+ T cell activation and infiltration via activation of regulation of STAT1 signaling in breast cancer patients (Mattiuz et al. 2021).

6.2.5 Myeloid-Derived Suppressor Cells (MDSCs)

MDSCs, with CD33 as one of the key cell surface markers, are a group of heterogeneous cells (with granulocytic and monocytic morphology) that belong to the myeloid lineage and are frozen at different stages of differentiation. During normal conditions, the immature myeloid cells (IMCs) are known to differentiate into granulocytes and monocytes, hence generating an immune response. However, during cancer, the differentiation of myeloid cells is blocked, thus leading to the expansion of MDSCs (Gabrilovich and Nagaraj 2009). MDSCs are known to produce high levels of reactive oxygen species (ROS), thus inhibiting the proliferation and inducing apoptosis of the T cells (Yang et al. 2020). Expansion of MDSCs and breast cancer cell growth is shown to be influenced by a combination of cytokines and growth factors. For example, Flt3L, TGF-β, IL-6, IL-1, and GM-CSF contribute to the expansion of MDSCs via breast cancer cell lines. Moreover, VEGF was also shown to enhance MDSC expansion in breast cancer mice models, where anti-VEGF antibody treatment caused a decrease in MDSC accumulation and increased the number of mature DCs (Markowitz et al. 2013). In addition, IL-12 treatment reduced MDSCs expansion and increased its differentiation to macrophages and DCs in tumor-bearing mice (Markowitz et al. 2013). In another study, MDSCs were demonstrated to enhance cancer stem cell-like properties via secretion of IL-6, which led to the phosphorylation of STAT3, and also activated NOTCH signaling through nitric oxide in breast cancer cells (Peng et al. 2016). CCL20, a chemokine secreted by breast cancer cells, binds to its receptor, CCR6, thus causing MDSCs expansion, activating the CXCR2/NOTCH1/HEY1 signaling pathway in cancer cells, thus promoting breast cancer stemness (Zhang et al. 2023). Moreover, a high expression of ALDH1A in tumor-initiating cells (TICs), or breast cancer stem cells, activates NF-κB signaling, thus causing the secretion of GM-CSF, which leads to the expansion of MDSCs and immunosuppression (Liu et al. 2021a).

6.2.6 Cancer-Associated Fibroblasts (CAFs)

Normal fibroblasts are spread throughout the connective tissue and are known to produce collagen and extracellular matrix, providing a structural framework to the tissues. They also play an important role in wound healing, where they are recruited to the site of injury and work together with other cells to repair the tissue and maintain homeostasis (Sahai et al. 2020). However, in cancer, various stimulating factors from the cancer cells lead to the activation of CAFs from fibroblasts. CAFs are a heterogeneous population of stromal cells or mesenchymal stromal cells and are the most abundantly present component of the TME (Monteran and Erez 2019). CAFs secrete CXCL12, which helps downregulate mammalian diaphanous-related formin 2 (mDia2) in breast cancer cells, thereby enhancing breast cancer cell migration and invasion (Dvorak et al. 2018). Moreover, CAFs also increased the expression of *S100A4, TGFβ, FGF7, PDGFA, uPA, IL-6, IL-8, MMP2, MMP11, TIMP1,* and *VEGFA* in breast cancer cell lines. This led to an increase in breast cancer cell invasion and also induced angiogenesis via increased tube formation by endothelial cells through an increase in the expression of phosphorylated VEGFR-2, ERK1/2, and p-38 (Eiro et al. 2018). Compressive stress is an outcome of increased cancer cell proliferation within a limited tissue space. This compressive stress has been shown to increase lactate production via increased aerobic glycolysis in CAFs by inducing enolase 2 (*ENO2*), hexokinase 2 (*HK2*), and 6-phosphofructo-2-kinase/fructose-2,6-biphosphatase 3 (*PFKFB3*) gene expression through activation of the c-Jun terminal kinase signaling pathway. This led to an increase in the expression of genes associated with EMT (*TWIST1, SNAI1, ZEB1, ZEB2, CDH1, CDH2,* and *MMP2*) and angiogenesis (VEGFA/B) in breast cancer cells, thus contributing to cancer progression and metastasis (Kim et al. 2019a). Activation of the AGE-RAGE signaling pathway in CAFs led to a high secretion of IL-8, which induced IL-8/CXCR1/2-mediated interactions between CAFs and breast cancer cells, thus contributing to migratory and invasive characteristics of breast cancer cells (Santolla et al. 2022). Moreover, activation of GPR30 increased HMGB1 secretion via induction of PI3K/AKT signaling in CAFs. This further triggered MEK/ERK signaling, causing autophagy and tamoxifen resistance in breast cancer cells (Liu et al. 2021b).

6.2.7 Endothelial Cells

Endothelial cells are known to be involved in the process of angiogenesis, where they line up the vascular system and regulate the contraction and relaxation of the blood vessels. Solid tumors, like breast cancer, are known to be highly vascularized as cancer cells constantly need oxygen and other nutrients for their growth, proliferation, and also invasion (Yang et al. 2021a). A high expression of vascular endothelial growth factor (VEGF) through the cross talk between breast cancer cells and the endothelial cells has been shown to promote angiogenesis (Buchanan et al. 2012). In another study, it was demonstrated that radiation and doxorubicin treatment induce senescence in endothelial cells, which increases the secretion of CXCL11 and promotes migratory and invasive properties via induction of EMT through binding to its receptor CXCR3 and activating ERK signaling pathway in breast cancer cells (Hwang et al. 2020). Ets-1, a transcription factor, is found to be upregulated in breast cancer cells, which causes the activation of MMP-9, which is then recruited by endothelial cells, thus inducing endothelial cell capillary-like morphogenesis, which thereby increases the invasive morphogenetic properties of breast cancer cells (Furlan et al. 2019). Transforming growth factor beta (TGF-β), a multifunctional cytokine is produced by cells of the immune system, mainly white blood cell lineages, and plays a crucial role in the induction of EMT via activation of TGF-β signaling pathway in cancer cells, including breast cancer cells. These TGF-β-induced EMT breast cancer cells overexpress CCR7 by p38 MAPK signaling via JunB transcription factor, which enhances chemotaxis potential and helps migrate cancer cells toward CCL21-producing endothelial cells, which act as a ligand for CCL21 (Pang et al. 2016). In another study, TGF-β-induced EMT TNBC cells were shown to produce plasminogen activator inhibitor-1 (PAI-1), which further stimulates the secretion of CCL5 from endothelial cells, thus increasing the chemotaxis potential of TNBC cells. CCL5 secreted by endothelial cells forms a positive feedback loop, enhancing the TNBC potential to secrete PAI-1, thus promoting TNBC metastasis (Zhang et al. 2018).

6.2.8 Cytokines and Chemokines

Cytokines and chemokines are cellular secreted proteins known to regulate several processes in modulating the immune response. They are secreted by both the cancer cells and the immune system cells and help in communication between the two. In the TME, cytokines promote the formation of an immunosuppressive environment, thus promoting cancer progression. Interleukin-4 (IL-4), a cytokine that is mainly produced by the cells of the immune system and the cancer cells itself, is involved in the polarization of macrophages into the M2 phenotype (Wang and Joyce 2010). Further, in breast cancer, these M2 macrophages (TAMs) are known to secrete cytokine transforming growth factor (TGF-β), which activates epithelial to mesenchymal transition (EMT), thus causing lung metastasis (Liu et al. 2020). Chemokines, also called chemotactic cytokines, are a type of cytokines that belong to the family of small secreted proteins that are involved in cell migration and motility (Hughes and Nibbs 2018). VEGF is known to act as a chemokine that attracts the endothelial cells thus leading to angiogenesis and vascularization of the tumor (Buchanan et al. 2012). High secretion of IL-6, IL1β, and TNF-α was shown to cause cognitive impairment in breast cancer patients receiving chemotherapy (Tyagi et al. 2023). Moreover, a proinflammatory cytokine, IL-17, plays a crucial role in breast cancer proliferation, invasion, and metastasis and correlates with a poor prognosis of the disease (Song et al. 2021).

6.3 Epigenetic Modifications in Breast Cancer Tumor Microenvironment

As we discussed, immune system cells have high plasticity, where their differentiation (activation or suppression) depends on the regulation of gene expression. There is mounting evidence demonstrating the role of aberrant epigenetic machinery in the deregulation of gene expression, contributing to breast cancer progression (Mondal et al. 2020). Interestingly, recent studies have shown that aberrant epigenetic mechanisms also play a role in reshaping the TME of breast cancer and thus changing the antitumorigenic phenotype to pro-tumorigenic or immunosuppressive phenotype, some of which have been summarized in Table 6.2 and depicted the key mechanism in Fig. 6.1.

6.3.1 DNA Methylation

DNA methylation, catalyzed by a family of enzymes called the DNA methyltransferases (DNMTs), is a biological process, where a methyl group is added to the CpG dinucleotide in the DNA, which thereby hinders the binding of the transcriptional factors causing transcriptional repression of the gene. Studies have demonstrated that breast cancer cells can induce DNA methylation changes in the TME. The deregulation of DNA methylation machinery has been further shown to affect breast cancer tumorigenesis by regulating the components of the TME.

The function of TAMs or M2 macrophages has been established in the progression of breast cancer; however, the role of DNMTs in the polarization of

Table 6.2 Epigenetic dysregulation studies in breast cancer tumor microenvironment

Epigenetic dysregulation	Target gene	Target cell	Contribution	References
↑ P300 histone acetyltransferase	↑ DNMT1 via IL-6-pSTAT3-ZEB1-DNMT1 axis	Breast cancer cells	High infiltration of TAMs in breast tumor	(Li et al. 2022a, b)
↑ DNMT1	↓ STING causing suppression of the cGAS-STING pathway	Breast cancer cells	Impairment of T cell infiltration and cytolytic function, thus promoting breast cancer growth	(Wu et al. 2021)
↑ HDAC6	↑ COX2 via regulation of STAT3 signaling	CAFs	The immunosuppressive phenotype of CAFs, associated with poor survival outcomes in breast cancer patients	(Li et al. 2018)
↑ miR-20a	↓ LBP2 and MICA/B via inhibition of MAPK/ERK pathway	Breast cancer cells	NK cell exhaustion	(Shen et al. 2017)
↑ exosomal miR-181d-5p	↓ CDX2/HOXA5	CAFs	Induction of EMT in breast cancer cells	(Wang et al. 2020)
↑ exosomal miR-660	↓ KLHL21	TAMs	Promote breast cancer cell invasion and migration	(Li et al. 2022a, b)
↑ exosomal miR-9 and miR-181a	↓ SOCS3 and PIAS3	Breast cancer cells	MDSC expansion and T cell exhaustion in breast cancer	(Jiang et al. 2020)

↑ upregulation; ↓ downregulation

macrophages in breast cancer is largely unknown. A low expression of DNMT3b was observed in adipose tissue macrophages (ATMs), which were involved in the anti-inflammatory M2 polarization of the macrophages through modulating PPARγ expression (Yang et al. 2014b). Further, a high expression of DNMT1 has been shown to cause M1 polarization and inflammation (Zhou et al. 2017). In contrast, DNMT3a and DNMT3al expression were found to be upregulated in M2-polarized porcine macrophages (Sang et al. 2014). However, extensive research is required to decipher the role of epigenetic modification in the macrophage polarization and epigenetically altered macrophages' role in breast cancer.

In a computational study, high expression of FLAD1 in breast cancer was shown to have a positive correlation with DNMT expression and a negative correlation with CD8$^+$ T cell infiltration (Zhang et al. 2021). Methylation-derived neutrophil-to-lymphocyte ratios (mdNLRs) are a prognostic biomarker and are estimated using array-based DNA methylation data. The mdNLR was higher in TNBC patients compared to control, which was associated with lower ratios of NK cells, CD4 + T cells, CD8 + T cells, monocytes, and B cells, with the strongest association with lower NK cell ratio (Manoochehri et al. 2021).

IRF8 is a transcriptional factor that is crucial in differentiating myeloid cells from monocyte precursor cells. The promoter region for IRF8 was found to be

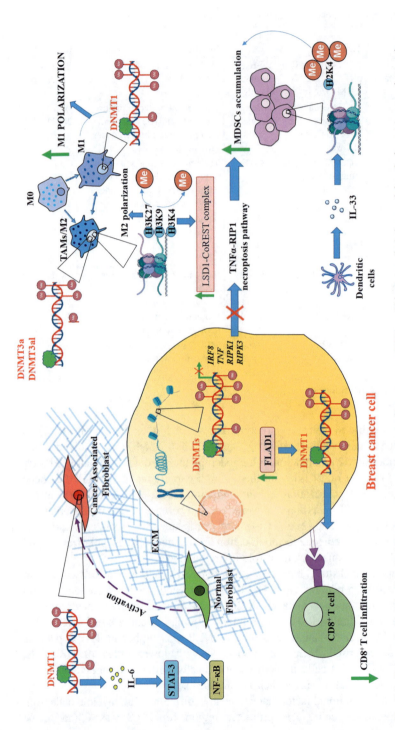

Fig. 6.1 Epigenetic modifications in the tumor microenvironment of breast cancer DNA and histone methylation modulate the breast cancer microenvironment by altering the expression of immune-related genes. Hyperactivity of DNMT1 reduced CD8+ T cell infiltration and also caused activation of cancer-associated fibroblasts (CAFs) from normal fibroblasts in breast cancer by suppressing the IL-6/STAT-3/NF-κB feedback loop. In M1 macrophages, DNMT1 enhanced the methylation, which induces M1 polarization. In contrast, demethylation of histone induces M2 macrophage polarization. Hypermethylation interrupts the TNFα-RIP1 necroptosis pathway, which induces MDSC accumulation. On the other side, dendritic cells release IL-33, which enhances histone trimethylation and thereby increases MDSC accumulation in TME. The symbol (↓) signifies downregulated and (↑) signifies upregulated in this figure

hypermethylated, along with the promoter regions of TNF, RIPK1, and RIPK3. This led to the impairment of the TNFα-RIP1 necroptosis pathway, which caused a high level of MDSC accumulation in mammary tumor-bearing mice (Smith et al. 2020).

High expression of DNMT1 resulted in the inhibition of the IL-6/STAT-3/NF-κB feedback loop, which promoted the activation of CAFs from normal fibroblasts in breast cancer (Al-Kharashi et al. 2018). In another study, it was demonstrated that high expression of DNMT3b led to the low expression of miR-200b/c, resulting in the increase of miR-221 upon TGF-β treatment which was shown to induce the activation of CAFs in breast cancer (Tang et al. 2019). Epigenetic deregulation also played a key role in altering the gene expression related to endothelial cell proliferation and migration to form new blood vessels. Hypermethylation of miR-148a and miR-152 via high expression of DNMT1 was observed to enhance breast tumor angiogenesis (Xu et al. 2013).

6.3.2 Histone Modifications

Histones are a family of proteins that help in the condensation of chromatin, thus providing structural support to the chromosome. Histone modifications are a form of posttranslational modifications (PTMs) that include histone acetylation, methylation and phosphorylation, less known ubiquitylation, deamination, and sumoylation, which play a role in chromatin packing and thus affecting the availability of the DNA to the transcription factors.

HDAC1/2 and lysine-specific histone demethylase 1 are known to form a complex with CoREST protein and have been demonstrated to play a critical role in gene silencing in many diseases, including cancer (Kalin et al. 2018). The high expression of the LSD1-CoREST complex has been observed in the M2 macrophages caused due to demethylation of the lysine residue at the H3K4 and H3K9 positions in the triple-negative breast cancer TME (Tan et al. 2019). In addition, the removal of H3K27 trimethylation of the lysine residues by histone demethylase JMJD3 has been shown to play a role in the M2 polarization (Iwanowycz et al. 2016). In another study, vorinostat (HDAC inhibitor) was used to decrease the tumor size in breast cancer-induced mice by decreasing the TAM infiltration in the tumor, thus suggesting a role of HDAC in macrophage polarization in breast cancer progression (Tran et al. 2013).

High expression of IL-33 in the TME of breast cancer reduced apoptosis and facilitated the expansion of MDSCs in breast tumors. This expansion was caused due to the secretion of GM-CSF via activation of NF-κB and MAPK signaling through increased histone trimethylation of H2K4 residue in MDSCs (Xiao et al. 2016).

6.3.3 NcRNA-mediated Gene Silencing

Noncoding RNAs (ncRNAs) are a large class of transcribed functional RNA molecules that regulate gene expression at a transcriptional and posttranscriptional level. NcRNAs can be divided into three categories, based on their length (1) long noncoding RNA (lncRNA) (>200 nts), (2) small nuclear ribonucleic acid RNA (snRNA) (200–40 nts), and (3) microRNA (miRNA) (<40 nts) (Mondal and Meeran 2020).

TAMs have been demonstrated to secrete microvesicles containing miRNA that aid breast cancer progression and invasion (Yang et al. 2011). Downregulation of miR-19a-3p led to the upregulation of proto-oncogene Fra1 in TAMs in mouse breast cancer models. This upregulation of Fra1 was reversed upon transfection with miR-19a-3p mimic, which also reduced the expression of its downstream genes VEGF, STAT3, and pSTAT3 (Yang et al. 2014a). In another study, the restoration of miR-200c expression in breast cancer cells induced GM-CSF secretion, which promoted M1 polarization in the breast tumor (Williams et al. 2021). Similarly, loading of tumor exosomes with miR-130 caused the upregulation of M1-specific markers and cytokines, such as CD86, Irf5, Nos2, TNF-α, and IL-1β (Moradi-Chaleshtori et al. 2021). M2-derived exosomes contain miR-503-3p, which is shown to downregulate DACT2, which activates the Wnt/β-catenin signaling pathway and also increased glucose intake while repressing oxygen consumption in breast cancer cells (Huang et al. 2021).

A high expression of lncRNA SNHG1 has been shown to increase the population of FOXP3$^+$ Treg cells and also increased CD4$^+$ tumor-infiltrating lymphocytes (TILs) infiltration in breast tumors via downregulating miR-448 expression and thus increasing IDO and IL-10 expression (Pei et al. 2018). In another study, overexpression of lncRNA NKILA has been shown to sensitize the cytotoxic T cells and Th1 cells to activation-induced cell death (AICD), which thereby helps in the immune evasion of the breast cancer cells (Huang et al. 2018).

A high expression of miR-519a-3p has been observed in the immune evasion of NK cells in breast cancer cells, thus conferring resistance to apoptosis of the cancer cells (Breunig et al. 2017). NKG2DLs present on the surface of malignant cells, which, upon interaction with NKG2D on NK cells, increase NK cell activity, were found to be in an inverse correlation with miR-20a. Mir-20a/b directly targeted the 3'-UTR of MICA/B, thus downregulating their expression on breast cancer cells. Mir-20a also indirectly downregulated ULBP2 expression in breast cancer cells by targeting MAPK/ERK signaling pathways (Shen et al. 2017). In another study, it was reported that hepatocellular carcinoma (HCC) cell-derived exosomal circular ubiquitin-like, with PHD and ring finger domain 1 RNA (circUHRF1), leads to the upregulation of TIM-3 via the degradation of miR-449c-5p, which ultimately causes NK cells exhaustion and decreases TNFα and IFNγ secretion (Zhang et al. 2020a).

DCs in the TME of breast cancer have a low expression of miR-155, which has been observed to cause DCs dysfunction (Wang et al. 2016). Upregulation of miR-155 was shown to increase the migration, antigen uptake, and maturation of

DC cells, which also significantly induced T cell proliferation in the 3-D TME of breast cancer (Yang et al. 2021b). Further, overexpression of miR-155, miR-142, and let-7i was shown to be involved in the maturation of DCs in breast cancer (Taghikhani et al. 2019). Also, a computational investigation showed that a low expression of lncRNA TCL6 positively correlated with immune infiltrating cells, including DCs, and was associated with worse overall survival of breast cancer patients (Zhang et al. 2020b).

In addition, doxorubicin treatment in breast cancer led to the induction of MDSC-derived exosomal miR-126a, which enhanced chemoresistance in the cancer cells and also promoted lung metastasis (Deng et al. 2017). Also, CAFs-derived exosomes contain miR-181d-5p, which targets CDX2, a transcription factor that drives HOXA5 expression. Breast cancer cells expressed poor levels of HOXA5, which enhanced cancer cell proliferation, invasion, migration, and epithelial to mesenchymal transition (EMT) and reduced apoptosis (Wang et al. 2020). High expression levels of miR-155, miR-526b, and miR-655 have been demonstrated to promote breast tumor angiogenesis (Hunter et al. 2019; Kong et al. 2014).

6.4 Epigenetic Modulators in the Tumor Microenvironment of Breast Cancer

Given the high level of heterogeneity in breast cancer, owing to genetic and epigenetic alterations in both cancer cells and cells of the TME, the use of the single agent chemotherapy to achieve complete (CR) or even a partial (PR) becomes a strenuous task (Carrick et al. 2009). Interestingly, epigenetic alterations in the TME play a crucial role in the cellular differentiation of immune cells from an immunomodulatory to an immunosuppressive phenotype. Hence, the use of epigenetic modulators may be beneficial in cancer immunotherapy by modulating the cellular differentiation and function of the components of the TME, as depicted in Fig. 6.2. The use of immunotherapy, along with other types of cancer treatments, such as chemotherapy and radiation, is one of the combination therapy treatments widely used, as it restores or enhances the immune system's ability to fight cancer. In addition, epigenetic modulators, along with immunotherapy, may further help enhance the efficacy of immunomodulators, the idea of which is being tested in many ongoing clinical trials.

5-Azacytidine and 5-aza-2′-deoxycytidine (decitabine) are FDA-approved nucleoside analog DNMTi for the treatment of myelodysplastic syndrome; however, they have been used in the treatment of solid tumors as well. As these drugs belong to the family of nucleoside analogs, they demonstrate their DNMT inhibitory activity by incorporating themselves into the DNA sequence, hence inhibiting the DNA methylation activity of DNMTs (Christman 2002). In a study demonstrated by Li et al. (2014), 63 cancer cell lines belonging to breast, colorectal, and ovarian cancer were treated with low doses of azacytidine for 3 days. Gene set enrichment analysis

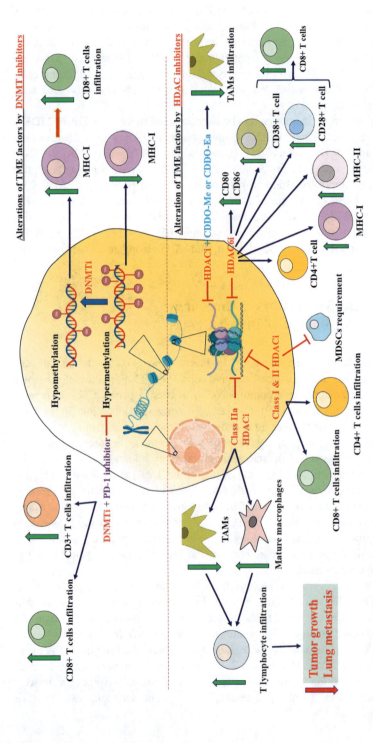

Fig. 6.2 Epigenetic modifiers alter the tumor microenvironment in breast cancer Epigenetic modifiers, especially DNMTi and HDACi, play a key role in modulating the TME in breast cancer. DNMTi suppresses DNA hypermethylation, thereby enhancing the expression of MHC-1, which induces CD8+ T cell infiltration in the tumor site. DNMTi and PD-1 inhibitors also inhibit hypermethylation and enhance the infiltration of CD8+ and CD3+ T cells. Similar to DNMTi, HDACi also increases CD8+ and CD4+ T cells infiltration. By inhibiting HDAC 6 activity, HDAC 6i enhances the expression of several immune

Fig. 6.2 (continued) modulators like MHC-I and II and induces CD8 + T cell proliferation by increasing CD38+ and CD28 + T cells. HDACi also impedes the MDSC requirement. The combination of HDACi and 2-cyano-3,12-dioxooleana-1,9(11)-dien-28-oic acid (CDDO)-methyl ester (CDDO-Me) or CDDO-ethylamide (CDDO-Ea) can inhibit TAMs filtration. Even HDACi alone has been shown to decrease the TAM population by augmenting major macrophages, thereby increasing T lymphocyte infiltration. This activity suppresses tumor growth and also inhibits metastasis. The symbol (↓) signifies downregulated and (↑) signifies upregulated in this figure

(GSEA) of the dysregulated genes upon azacytidine treatment revealed an upregulation of immune gene sets belonging to interferon signaling, antigen presentation, chemokine and cytokine signaling, inflammation, and influenza (Li et al. 2014). This suggests that solid tumors, such as breast cancer, can benefit from epigenetic modulators through the modulation of immune pathways. MHC-1 is found to be downregulated with high levels of methylation on its promoter region in breast cancer cells. Upon treatment with guadecitabine, also called SGI-110, a second-generation DNMTi (DNMT inhibitor), DNMT1 expression was observed to be inhibited, which enhanced MHC-1 expression on breast cancer cell surface. Since the function of MHC-1 is to present its antigen to cytotoxic CD8+ T cells, an increase in the population of CD8+ T cells infiltrating the tumor site was also observed. In addition, an increase in the mRNA expression of genes NF-κB activity and interferon signaling (*Irf7*, *Tlr3*, *Ifit1*, *Ifit3*, *Ifi44*, *Ifi35*, and *Tnssf10*) (Luo et al. 2018).

MYC, a well-known transcription factor, is shown to be upregulated in TNBC, which negatively correlates with reduced T cell infiltration of CD3+, CD8+, and granzyme B+ cells in vivo. Upregulation of MYC further increases the transcription levels of DNMT1, which in turn increases the methylation at the promoter region of *STING*, thus suppressing the STNG-dependent IFN response and also decreasing MHC-1 expression on TNBC cells. Decitabine treatment restored STNG expression in MYC overexpressing TNBC cells. Further, decitabine, in combination with a PD-1 inhibitor, significantly decreased tumor growth compared to the treatment alone group by increasing CD3+ and CD8+ T cell infiltration. Combination treatment also increased granzyme B expression in CD8+ cells, which is a serine protease with pro-apoptotic activity (Wu et al. 2021). Aberrant expression of DNMTs has been observed in the tumor-exposed NK (teNK) cells of breast cancer, where they promote colony formation. Pretreatment of teNK cells with DNMT inhibitors (azacytidine and decitabine) led to a decreased number of colonies formation by neutralizing teNK cell phenotype and also augmented NK-directed immunotherapy efficacy (Chan et al. 2020).

TMP195 is a selective competitive class IIa HDAC inhibitor that competes against the binding of HDAC to various side chain modifications by occupying the acetyllysine-binding site of class IIa HDACs. TMP195 treatment significantly increases mature macrophages (Mac-2$^+$, CD115$^+$, F4/80$^+$) population and reduced TAMs (CD45$^+$MHCII$^+$CD11b^{lo}) in transgenic mice tumors which promotes phagocytic and immunostimulatory functions in macrophages, which thereby also increase cytotoxic T lymphocyte infiltration. This led to decreased macrophage-dependent tumor growth reduction and reduced lung metastasis. Further, TMP195 combined with carboplatin or paclitaxel and with anti-PD-1 significantly reduced tumor burden compared to treatment alone, suggesting that TMP195 augments the efficacy of chemotherapeutic agents and immune checkpoint blockade (Guerriero et al. 2017). Trichostatin A (TSA), a class I and II HDAC inhibitor, was also demonstrated to increase CD4+ and CD8+ T cell infiltration along with CD45+ inflammatory cells while also inhibiting the recruitment of MDSCs in breast tumor-bearing mice. Low-dose TSA treatment significantly decreased expression levels of M2 markers

Arg1, *Cd206*, and *Fizz1* while significantly increasing M1 markers *Nos2* and *IL6*, which induced the pro-inflammatory action of macrophages and also enhanced T cell proliferation. Further, TSA treatment in combination with anti-PD-L1 immunotherapy acts synergistically in inhibiting tumor growth and increasing overall survival in a syngeneic mouse model (Li et al. 2021).

Vorinostat (suberoylanilide hydroxamic acid (SAHA)), an FDA-approved HDAC inhibitor, when fed to breast tumor-bearing mice in combination with CDDO-Me or CDDO-Ea, which are synthetic triterpenoids, was shown to inhibit TAMs infiltration by reducing macrophage colony-stimulating factor (M-CSF) and MMP-9 levels in mammary tumors (Tran et al. 2013). Further, SAHA treatment also increased the expression levels of NKG2DLs, MICA/B, and ULBP2 in breast cancer cells by inhibiting members of the miR-17-92 cluster, miR-20a, miR-20b, miR-93 and miR-106b, which enhanced NK cell activity and sensitized breast cancer cells to NK cell-mediated cytotoxicity (Shen et al. 2017). ACY241, a selective HDAC6 inhibitor, was also demonstrated to increase the expression of CD80 and CD86, as well as MHC class I and class II proteins on myeloma, breast cancer, colon cancer, and dendritic cells. ACY241 treatment also induced CD4+ and CD28+, CD38+, and CD28 + CD38+, expressing CD8+ T cell proliferation within circulating T lymphocytes (CTLs). This increased XBP1 peptide-specific T cell proliferation and augmented the antitumor activity of XBP1-CTLs against cancer cells. This was achieved due to the activation of AKT/mTOR/NF-κB p65 pathways within CD8+ T cells via upregulation of key transcriptional regulators Bcl-6, HIF-1, Eomes, and T-bet, which contribute to the induction and maintenance of antigen-specific memory CD8+ T cells with antitumor activities (Bae et al. 2018).

Eugenol, a phenolic compound found in the essential oils of different spices, such as *Syzygium aromaticum* (clove), *Pimenta racemosa* (bay leaves), and *Cinnamomum verum* (cinnamon leaf), has been reported to have many properties, such as antioxidant, anti-inflammatory, antibacterial, antiseptic, and also anticancer properties (Zari et al. 2021). Recently, its epigenetic modulatory efficacy against active breast CAFs was demonstrated in a study by Al-Kharashi et al. (2021). Eugenol, like decitabine, inhibited the expression of DNMT1 and DNMT3A by inhibiting E2F1 expression in breast CAFs, significantly reducing myofibroblasts biomarkers protein expression, α-SMA, SDF-1, TGFβ1, and IL-6. This led to the suppression of paracrine and autocrine pro-carcinogenic features of the breast stromal fibroblasts (BSFs) (Al-Kharashi et al. 2021).

6.5 Conclusion

Over the past decades, the role of epigenetic aberrations has mainly been focused on the malignant cells, where it drives cancer growth and progression (Shukla et al. 2019). However, it has been well established that a solid tumor, including breast cancer, is a group of the heterogeneous population, along with the presence of immune cells and other growth factors, which form the tumor microenvironment, that plays a vital role in cancer progression and is also responsible for the development of resistance against chemotherapeutic drugs. Recent evidence suggests that epigenetic alterations are also present in the cells of the TME, which drives the differentiation of the immune cells from an antitumorigenic environment to a pro-tumorigenic or immunosuppressive environment, ideal for breast cancer development. Thus, epigenetic drugs or dietary epigenetic modulators to alter the components of the TME holds great potential in breast cancer therapy. However, single-agent chemotherapy has not received much positive feedback in clinical trials. Thus, epigenetic modulators in combination with immunomodulators have gained the limelight over the last few years. Thus, there is a need to better understand the mechanism behind the plasticity of these cells and how epigenetic deregulation contributes to it.

Acknowledgments BK and PM acknowledge research fellowships from the CSIR and UGC, Government of India, respectively. We thank the Director, CSIR-CFTRI, Mysore, India, for providing all the required facilities.

Funding The work was supported by CSIR-FIRST grant, MLP-0299.

Conflict of interest The authors declare that they have no conflict of interest.

Ethical approval Not applicable.

References

Al-Kharashi LA, Al-Mohanna FH, Tulbah A, Aboussekhra A (2018) The DNA methyl-transferase protein DNMT1 enhances tumor promoting properties of breast stromal fibroblasts. Oncotarget 9(2):2329–2343. https://doi.org/10.18632/oncotarget.23411

Al-Kharashi LA, Bakheet T, AlHarbi WA, Al-Moghrabi N, Aboussekhra A (2021) Eugenol modulates genomic methylation and inactivates breast cancer-associated fibroblasts through E2F1-dependent downregulation of DNMT1/DNMT3A. Mol Carcinog 60(11):784. https://doi.org/10.1002/mc.23344

Ali HR, Provenzano E, Dawson SJ, Blows FM, Liu B, Shah M, Earl HM, Poole CJ, Hiller L, Dunn JA, Bowden SJ, Twelves C, Bartlett JMS, Mahmoud SMA, Rakha E, Ellis IO, Liu S, Gao D, Nielsen TO, Caldas C (2014) Association between CD8+ T-cell infiltration and breast cancer survival in 12 439 patients. Ann Oncol 25(8). https://doi.org/10.1093/annonc/mdu191

An J, Du Y, Fan X, Wang Y, Ivan C, Zhang XG, Sood AK, An Z, Zhang N (2019) EGFL6 promotes breast cancer by simultaneously enhancing cancer cell metastasis and stimulating tumor angiogenesis. Oncogene 38(12):2123. https://doi.org/10.1038/s41388-018-0565-9

Bae J, Hideshima T, Tai YT, Song Y, Richardson P, Raje N, Munshi NC, Anderson KC (2018) Histone deacetylase (HDAC) inhibitor ACY241 enhances anti-tumor activities of antigen-specific central memory cytotoxic T lymphocytes against multiple myeloma and solid tumors. Leukemia 32(9). https://doi.org/10.1038/s41375-018-0062-8

Baghban R, Roshangar L, Jahanban-Esfahlan R, Seidi K, Ebrahimi-Kalan A, Jaymand M, Kolahian S, Javaheri T, Zare P (2020) Tumor microenvironment complexity and therapeutic implications at a glance. Cell Commun Signal 18:1. https://doi.org/10.1186/s12964-020-0530-4

Baginska J, Viry E, Berchem G, Poli A, Noman MZ, Van Moer K, Medves S, Zimmer J, Oudin A, Niclou SP, Bleackley RC, Goping IS, Chouaib S, Janji B (2013) Granzyme B degradation by autophagy decreases tumor cell susceptibility to natural killer-mediated lysis under hypoxia. Proc Natl Acad Sci U S A 110(43):17450. https://doi.org/10.1073/pnas.1304790110

Bergenfelz C, Larsson AM, Von Stedingk K, Gruvberger-Saal S, Aaltonen K, Jansson S, Jernström H, Janols H, Wullt M, Bredberg A, Rydén L, Leandersson K (2015) Systemic monocytic-MDSCs are generated from monocytes and correlate with disease progression in breast cancer patients. PLoS One 10(5). https://doi.org/10.1371/journal.pone.0127028

Bergenfelz C, Roxå A, Mehmeti M, Leandersson K, Larsson AM (2020) Clinical relevance of systemic monocytic-MDSCs in patients with metastatic breast cancer. Cancer Immunol Immunother 69(3):435. https://doi.org/10.1007/s00262-019-02472-z

Boieri M, Malishkevich A, Guennoun R, Marchese E, Kroon S, Trerice KE, Awad M, Park JH, Iyer S, Kreuzer J, Haas W, Rivera MN, Demehri S (2022) CD4+ T helper 2 cells suppress breast cancer by inducing terminal differentiation. J Exp Med 219(7). https://doi.org/10.1084/jem.20201963

Breunig C, Pahl J, Küblbeck M, Miller M, Antonelli D, Erdem N, Wirth C, Will R, Bott A, Cerwenka A, Wiemann S (2017) Micro RNA-519a-3p mediates apoptosis resistance in breast cancer cells and their escape from recognition by natural killer cells. Cell Death and Disease 8(8). https://doi.org/10.1038/cddis.2017.364

Buchanan CF, Szot CS, Wilson TD, Akman S, Metheny-Barlow LJ, Robertson JL, Freeman JW, Rylander MN (2012) Cross-talk between endothelial and breast cancer cells regulates reciprocal expression of angiogenic factors in vitro. J Cell Biochem 113(4):1142–1151. https://doi.org/10.1002/jcb.23447

Carrick S, Parker S, Thornton CE, Ghersi D, Simes J, Wilcken N (2009) Single agent versus combination chemotherapy for metastatic breast cancer. Cochrane Database Syst Rev 2009: CD003372. https://doi.org/10.1002/14651858.CD003372.pub3

Chan IS, Knutsdóttir H, Ramakrishnan G, Padmanaban V, Warrier M, Ramirez JC, Dunworth M, Zhang H, Jaffee EM, Bader JS, Ewald AJ (2020) Cancer cells educate natural killer cells to a metastasis-promoting cell state. J Cell Biol 219(9). https://doi.org/10.1083/jcb.202002077

Chatterjee S, Chatterjee A, Jana S, Dey S, Roy H, Das MK, Alam J, Adhikary A, Chowdhury A, Biswas A, Manna D, Bhattacharyya A (2021) Transforming growth factor beta orchestrates PD-L1 enrichment in tumor-derived exosomes and mediates CD8 T-cell dysfunction regulating early phosphorylation of TCR signalome in breast cancer. Carcinogenesis 42(1):38. https://doi.org/10.1093/carcin/bgaa092

Chávez-Galán L, Olleros ML, Vesin D, Garcia I (2015) Much more than M1 and M2 macrophages, there are also CD169+ and TCR+ macrophages. Front Immunol 6. https://doi.org/10.3389/fimmu.2015.00263

Chen J, Yao Y, Gong C, Yu F, Su S, Chen J, Liu B, Deng H, Wang F, Lin L, Yao H, Su F, Anderson KS, Liu Q, Ewen ME, Yao X, Song E (2011) CCL18 from tumor-associated macrophages promotes breast cancer metastasis via PITPNM3. Cancer Cell 19(4):814. https://doi.org/10.1016/j.ccr.2011.02.006

Chen X, Shao Q, Hao S, Zhao Z, Wang Y, Guo X, He Y, Gao W, Mao H (2017) CTLA-4 positive breast cancer cells suppress dendritic cells maturation and function. Oncotarget 8(8): 13703–13715. https://doi.org/10.18632/oncotarget.14626

Chen Z, Wu J, Wang L, Zhao H, He J (2022) Tumor-associated macrophages of the M1/M2 phenotype are involved in the regulation of malignant biological behavior of breast cancer cells through the EMT pathway. Med Oncol 39(5):83. https://doi.org/10.1007/s12032-022-01670-7

Christman JK (2002) 5-Azacytidine and 5-aza-2′-deoxycytidine as inhibitors of DNA methylation: mechanistic studies and their implications for cancer therapy. Oncogene 21(35). https://doi.org/10.1038/sj.onc.1205699

Coffelt SB, Kersten K, Doornebal CW, Weiden J, Vrijland K, Hau CS, Verstegen NJM, Ciampricotti M, Hawinkels LJAC, Jonkers J, De Visser KE (2015) IL-17-producing γδ T cells and neutrophils conspire to promote breast cancer metastasis. Nature 522(7556):345. https://doi.org/10.1038/nature14282

De Francesco EM, Lappano R, Santolla MF, Marsico S, Caruso A, Maggiolini M (2013) HIF-1α/GPER signaling mediates the expression of VEGF induced by hypoxia in breast cancer associated fibroblasts (CAFs). Breast Cancer Res 15(4). https://doi.org/10.1186/bcr3458

De Paolis V, Maiullari F, Chirivì M, Milan M, Cordiglieri C, Pagano F, La Manna AR, De Falco E, Bearzi C, Rizzi R, Parisi C (2022) Unusual association of NF-κB components in tumor-associated macrophages (TAMs) promotes HSPG2-mediated immune-escaping mechanism in breast cancer. Int J Mol Sci 23(14). https://doi.org/10.3390/ijms23147902

de Winde CM, Munday C, Acton SE (2020) Molecular mechanisms of dendritic cell migration in immunity and cancer. Med Microbiol Immunol 209:515. https://doi.org/10.1007/s00430-020-00680-4

Della Bella S, Gennaro M, Vaccari M, Ferraris C, Nicola S, Riva A, Clerici M, Greco M, Villa ML (2003) Altered maturation of peripheral blood dendritic cells in patients with breast cancer. Br J Cancer 89(8):1463. https://doi.org/10.1038/sj.bjc.6601243

Deng Z, Rong Y, Teng Y, Zhuang X, Samykutty A, Mu J, Zhang L, Cao P, Yan J, Miller D, Zhang HG (2017) Exosomes miR-126a released from MDSC induced by DOX treatment promotes lung metastasis. Oncogene 36(5):639–651. https://doi.org/10.1038/onc.2016.229

Dong H, Diao H, Zhao Y, Xu H, Pei S, Gao J, Wang J, Hussain T, Zhao D, Zhou X, Lin D (2019) Overexpression of matrix metalloproteinase-9 in breast cancer cell lines remarkably increases the cell malignancy largely via activation of transforming growth factor beta/SMAD signalling. Cell Prolif 52(5):e12633. https://doi.org/10.1111/cpr.12633

Duque GA, Descoteaux A (2014) Macrophage cytokines: involvement in immunity and infectious diseases. Front Immunol 5:491. https://doi.org/10.3389/fimmu.2014.00491

Dvorak KM, Pettee KM, Rubinic-Minotti K, Su R, Nestor-Kalinoski A, Eisenmann KM (2018) Carcinoma associated fibroblasts (CAFs) promote breast cancer motility by suppressing mammalian diaphanous-related formin-2 (mDia2). PLoS One 13(3):e0195278. https://doi.org/10.1371/journal.pone.0195278

Eiro N, González L, Martínez-Ordoñez A, Fernandez-Garcia B, González LO, Cid S, Dominguez F, Perez-Fernandez R, Vizoso FJ (2018) Cancer-associated fibroblasts affect breast cancer cell gene expression, invasion and angiogenesis. Cell Oncol 41(4). https://doi.org/10.1007/s13402-018-0371-y

Furlan A, Vercamer C, Heliot L, Wernert N, Desbiens X, Pourtier A (2019) Ets-1 drives breast cancer cell angiogenic potential and interactions between breast cancer and endothelial cells. Int J Oncol 54(1):29. https://doi.org/10.3892/ijo.2018.4605

Gabrilovich DI, Corak J, Ciernik IF, Kavanaugh D, Carbone DP (1997) Decreased antigen presentation by dendritic cells in patients with breast cancer. Clin Cancer Res 3(3)

Gabrilovich DI, Nagaraj S (2009) Myeloid-derived suppressor cells as regulators of the immune system. Nat Rev Immunol 9(3):162–174. https://doi.org/10.1038/nri2506

Giorello MB, Matas A, Marenco P, Davies KM, Borzone FR, de Calcagno ML, García-Rivello H, Wernicke A, Martinez LM, Labovsky V, Chasseing NA (2021) CD1a- and CD83-positive dendritic cells as prognostic markers of metastasis development in early breast cancer patients. Breast Cancer 28(6):1328. https://doi.org/10.1007/s12282-021-01270-9

Greene LI, Bruno TC, Christenson JL, D'Alessandro A, Culp-Hill R, Torkko K, Borges VF, Slansky JE, Richer JK (2019) A role for tryptophan-2,3-dioxygenase in CD8 T-cell suppression and evidence of tryptophan catabolism in breast cancer patient plasma. Mol Cancer Res 17(1):131. https://doi.org/10.1158/1541-7786.MCR-18-0362

Guerriero JL, Sotayo A, Ponichtera HE, Castrillon JA, Pourzia AL, Schad S, Johnson SF, Carrasco RD, Lazo S, Bronson RT, Davis SP, Lobera M, Nolan MA, Letai A (2017) Class IIa HDAC inhibition reduces breast tumours and metastases through anti-tumour macrophages. Nature 543(7645). https://doi.org/10.1038/nature21409

Henze AT, Mazzone M (2016) The impact of hypoxia on tumor-associated macrophages. J Clin Investig 126:3672. https://doi.org/10.1172/JCI84427

Huang D, Chen J, Yang L, Ouyang Q, Li J, Lao L, Zhao J, Liu J, Lu Y, Xing Y, Chen F, Su F, Yao H, Liu Q, Su S, Song E (2018) NKILA lncRNA promotes tumor immune evasion by sensitizing T cells to activation-induced cell death. Nat Immunol 19(10):1112–1125. https://doi.org/10.1038/s41590-018-0207-y

Huang S, Fan P, Zhang C, Xie J, Gu X, Lei S, Chen Z, Huang Z (2021) Exosomal microRNA-503-3p derived from macrophages represses glycolysis and promotes mitochondrial oxidative phosphorylation in breast cancer cells by elevating DACT2. Cell Death Discov 7(1):119. https://doi.org/10.1038/s41420-021-00492-2

Hughes CE, Nibbs RJB (2018) A guide to chemokines and their receptors. FEBS J 285(16): 2944–2971. https://doi.org/10.1111/febs.14466

Hunter S, Nault B, Ugwuagbo KC, Maiti S, Majumder M (2019) Mir526b and mir655 promote tumour associated angiogenesis and lymphangiogenesis in breast cancer. Cancers 11(7). https://doi.org/10.3390/cancers11070938

Hwang HJ, Lee YR, Kang D, Lee HC, Seo HR, Ryu JK, Kim YN, Ko YG, Park HJ, Lee JS (2020) Endothelial cells under therapy-induced senescence secrete CXCL11, which increases aggressiveness of breast cancer cells. Cancer Lett 490:100. https://doi.org/10.1016/j.canlet.2020.06.019

Iwanowycz S, Wang J, Hodge J, Wang Y, Yu F, Fan D (2016) Emodin inhibits breast cancer growth by blocking the tumor-promoting feedforward loop between cancer cells and macrophages. Mol Cancer Ther 15(8):1931–1942. https://doi.org/10.1158/1535-7163.MCT-15-0987

Jayasingam SD, Citartan M, Thang TH, Mat Zin AA, Ang KC, Ch'ng, E. S. (2020) Evaluating the polarization of tumor-associated macrophages into M1 and M2 phenotypes in human cancer tissue: technicalities and challenges in routine clinical practice. Front Oncol 9(January):1–9. https://doi.org/10.3389/fonc.2019.01512

Jeschke J, Collignon E, Fuks F (2015) DNA methylome profiling beyond promoters–taking an epigenetic snapshot of the breast tumor microenvironment. FEBS J 282(9). https://doi.org/10.1111/febs.13125

Jiang M, Zhang W, Zhang R, Liu P, Ye Y, Yu W, Guo X, Yu J (2020) Cancer exosome-derived miR-9 and miR-181a promote the development of early-stage MDSCs via interfering with SOCS3 and PIAS3 respectively in breast cancer. Oncogene 39(24):4681. https://doi.org/10.1038/s41388-020-1322-4

Juliá EP, Amante A, Pampena MB, Mordoh J, Levy EM (2018) Avelumab, an IgG1 anti-PD-L1 immune checkpoint inhibitor, triggers NK cell-mediated cytotoxicity and cytokine production against triple negative breast cancer cells. Front Immunol 9(SEP). https://doi.org/10.3389/fimmu.2018.02140

Kalin JH, Wu M, Gomez AV, Song Y, Das J, Hayward D, Adejola N, Wu M, Panova I, Chung HJ, Kim E, Roberts HJ, Roberts JM, Prusevich P, Jeliazkov JR, Roy Burman SS, Fairall L, Milano C, Eroglu A, Cole PA (2018) Targeting the CoREST complex with dual histone deacetylase and demethylase inhibitors. Nat Comm 9(1). https://doi.org/10.1038/s41467-017-02242-4

Kim BG, Sung JS, Jang Y, Cha YJ, Kang S, Han HH, Lee JH, Cho NH (2019a) Compression-induced expression of glycolysis genes in CAFs correlates with EMT and angiogenesis gene expression in breast cancer. Comm Biol 2(1):313. https://doi.org/10.1038/s42003-019-0553-9

Kim R, Kawai A, Wakisaka M, Funaoka Y, Yasuda N, Hidaka M, Morita Y, Ohtani S, Ito M, Arihiro K (2019b) A potential role for peripheral natural killer cell activity induced by preoperative chemotherapy in breast cancer patients. Cancer Immunol Immunother 68(4):577. https://doi.org/10.1007/s00262-019-02305-z

Kong W, He L, Richards EJ, Challa S, Xu CX, Permuth-Wey J, Lancaster JM, Coppola D, Sellers TA, Djeu JY, Cheng JQ (2014) Upregulation of miRNA-155 promotes tumour angiogenesis by targeting VHL and is associated with poor prognosis and triple-negative breast cancer. Oncogene 33(6):679. https://doi.org/10.1038/onc.2012.636

Krneta T, Gillgrass A, Chew M, Ashkar AA (2016) The breast tumor microenvironment alters the phenotype and function of natural killer cells. Cell Mol Immunol 13(5):628. https://doi.org/10.1038/cmi.2015.42

Lee H, Lee HJ, Song IH, Bang WS, Heo SH, Gong G, Park IA (2018) CD11c-positive dendritic cells in triple-negative breast cancer. In Vivo 32(6). https://doi.org/10.21873/invivo.11415

Li A, Chen P, Leng Y, Kang J (2018) Histone deacetylase 6 regulates the immunosuppressive properties of cancer-associated fibroblasts in breast cancer through the STAT3–COX2-dependent pathway. Oncogene 37(45):5952–5966. https://doi.org/10.1038/s41388-018-0379-9

Li C, Li R, Hu X, Zhou G, Jiang G (2022a) Tumor-promoting mechanisms of macrophage-derived extracellular vesicles-enclosed microRNA-660 in breast cancer progression. Breast Cancer Res Treat 192(2):353. https://doi.org/10.1007/s10549-021-06433-y

Li H, Chiappinelli KB, Guzzetta AA, Easwaran H, Yen RWC, Vatapalli R, Topper MJ, Luo J, Connolly RM, Azad NS, Stearns V, Pardoll DM, Davidson N, Jones PA, Slamon DJ, Baylin SB, Zahnow CA, Ahuja N (2014) Immune regulation by low doses of the DNA methyltransferase inhibitor 5-azacitidine in common human epithelial cancers. Oncotarget 5(3):587. https://doi.org/10.18632/oncotarget.1782

Li X, Su X, Liu R, Pan Y, Fang J, Cao L, Feng C, Shang Q, Chen Y, Shao C, Shi Y (2021) HDAC inhibition potentiates anti-tumor activity of macrophages and enhances anti-PD-L1-mediated tumor suppression. Oncogene 40(10). https://doi.org/10.1038/s41388-020-01636-x

Li Z, Wang P, Cui W, Yong H, Wang D, Zhao T, Wang W, Shi M, Zheng J, Bai J (2022b) Tumour-associated breast cancer macrophages enhance breast cancer malignancy via inducing ZEB1-mediated DNMT1 transcriptional activation. Cell Biosci 12(1):176. https://doi.org/10.1186/s13578-022-00913-4

Linde N, Casanova-Acebes M, Sosa MS, Mortha A, Rahman A, Farias E, Harper K, Tardio E, Reyes Torres I, Jones J, Condeelis J, Merad M, Aguirre-Ghiso JA (2018) Macrophages orchestrate breast cancer early dissemination and metastasis. *Nature*. Communications 9(1). https://doi.org/10.1038/s41467-017-02481-5

Liu C, Qiang J, Deng Q, Xia J, Deng L, Zhou L, Wang D, He X, Liu Y, Zhao B, Lv J, Yu Z, Lei QY, Shao ZM, Zhang XY, Zhang L, Liu S (2021a) ALDH1A1 activity in tumor-initiating cells remodels myeloid-derived suppressor cells to promote breast cancer progression. Cancer Res 81(23):5919. https://doi.org/10.1158/0008-5472.CAN-21-1337

Liu L, Liu S, Luo H, Chen C, Zhang X, He L, Tu G (2021b) GPR30-mediated HMGB1 upregulation in CAFs induces autophagy and tamoxifen resistance in ERα-positive breast cancer cells. Aging 13(12):16178. https://doi.org/10.18632/aging.203145

Liu Q, Hodge J, Wang J, Wang Y, Wang L, Singh U, Li Y, Yao Y, Wang D, Ai W, Nagarkatti P, Chen H, Xu P, Angela Murphy E, Fan D (2020) Emodin reduces breast cancer lung metastasis by suppressing macrophage-induced breast cancer cell epithelial-mesenchymal transition and cancer stem cell formation. Theranostics 10(18):8365–8381. https://doi.org/10.7150/thno.45395

Lodewijk I, Nunes SP, Henrique R, Jerónimo C, Dueñas M, Paramio JM (2021) Tackling tumor microenvironment through epigenetic tools to improve cancer immunotherapy. Clin Epigenetics 13:63. https://doi.org/10.1186/s13148-021-01046-0

Luckheeram RV, Zhou R, Verma AD, Xia B (2012) CD4 +T cells: differentiation and functions. Clin Dev Immunol 2012:925135. https://doi.org/10.1155/2012/925135

Luo N, Nixon MJ, Gonzalez-Ericsson PI, Sanchez V, Opalenik SR, Li H, Zahnow CA, Nickels ML, Liu F, Tantawy MN, Sanders ME, Manning HC, Balko JM (2018) DNA methyltransferase inhibition upregulates MHC-I to potentiate cytotoxic T lymphocyte responses in breast cancer. Nat Comm 9(1). https://doi.org/10.1038/s41467-017-02630-w

Yang M, Chen J, Su F, Yu B, Su F, Lin L, Liu Y, Huang J-D, Song E (2011) Microvesicles secreted by macrophages shuttle invasion-potentiating microRNAs into breast cancer cells. Mol Cancer *10*:6–10. https://doi.org/10.1186/1476-4598-10-117

Mahmoud SMA, Paish EC, Powe DG, Macmillan RD, Grainge MJ, Lee AHS, Ellis IO, Green AR (2011) Tumor-infiltrating CD8+ lymphocytes predict clinical outcome in breast cancer. J Clin Oncol 29(15):1949. https://doi.org/10.1200/JCO.2010.30.5037

Manoochehri M, Hielscher T, Borhani N, Gerhäuser C, Fletcher O, Swerdlow AJ, Ko YD, Brauch H, Brüning T, Hamann U (2021) Epigenetic quantification of circulating immune cells in peripheral blood of triple-negative breast cancer patients. Clin Epigenetics 13(1):207. https://doi.org/10.1186/s13148-021-01196-1

Markowitz J, Wesolowski R, Papenfuss T, Brooks TR, Carson WE (2013) Myeloid-derived suppressor cells in breast cancer. Breast Cancer Res Treat 140:13. https://doi.org/10.1007/s10549-013-2618-7

Mattiuz R, Brousse C, Ambrosini M, Cancel JC, Bessou G, Mussard J, Sanlaville A, Caux C, Bendriss-Vermare N, Valladeau-Guilemond J, Dalod M, Crozat K (2021) Type 1 conventional dendritic cells and interferons are required for spontaneous CD4+ and CD8+ T-cell protective responses to breast cancer. Clin Transl Immunology 10(7):e1305. https://doi.org/10.1002/cti2.1305

Mercier I, Casimiro MC, Wang C, Rosenberg AL, Quong J, Minkeu A, Allen KG, Danilo C, Sotgia F, Bonuccelli G, Jasmin JF, Xu H, Bosco E, Aronow B, Witkiewicz A, Pestell RG, Knudsen ES, Lisanti MP (2008) Human breast cancer-associated fibroblasts (CAFs) show caveolin-1 downregulation and RB tumor suppressor functional inactivation: implications for the response to hormonal therapy. Cancer Biol Ther 7(8):1212–1225. https://doi.org/10.4161/cbt.7.8.6220

Mills CD (2015) Anatomy of a discovery: M1 and M2 macrophages. Front Immunol 6:212. https://doi.org/10.3389/fimmu.2015.00212

Mittal D, Gubin MM, Schreiber RD, Smyth MJ (2014) New insights into cancer immunoediting and its three component phases-elimination, equilibrium and escape. Curr Opin Immunol 27: 16–25. https://doi.org/10.1016/j.coi.2014.01.004

Mondal P, Kaur B, Natesh J, Meeran SM (2021) The emerging role of miRNA in the perturbation of tumor immune microenvironment in chemoresistance: therapeutic implications. Semin Cell Dev Biol

Mondal P, Meeran SM (2020) Long non-coding RNAs in breast cancer metastasis. Non-Coding RNA Res 5(4):208–218. https://doi.org/10.1016/j.ncrna.2020.11.004

Mondal P, Natesh J, Penta D, Meeran SM (2020) Progress and promises of epigenetic drugs and epigenetic diets in cancer prevention and therapy: a clinical update. Semin Cancer Biol 83:503. https://doi.org/10.1016/j.semcancer.2020.12.006

Monteran L, Erez N (2019) The dark side of fibroblasts: cancer-associated fibroblasts as mediators of immunosuppression in the tumor microenvironment. Front Immunol 10(AUG):1–15. https://doi.org/10.3389/fimmu.2019.01835

Moradi-Chaleshtori M, Shojaei S, Mohammadi-Yeganeh S, Hashemi SM (2021) Transfer of miRNA in tumor-derived exosomes suppresses breast tumor cell invasion and migration by inducing M1 polarization in macrophages. Life Sci 282. https://doi.org/10.1016/j.lfs.2021.119800

Mou W, Xu Y, Ye Y, Chen S, Li X, Gong K, Liu Y, Chen Y, Li X, Tian Y, Xiang R, Li N (2015) Expression of sox 2 in breast cancer cells promotes the recruitment of M2 macrophages to tumor microenvironment. Cancer Lett 358(2):115. https://doi.org/10.1016/j.canlet.2014.11.004

Mu X, Shi W, Xu Y, Xu C, Zhao T, Geng B, Yang J, Pan J, Hu S, Zhang C, Zhang J, Wang C, Shen J, Che Y, Liu Z, Lv Y, Wen H, You Q (2018) Tumor-derived lactate induces M2 macrophage polarization via the activation of the ERK/STAT3 signaling pathway in breast cancer. Cell Cycle 17(4):428. https://doi.org/10.1080/15384101.2018.1444305

Obeid E, Nanda R, Fu YX, Olopade OI (2013) The role of tumor-associated macrophages in breast cancer progression (review). Int J Oncol 43. https://doi.org/10.3892/ijo.2013.1938

Olkhanud PB, Damdinsuren B, Bodogai M, Gress RE, Sen R, Wejksza K, Malchinkhuu E, Wersto RP, Biragyn A (2011) Tumor-evoked regulatory B cells promote breast cancer metastasis by converting resting CD4+ T cells to T-regulatory cells. Cancer Res 71(10). https://doi.org/10.1158/0008-5472.CAN-10-4316

Orecchioni M, Ghosheh Y, Pramod AB, Ley K (2019) Macrophage polarization: different gene signatures in M1(Lps+) vs. classically and M2(LPS-) vs. alternatively activated macrophages. Front Immunol 10. https://doi.org/10.3389/fimmu.2019.01084

Pang MF, Georgoudaki AM, Lambut L, Johansson J, Tabor V, Hagikura K, Jin Y, Jansson M, Alexander JS, Nelson CM, Jakobsson L, Betsholtz C, Sund M, Karlsson MCI, Fuxe J (2016) TGF-β1-induced EMT promotes targeted migration of breast cancer cells through the lymphatic system by the activation of CCR7/CCL21-mediated chemotaxis. Oncogene 35(6). https://doi.org/10.1038/onc.2015.133

Park IH, Yang HN, Lee KJ, Kim TS, Lee ES, Jung SY, Kwon Y, Kong SY (2017) Tumor-derived IL-18 induces PD-1 expression on immunosuppressive NK cells in triple-negative breast cancer. Oncotarget 8(20). https://doi.org/10.18632/oncotarget.16281

Park MH, Lee JS, Yoon JH (2012) High expression of CX3CL1 by tumor cells correlates with a good prognosis and increased tumor-infiltrating CD8+ T cells, natural killer cells, and dendritic cells in breast carcinoma. J Surg Oncol 106(4). https://doi.org/10.1002/jso.23095

Pei X, Wang X, Li H (2018) Lnc RNA SNHG1 regulates the differentiation of Treg cells and affects the immune escape of breast cancer via regulating miR-448/IDO. Int J Biol Macromol 118:24–30. https://doi.org/10.1016/j.ijbiomac.2018.06.033

Peng D, Tanikawa T, Li W, Zhao L, Vatan L, Szeliga W, Wan S, Wei S, Wang Y, Liu Y, Staroslawska E, Szubstarski F, Rolinski J, Grywalska E, Stanisławek A, Polkowski W, Kurylcio A, Kleer C, Chang AE, Kryczek I (2016) Myeloid-derived suppressor cells endow stem-like qualities to breast cancer cells through IL6/STAT3 and NO/NOTCH cross-talk signaling. Cancer Res 76(11):3156. https://doi.org/10.1158/0008-5472.CAN-15-2528

Plitas G, Konopacki C, Wu K, Bos PD, Morrow M, Putintseva EV, Chudakov DM, Rudensky AY (2016) Regulatory T cells exhibit distinct features in human breast cancer. Immunity 45(5):1122. https://doi.org/10.1016/j.immuni.2016.10.032

Sahai E, Astsaturov I, Cukierman E, DeNardo DG, Egeblad M, Evans RM, Fearon D, Greten FR, Hingorani SR, Hunter T, Hynes RO, Jain RK, Janowitz T, Jorgensen C, Kimmelman AC, Kolonin MG, Maki RG, Powers RS, Puré E et al (2020) A framework for advancing our understanding of cancer-associated fibroblasts. Nat Rev Cancer 20(3):174–186. https://doi.org/10.1038/s41568-019-0238-1

Saleh R, Toor SM, Khalaf S, Elkord E (2019) Breast cancer cells and PD-1/PD-L1 blockade upregulate the expression of PD-1, CTLA-4, TIM-3 and LAG-3 immune checkpoints in CD4+ T cells. Vaccine 7(4). https://doi.org/10.3390/vaccines7040149

Sang Y, Brichalli W, Rowland RRR, Blecha F (2014) Genome-wide analysis of antiviral signature genes in porcine macrophages at different activation statuses. PLoS One 9(2):1–12. https://doi.org/10.1371/journal.pone.0087613

Santolla MF, Talia M, Cirillo F, Scordamaglia D, De Rosis S, Spinelli A, Miglietta AM, Nardo B, Filippelli G, De Francesco EM, Belfiore A, Lappano R, Maggiolini M (2022) The AGEs/RAGE transduction signaling prompts IL-8/CXCR1/2-mediated interaction between cancer-associated fibroblasts (CAFs) and breast cancer cells. Cells 11(15). https://doi.org/10.3390/cells11152402

Satthaporn S, Robins A, Vassanasiri W, El-Sheemy M, Jibril JA, Clark D, Valerio D, Eremin O (2004) Dendritic cells are dysfunctional in patients with operable breast cancer. Cancer Immunol Immunother 53(6). https://doi.org/10.1007/s00262-003-0485-5

Shen J, Pan J, Du C, Si W, Yao M, Xu L, Zheng H, Xu M, Chen D, Wang S, Fu P, Fan W (2017) Silencing NKG2D ligand-targeting miRNAs enhances natural killer cell-mediated cytotoxicity in breast cancer. Cell Death and Disease 8(4):e2740. https://doi.org/10.1038/cddis.2017.158

Shukla S, Penta D, Mondal P, Meeran SM (2019) Epigenetics of breast cancer: clinical status of epi-drugs and phytochemicals. Adv Exp Med Biol 1152:293–310. https://doi.org/10.1007/978-3-030-20301-6_16

Sisirak V, Faget J, Gobert M, Goutagny N, Vey N, Treilleux I, Renaudineau S, Poyet G, Labidi-Galy SI, Goddard-Leon S, Durand I, Le Mercier I, Bajard A, Bachelot T, Puisieux A, Puisieux I,

Blay JY, Ménétrier-Caux C, Caux C, Bendriss-Vermare N (2012) Impaired IFN-α production by plasmacytoid dendritic cells favors regulatory T-cell expansion that may contribute to breast cancer progression. Cancer Res 72(20):5188. https://doi.org/10.1158/0008-5472.CAN-11-3468

Slattery K, Woods E, Zaiatz-Bittencourt V, Marks S, Chew S, Conroy M, Goggin C, Maceochagain C, Kennedy J, Lucas S, Finlay DK, Gardiner CM (2021) TGFβ drives NK cell metabolic dysfunction in human metastatic breast cancer. J Immunother Cancer 9(2). https://doi.org/10.1136/jitc-2020-002044

Smith AD, Lu C, Payne D, Paschall AV, Klement JD, Redd PS, Ibrahim ML, Yang D, Han Q, Liu Z, Shi H, Hartney TJ, Nayak-Kapoor A, Liu K (2020) Autocrine IL6-mediated activation of the STAT3-DNMT axis silences the TNFa-RIP1 necroptosis pathway to sustain survival and accumulation of myeloid-derived suppressor cells. Cancer Res 80(15):3145–3156. https://doi.org/10.1158/0008-5472.CAN-19-3670

Song X, Wei C, Li X (2021) The potential role and status of IL-17 family cytokines in breast cancer. Int Immunopharmacol 95:107544. https://doi.org/10.1016/j.intimp.2021.107544

Sousa S, Brion R, Lintunen M, Kronqvist P, Sandholm J, Mönkkönen J, Kellokumpu-Lehtinen PL, Lauttia S, Tynninen O, Joensuu H, Heymann D, Määttä JA (2015) Human breast cancer cells educate macrophages toward the M2 activation status. Breast Cancer Res 17(1):101. https://doi.org/10.1186/s13058-015-0621-0

Su S, Liao J, Liu J, Huang D, He C, Chen F, Yang LB, Wu W, Chen J, Lin L, Zeng Y, Ouyang N, Cui X, Yao H, Su F, Huang JD, Lieberman J, Liu Q, Song E (2017) Blocking the recruitment of naive CD4+ T cells reverses immunosuppression in breast cancer. Cell Res 27(4):461. https://doi.org/10.1038/cr.2017.34

Sung H, Ferlay J, Siegel RL, Laversanne M, Soerjomataram I, Jemal A, Bray F (2021) Global cancer statistics 2020: GLOBOCAN estimates of incidence and mortality worldwide for 36 cancers in 185 countries. CA Cancer J Clin 71(3):209–249. https://doi.org/10.3322/caac.21660

Taghikhani A, Hassan ZM, Ebrahimi M, Moazzeni SM (2019) microRNA modified tumor-derived exosomes as novel tools for maturation of dendritic cells. J Cell Physiol 234(6):9417. https://doi.org/10.1002/jcp.27626

Tan AHY, Tu WJ, McCuaig R, Hardy K, Donovan T, Tsimbalyuk S, Forwood JK, Rao S (2019) Lysine-specific histone demethylase 1A regulates macrophage polarization and checkpoint molecules in the tumor microenvironment of triple-negative breast cancer. Front Immunol 10 (JUN):1–17. https://doi.org/10.3389/fimmu.2019.01351

Tang M, Diao J, Cattral MS (2017) Molecular mechanisms involved in dendritic cell dysfunction in cancer. Cell Mol Life Sci 74(5):761–776. https://doi.org/10.1007/s00018-016-2317-8

Tang X, Tu G, Yang G, Wang X, Kang L, Yang L, Zeng H, Wan X, Qiao Y, Cui X, Liu M, Hou Y (2019) Autocrine TGF-β1/miR-200s/miR-221/DNMT3B regulatory loop maintains CAF status to fuel breast cancer cell proliferation. Cancer Lett 452:79. https://doi.org/10.1016/j.canlet.2019.02.044

Thacker G, Henry S, Nandi A, Debnath R, Singh S, Nayak A, Susnik B, Boone MM, Zhang Q, Kesmodel SB, Gumber S, Das GM, Kambayashi T, Dos Santos CO, Chakrabarti R (2023) Immature natural killer cells promote progression of triple-negative breast cancer. Sci Transl Med 15(686):eabl4414. https://doi.org/10.1126/scitranslmed.abl4414

Tiainen S, Tumelius R, Rilla K, Hämäläinen K, Tammi M, Tammi R, Kosma VM, Oikari S, Auvinen P (2015) High numbers of macrophages, especially M2-like (CD163-positive), correlate with hyaluronan accumulation and poor outcome in breast cancer. Histopathology 66(6): 873. https://doi.org/10.1111/his.12607

Tran K, Risingsong R, Royce DB, Williams CR, Sporn MB, Pioli PA, Gediya LK, Njar VC, Liby KT (2013) The combination of the histone deacetylase inhibitor vorinostat and synthetic triterpenoids reduces tumorigenesis in mouse models of cancer. Carcinogenesis 34(1): 199–210. https://doi.org/10.1093/carcin/bgs319

Tyagi K, Masoom M, Majid H, Garg A, Bhurani D, Agarwal NB, Khan MA (2023) Role of cytokines in chemotherapy-related cognitive impairment of breast cancer patients: a systematic

review. Curr Rev Clin Exp Pharmacol 18:110. https://doi.org/10.2174/ 2772432817666220304212456
Vesely MD, Schreiber RD (2013) Cancer immunoediting: antigens, mechanisms, and implications to cancer immunotherapy. Ann N Y Acad Sci 1284(1):1. https://doi.org/10.1111/nyas.12105
Wang HW, Joyce JA (2010) Alternative activation of tumor-associated macrophages by IL-4: priming for protumoral functions. Cell Cycle 9(24):4824–4835. https://doi.org/10.4161/cc.9.24. 14322
Wang H, Wei H, Wang J, Li L, Chen A, Li Z (2020) MicroRNA-181d-5p-containing exosomes derived from CAFs promote EMT by regulating CDX2/HOXA5 in breast cancer. Molecular Therapy-Nucleic Acids 19:654. https://doi.org/10.1016/j.omtn.2019.11.024
Wang J, Iwanowycz S, Yu F, Jia X, Leng S, Wang Y, Li W, Huang S, Ai W, Fan D (2016) microRNA-155 deficiency impairs dendritic cell function in breast cancer. Onco Immunol 5(11):1–14. https://doi.org/10.1080/2162402X.2016.1232223
Wang ZK, Yang B, Liu H, Hu Y, Yang JL, Wu LL, Zhou ZH, Jiao SC (2012) Regulatory T cells increase in breast cancer and in stage IV breast cancer. Cancer Immunol Immunother 61(6): 911–916. https://doi.org/10.1007/s00262-011-1158-4
Williams MM, Christenson JL, O'Neill KI, Hafeez SA, Ihle CL, Spoelstra NS, Slansky JE, Richer JK (2021) MicroRNA-200c restoration reveals a cytokine profile to enhance M1 macrophage polarization in breast cancer. Npj Breast Cancer 7(1):64. https://doi.org/10.1038/s41523-021-00273-1
Wu SY, Xiao Y, Wei JL, Xu XE, Jin X, Hu X, Li DQ, Jiang YZ, Shao ZM (2021) MYC suppresses STING-dependent innate immunity by transcriptionally upregulating DNMT1 in triple-negative breast cancer. J Immunother Cancer 9(7). https://doi.org/10.1136/jitc-2021-002528
Xiao P, Wan X, Cui B, Liu Y, Qiu C, Rong J, Zheng M, Song Y, Chen L, He J, Tan Q, Wang X, Shao X, Liu Y, Cao X, Wang Q (2016) Interleukin 33 in tumor microenvironment is crucial for the accumulation and function of myeloid-derived suppressor cells. Onco Immunology 5(1): 1–12. https://doi.org/10.1080/2162402X.2015.1063772
Xu Q, Jiang Y, Yin Y, Li Q, He J, Jing Y, Qi YT, Xu Q, Li W, Lu B, Peiper SS, Jiang BH, Liu LZ (2013) A regulatory circuit of miR-148a/152 and DNMT1 in modulating cell transformation and tumor angiogenesis through IGF-IR and IRS1. J Mol Cell Biol 5(1):3. https://doi.org/10.1093/jmcb/mjs049
Yadav P, Shankar BS (2019) Radio resistance in breast cancer cells is mediated through TGF-β signalling, hybrid epithelial-mesenchymal phenotype and cancer stem cells. Biomed Pharmacother 111:119. https://doi.org/10.1016/j.biopha.2018.12.055
Yan Y, Song Q, Yao L, Zhao L, Cai H (2022) YAP overexpression in breast cancer cells promotes angiogenesis through activating YAP signaling in vascular endothelial cells. Anal Cell Pathol 2022. https://doi.org/10.1155/2022/5942379
Yang D, Guo P, He T, Powell CA (2021a) Role of endothelial cells in tumor microenvironment. Clin Transl Med 11(6):19–22. https://doi.org/10.1002/ctm2.450
Yang J, Zhang Z, Chen C, Liu Y, Si Q, Chuang TH, Li N, Gomez-Cabrero A, Reisfeld RA, Xiang R, Luo Y (2014a) MicroRNA-19a-3p inhibits breast cancer progression and metastasis by inducing macrophage polarization through downregulated expression of Fra-1 proto-oncogene. Oncogene 33(23):3014–3023. https://doi.org/10.1038/onc.2013.258
Yang P, Cao X, Cai H, Chen X, Zhu Y, Yang Y, An W, Jie J (2021b) Upregulation of microRNA-155 enhanced migration and function of dendritic cells in three-dimensional breast cancer microenvironment. Immunol Investig 50(8):1058–1071. https://doi.org/10.1080/08820139. 2020.1801721
Yang X, Wang X, Liu D, Yu L, Xue B, Shi H (2014b) Epigenetic regulation of macrophage polarization by DNA methyltransferase 3b. Mol Endocrinol 28(4):565. https://doi.org/10.1210/me.2013-1293
Yang Y, Li C, Liu T, Dai X, Bazhin AV (2020) Myeloid-derived suppressor cells in tumors: from mechanisms to antigen specificity and microenvironmental regulation. Front Immunol 11 (July):1–22. https://doi.org/10.3389/fimmu.2020.01371

Yu J, Du W, Yan F, Wang Y, Li H, Cao S, Yu W, Shen C, Liu J, Ren X (2013) Myeloid-derived suppressor cells suppress antitumor immune responses through IDO expression and correlate with lymph node metastasis in patients with breast cancer. J Immunol 190(7). https://doi.org/10.4049/jimmunol.1201449

Zari AT, Zari TA, Hakeem KR (2021) Anticancer properties of eugenol: a review. Molecules 26(23):7407. https://doi.org/10.3390/molecules26237407

Zhang L, Deng J, Tang R, Sun J, Chi F, Wu S (2021) Flavin adenine dinucleotide synthetase 1 is an up-regulated prognostic marker correlated with immune infiltrates in breast cancer. 1–23

Zhang PF, Zhang PF, Zhang PF, Gao C, Gao C, Huang XY, Huang XY, Lu JC, Lu JC, Guo XJ, Guo XJ, Shi GM, Shi GM, Cai JB, Cai JB, Ke AW, Ke AW (2020a) Cancer cell-derived exosomal circUHRF1 induces natural killer cell exhaustion and may cause resistance to anti-PD1 therapy in hepatocellular carcinoma. Mol Cancer 19(1). https://doi.org/10.1186/s12943-020-01222-5

Zhang R, Dong M, Tu J, Li F, Deng Q, Xu J, He X, Ding J, Xia J, Sheng D, Chang Z, Ma W, Dong H, Zhang Y, Zhang L, Zhang L, Liu S (2023) PMN-MDSCs modulated by CCL20 from cancer cells promoted breast cancer cell stemness through CXCL2-CXCR2 pathway. Signal Transduct Target Ther 8(1):97. https://doi.org/10.1038/s41392-023-01337-3

Zhang W, Xu J, Fang H, Tang L, Chen W, Sun Q, Zhang Q, Yang F, Sun Z, Cao L, Wang Y, Guan X (2018) Endothelial cells promote triple-negative breast cancer cell metastasis via PAI-1 and CCL5 signaling. FASEB J 32(1). https://doi.org/10.1096/fj.201700237RR

Zhang Y, Li Z, Chen M, Chen H, Zhong Q, Liang L, Li B (2020b) lncRNA TCL6 correlates with immune cell infiltration and indicates worse survival in breast cancer. Breast Cancer 27(4): 573–585. https://doi.org/10.1007/s12282-020-01048-5

Zhao X, Qu J, Sun Y, Wang J, Liu X, Wang F, Zhang H, Wang W, Ma X, Gao X, Zhang S (2017) Prognostic significance of tumor-associated macrophages in breast cancer: a meta-analysis of the literature. Oncotarget 8(18):30576. https://doi.org/10.18632/oncotarget.15736

Zhou D, Yang K, Chen L, Zhang W, Xu Z, Zuo J, Jiang H, Luan J (2017) Promising landscape for regulating macrophage polarization: epigenetic viewpoint. Oncotarget 8(34):57693–57706. https://doi.org/10.18632/oncotarget.17027

Zhu H, Gu Y, Xue Y, Yuan M, Cao X, Liu Q (2017) CXCR2+ MDSCs promote breast cancer progression by inducing EMT and activated T cell exhaustion. Oncotarget 8(70). https://doi.org/10.18632/oncotarget.23020

Zhu J, Paul WE (2008) CD4 T cells: fates, functions, and faults. Blood 112(5):1557. https://doi.org/10.1182/blood-2008-05-078154

Chapter 7
The Epigenetics of Brain Tumors: Fundamental Aspects of Epigenetics in Glioma

Sevilhan Artan and Ali Arslantas

Abstract A brain tumor is an abnormal growth of heterogeneous cells around the central nervous system and spinal cord. It is seen in different types and frequencies in children and adults. Today, surgery, radiotherapy, and chemotherapy are common treatment approaches. Comprehensive molecular profiling has dramatically changed the diagnostic neuropathology of brain tumors. Diffuse gliomas, the most common and fatal brain tumor variants, are now classified by highly repetitive biomarkers rather than histomorphological features. Many critical molecular changes that drive glioma classification involve fundamental epigenetic dysregulation, an area not previously thought to play important roles in glioma pathogenesis. Considering tumor heterogeneity in the classification of brain tumors, molecular markers provide more accurate results in diagnosis, prognosis, and selection of treatment approaches. Recently, epigenetic changes have received increasing attention as they aid in understanding the mechanism of chromatin-mediated disease. Epigenetic modification alters the chromatin structure, which affects the docking site of many drugs that cause chemoresistance in cancer therapy. This chapter will review the main epigenetic changes underlying malignant gliomas and their possible mechanisms of action, based on the WHO 2021 classification.

Keywords Epigenetic mechanisms · DNA demethylation · Glioma CpG island methylator phenotype (G-CIMP) · IDH genes · MGMT · ATRX gene · H3F3AK27 · H3F3A G34R/V

S. Artan (✉)
Department of Medical Genetics, Eskisehir Osmangazi University, Medical Faculty, Eskişehir, Turkey
e-mail: sartan@ogu.edu.tr

A. Arslantas
Department of Neurosurgery, Eskisehir Osmangazi University, Medical Faculty, Eskişehir, Turkey

Abbreviations

5caC	5-Carboxylcytosine
5fC	5-Formylcytosine
APNG	Alkylpurine-DNA-N-glycosylase
ASIR	Age-standardized incidence rate
ASMR	Age-standardized mortality rate
ATPase	Adenosine triphosphatase
CGI	CpG islands
CNS	Central nervous system
CTCF	CCCTC-binding factor
DIPGs	Diffuse intrinsic pontine gliomas
DNMTs	DNA methyltransferases
G-CIMP	Glioma CpG island methylator phenotype
HATs	Histone acetyltransferases
HDACi	HDAC inhibitors
HDACs	Histone deacetylases
mRNAs	Messenger RNAs
NADP(+)	Nicotinamide adenine dinucleotide phosphate
ncRNAs	Noncoding RNAs
TCGA	The Cancer Genome Atlas
TMZ	Temozolomide
α-KG	α-Ketoglutarate

7.1 Introduction

Primary brain and central nervous system (CNS) cancers refer to a heterogeneous group of tumors arising from cells within the CNS (Guo et al. 2019; Louis et al. 2021). They affect both children and adults and are diagnosed in all CNS anatomical regions. While nearly histologically distinct over 150 malignant and nonmalignant brain and CNS tumors exist, each has its own epidemiology, clinical treatments, and prognosis. Primary brain tumor is a disease that occurs when the brain parenchyma cells differentiate and multiply uncontrollably. They are stratified based on age, tumor histology, and growth. The majority of CNS malignancies (>90%) (Koivunen and Laukka 2018) occur in the brain, whereas the rest are seen in the spinal cord, meninges, and cranial nerves. They include a wide range of diseases affecting children and adults, some of which are the most aggressive and fatal forms of cancer (Patel et al. 2019; Kukreja et al. 2021; Miller et al. 2021).

According to Global Cancer Statistics 2020, the global brain and CNS cancer age-standardized incidence rate (ASIR) was 4.34 (3.27–4.86) per 100,000 population, and the global age-standardized mortality rate (ASMR) was 3.05(2.29–3.36) per 100,000 population. Since there is an increasing global incidence and mortality rate of brain and CNS cancers, these cancers remain a major public health burden worldwide (Fan et al. 2022).

Brain tumors are a highly heterogeneous group of tumors with a robust variety of incidence by age but also by sex (Louis et al. 2016). Although brain and CNS tumors are more common in males (58% vs. 41%), it varies according to the tumor type. While malignant tumors are more common in men than women (with an annual rate of 8.3 vs. 6.0), the opposite is true for nonmalignant tumors (with a rate of 19.8 vs. 12.5) (Miller et al. 2021; Le Rhun and Weller 2020).

Besides, brain tumors can be seen in almost any age range but are more common in children under the age of 10 and people over the age of 65. Primary malignant CNS tumors are the most frequent solid cancer observed in children and adolescents, with an incidence reported of 5.7 per 100,000 children, and have the highest mortality of all childhood cancers. In contrast, the incidence is much higher (29.9 per 100,000 persons) in adults (Greuter et al. 2021; Ross et al. 2021; Miller et al. 2021; Duke and Packer 2020). Five-year survival for patients diagnosed with a nonmalignant tumor was 91.5%, but it is about 36% overall in the patients after a diagnosis of a malignant brain and other CNS tumor. However, survival rates of the patients with malignant tumors vary drastically based on the type of brain tumor and by the age. Relative survival is still lowest for glioblastoma (6.8%), and survival following diagnosis seems to be highest in children and adolescents (about 80%) if compared to those ages ≥40+ years (21.3%) (Ostrom et al. 2019; Kukreja et al. 2021; Miller et al. 2021).

Thanks to the recent developments in molecular technology, understanding of cancer biology, and the discovery of signaling pathways and molecules that play a role in the development and progression of many cancer types, these improvements have dramatically altered both neoplastic and nonneoplastic disease classification, prognostic evaluation, and the individualization of cancer treatment. While precision medicine approaches, defined as delivering the right drug to the right patient at the right time by precisely targeting disease-specific molecular pathways, are well established in a few extracranial solid tumors, prospective evidence for primary CNS tumors is still in progress (Kheder and Hong 2018; Özdemir et al. 2022; Leibetseder et al. 2022). However, with the development of new molecular techniques and their increased availability worldwide, advances have been made in tumor diagnosis and classification through nucleic acid-based technologies. These developments formed the basis of the updated WHO 2021 classification of CNS tumors (WHO CNS5). Numerous genomic and epigenetic molecular modifications with clinicopathological utility have been included in the WHO CNS5 (Gritsch et al. 2022) that are important for defining more accurate diagnosis and classification of CNS neoplasms, determining prognosis and treatment approaches, and providing ancillary information (Leibetseder et al. 2022; Mcnamara et al. 2022; Gritsch et al. 2022). As can be seen in the table, not only genomic but also epigenetic markers are guiding in clinicopathological CNS neoplasia classification. In this chapter, following a general review of epigenetic mechanisms involved in cancer, epigenetic biomarkers included in WHO 2021 (WHO CNS5 2021) as diagnostic and prognostic criteria in CNS tumors will be reviewed in adult and childhood tumors separately.

7.2 Overview of Epigenetic Mechanisms

The expression pattern of a particular gene is determined by the tight structure of chromatin, the ability of transcription factors to bind to DNA regulatory regions in promoters, as well as the presence of chromatin-modifying enzymes. Since chromatin modifications play an instructive role in regulating all DNA-based processes, including transcription, repair, and replication, epigenetic regulation of the genetic code plays a role in determining the fate of the cell (You and Jones 2012; Cavalli and Heard 2019).

The term *epigenetics* is described as molecules and mechanisms that can maintain mitotically heritable alterations in gene expression without changing the sequence of DNA. DNA methylation, histone modifications, chromatin remodeling, and non-coding RNAs, especially microRNAs (miRNAs) and long noncoding RNAs, are known epigenetic mechanisms. Since the temporal and spatial controls of gene expression in all biological processes are controlled by epigenetic mechanisms, accumulating evidence suggests both global epigenetic signature changes and genetic alterations are driving events in several diseases, including cancer. The first human disease associated with epigenetic alterations was cancer. Feinberg and Vogelstein (1983) reported global DNA hypomethylation if compared to adjacent analogous normal tissues from which tumors are derived (Feinberg and Vogelstein 1983; Cheng et al. 2019).

7.2.1 DNA Methylation

DNA methylation is a well-known and characterized reversible covalent DNA modification by the addition of a methyl group (CH3) at the cytosine residue of the CpG dinucleotide at position 5′ (5mC), which is mediated by dedicated enzymes called *DNA methyltransferases* (DNMTs). Around 60% of the human genes have promoters with a high frequency of CpG sites called CpG islands (CGIs) which are hypermethylated in a tissue-specific manner during various phases of development and differentiation. These regions are thought to be crucial for gene regulation, and hypermethylation of promoters typically results in the silencing of gene expression. Despite the significance of transcriptional control, the methylation status of CpG islands in gene promoters is only a minor portion of the genome. Contrarily, global methylome analyses revealed a bulk of CpG dinucleotides scattered in the genome, particularly in repetitive sequences, which are methylated (cell type-specific) with a frequency of 60–90% (Asmar et al. 2015; Dabrowski et al. 2019; Perez and Capper 2020).

DNA methylation is a reversible process carried out by a highly conserved protein family referred to as DNA methyltransferases (*DNMT*s) (Sharma and Aazmi). These *DNMT*s mediate the transfer of a methyl group from S-adenosyl methionine as a methyl donor to DNA, and the family members, including *DNMT1*,

DNMT2, *DNMT3A*, *DNMT3B*, and *DNMT3L*, are classified as de novo *DNMTs* (DNMT3A and DNMT3B) and maintenance *DNMTs* (*DNMT1*) (Dabrowski et al. 2019; Greenberg and Bourćhis 2019).

7.2.1.1 DNA Demethylation and Hypermethylated Phenotype

Normal tissue homeostasis depends on epigenetic dynamics, and disruption of these dynamics affects gene expression networks associated with different cancer types. The high responsiveness of the epigenome to environmental changes and cellular signaling systems is one of its fundamental characteristics. It is noteworthy that epigenetic dynamics are not unidirectional but possess a high level of plasticity to allow changes in cell destiny in response to obstacles within various cellular microenvironments. Among the epigenetic modifications, especially DNA demethylation is significant for the induction of pluripotent gene expressions and reprogramming of the epigenome during the formation of cancer stem cells.

DNA demethylation mechanisms in mammals are divided into passive or active demethylation, depending on whether they rely on DNA replication. DNA methylation can be reversed by passive demethylation during DNA replication cycles if the specific methylation sites cannot be maintained by DNMT1 and UHR1 activity. Active demethylation is the removal of 5mC from DNA without being dependent on replication driven by the ten-eleven translocation family of 2-oxoglutarate-dependent dioxygenase enzymes (TET1, TET2, TET3) (Greenberg and Bourćhis 2019; Ferrer et al. 2020; Nishiyama and Nakanishi 2021; Asmar et al. 2015; Sharma and Aazmi 2019). Active DNA demethylation in cells is based on the mechanism of oxidation of 5mC instead of breaking the covalent bond and removing the methyl group (Ko et al. 2015). In the TET-dependent pathway, a series of sequential oxidation reactions leads to conversion of 5mC into 5-hydroxymethyl cytosine (5hC), then into 5-formylcytosine (5fC), and finally into 5-carboxylcytosine (5caC) (Iurlaro et al. 2013).

5hmC, which is the most prevalent form of oxidized 5mCs, has been the subject of the most research attention. High amounts of 5hmC are present in many different stem-cell types as well as in neural cell lines. The brain has levels of 5hmC that are roughly ten-fold higher than those of other tissues, indicating that 5hmC may serve as a crucial epigenetic marker for the nervous system (Globisch et al. 2010; Münzel et al. 2010), and in glioblastoma cells, it reaches about 1% (Takai et al. 2014). Since these cell types also exhibit high TET enzyme expression, it has been hypothesized that TETs and 5hmC together regulate pluripotency and cell differentiation and that a deficiency in 5hmC is also linked to the malignant development of cancer cells (Dabrowski et al. 2019). Moreover, it has been reported that all three TET enzymes are expressed in the brain and are associated with the development of neural progenitor cells and neuronal differentiation (Hahn et al. 2013).

It is well known that promoter CGIs of many genes are usually unmethylated in non-cancerous cells. Transcriptional repression is linked to the methylation state of CGIs close to promoter regions. Since the identification of the CpG island methylator phenotype (CIMP) in colorectal cancer, abnormal CGI methylation in promoter regions of tumor suppressor genes has become a major focus of DNA methylation research. Since TET enzymes convert 5mC to 5hmC, aberrant TET function that results in decreased 5hmC levels may be linked to dysregulated DNA demethylation, which can lead to elevated 5mC levels, CIMP, and the hypermethylated phenotype (Pfeifer et al. 2014; Thienpont et al. 2016; An et al. 2017). A number of solid tumors such as glial tumors have decreased global 5hmC levels, which may be related to a chromatin hypermethylator phenotype or CGIs (Chen et al. 2017; Orr et al. 2012; Van Damme et al. 2016; Nishiyama and Nakanishi 2021). It has also been proposed that TETs and 5hmC levels may govern cell differentiation and the epithelial-to-mesenchymal transition (EMT), and low 5hmC has been linked to a worse prognosis since lower 5hmC levels are an epigenetic signature of a variety of malignancies (Zhao et al. 2021; Asmar et al. 2015; Koivunen and Laukka 2018; Carella et al. 2020; Ehrlich 2019; Ferrer et al. 2020).

It is well established that isocitrate dehydrogenases (IDH1 and IDH2) affect DNA methylation, and the mutations of these enzymes are seen in several human malignancies, particularly gliomas and AML. These genes are well-known significant molecular prognostic marker of glioma patients. IDH mutants, as described in glioblastoma, produce the oncometabolite 2-hydroxyglutarate (2HG) that are involved in inactivation of TET enzymes leading to DNA hypermethylation (Gusyatiner and Hegi 2018; Turcan et al. 2012) (Koivunen and Laukka 2018; Scourzic et al. 2015).

7.2.2 Posttranslational Modifications of Histones

DNA methyltransferases (DNMTs), the enzymes responsible for writing DNA methylation, can be assembled by covalent posttranslational modifications of histones, which emphasizes how epigenetic markers interact with one another. Chromatin structure is crucial in processes such as gene transcription, alternative splicing, chromosome condensation, DNA replication, and repair. Posttranslational modifications in histone proteins play a role in the dynamic changes that provide regional relaxation/rearrangement in the chromatin structure. The positioning and composition of chromatin components, as well as chromatin alterations, impact the state of chromatin. Chromatin states can be affected by variations in the histone makeup of nucleosomes. The structure of the core proteins in nucleosomes is predominately globular, with an unstructured N-terminal tail. Most histone posttranslational modifications (PTMs), such as acetylation, methylation, and phosphorylation, primarily affect the amino acids in the N-terminal tail domains (Mancarella and Plass 2021). These modifications cause the alteration of the electrostatic charge, thus changing in histone shape and DNA-binding affinity and, consequently, resulting in gene expression suppression or activation.

Methylation, acetylation, and phosphorylation are the best-characterized histone modifications. High levels of acetylation (e.g., histone H3 lysine acetylation at

H3K9, H3K14, and H3K27) and trimethylated H3K4, H3K36, and H3K79 have been observed in actively transcribed euchromatin. In contrast, the transcriptionally inactive chromatin has high amounts of H3K9, H3K27, and H4K20 methylation and low levels of acetylation. Histones can undergo simultaneous modifications at many locations. As a result, a histone code incorporating posttranslational modifications that would control the chromatin's state at a certain instant has been proposed (Blakey and Litt 2015; Lu et al. 2020).

Histone posttranslational modifications are mediated or reversed by different enzyme groups. An example is the transfer of an acetyl group to the N-terminal tails of histones by *histone acetyltransferases* (HATs). *HAT*s add a negative charge by transferring the acetyl group, resulting in an electrostatic contact between amino acid residues. Additionally, this causes the chromatin loop to open, making it easier for transcription factors to bind and stimulate gene expression (Dawson 2017). On the other hand, the acetyl group is removed by *histone deacetylases* (HDACs), which give the molecule a positive charge by removing one of its negative charges. Maintaining a stable closed state of the chromatin loop and bringing out the electrostatic interactions between amino acids silence the gene. Since overexpression or downregulation of *HAT*s is linked to cancer and poor prognosis, proper acetylation within cells is crucial. Many malignancies, including glioblastoma (GBM), have increased *HDAC* expressions, and HDAC inhibitors (*HDACi*) have been thoroughly investigated as GBM therapies (Kukreja et al. 2021; Romani et al. 2018).

In contrast to histone acetylation, the effect of histone methylation is more complex and based on the targeted residues, and methylation can either promote or inhibit gene expression. For instance, methylation at histone H3 lysine 4/36/79 (H3K4/36/79) often promotes transcription, whereas methylation at histone H3 lysine 9/27 (H3K9Me2/3/H3K27Me3) and histone H4 lysine 20 (H4K20me) is typically regarded as repressive epigenetic markers (Shoaib et al. 2018; Wiles and Selker 2017; Black et al. 2012). Different histone methyltransferases (*HMT*s), the majority of which contain a SET domain, are the only enzymes that can catalyze them. For instance, the transcriptional silencing protein enhancer of zeste 2 (EZH2), catalytic components of polycomb repressive complexes (PRCs), is specific for the H3K27 trimethylation (H3K27me3). Inflammatory genes are activated to express themselves when H3K4me is catalyzed by SET7/9 (Lu et al. 2020). Lysine 27 to methionine (H3K27M) and H3K36M mutations are two significant oncogenic events, and both act as the primary drivers of juvenile gliomas and sarcomas, respectively. More than 70% of diffuse intrinsic pontine gliomas (DIPGs) and 20% of juvenile glioblastomas have been shown to have H3K27M, which causes a global decrease in the trimethylation of H3K27 (H3K27me3) (Cheng et al. 2019; Duchatel et al. 2019).

7.2.3 Noncoding RNAs

The alterations in the expressions of noncoding RNAs (ncRNAs) resulting from amplifications, deletions, and mutations, consequently, affect the functionality of their particular targets. ncRNAs can behave in a tumor-promoting, tumor-suppressing, or context-dependent manner. They comprise more than 70% of the human genome and influence regulation. microRNAs (miRNAs) have been further investigated in cancers and are promising in the development of preclinical therapeutics, and long noncoding RNAs (lncRNAs) are known as important epigenetic regulators in cancer. In humans, miRNAs regulate over 60% of the protein-coding genes and increases to over 80% among cancer genes (Tian et al. 2022). They complementarily bind to the 3′ UTR of the target mRNA to reduce gene expression. They can behave as oncomiRNAs or tumor suppressors depending on the function of target genes. More than 50% of miRNA genes are found near CGIs, making them sensitive to further epigenetic changes (Lu et al. 2020; Patil et al. 2021).

LncRNAs are a broad family of long transcripts (longer than 200 nucleotides) produced at various genomic sites. They can act on target sites in the cytoplasm or the nucleus and are not only involved in gene expression but also have roles as chromatin regulators, enhancers, sponges for ncRNAs, molecular scaffolds, etc. In general, lncRNAs and mRNAs that undergo splicing, 5′-cap formation, and polyadenylation are comparable, except for recently found circular RNAs (circRNAs), which lack a cap and have a poly-A tail. The dysregulation of lncRNAs is increasingly associated with many human diseases, especially cancers.

Downregulation of various tumor suppressor miRNAs such as miR-7, miR-124, miR-128, miR-137, miR-181a/b, and mir-138 in GBM or elevated miRNAs functioning as oncogenes has been reported in adult brain tumors. These miRNAs play essential roles in regulating cell-cycle progression, apoptosis, DNA repair, or angiogenesis, but expression alterations are consequently associated with tumor growth and metastasis. In addition, studies on using miRNAs as prognostic and predictive biomarkers have increased in recent years, and meta-analyses have provided promising data. For example, overexpression of plasma miRNA-222, miRNA-155, miRNA-221, and miRNA-21 has been associated with worse prognosis in glioblastoma patients (Puduvalli 2014; Śledzińska et al. 2021; Kukreja et al. 2021).

In the following sections, the most significant developments in molecular diagnosis of adult- and child-type primary tumors of the CNS are highlighted, with a strong focus on mechanisms of epigenetic markers used in diagnostic and/or prognostic information based on the WHO 2021 classification.

7.3 Epigenetic Changes in Gliomas

High-throughput genomic and molecular profiling technologies have dramatically altered the classification of neoplastic and nonneoplastic disorders, which frequently identify more uniformly described biological and clinical entities beyond histological classification. A better understanding of the molecular changes that support the central nervous system is now reflected in the molecular profile-based classification of brain tumors and clinical practice, leading to advances in tumor diagnosis, prognostic evaluation, and even therapeutic strategy approaches.

Gliomas, categorized based on the cell from whence they originated, are the most prevalent CNS tumors. About 30% of primary brain tumors are gliomas, which account for 80% of malignant ones and most primary brain tumor-related mortality (Louis et al. 2021). Adults can develop gliomas at a rate of 1.9 to 9.6 per 100,000, depending on age, sex, race, and location (Ostrom et al. 2019). According to the histological subtype, the median age at diagnosis varies, with pilocytic astrocytomas more frequently affecting children and adolescents, low-grade oligodendrogliomas peaking in the third and fourth decades, and glioblastomas typically presenting in patients over the age of 50 (Patel et al. 2019; Molinaro et al. 2019). Although the four lobes of the brain – frontal (23.6%), temporal (17.4%), parietal (10.6%), and occipital – are where most gliomas develop, gliomas in the brain stem, cerebellum, and spinal cord are observed less frequently (Ostrom et al. 2014).

Various classifications have been proposed to better understand the glioma genetics, ranging from the earliest microscopic findings to the most recent analyses of gene expression (Arslantas et al. 2007, Arslantas et al. 2004). The most frequent and lethal primary brain tumors are diffusely infiltrating gliomas, also known as diffuse gliomas. Neoplastic categories based on the histopathological traits of astrocytes and oligodendrocytes have been revised and optimized with the integration of biologically, clinically, and prognostically distinct disease-specific molecular markers in both adult and pediatric populations. In the past, diffuse gliomas were only categorized by their histological features as oligodendrogliomas, lower-grade astrocytomas, and high-grade glioblastomas, with some lower-grade tumors displaying both glial phenotypes (oligoastrocytomas). However, it has become increasingly apparent that tumor morphology alone cannot fully predict clinical behavior because tumor growth varies significantly depending on the histological subtype. More recently, the WHO 2021 extensively updated its classification of diffuse gliomas to include highly penetrant molecular anomalies. Not only genetic but also epigenetic criteria are integrated into the WHO classification.

The mutations in the genes encoding isocitrate dehydrogenase enzymes (IDH1 and IDH2), H3 histone monomers (H3F3A and HIST1H3B), and histone chaperone-thalassemia mental retardation X-linked (ATRX) are associated with disruption of epigenetic mechanisms (Dharmaiah and Huse 2022, Turcan et al. 2012, La Madrid and Kieran 2018, Perez and Capper 2020, Kristensen et al. 2019, Dabrowski et al. 2019). Since they highlight the significance of epigenetic changes as drivers in the evolution and biology of gliomas, they have become informative biomarkers for tumor classification.

7.3.1 Glioma CpG Island Methylator Phenotype (G-CIMP)

Following the discovery of a hypermethylated CGI pattern within promoter regions of the tumor suppressor genes in colorectal cancer (Toyota et al. 1999), a similar pattern has also reported in gliomas and is named glioma-CpG island methylator phenotype (G-CIMP). The G-CIMP phenotype is most frequently found in gliomas of grades 2 and 3 (Noushmehr et al. 2010) but is seen in recurrent glioblastomas with IDH gene mutations, as well. In the study by Verhaak et al., they discovered a significantly high prevalence of G-CIMP+ subtypes in younger patients and among the proneural subtypes than in G-CIMP- tumors (Verhaak et al. 2010). Later, the G-CIMP subtype in pediatric glioma patients was also described, but in lower frequency (La Madrid and Kieran 2018). Notably, numerous investigations revealed that G-CIMP+ was strongly correlated with IDH-mutant gliomas (Malta et al. 2018, Dabrowski et al. 2019, Romani et al. 2018). As IDH mutations are present in 75% of grade 2/3 tumors, it is safe to say that IDH mutations are a factor in the development of the G-CIMP phenotype.

If we summarize the relationship of G-CIMP to the prognostic features of glial tumors, previous studies have emphasized that IDH-mutant G-CIMP+ GBM and low-grade gliomas (LGG) share multiple molecules and extended survival properties (Gusyatiner and Hegi 2018; Turcan et al. 2012; Brennan et al. 2013; Malta et al. 2018; Noushmehr et al. 2010). In contrast, IDH wild-type, G-CIMP- LGG tumors are highly similar to the molecular and clinical features of GBM. On the other hand, not all IDH-mutant G-CIMP+ tumors have the same prognostic features. The overall survival rate in glial tumors with low G-CIMP levels (median survival G-CIMP-low $= 2.7$y) is highly similar to that of IDH wild-type gliomas (median survival 1.2 y) (Ceccarelli et al. 2016).

The known prognostic biomarkers for glial tumors are mainly IDH mutations, 1p/19q codeletion, MGMT promoter methylation, and G-CIMP+, all independent markers related to a good prognosis (Śledzińska et al. 2021; Mur et al. 2015). In the following sections, the molecular mechanisms of each of the known prognostic and therapeutic epigenetic based markers in gliomas and the reasons for being a marker will be detailed.

7.3.2 IDH1 Mutations and Relationship to Epigenetic Changes in Gliomas

IDH mutations are the most significant glioma classification and prognostic biomarkers. According to the WHO 2016 and WHO 2021 guidelines (Gritsch et al. 2022), IDH mutations are primarily associated with lower-grade (WHO 2/3) cancers. Compared to IDH wild-type glioblastoma, the typical WHO grade 4 primary brain tumor, diffuse gliomas in grades 2 and 3 are different. IDH wild-type GBM nearly always originates de novo in a completely malignant state marked by the aggressive histological hallmarks of microvascular proliferation and necrosis (Romani et al. 2018).

Adults with lower-grade astrocytomas and oligodendrogliomas are generally defined by heterozygous mutations in IDH1 and, less frequently, IDH2, with a more better clinical outcome (median survival of 65 months) than those of IDH wild-type GBM (median survival of 15 months). IDH mutations that have a tremendous effect on global DNA methylation patterns are considered the initiating event in the oncogenesis of IDH-mutant gliomas (Kristensen et al. 2019; Dharmaiah and Huse 2022). IDH-mutant gliomas in adults are further categorized depending on the presence of coincident chromosomes 1p and 19q codeletions (1p/19q). According to WHO 2021, oligodendrogliomas (WHO 2/3) are now characterized by the concurrent IDH mutation status and presence of 1p/19q codeletion. The median survival times of the patients whose tumors have these contemporary aberrations are relatively extended. Therefore, patients who exhibit the G-CIMP phenotype, particularly those who also have a 1p19q codeletion, have a better prognosis than patients who do not (Gritsch et al. 2022). However, by contrast, IDH-mutant astrocytomas (WHO 2/3) have mainly combined loss-of-function mutations in TP53 and ATRX genes, instead of 1p/19q deletion, and exhibit a relatively more aggressive biological behavior than their oligodendroglial equivalents. Besides, because the median overall survival times for high-grade (WHO4) IDH1/2 wild-type astrocytoma patients and IDH1/2 wild-type GBMs were similar, Tesileanu et al. (Tesileanu et al. 2020) noted that prognostic and therapeutic evaluations of high-grade astrocytomas without other qualifying molecular changes could be managed as IDH1/2 wild-type glioblastomas. In conclusion, IDH1/2 mutations, and hence G-CIMP tumor models, are the most critical and informative classification biomarkers for gliomas and also serve as a prognostic marker and a potential drug target.

Due to its importance in classification, prognostic evaluations, and possible therapeutic targets in glioma, the role of IDH1 is one of the most important clinical discoveries in neurooncology and deserves detailed scrutiny.

7.3.2.1 IDH1 and IDH2 Genes

IDH1 and IDH2 genes encode two of the three IDH enzymes and are involved in critical metabolic processes such as the Krebs cycle, lipid metabolism, and the

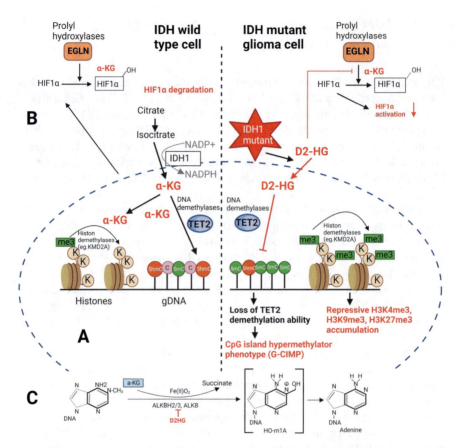

Fig. 7.1 Schematic view about how IDH mutations affect epigenetic patterns of the cell. A. D2-HG resulting from IDH mutation causes inhibition of TET DNA and histone demethylases and increases the cell's DNA and histone methylation levels. B. Histone demethylases like KDM regulate histone methylation state and gene expression in concert. Because of IDH mutations, increased D2-HG inhibits histone demethylase activities, causing the accumulation of repressive H3K4me3, H3K9me3, and H3K27me3 histone methylation markers. C. The ketoglutarate-dependent DNA repair enzymes remove methyl groups and some alkylation lesions from purine and pyrimidine bases. In the case of IDH mutations, the presence of 2HG causes changes in the DNA repair process

control of oxidative damage (Sun et al. 2021). IDH1 is primarily expressed in the cytoplasm and peroxisomes, whereas IDH2 is in the mitochondrial matrix. Both are obligate homodimers and use nicotinamide adenine dinucleotide phosphate (NADP+) as a cofactor and catalyze the production of α-ketoglutarate (α-KG) from isocitrate in the Krebs cycle (Fig. 7.1). IDH1/IDH2 enzymes catalyze the reaction from the NADP+ to generate NADPH. NADPH is a critical reducing agent that regulates cellular defense systems against oxidative damage through the reduction of glutathione and thioredoxins and the synthesis of activated catalase as well. As a

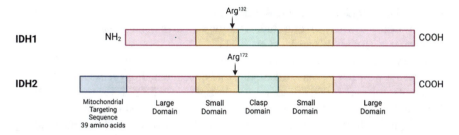

Fig. 7.2 IDH1 and IDH2 genes have three different domains: large domain, small domain, and clasp domain. IDH2 also contains an additional mitochondrial targeting sequence of 39 amino acids at its NH2-terminal

result, it is plausible that decreased or lack of IDH1/IDH2 function could compromise detoxification processes, resulting in DNA damage and genomic instability, which are typical features of cancerous cells. About a twofold decrease in NADP+-dependent IDH activity and a significant reduction in NADPH generation have been shown in GBM tumors (Dang et al. 2016; Waitkus et al. 2015).

IDH1 and IDH2 genes are localized on chromosomes 2q33 and 15q26, respectively, and both have three different domains: large domain, small domain, and clasp domain. IDH2 also contains an additional mitochondrial targeting sequence of 39 amino acids at its NH2-terminal (Fig. 7.2). IDH mutations are early lesions in the development of gliomas. Although other locations have occasionally been observed (Gupta et al. 2013), remarkably, missense mutations in IDH1 and IDH2, which are universal and almost in heterozygous form, cluster at specific arginine residues at the active isocitrate binding sites of the genes. IDH1 codon 132 and codons 140 and 172 for IDH2 are hotspot regions, and IDH1 Arg132 and IDH2 Arg172 are evolutionarily conserved residues. About 90% of IDH alterations in gliomas are caused by a single mutation in IDH1, which changes arginine 132 to histidine (R132H) (Dharmaiah and Huse 2022; Gusyatiner and Hegi 2018; Waitkus et al. 2015).

In heterozygous IDH-mutant glioma, the product of wild-type allele of the IDH gene catalyzes the production of α-ketoglutarate (α-KG) from isocitrate, but the mutant allele works with NADPH for conversion of α-KG into R(−)-2-hydroxyglutarate (D2HG) (Waitkus et al. 2015). Although mutations in IDH result in the loss of the ability for conversion of isocitrate to α-KG, the substrate specificity of mutant IDH has changed; unlike wild-type IDH1, it catalyzes NADPH dependent reduction of α-KG to D2HG, that is, a gain of function mutation. Detection of increased D2HG levels in glioma samples also supports that D2HG is an oncometabolite that plays a role in the early stages of gliomagenesis. Because α-KG and D2HG are structurally similar, the only difference between them is the presence of the C2 hydroxyl group in 2HG, but the C2 carbonyl group in α-KG, accumulated 2HG, behaves as a competitive inhibitor of α-KG-dependent dioxygenases, including the histone lysine demethylases, the TET family of methylcytosine hydroxylase, and the AlkB family of oxidative demethylases.

Previously, inhibition of these α-KG-dependent dioxygenases by D2HG accumulation has been shown by several studies (Puduvalli 2014; Dharmaiah and Huse 2022; Romani et al. 2018; Śledzińska et al. 2021).

As reviewed before, oxidation reactions of 5mC to 5hmC, to 5fC, and to 5caC are catalyzed by Fe2 + − and α-ketoglutarate-dependent dioxygenases of the Ten Eleven Translocation (TET) family. All three TETs (TET1, TET2, and TET3) are important regulators of epigenetic controls of gene expression through removing methyl groups from nucleic acid and protein. Elevated levels of D2HG resulting from IDH mutations block the TET family functions and cause abnormally high levels of methylation throughout the genome, hence giving rise to the glioma CpG island hypermethylator phenotype (G-CIMP) (Fig. 7.1a). These findings suggest that IDH mutations, lowered α-KG levels, elevated D2HG levels, and the epigenetic regulator mechanisms of glioma cells are in an intimate association. Several lines of evidence now point to the sufficiency of IDH mutations to cause the G-CIMP phenotype in glioma cells (Turcan et al. 2012; Dharmaiah and Huse 2022; Waitkus et al. 2015).

Accumulated 2HG in IDH-mutant gliomas also affects the histone methylation status by inhibiting histone lysine demethylases, such as lysine-specific demethylase (KDM) (Xu et al. 2011). As well known, histone methyltransferases, including EZH2, SET, GLP, and G9a, and histone demethylases such as KDM and JARID control histone methylation status and, consequently, gene expression in coordination with each other. It has been shown that KDM4 and KDM5 histone demethylases are inhibited by the elevated 2HG, resulting in the accumulation of repressive histone marks H3K4me3, H3K9me3, and H3K27me3, histone methylation markers in IDH-mutant tumors (Fig. 7.1b) (Dabrowski et al. 2019, Ceccarelli et al. 2016, Han et al. 2020).

Basic mechanisms of carcinogenesis, including genomic instability, cellular motility, and growth control deficiencies, are directly related to the production level of reactive oxygen species (ROS), whose excessive production damages nucleic acids, proteins, and lipids. Therefore, balancing appropriate levels is essential for both carcinogenesis and therapeutic resistance. Because tumors with IDH mutations consume cellular NADPH instead of NADP+ and isocitrate, NADPH usage results in ROS accumulation and, consequently, oxidative damage. During hypoxia, the HIF transcription factor activates the genes promoting cellular survival. The prolyl hydroxylation by α-ketoglutarate (αKG) dependent dioxygenases known as EGLNs work as oxygen sensors and regulate the subunit of HIF. In the presence of oxygen, EglN hydroxylates HIFα, and then von Hippel-Lindau (VHL) tumor suppressor protein polyubiquitylates the hydroxylated HIF for degradation. However, in the case of IDH-mutant gliomas, elevated 2HG has been shown to increase the EglN activity and decrease the HIF activation, resulting in the progression of oncogenesis (Losman and Kaelin 2013; Dharmaiah and Huse 2022).

α-Ketoglutarate-dependent enzymes are also involved in DNA repairing, and the ALKB enzyme, which is a DNA repair enzyme, removes methyl groups and some larger alkylation lesions from endocyclic positions on purine and pyrimidine bases. The enzyme is a dioxygenase and requires α-ketoglutarate and iron for its activity

(Fig. 7.1c). Therefore, in IDH-mutant gliomas, changes occur in the DNA repair process due to enzyme inhibition in the presence of 2HG, which plays a role in the chemotherapy response.

7.3.2.2 Classifying Marker 1p/19q Codeletion

Although IDH mutations are the most important classifying marker for gliomas, they are further divided into subgroups based on the simultaneous deletion of the 1p and 19q chromosomal arms (1p/19q codeletion). The 1p/19q codel is exclusively seen in IDH-mutant oligodendrogliomas, WHO grades 2 and 3, and has been shown to have a positive prognostic marker in gliomas. According to The Cancer Genome Atlas Research Network data, the median survival time for patients with lower-grade gliomas with an IDH mutation and a 1p/19q codeletion was 8.0 years, compared to 6.3 years for patients with an IDH mutation but no codeletion (Network 2015). However, molecular-based features of IDH-mutant astrocytomas, WHO grades 2 and 3, are ATRX and TP53 loss-of-function mutations instead of 1p/19q codel. Their clinical progression is relatively more aggressive than oligodendroglial counterparts. Besides, homozygous CDKN2A/B deletion is an independent, negative prognostic factor in this tumor group and has been repeatedly reported as associated with poor prognosis (Gritsch et al. 2022).

7.3.2.3 MGMT Promoter Methylation

Prognostic markers help to determine differences in the clinical course due to intrinsic differences in patients' own tumors, regardless of treatment. But in the case of predictive biomarkers, the goal is to predict responses to specific therapies (Puduvalli 2014). In previous sections, proposed epigenetic biomarkers for the classification of glial tumors according to WHO 2021 were discussed. However, we will now examine the epigenetic marker MGMT (6-O-methylguanine-DNA methyltransferase), which has critical importance not in classifying glial tumors but in the treatment response (Kalkan et al. 2015).

Glioblastoma is a high-grade brain tumor with a poor prognosis. The standard sequential treatment approach is surgical resection followed by radiotherapy and concurrent and adjuvant temozolomide (TMZ) therapy. One of the most important prognostic and predictive biomarkers for GBM patients treated with TMZ is the methylation status of the MGMT promoter. In glial tumors, the MGMT gene promoter methylation status may specify the responsiveness of the patient to alkylating agent chemotherapy (Raviraj et al. 2020; Uddin et al. 2022; Liu et al. 2022).

The MGMT gene is located on chromosome 10q26 and encodes a DNA repair protein that removes alkyl groups from the O6 position of guanine. The promoter region of the gene is CpG rich and contains about 97 CpG sites for methylation and, consequently, inhibition of the gene transcription. DNA-alkylating agent

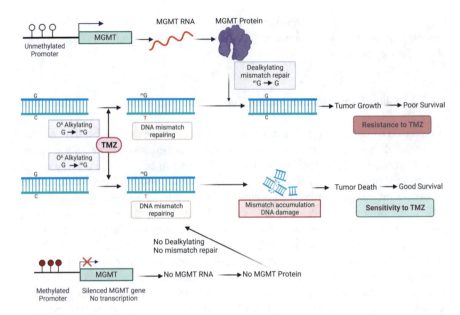

Fig. 7.3 MGMT methylation status is a potential predictive marker for temozolomide response. TMZ treatment leads to DNA mismatch, and high-level active MGMT protein is involved in mismatch repairing, resulting in TMZ resistance and hence poor prognosis. If MGMT gene promoter CGIs are hypermethylated, then the tumor cells are sensitive to TMZ because of the loss of MGMT protein

temozolomide (TMZ) is known to cause cell-cycle arrest at G2/M, which eventually results in apoptosis. The addition of methyl groups at the N7 and O6 sites on guanines and the O3 site on adenines in genomic DNA is the mechanism through which TMZ causes cytotoxicity. Alkylation of the O6 atom on guanine causes the coding specificity of the base to change in subsequent replication, causing methylguanine to mispair with thymine instead of cytosine. The mismatches of methylated DNA can be repaired by base excision or DNA mismatch repair pathways, through the involvement of a DNA glycosylase like alkylpurine-DNA-N-glycosylase (APNG) or a demethylating enzyme like MGMT. As a result, high MGMT mRNA and protein levels have been associated with resistance to DNA-alkylating chemicals, whereas methylation of CpG islands in the MGMT promoter region, inhibiting the protein expression, enhances chemosensitivity to these agents (Fig. 7.3). Numerous clinical trials and studies have demonstrated the significance of MGMT promoter methylation as a prognostic and predictive biomarker. According to multiple trials, temozolomide obviously has a lower benefit for glioblastoma patients over the age of 65 to 70 whose tumors lack MGMT promoter methylation (Kukreja et al. 2021; Śledzińska et al. 2021; Liu et al. 2020; Dabrowski et al. 2019).

7.3.2.4 IDHs, G-CIMP Status, and MGMT Promoter Methylation Relations

Because IDH1 mutation is related to the globally hypermethylated G-CIMP phenotype, regardless of the glioma grade, IDH mutation and G-CIMP status correlate with MGMT promoter methylation. In fact, it was discovered that the context of the IDH mutation and G-CIMP status was at least somewhat correlated with the prognostic vs. predictive significance of MGMT promoter methylation. According to The Cancer Genome Atlas (TCGA) data of Ceccarelli et al., the MGMT promoter hypermethylation rates ranged from 40.0% among IDH wild-type tumors to 91.8% of IDH-mutant glioma tumors (Ceccarelli et al. 2016; Malta et al. 2018). The MGMT promoter methylation has been reported to be a prognostic indicator for better survival and slowed tumor progression in gliomas harboring IDH mutation, including G-CIMP+ cases, regardless of treatment with radiotherapy and alkylating agent chemotherapy or with radiotherapy alone. On the other hand, MGMT promoter methylation was reported as a prognostic sign of a positive response to alkylating drug treatment in IDH wild-type gliomas, often known as G-CIMP negative tumors (Turcan et al. 2012).

Patients with MGMT promoter methylation may receive chemotherapy alone, while those without MGMT promoter methylation may receive radiotherapy alone due to the higher toxicity of radiotherapy and chemotherapy when administered together in senior patients (Liu et al. 2022). Data from four large clinical trials' control arms, which included over 4000 glioblastoma patients who received radiotherapy and temozolomide and underwent centralized MGMT promoter methylation testing, confirmed that those without MGMT promoter methylation had a worse prognosis than those who had it. Additionally, 10% of the patients in this study had minimal MGMT promoter methylation. The results for these "gray zone" patients matched more closely with the methylated group than the unmethylated group. These findings confirm those from ongoing studies that suggest temozolomide medication is beneficial for both MGMT promoter methylation-positive and methylation-negative individuals. Perhaps only patients who actually lack MGMT promoter methylation should be denied this medication (Molinaro et al. 2019; Hegi et al. 2019).

In conclusion, samples with low G-CIMP ratio also have a low MGMT promoter methylation ratio compared to high G-CIMP samples. These findings suggest that in determining the predictive significance of MGMT promoter methylation for predicting the benefit of alkylating chemotherapy in glioma samples, the G-CIMP positivity rate and IDH mutation status of the tumors as well should be taken into consideration.

Although MGMT is a reliable prognostic biomarker for GBM in adults and the elderly, the significance in children is less noteworthy. Glioblastoma falls under the adult diffuse glioma category in the WHO 2021 classification. However, pediatric gliomas are categorized individually along with other significant genetic markers, such as the H3K27 alteration and H3G34 mutation, which are also discussed below.

7.4 Epigenetic Biomarkers Related with Chromatin Modifications

Chromatin remodeling complexes have adenosine triphosphatase (ATPase) activity and rely on ATP hydrolysis to provide energy for completing chromatin structure changes (Stanton et al. 2017). The complexes can be divided into ISW I, SWI/SNF, and other types based on the various subunits that can hydrolyze ATP. These complexes and their associated proteins are also linked to cell-cycle activation and suppression, DNA repair, DNA methylation, and DNA transcription.

Many human diseases are caused by mutations in the chromatin remodelings main proteins. These mutations are also responsible for chromatin remodeling failures in which nucleosomes cannot be properly positioned, halting transcriptional machinery and preventing complexes that could repair DNA damage from accessing DNA. This may result in the expression of the abnormal gene. When these mutations cause changes in tumor suppressor genes or proteins that control the cell cycle, they may eventually contribute to cancer incidence (Uddin et al. 2022).

7.4.1 CTCF (CCCTC-Binding Factor)

CTCF is another crucial transcription factor that has been demonstrated to be impacted by DNA methylation. Gene expression patterns are strongly affected by changes in global CTCF binding, mainly through upsetting the three-dimensional chromatin structure. CTCF is not a typical transcription factor; it plays critical roles in chromatin loop formation and chromatin compartmentalized borders. Since particular genes should be silenced while others are kept active at a given time and/or tissue location in normal physiological conditions in differentiating cells and terminally differentiated cells, disruption of this mechanism may have serious responses. Additionally, IDH mutations causing DNA hypermethylation can also affect CTCF binding. Because of the widespread hypermethylation of CTCF binding sites in IDH-mutated gliomas, disruption of border elements entirely alter the topological architecture of chromatin. In the study by Flavahan et al., they reported the detection of hypermethylated sites at cohesin and CCCTC-binding factor (CTCF)-binding sites in IDH-mutant gliomas (Flavahan et al. 2016). Because CTCF is a methylation-sensitive insulator and essential in maintaining the barrier between normal chromatin topology and preventing abnormal gene activation, hypermethylation of these sites impairs the interaction ability of this crucial insulator protein. The receptor tyrosine kinase gene PDGFRA is expressed constitutively when CTCF binding is lost, which encourages the development of gliomas (Liu et al. 2020).

7.4.2 Nuclear Alpha-Thalassemia/Mental Retardation X-Linked Syndrome (ATRX) Gene

The Nuclear Alpha-Thalassemia/Mental Retardation X-linked syndrome (ATRX) gene, located on Xq21.1, was first discovered by Gibbons et al. in the analysis of patients with ATR-X syndrome (Gibbons et al. 1995). The protein encoded by the ATRX gene is a histone chaperone that assists in the loading of histones onto telomeres and the maintenance of heterochromatin environments. ATRX is a SNF2-type chromatin remodeling protein enriched at GC-rich and repetitive sequences, including inactivated X chromosome, telomere, and pericentromeric heterochromatin. It has two highly conserved domains (Valle-García et al. 2016). An ADD (ATRX-DNMT3-DNMT3L) domain in the N-terminal of the protein binds to and regulates H3K9me3-modified chromatin. ADD domain is similar to DNA methyltransferase DNMT3 with cysteine-rich motifs. The SWI/SNF domain in the C-terminal of the protein uses ATP to carry out ATRX's chromatin remodeling functions. ATRX also binds to Death Domain Associated Protein (DAXX) to deposit the histone variant H3.3 at repeat GC-rich sequences, including pericentromeric and telomeric regions and other transcriptionally silent genomic regions, known to be enriched with H3K9me3, H4K20me3, and DNA methylation (Qin et al. 2022; Nandakumar et al. 2017)

Recent research in human gliomas has revealed recurrent ATRX loss-of-function mutations. ATRX mutation has been reported in 31% of pediatric patients with high-grade tumor, almost always with concurrent TP53 mutation and frequently with mutations in the histone variant gene H3F3A (Miklja et al. 2019). ATRX is mutated in WHO grade 2/3 astrocytic glioma in adults, with 75% of astrocytic gliomas with TP53 and IDH1 mutations also carrying ATRX mutations (Śledzińska et al. 2021; Dharmaiah and Huse 2022). Alternative lengthening of telomeres, a telomerase-independent mechanism, is used by ATRX-deficient malignancies to preserve telomere length (ALT). Since the ATRX protein suppresses ALT, cancer cells with the ATRX mutation can continue to extend their telomeres by homologous recombination (HR) (Network 2015).

ATRX deficiency results in the formation of abnormal DNA secondary structures known as G-quadruplexes (G4s) at GC-rich sites of the genome (Fig. 7.4). For the maintaining of normal DNA conformation, ATRX binds at these GC-rich sites by DAXX-dependent H3.3 monomer incorporation. In tumors with loss of ATRX, probably accumulation of G4s causes increasing replication stress and DNA damage throughout the genome; consequently, genomic instability and abnormal transcriptional expression can occur (Dharmaiah and Huse 2022).

Due to the role of the ATRX protein in different cellular processes, the deficiency of this protein has many effects, especially genomic instability and telomere maintenance through the HR mechanism. Especially in diffuse astrocytomas, loss of P53 and ATRX genes with IDH mutation is observed in approximately 90% of tumors, so ATRX was included in the WHO 2021 classification as a biomarker (Gritsch et al. 2022).

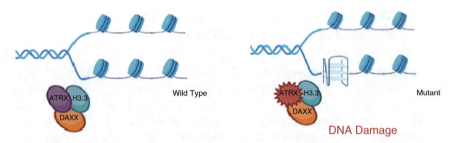

Fig. 7.4 The protein encoded by the ATRX gene is a multidomain binding protein, containing DAXX binding region to deposit the histone variant H3.3 at GC-rich repeat sequences, including telomeric, pericentromeric, and H3K9me3, H4K20me3, and DNA methylation-rich transcriptionally silent genomic regions. Loss-of-function mutations in ATRX results in disrupting the function of the ATRX/DAXX complex, including collection and deposition of histone variant H3.3 at sites of replication stress and DNA damage, hence ultimately leading to genomic instability in ATRX-deficient glioma. ATRX is frequently mutated in WHO grade 2/3 astrocytic glioma in adults with TP53 and IDH1 mutations and about 31% of pediatric patients with high-grade tumors, almost always with concurrent TP53 and H3F3A mutations

7.5 Histone Modifications

As discussed before, histones can undergo changes that either activate or suppress transcription. The DNA is packaged into nucleosome octamers, which each contains two copies, by the four core histone monomers H2A, H2B, H3, and H4, which are then further compressed into chromatin. These histones are essential for transcription and genomic stability, and a number of posttranslational changes are necessary for them to send regulatory signals to the transcriptional apparatus. The transcriptional features of a gene are defined by histone modifications, which are carried out by more than 100 enzymes that can interact with one another and contribute to the development of cancer and other disorders as well as treatment response.

7.5.1 Histone Acetylation

The relaxation of the connection between DNA and histones is made possible by the addition of acetyl groups to specific lysines in histones H3 and H4. This enables gene transcription. Acetyl groups that cause chromatin condensation and gene inactivation are eliminated by deacetylation. Histone acetyltransferases (HATs) and histone deacetylases mediate the dynamic processes of acetylation and deacetylation (HDAC). Gains in HDAC expression have been shown in a variety of malignancies, including GBM, and the therapeutic potential of HDAC inhibitors (HDACi) for GBM has been thoroughly studied. HDAC inhibitors are currently FDA-approved and have a wide range of anticancer efficacy (Romani et al. 2018).

7.5.2 Histone Methylation and Pediatric Gliomas

Histone methylation is a reversible process mediated by about 30 enzymes and is associated with various physiological and pathological conditions such as cancer, neurological disorders, and normal immune response. By the methylation of lysine and arginine amino acids of H3 and H4, gene expression can be activated or repressed.

Brain tumors are the most commonly seen solid tumors in children, and they are a major source of morbidity and mortality in this age range. A separate group of tumors known as pediatric gliomas differs significantly from their adult counterparts in terms of both clinical traits and therapeutic responses. Genomic research on juvenile tumors has only lately been conducted, thanks to the development of genome-wide assessments of changes in malignancies. Pediatric individuals are more susceptible to the aggressive malignancies glioblastoma and diffuse intrinsic pontine glioma (DIPG) (Broniscer and Gajjar 2004). Whole-exome sequencing research on juvenile glioblastoma provided the first look at the genomic makeup of these tumors by demonstrating for the first time that histone alterations can be linked to cancer.

Chromatin dysregulation in pediatric gliomas, as opposed to adult diffuse gliomas, results from histone H3 mutations. H3 subtypes include H3.1 and H3.2, which are controlled by the cell cycle and only deposited during S-phase and DNA repair, as well as H3.3, which is deposited at GC-rich heterochromatic areas of the genome by the ATRX/DAXX complex in a replication-independent manner. In pediatric diffuse gliomas, histone H3 mutations – highly conserved somatic alterations – drive carcinogenesis. The two most frequent of these mutations consist of guanine-to-adenine or adenine-to-thymine transversion resulting, respectively, in replacements for glycine at position 34 to arginine or valine (G34R/V) and that of lysine at position 27 to methionine (K27M) in the H3 variant genes H3F3A and HIST1H3B. Besides, ATRX and DAXX mutations have been reported in about 31% of all samples overall and tumors with G34R/G34V H3.3 mutations (Lowe et al. 2019).

In glioblastoma, histone methylation affects pediatric and adult patients differently. Lysine 27 (K27M) and glycine 34 (G34R/V) are two places where the histone variation H3.3 (H3F3A), which designates active chromatin domains, can be altered in pediatric cancers. Although aggressive K27M-mutant diffuse midline gliomas are most frequently found in the pons of the brainstem, they can also be seen from the base of the spinal cord to the thalamus and basal ganglia. Another thing to keep in mind is that K27M mutations do not just occur in diffuse midline gliomas; they can also occur in a variety of tumors, such as pilocytic astrocytomas, gangliogliomas, and posterior fossa ependymomas.

K27M causes transcriptional activation by reducing methylation at K27. The H3F3A K27M inhibits the enzymatic activity of histone methyltransferase PRC2-EZH2 and results in decreasing of normally repressive H3K27me3 chromatin mark deposition and the CpG hypomethylation phenotype (CHOP), which leads to abnormally activated gene expression. As detailed above, IDH mutations and elevated

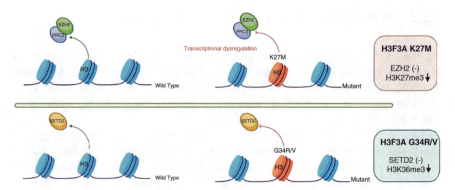

Fig. 7.5 H3.3, generally presented in transcription start sites and telomeric regions, is a ubiquitous, replication-independent histone related to open and active chromatin. The mutant form, H3F3A K27M, inhibits the enzymatic activity of histone methyltransferase PRC2-EZH2 and results in decreasing of typically repressive H3K27me3 chromatin mark deposition and the CpG hypomethylation phenotype (CHOP), which leads to abnormally activated gene expression. H3G34 mutations prevent the essential histone methylase SETD2 from catalyzing, which results in differential K36 binding and disruption of H3K36me3 deposition

D2-HG are associated with G-CIMP, which also interferes with the normal methylation patterns of H3K27 residues (Aldera and Govender 2022; La Madrid and Kieran 2018).

In contrast to K27M-mutant tumors, which are virtually exclusively located in the cerebral hemispheres of the brain, G34R/V mutation-carrying tumors have a somewhat longer overall survival (median 18.0 months). These mutations affect the adjacent K36 residue on the H3 tail and are frequently seen in conjunction with ATRX/DAXX and TP53 mutations. H3G34 mutations prevent the essential histone methylase SETD2 from catalyzing, which results in differential K36 binding and disruption of H3K36me3 deposition (Fig. 7.5). The expression of MYCN, an oncogenic driver of GBM89, as well as markers of stem-cell maintenance, neural differentiation, and cellular proliferation are among the epigenetic and transcriptional landscapes that are substantially altered by this process. It may be possible to treat G34R/V-mutant glioma with kinase inhibitors that stabilize MYCN (Lowe et al. 2019; Dharmaiah and Huse 2022; Duchatel et al. 2019).

Recently, Fang et al. have shown that G34R/V mutations impair SETD2's catalytic activity and cause mismatch repair deficit and a hypermutator phenotype by preventing the H3K36me3 mark from interacting with the mismatch repair protein MutS/MSH6 and K36-specific methyltransferases. Additionally, they have shown the relation between the G34R/V mutations and increasing mutational frequency and a decrease in binding of MSH6 to chromatin because MSH6 had a decreased affinity for binding the H3-mutant tail and G34R/V cells had less H3K36me3 (Fang et al. 2018).

7.6 Noncoding RNA and Glial Tumors

Of noncoding RNA molecules, microRNAs (miRNAs), which are around 22 nucleotides long, regulate gene expression by interacting with messenger RNAs (mRNAs). MiRNAs may impact up to 60% of protein-coding genes (Kamińska et al. 2019). Numerous recent researches have shown that different pathways regulate miRNA metabolism and function. Although miRNAs can be detected in tissues as well as body fluids like blood, CSF, or urine, the recent analysis suggested a suitable amount of genetic testing material for the evaluation of somatic changes such as point mutations or 1p/19q codeletions is necessary. Therefore, peripheral blood is a relatively noninvasive process, and peripheral blood miRNA expression could offer an inventive approach to determining diagnosis, prognosis, and predicting responses to therapy with so-called liquid biopsies.

MiRNAs are potential classifying markers. They can act as distinctive indicators for the minimally invasive diagnosis of glioblastoma because of their role in carcinogenesis and stability. Since miRNAs can be thought of as biomarkers and their identification in the blood explains the need for additional testing, Roth et al. studied whether a particular blood-derived miRNA fingerprint could be determined in glioblastoma patients and concluded it as potential biomarkers with 79% specificity and 83% sensitivity (Roth et al. 2011). According to the meta-analysis by Wang et al., cell-free miRNA-21 is the most promising diagnostic miRNA for glioma detection, followed by miRNA-125 and miRNA-222 (Wang et al. 2019). Additionally, miRNA-21 and miRNA-26 were both highly elevated in pre- and postoperative serum samples from glioblastoma patients, and therefore they suggested that these miRNAs have a potential to be used as serum-derived biomarkers. Akers et al. reported a sensitivity of 28% and a specificity of 95% for glioblastoma diagnosis in lumbar CSF compared to cisternal CSF's 80% sensitivity and 76% specificity (Akers et al. 2017).

In the last decade, attention has been focused on the functional importance of miRNAs as prognostic and predictive biomarkers and has been extensively analyzed. Several meta-analyses have been carried out to examine prognostic importance of miRNAs. The upregulation of the plasma miRNAs 155, 221, and 222, as well as miRNA-21, is associated with a worse prognosis. Additionally, a considerable unfavorable connection between high blood levels of miRNA-21 expression and OS and PFS has been reported. Patients with glioblastoma who showed high levels of the miRNA-10 family members in their tissue had also a significantly worse prognosis. MiRNAs may also have a potential significance as prognostic biomarkers because miRNA expression is frequently linked to therapeutic responses. When cells were treated with TMZ, it has been reported that highly expressed plasma levels of miRNA-223 and miRNA-125b-2 increased cell survival, as well (Siegal et al. 2016; Śledzińska et al. 2021).

Although many studies related to up- or downregulation of miRNA expressions in glioma patients have been published, definite miRNA signatures for glioblastoma classification or as prognostic or predictive biomarkers are still being investigated.

7.7 Conclusion

Despite all the advances in medicine, the median survival of patients with GBM has not increased much. This is probably because the tumor rapidly develops to be radio- and chemoresistance and infiltrates the surrounding brain tissue, making surgical removal of the tumor impossible. Numerous experimental solutions have been developed to break this stalemate, but none have had the desired effects. The fact that glioma classification and other brain tumor diagnoses are now based on epigenetic markers of dysfunction is just the beginning. Uncovering the underlying studies of glioma biology will undoubtedly be possible by adding proteomic and metabolomic methods to the important epigenomic and transcriptome research already completed. This will allow for the development of more potent drugs.

Compliance with Ethical Standards A.A. and S.A. wrote the chapter. All authors read and approved the final manuscript.

The authors declare that they have no competing interests. Although some figures of the chapter are similar to the figures in the literature in the schematic description of the molecular mechanisms, all the figures in this section were produced and drawn by the authors themselves.

References

Akers JC, Hua W, Li H, Ramakrishnan V, Yang Z, Quan K, Zhu W, Li J, Figueroa J, Hirshman BR, Miller B, Piccioni D, Ringel F, Komotar R, Messer K, Galasko DR, Hochberg F, Mao Y, Carter BS, Chen CC (2017 Jun 1) A cerebrospinal fluid microRNA signature as biomarker for glioblastoma. Oncotarget 8(40):68769–68779. https://doi.org/10.18632/oncotarget.18332. PMID: 28978155; PMCID: PMC5620295.

Aldera AP, Govender D (2022) Gene of the month: H3F3A and H3F3B. J Clin Pathol 75:1–4

An J, Rao A, Ko M (2017) TET family dioxygenases and DNA demethylation in stem cells and cancers. Exp Mol Med 49:e323–e323

Arslantas A, Artan S, Öner Ü, Müslümanoğlu H, Durmaz R, Cosan E, Atasoy MA, Başaran N, Tel E (2004) The importance of genomic copy number changes in the prognosis of glioblastoma multiforme. Neurosurg Rev 27:58–64

Arslantas A, Artan S, Öner Ü, Müslümanoğlu MH, Özdemir M, Durmaz R, Arslantas D, Vural M, Cosan E, Atasoy MA (2007) Genomic alterations in low-grade, anaplastic astrocytomas and glioblastomas. Pathol Oncol Res 13:39–46

Asmar F, Søgaard A, Grønbæk K (2015) DNA methylation and hydroxymethylation in cancer. Elsevier, Epigenetic Cancer Therapy

Black JC, Van Rechem C, Whetstine JR (2012) Histone lysine methylation dynamics: establishment, regulation, and biological impact. Mol Cell 48:491–507

Blakey CA, Litt MD (2015) Histone modifications—models and mechanisms. In: Epigenetic gene expression and regulation. Elsevier

Brennan CW, Verhaak RG, Mckenna A, Campos B, Noushmehr H, Salama SR, Zheng S, Chakravarty D, Sanborn JZ, BERMAN, S. H. (2013) The somatic genomic landscape of glioblastoma. Cell 155:462–477

Broniscer A, Gajjar A (2004) Supratentorial high-grade astrocytoma and diffuse brainstem glioma: two challenges for the pediatric oncologist. Oncologist 9(2):197–206. https://doi.org/10.1634/theoncologist.9-2-197

Carella A, Tejedor JR, García MG, Urdinguio RG, Bayón GF, Sierra M, López V, García-Toraño E, Santamarina-Ojeda P, Pérez RF (2020) Epigenetic downregulation of TET3 reduces genome-wide 5hmC levels and promotes glioblastoma tumorigenesis. Int J Cancer 146:373–387

Cavalli G, Heard E (2019) Advances in epigenetics link genetics to the environment and disease. Nature 571:489–499

Ceccarelli M, Barthel FP, Malta TM, Sabedot TS, Salama SR, Murray BA, Morozova O, Newton Y, Radenbaugh A, Pagnotta SM (2016) Molecular profiling reveals biologically discrete subsets and pathways of progression in diffuse glioma. Cell 164:550–563

Chen Z, Shi X, Guo L, Li Y, Luo M, He J (2017) Decreased 5-hydroxymethylcytosine levels correlate with cancer progression and poor survival: a systematic review and meta-analysis. Oncotarget 8:1944

Cheng Y, He C, Wang M, Ma X, Mo F, Yang S, Han J, Wei X (2019) Targeting epigenetic regulators for cancer therapy: mechanisms and advances in clinical trials. Signal Transduct Target Ther 4:1–39

Dabrowski J, Michal, Wojtas B (2019) Global DNA methylation patterns in human gliomas and their interplay with other epigenetic modifications. Int J Mol Sci 20:3478

Dang L, Yen K, Attar E (2016) IDH mutations in cancer and progress toward development of targeted therapeutics. Ann Oncol 27:599–608

Dawson MA (2017) The cancer epigenome: concepts, challenges, and therapeutic opportunities. Science 355:1147–1152

Dharmaiah S, Huse JT (2022) The epigenetic dysfunction underlying malignant glioma pathogenesis. Lab Investig 102:1–9

Duchatel RJ, Jackson ER, Alvaro F, Nixon B, Hondermarck H, Dun MD (2019) Signal transduction in diffuse intrinsic pontine glioma. Proteomics 19:1800479

Duke ES, Packer RJ (2020) Update on pediatric brain tumors: the molecular era and neuro-immunologic beginnings. Curr Neurol Neurosci Rep 20:1–8

Ehrlich M (2019) DNA hypermethylation in disease: mechanisms and clinical relevance. Epigenetics 14:1141–1163

Fan Y, Zhang X, Gao C, Jiang S, Wu H, Liu Z, Dou T (2022) Burden and trends of brain and central nervous system cancer from 1990 to 2019 at the global, regional, and country levels. Arch Publ Health 80:1–14

Fang J, Huang Y, Mao G, Yang S, Rennert G, Gu L, Li H, Li G-M (2018) Cancer-driving H3G34V/R/D mutations block H3K36 methylation and H3K36me3–MutSα interaction. Proc Natl Acad Sci 115:9598–9603

Feinberg AP, Vogelstein B (1983) Hypomethylation distinguishes genes of some human cancers from their normal counterparts. Nature 301:89–92

Ferrer AI, Trinidad JR, Sandiford O, Etchegaray J-P, Rameshwar P (2020) Epigenetic dynamics in cancer stem cell dormancy. Cancer Metastasis Rev 39:721–738

Flavahan WA, Drier Y, Liau BB, Gillespie SM, Venteicher AS, Stemmer-Rachamimov AO, Suvà ML, Bernstein BE (2016) Insulator dysfunction and oncogene activation in IDH mutant gliomas. Nature 529(7584):110–114. https://doi.org/10.1038/nature16490

Gibbons RJ, Picketts DJ, Villard L, Higgs DR (1995) Mutations in a putative global transcriptional regulator cause X-linked mental retardation with α-thalassemia (ATR-X syndrome). Cell 80:837–845

Globisch D, Münzel M, Müller M, Michalakis S, Wagner M, Koch S, Brückl T, Biel M, Carell T (2010) Tissue distribution of 5-hydroxymethylcytosine and search for active demethylation intermediates. PLoS One 5:e15367

Greenberg MV, Bourćhis D (2019) The diverse roles of DNA methylation in mammalian development and disease. Nat Rev Mol Cell Biol 20:590–607

Greuter L, Guzman R, Soleman J (2021) Typical pediatric brain tumors occurring in adults—differences in management and outcome. Biomedicine 9:356

Gritsch S, Batchelor TT, Gonzalez Castro LN (2022) Diagnostic, therapeutic, and prognostic implications of the 2021 World Health Organization classification of tumors of the central nervous system. Cancer 128:47–58

Guo M, Peng Y, Gao A, Du C, Herman JG (2019) Epigenetic heterogeneity in cancer. *Biomarker Res* 7:1–19

Gupta R, Flanagan S, Li CC, Lee M, Shivalingham B, Maleki S, Wheeler HR, Buckland ME (2013) Expanding the spectrum of IDH1 mutations in gliomas. Modern pathology: an official journal of the United States and Canadian Academy of Pathology. Inc 26(5):619–625. https://doi.org/10.1038/modpathol.2012.210

Gusyatiner O, Hegi ME (2018) Glioma epigenetics: from subclassification to novel treatment options. In: Seminars in cancer biology. Elsevier, pp 50–58

Hahn MA, Qiu R, Wu X, Li AX, Zhang H, Wang J, Jui J, Jin S-G, Jiang Y, Pfeifer GP (2013) Dynamics of 5-hydroxymethylcytosine and chromatin marks in mammalian neurogenesis. Cell Rep 3:291–300

Han S, Liu Y, Cai SJ, Qian M, Ding J, Larion M, Gilbert MR, Yang C (2020) IDH mutation in glioma: molecular mechanisms and potential therapeutic targets. Br J Cancer 122:1580–1589

Hegi ME, Genbrugge E, Gorlia T, Stupp R, Gilbert MR, Chinot OL, Nabors LB, Jones G, Van Criekinge W, Straub J (2019) MGMT promoter methylation cutoff with safety margin for selecting glioblastoma patients into trials omitting Temozolomide: A pooled analysis of four clinical TrialsMGMT safety margin for glioblastoma. Clin Cancer Res 25:1809–1816

Iurlaro M, Ficz G, Oxley D, Raiber E-A, Bachman M, Booth MJ, Andrews S, Balasubramanian S, Reik W (2013) A screen for hydroxymethylcytosine and formylcytosine binding proteins suggests functions in transcription and chromatin regulation. Genome Biol 14:1–11

Kalkan R, Atli Eİ, Özdemir M, Çiftçi E, Aydin HE, Artan S, Arslantaş A (2015) IDH1 mutations is prognostic marker for primary glioblastoma multiforme but MGMT hypermethylation is not prognostic for primary glioblastoma multiforme. Gene 554:81–86

Kamińska K, Nalejska E, Kubiak M, Wojtysiak J, Żołna Ł, Kowalewski J, Lewandowska MA (2019) Prognostic and predictive epigenetic biomarkers in oncology. Mol Diagn Ther 23: 83–95

Kheder ES, Hong DS (2018) Emerging targeted therapy for tumors with NTRK fusion ProteinsNovel targeted therapy for NTRK-rearranged tumors. Clin Cancer Res 24:5807–5814

Ko M, An J, Pastor WA, Koralov SB, Rajewsky K, Rao A (2015) TET proteins and 5-methylcytosine oxidation in hematological cancers. Immunol Rev 263:6–21

Koivunen P, Laukka T (2018) The TET enzymes. Cell Mol Life Sci 75:1339–1348

Kristensen B, Priesterbach-Ackley L, Petersen J, Wesseling P (2019) Molecular pathology of tumors of the central nervous system. Ann Oncol 30:1265–1278

Kukreja L, Li CJ, Ezhilan S, Iyer VR, Kuo JS (2021) Emerging epigenetic therapies for brain tumors. NeuroMolecular Med 24:1–9

La Madrid AM, Kieran MW (2018) Epigenetics in clinical management of children and adolescents with brain tumors. Curr Cancer Drug Targets 18:57–64

Le Rhun E, Weller M (2020) Sex-specific aspects of epidemiology, molecular genetics and outcome: primary brain tumours. ESMO Open 5:e001034

Leibetseder A, Preusser M, Berghoff AS (2022) New approaches with precision medicine in adult brain tumors. Cancers 14:712

Liu D, Yang T, Ma W, Wang Y (2022) Clinical strategies to manage adult glioblastoma patients without MGMT hypermethylation. J Cancer 13:354

Liu Y, Lang F, Chou F-J, Zaghloul KA, Yang C (2020) Isocitrate dehydrogenase mutations in glioma: genetics, biochemistry, and clinical indications. Biomedicine 8:294

Losman J-A, Kaelin WG (2013) What a difference a hydroxyl makes: mutant IDH,(R)-2-hydroxyglutarate, and cancer. Genes Dev 27:836–852

Louis DN, Perry A, Reifenberger G, Von Deimling A, Figarella-Branger D, Cavenee WK, Ohgaki H, Wiestler OD, Kleihues P, Ellison DW (2016) The 2016 World Health Organization classification of tumors of the central nervous system: a summary. Acta Neuropathol 131:803–820

Louis DN, Perry A, Wesseling P, Brat DJ, Cree IA, Figarella-Branger D, Hawkins C, Ng H, Pfister SM, Reifenberger G (2021) The 2021 WHO classification of tumors of the central nervous system: a summary. Neuro-Oncology 23:1231–1251

Lowe BR, Maxham LA, Hamey JJ, Wilkins MR, Partridge JF (2019) Histone H3 mutations: an updated view of their role in chromatin deregulation and cancer. Cancers 11:660

Lu Y, Chan Y-T, Tan H-Y, Li S, Wang N, Feng Y (2020) Epigenetic regulation in human cancer: the potential role of epi-drug in cancer therapy. Mol Cancer 19:1–16

Malta TM, De Souza CF, Sabedot TS, Silva TC, Mosella MS, Kalkanis SN, Snyder J, Castro AVB, Noushmehr H (2018) Glioma CpG Island methylator phenotype (G-CIMP): biological and clinical implications. Neuro-Oncology 20:608–620

Mancarella D, Plass C (2021) Epigenetic signatures in cancer: proper controls, current challenges and the potential for clinical translation. Genome Med 13:1–12

Mcnamara C, Mankad K, Thust S, Dixon L, Limback-Stanic C, D'arco F, Jacques TS, Löbel U (2022) 2021 WHO classification of tumours of the central nervous system: a review for the neuroradiologist. Neuroradiology 64:1–32

Miklja Z, Pasternak A, Stallard S, Nicolaides T, Kline-Nunnally C, Cole B, Beroukhim R, Bandopadhayay P, Chi S, Ramkissoon SH (2019) Molecular profiling and targeted therapy in pediatric gliomas: review and consensus recommendations. Neuro-Oncology 21:968–980

Miller KD, Ostrom QT, Kruchko C, Patil N, Tihan T, Cioffi G, Fuchs HE, Waite KA, Jemal A, Siegel RL (2021) Brain and other central nervous system tumor statistics, 2021. CA Cancer J Clin 71:381–406

Molinaro AM, Taylor JW, Wiencke JK, Wrensch MR (2019) Genetic and molecular epidemiology of adult diffuse glioma. Nat Rev Neurol 15:405–417

Münzel M, Globisch D, Brückl T, Wagner M, Welzmiller V, Michalakis S, Müller M, Biel M, Carell T (2010) Quantification of the sixth DNA base hydroxymethylcytosine in the brain. Angew Chem Int Ed 49:5375–5377

Mur P, Rodríguez de Lope Á, Díaz-Crespo FJ, Hernández-Iglesias T, Ribalta T, Fiaño C, García JF, Rey JA, Mollejo M, Meléndez B (2015) Impact on prognosis of the regional distribution of MGMT methylation with respect to the CpG Island methylator phenotype and age in glioma patients. J Neuro-Oncol 122:441–450

Nandakumar P, Mansouri A, DAS, S. (2017) The role of ATRX in glioma biology. Front Oncol 7:236

Network CGAR (2015) Comprehensive, integrative genomic analysis of diffuse lower-grade gliomas. N Engl J Med 372:2481–2498

Nishiyama A, Nakanishi M (2021) Navigating the DNA methylation landscape of cancer. Trends Genet 37:1012–1027

Noushmehr H, Weisenberger DJ, Diefes K, Phillips HS, Pujara K, Berman BP, Pan F, Pelloski CE, Sulman EP, Bhat KP (2010) Identification of a CpG Island methylator phenotype that defines a distinct subgroup of glioma. Cancer Cell 17:510–522

Orr BA, Haffner MC, Nelson WG, Yegnasubramanian S, Eberhart CG (2012) Decreased 5-hydroxymethylcytosine is associated with neural progenitor phenotype in normal brain and shorter survival in malignant glioma. PLoS One 7:e41036

Ostrom QT, Bauchet L, Davis FG, Deltour I, Fisher JL, Langer CE, Pekmezci M, Schwartzbaum JA, Turner MC, Walsh KM (2014) The epidemiology of glioma in adults: a "state of the science" review. Neuro-Oncology 16:896–913

Ostrom QT, Cioffi G, Gittleman H, Patil N, Waite K, Kruchko C, Barnholtz-Sloan JS (2019) CBTRUS statistical report: primary brain and other central nervous system tumors diagnosed in the United States in 2012–2016. Neuro-Oncology 21:v1–v100

Özdemir Sİ, Şimşek AY, Emel Ü (2022) NTRK somatic fusions and tumor agnostic treatment in pediatric cancers. J Contemp Med:12(6):1019-1024.

Patel AP, Fisher JL, Nichols E, Abd-Allah F, Abdela J, Abdelalim A, Abraha HN, Agius D, Alahdab F, Alam T (2019) Global, regional, and national burden of brain and other CNS cancer,

1990–2016: a systematic analysis for the global burden of disease study 2016. Lancet Neurol 18:376–393

Patil N, Abba ML, Zhou C, Chang S, Gaiser T, Leupold JH, Allgayer H (2021) Changes in methylation across structural and MicroRNA genes relevant for progression and metastasis in colorectal cancer. Cancers 13:5951

Perez E, Capper D (2020) Invited review: DNA methylation-based classification of paediatric brain tumours. Neuropathol Appl Neurobiol 46:28–47

Pfeifer GP, Xiong W, Hahn MA, Jin S-G (2014) The role of 5-hydroxymethylcytosine in human cancer. Cell Tissue Res 356:631–641

Puduvalli VK (2014) Epigenetic changes in gliomas. In: Glioma cell biology. Springer

Qin T, Mullan B, Ravindran R, Messinger D, Siada R, Cummings JR, Harris M, Muruganand A, Pyaram K, Miklja Z (2022) ATRX loss in glioma results in dysregulation of cell-cycle phase transition and ATM inhibitor radio-sensitization. Cell Rep 38:110216

Raviraj R, Nagaraja SS, Selvakumar I, Mohan S, Nagarajan D (2020) The epigenetics of brain tumors and its modulation during radiation: A review. Life Sci 256:117974

Romani M, Pistillo MP, Banelli B (2018) Epigenetic targeting of glioblastoma. Front Oncol 8:448

Ross JL, Velazquez Vega J, Plant A, Macdonald TJ, Becher OJ, Hambardzumyan D (2021) Tumour immune landscape of paediatric high-grade gliomas. Brain 144:2594–2609

Roth P, Wischhusen J, Happold C, Chandran PA, Hofer S, Eisele G, Weller M, Keller A (2011) A specific miRNA signature in the peripheral blood of glioblastoma patients. J Neurochem 118(3): 449–457. https://doi.org/10.1111/j.1471-4159.2011.07307.x

Scourzic L, Mouly E, Bernard OA (2015) TET proteins and the control of cytosine demethylation in cancer. Genome Med 7:1–16

Sharma S, Aazmi O (2019) Basics of epigenetics: it is more than simple changes in sequence that govern gene expression. Elsevier, Prognostic Epigenetics

Shoaib M, Walter D, Gillespie PJ, Izard F, Fahrenkrog B, Lleres D, Lerdrup M, Johansen JV, Hansen K, Julien E (2018) Histone H4K20 methylation mediated chromatin compaction threshold ensures genome integrity by limiting DNA replication licensing. Nat Commun 9:1–11

Siegal T, Charbit H, Paldor I, Zelikovitch B, Canello T, Benis A, Wong ML, Morokoff A, Kaye AH, Lavon I (2016) Dynamics of circulating hypoxia-mediated miRNAs and tumor response in patients with high-grade glioma treated with bevacizumab. J Neurosurg 125(4):1008–1015. https://doi.org/10.3171/2015.8.JNS15437

Śledzińska P, Bebyn MG, Furtak J, Kowalewski J, Lewandowska MA (2021) Prognostic and predictive biomarkers in gliomas. Int J Mol Sci 22:10373

Stanton BZ, Hodges C, Crabtree GR, Zhao K (2017) A general non-radioactive ATPase assay for chromatin remodeling complexes. Curr Protocols Chem Biol 9:1–10

Sun L, Zhang H, Gao P (2021) Metabolic reprogramming and epigenetic modifications on the path to cancer. Protein Cell:1–43

Takai H, Masuda K, Sato T, Sakaguchi Y, Suzuki T, Suzuki T, Koyama-Nasu R, Nasu-Nishimura-Y, Katou Y, Ogawa H (2014) 5-Hydroxymethylcytosine plays a critical role in glioblastomagenesis by recruiting the CHTOP-methylosome complex. Cell Rep 9:48–60

Tesileanu CMS, Dirven L, Wijnenga MM, Koekkoek JA, Vincent AJ, Dubbink HJ, Atmodimedjo PN, Kros JM, Van Duinen SG, Smits M (2020) Survival of diffuse astrocytic glioma, IDH1/2 wildtype, with molecular features of glioblastoma, WHO grade IV: a confirmation of the cIMPACT-NOW criteria. Neuro-Oncology 22:515–523

Thienpont B, Steinbacher J, Zhao H, D'anna F, Kuchnio A, Ploumakis A, Ghesquière B, Van Dyck L, Boeckx B, Schoonjans L (2016) Tumour hypoxia causes DNA hypermethylation by reducing TET activity. Nature 537:63–68

Tian S, Wang J, Zhang F, Wang D (2022) Comparative analysis of microRNA binding site distribution and microRNA-mediated gene expression repression of oncogenes and tumor suppressor genes. *Gene* 13:481

Toyota M, Ahuja N, Ohe-Toyota M, Herman JG, Baylin SB, Issa J-PJ (1999) CpG Island methylator phenotype in colorectal cancer. Proc Natl Acad Sci 96:8681–8686

Turcan S, Rohle D, Goenka A, Walsh LA, Fang F, Yilmaz E, Campos C, Fabius AW, Lu C, Ward PS (2012) IDH1 mutation is sufficient to establish the glioma hypermethylator phenotype. Nature 483:479–483

Uddin MS, Al Mamun A, Alghamdi BS, Tewari D, Jeandet P, Sarwar MS, Ashraf GM (2022) Epigenetics of glioblastoma multiforme: from molecular mechanisms to therapeutic approaches. In: Seminars in cancer biology. Elsevier

Valle-García D, Qadeer ZA, Mchugh DS, Ghiraldini FG, Chowdhury AH, Hasson D, Dyer MA, Recillas-Targa F, Bernstein E (2016) ATRX binds to atypical chromatin domains at the 3′ exons of zinc finger genes to preserve H3K9me3 enrichment. Epigenetics 11:398–414

Van Damme M, Crompot E, Meuleman N, Maerevoet M, Mineur P, Bron D, Lagneaux L, Stamatopoulos B (2016) Characterization of TET and IDH gene expression in chronic lymphocytic leukemia: comparison with normal B cells and prognostic significance. Clin Epigenetics 8:1–11

Verhaak RG, Hoadley KA, Purdom E, Wang V, Qi Y, Wilkerson MD, Miller CR, Ding L, Golub T, Mesirov JP (2010) Integrated genomic analysis identifies clinically relevant subtypes of glioblastoma characterized by abnormalities in PDGFRA, IDH1, EGFR, and NF1. Cancer Cell 17:98–110

Waitkus MS, Diplas BH, Yan H (2015) Isocitrate dehydrogenase mutations in gliomas. Neuro-Oncology 18:16–26

Wang J, Che F, Zhang J (2019) Cell-free microRNAs as non-invasive biomarkers in glioma: a diagnostic meta-analysis. Int J Biol Markers 34(3):232–242. https://doi.org/10.1177/1724600819840033

Wiles ET, Selker EU (2017) H3K27 methylation: a promiscuous repressive chromatin mark. Curr Opin Genet Dev 43:31–37

Xu W, Yang H, Liu Y, Yang Y, Wang P, Kim S-H, Ito S, Yang C, Wang P, Xiao M-T (2011) Oncometabolite 2-hydroxyglutarate is a competitive inhibitor of α-ketoglutarate-dependent dioxygenases. Cancer Cell 19:17–30

You JS, Jones PA (2012) Cancer genetics and epigenetics: two sides of the same coin? Cancer Cell 22:9–20

Zhao F, Zhang Z-W, Zhang J, Zhang S, Zhang H, Zhao C, Chen Y, Luo L, Tong W-M, Li C (2021) Loss of 5-hydroxymethylcytosine as an epigenetic signature that correlates with poor outcomes in patients with medulloblastoma. Front Oncol 11:603686

Chapter 8
Epigenetic Alterations in Pancreatic Cancer

Cincin Zeynep Bulbul, Bulbul Muhammed Volkan, and Sahin Soner

Abstract The occurrence of pancreatic cancer (PC) is presented to have risen in the past few years. Pancreatic cancer includes 5% of all cancer-related deaths and almost 2% of existing cancer types. Pancreatic tumors can be categorized as endocrine pancreatic tumors and non-endocrine pancreatic tumors. The significant symptoms will usually not be determined until the advanced metastasis stage. Research in understanding the mechanism of pancreatic cancer focuses on genetic and epigenetic changes using high-throughput genomic sequencing techniques. The epigenetic alterations and genetic abnormalities could lead to tumor progression through increasing oncogene expression, the proliferation of tumor cells, or suppressing tumor suppressor gene expressions with various adaptations. In pancreatic cancer, the progression of tumor metastasis could be related to epigenetic changes, including hypomethylation and hypermethylation of DNA and histone modifications. In today's world, which is also described as the post-genomic era, the epigenetic foundations of cancer development reveal revolutionary results in cancer genetics and provide the development of promising new methods in cancer treatment. This book chapter discusses epigenetic changes based on methylation, demethylation, and histone modification mechanisms in pancreatic cancer.

C. Z. Bulbul
Istanbul Nisantaşı University, Medical Faculty, Medical Biochemistry, Istanbul, Turkey

Istanbul Nisantaşı University, Health Ecosystem, Istanbul, Turkey
e-mail: zeynepbirsu.cincin@nisantasi.edu.tr

B. M. Volkan
Istanbul Nisantaşı University, Health Ecosystem, Istanbul, Turkey

Istanbul Nisantaşı University, Medical Faculty, Histology and Embryology, Istanbul, Turkey
e-mail: volkan.bulbul@nisantasi.edu.tr

S. Soner (✉)
Istanbul Nisantaşı University, Health Ecosystem, Istanbul, Turkey

Istanbul Nisantaşı University, Medical Faculty, Department of Neurosurgery, Istanbul, Turkey
e-mail: sahin.soner@nisantasi.edu.tr

© The Author(s), under exclusive license to Springer Nature Switzerland AG 2023
R. Kalkan (ed.), *Cancer Epigenetics*, Epigenetics and Human Health 11,
https://doi.org/10.1007/978-3-031-42365-9_8

Keywords Pancreatic cancer · Epigenetic · Metastasis · Methylation · Histone modification

Abbreviations

2-OG	2-Oxoglutarate
5-hmC	5-Hydroxymethylcytosine
5-mC	5-Methylcytosine
AKT	Protein kinase B
CDK	Cyclin-dependent kinase
CDKN2A	Cyclin-dependent tumor inhibitor 2A
c-Myc	Transcriptional regulator Myc-like
DCLK1	Doublecortin like kinase 1
DNMT	DNA methyltransferase
EGFR	Epidermal growth factor receptor
EMT	Epithelial-mesenchymal transition
EZH2	Enhancer of zeste homolog 2
FAD	Flavin adenine dinucleotide
FBW7	F-box and WD repeat domain containing 7
GC	Guanine-cytosine
GDP	Guanine diphosphate
GTP	Guanine triphosphate
H3	Histone 3
H4	Histone 4
JMJD	Jumonji domain containing protein
K	Lysine
KDM	Histone lysine demethylase
KLF	Kruppel-like factor 1
KMT	Lysine methyl transferase
KRAS	Kirsten rat sarcoma virus
LncRNA	Long-coding RNA
MBD	Methyl-CpG-binding domain
MPC1	Mitochondrial pyruvate carrier-I
MTAP	Methylthioadenosine phosphorylase
PanIN	Pancreatic intraepithelial neoplasms
PC	Pancreatic cancer
PDAC	Pancreatic ductal adenocarcinoma
PR C2	Polycomb repressive complex 2
PRMT	The protein arginine N-methyltransferase
R	Arginine
RFXAP	Regulatory X-related protein
SATB	Specific AT-rich sequence binding protein
SMAD4	SMAD family member 4

SMYD	Set and MYND domain protein
SWI/SNF	Switching defective/sucrosenon-fermenting complex
TDG	Thymine DNA glycosylase
TET	Ten eleven translocases
TF	Transcription factor
TGF-β	Transforming growth factor β
TP53	Tumor protein 53
WNK2	WNK lysine deficient protein kinase 2
ZEB1	Zinc Finger E-Box Binding Homeobox 1

8.1 Introduction

Pancreatic cancer ranked seventh in the list of cancer-related deaths in 2020 GLOBOCAN data, with 495,773 new patients and 466,003 deaths per year (Ferlay et al. 2021). Pancreatic cancer is among the deadliest call types, with a 5-year survival rate of 11% worldwide (Kung and Yu 2023). Ninety percent of these patients do not show clinical symptoms; therefore, there is no effective early diagnosis method for the disease (Chakma et al. 2020). Pancreatic cancer can be divided into various histopathological types. Among these types, the prognosis is the pancreatic ductal adenocarcinoma tissue type, with a rate of 90%. Despite the diagnostic methods, pre-and postsurgical treatments, and systematic approaches developed in recent years, the average survival time for the patients is approximately 1 year (Khan et al. 2021). Therefore, the rapid development of innovative and active biomarkers for the early treatment of pancreatic cancer is required (Rao et al. 2019).

High-throughput techniques in genomics revealed that genetic changes are directly related to the development of pancreatic tumors, especially in the early stage (Wang et al. 2021). The pancreatic ductal adenocarcinoma (PDAC) could be classified by using highly complex sequential techniques through the determination of genetic changes (Liu et al. 2022). Pancreatic intraepithelial neoplasms (PanINs) are precancerous lesions that can progress to PDAC (Li and Xie 2022). Four distinct genetic changes are observed in PanINs, mainly in the form of *KRAS* (Kirsten rat sarcoma virus) protooncogene activation and inactivation of Cyclin-dependent tumor inhibitor 2A (*CDKN2A*) (Kolbeinsson et al. 2023). The tumor protein 53 (*TP53*) and *SMAD4* tumor suppressor genes are responsible for the emergence and progression of PDAC (Botrus et al. 2021). PanINs are identified in nearly 90% of individuals with PDAC. However, as PDAC status progresses to advanced levels, additional gene variants also occur in these genes (Fig. 8.1) (Jiang et al. 2022).

The cumulation and aggregation of these genomic changes result in increased defects in tumor suppressor mechanisms and the emergence of dysregulated growth signals (Li et al. 2022a, b). Epigenetics determines the responses of the organism's genotype by the environmental conditions, independently from the mutation effects (Hanahan 2022). Through epigenetic mechanisms, chromatin structure changes

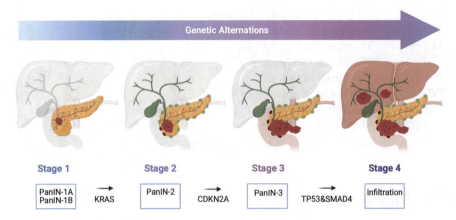

Fig. 8.1 The representative illustration of pancreatic cancer development from neoplasms to metastasis. Pancreatic intraepithelial neoplasms (PanINs) are precancerous lesions that can progress to PDAC. Four distinct genetic changes are directly related to the development of pancreatic tumors, especially in the early stage

without any change in DNA sequence, resulting in a different gene expression profile (Fitz-James and Cavalli 2022). The exact function of epigenetic mechanisms in a cell is the regulation of gene transcription and normal tissue development (Zhang et al. 2020). The epigenetic changes could evolve an undifferentiated cell into a differentiated state (Simpson et al. 2021). Furthermore, dysregulation in epigenetic mechanisms stimulates or suppresses gene transcription that can potentially result in cancer (Ilango et al. 2020). The point of view that epigenetic factors could play a critical role along with genetic factors in the development and progression of cancer has become the focal point nowadays (Sun et al. 2022). The epigenetic changes could induce the metastatic process, especially in pancreatic cancer (Montalvo-Javé et al. 2023).

The epigenetic foundations of cancer development are believed to reveal significant results in cancer genetics and improve promising new methods in cancer treatment (Miranda Furtado et al. 2019). This book chapter aims to explain broadly the mechanism of epigenetic changes based on DNA methylation, histone modification, and RNA in pancreatic cancer.

8.2 Genetic Alterations in Pancreatic Cancer

The most common genetic changes seen in PDACs are activating *KRAS* mutations in more than 90% of pancreatic cancer patients (Bannoura et al. 2021). The *KRAS* gene has a regulatory role in vital mechanisms as well as survival, growth, and distinction by catalyzing the GDP-GTP energy transfer and transforming the passive and active form of the GTPase enzyme (Gurreri et al. 2023). However, studies suggest that KRAS-activating mutations, which stimulate the onset of disease in the early stage of cancer, cannot show the same effect alone in the later stage (Timar and Kashofer 2020). The *CDKN2A* gene, located on the ninth chromosome, controls the cell cycle G1-S control mechanism by encoding the p16INK4A protein, which provides inhibition of the formation between CDK4/6 (cyclin-dependent kinase 4/6) and cyclin D proteins (Zhao et al. 2016). With the loss of function in the *CDKN2A* gene, KRAS-activating mutations in pancreatic cancer cells come into play, and cells escape senescence (Kimura et al. 2021). TP53, one of four crucial mutant leader genes, encodes the p53 transcription factor (Kastenhuber and Lowe 2017). This protein binds to DNA, so the transcription of genes that stimulate cell cycle arrest or death against DNA damage can proceed. TP53 mutations cause loss of function in the p53 protein (Smith et al. 2020). This condition generally occurs following the loss of CDKN2A in more than 50% of pancreatic cancer patients (Abe et al. 2021).

The SMAD4 gene is a significant mediator in the TGF-ß (transforming growth factor-ß) signal transduction system, which is thought to have different and essential roles in the pathogenesis of pancreatic cancer (Javle et al. 2014). In nonmalignant cells, TGF-β signaling directly affects the tumor suppression process by suppressing cyclin-dependent kinase (CDK) gene expression and increasing CDK inhibitor gene expression (Feng et al. 2023). However, TGF-ß signaling can create an immunosuppressive environment by showing tumor accelerating effect by stimulating the epithelial-mesenchymal transformation process in malignant cells (Ali et al. 2023). Loss of SMAD4 function suppresses the antiproliferative effect of TGF-ß signaling, causing the tumor to become more aggressive (Dardare et al. 2020).

Genetic mutations, seen less frequently (<10%) in hereditary pancreatic cancer cases, may cause the inactivation of proteins involved in the regulation mechanisms of critical cell signaling pathways such as cell cycle control systems, cell death, DNA damage control, and epithelial-mesenchymal transformation (Hayashi et al. 2021). Studies suggest that in these cases, proteins may be inactivated by the effect of germline shortening mutation in genes associated with DNA repair pathways (Crowley et al. 2021). It has been shown that in some patients carrying this mutation in epigenetic regulatory genes, abnormal changes in the epigenome cause a change in the transcriptional structure of the cells and predispose them to pancreatic cancer (Rah et al. 2021). In addition, all exon and genome sequencing studies and pancreatic cancer germline mutations reveal somatic mutations in pancreatic cancer patients' epigenetic regulators and chromatin remodeling structure (Bailey et al. 2016). Although there is an increase in the information on the genetic changes leading to the development of pancreatic cancer, there is limited literature information about the mechanisms of genetic change that are effective in the development of metastatic pancreatic cancer (Mishra and Guda 2017). This situation suggests epigenetic mechanisms are more effective than genetic processes in pancreatic cancer metastasis (Ganji and Farran 2022). Epigenetic programs, including transcription factor (TF)-driven histone modifications, chromatin remodeling mechanism, DNA methylation, and changes in transcriptional programs, are effective mechanisms in pancreatic cancer's progression and metastasis processes (Syren et al. 2017).

8.3 Epigenetic Events in Pancreatic Cancer

Epigenetic modifications aim to alter gene expression and function without changing the DNA sequence (Zhao et al. 2020). Histone modification is the most well-known type of epigenetic modification (Zhang et al. 2021). However, the effect of epigenetic reprogramming, particularly on pancreatic cancer development, remains unclear (Pandey et al. 2023). Understanding the changes in epigenetic modification mechanisms may provide biomarker data for the early detection and treatment of pancreatic cancer (Ciernikova et al. 2020).

Epigenetic events are significantly crucial in stimulating the processes of pancreatic carcinogenesis (Lomberk et al. 2019). The process of epigenetic regulation refers to controlling gene expression by changing chromatin structure and packaging without alteration of the DNA sequence (Peixoto et al. 2020). These changes may also occur as an adaptive mechanism response to evolving processes (Cheng et al. 2019). In addition, the epigenetic arrangement is essential for developing and maintaining different gene expression profiles (Dawson and Kouzarides 2012). Therefore, epigenetic changes are a critical feature that could stimulate the induction, initiation, and progression of all stages of carcinogenesis (Babar et al. 2022).

8.3.1 DNA Methylation

DNA methylation is the best-described mechanism of methylation in mammals (Mattei et al. 2022). In this mechanism, a 5-methylcytosine molecule is formed by attaching a methyl group to the fifth carbon location of a C nucleotide (cytosine) (Feng and Lou 2019). The DNA methyl transferase enzyme (DNMT) family is responsible for catalyzing methylation reactions (Lyko 2018). The DNMT1 enzyme ensures the methylation of ancestral genes and their transfer to offspring (Qureshi et al. 2022). De novo methylation mechanisms have extracted DNMT 3A/3B (Zhang et al. 2018). The increase in DNMT1 gene expression in 80% of pancreatic cancer cases is shown as the most crucial epigenetic change mechanism that causes hypermethylation (Fig. 8.2) (Wong 2020). The mechanism of DNA methylation in multicellular organisms has yet to be understood entirely (Law and Holland 2019). However, it has been suggested that transposons that can change their position, known as jumping genes or suppression of repetitive sequences, can maintain genomic stability.

The somatic mutations in chromatin remodeling complex regulators (SWF/SNF) and inactivation of histone modification enzymes occur together with oncogenic KRAS in identifying sleepy beauty transposon insertion mutagenesis (Takeda et al. 2016). It was detected that there is more than one mutation in the histone-modifying enzyme-encoding genes of all tumors (Wang and Tang 2023). Combined with the KRAS mutation, these mutations caused a change in the epigenome and caused the progression of pancreatic cancer (Li et al. 2021a, 2021b). The obtained data reveal

8 Epigenetic Alterations in Pancreatic Cancer

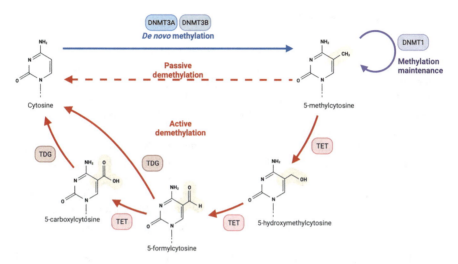

Fig. 8.2 The schematic illustration of genetic and epigenetic mechanisms commonly seen in PDACs. The DNA methyl transferase enzyme (DNMT) family catalyzes methylation reactions, while ten eleven translocases (TET) enzyme families work for demethylation. Thymine DNA glycosylase (TDG) excluded thymine groups from guanine-thymine mismatch points through hydrolysis of the C-N (carbon-nitrogen) bond between the DNA sugar-phosphate backbone and the mismatched thymine group

the significance of epigenetic modification in developing pancreatic cancer (Ushijima et al. 2021).

DNA demethylation processes accomplish the extraction or inversion of methylation signs (Wu and Zhang 2017). The DNA demethylation process can occur either actively or passively, depending on the action of the process (Wang et al. 2022). Passive demethylation begins when the methylation mechanism is not maintained to reduce the methylation content during DNA replication (Ross and Bogdanovic 2019). Active demethylation occurs in the presence of certain enzymes for the execution of processes. The most common and best-known demethylation process occurs with ten eleven translocases (TET) enzyme families (Hill et al. 2014). These enzymes first change 5-methylcytosine to 5-hydroxymethylcytosine, the second step to 5-formyl cytosine, the third step to 5-carboxy cytosine, and finally residue to unmodified cytosine nucleotide via MBD4 enzyme (Du et al. 2015a, 2015b).

DNA methylation rebalances tumor suppressor genes and protooncogenes during the beginning and improvement of pancreatic cancer (Xiao et al. 2022). Determining DNA methylation in peripheral blood samples can be used as an early diagnostic biomarker for the clinical laboratory of pancreatic cancer (Natale et al. 2019). Studies show that DNA methylation determined in the promoter regions of *ADAMTS1* and *BNC1* genes is directly related to pancreatic cancer (Ying et al. 2021). In recent years, the results of studies examining the relationship of DNA methylation with the improvement of pancreatic cancer have detected the presence of genes sensitive to the disruption of methylation processes (Liu and Pilarsky 2018).

DNA methylation usually occurs at C residues in areas rich in guanine-cytosine (GC) dinucleotides, known as CpG islands (Nishiyama and Nakanishi 2021). A significant portion of the promoter regions found in the human genome consists of unmethylated CpG islets (Edwards et al. 2017). These regions undergo methylation when gene transcription is active (Meng et al. 2015). The proportion of CpG dinucleotides found in the genome is less than that within the CpG islands because the 5-methylcytosine molecule is converted to thymine by deamination (Angeloni and Bogdanovic 2019). 5-Me(methyl) CpG translation induced by TpG mismatch is fixed by MBD4 (methyl-CpG-binding domain protein 4) protein, a thymine DNA glycosylase enzyme that replaces unmodified cytokine residues with modified nucleotides (Pidugu et al. 2021). However, MBD4 can bypass some parts of this mismatch and substitute guanine for adenine by identifying this nucleotide as usual instead of replacing the thymine nucleotide found in the TpG mismatch (Mangelinck and Mann 2021). This condition was first shown to cause the development of mutation-induced cancer (Galbraith and Snuderl 2022). Unrepaired deamination and conversion of thymine to cytosine cause somatic mutations in the p53 gene (Zemojtel et al. 2009). Studies indicate that cytosine residues in the methylated state may be a mutational attraction site and a predisposing factor in many types of cancer, especially pancreatic cancer (Lo et al. 2023).

The tumor suppressor *WNK2* gene shows a significant increase in tumor tissue when methylation levels of pancreatic cancer and inflamed tissues surrounding cancer are compared (Dutruel et al. 2014). *WNK2* is a serine-threonine kinase enzyme found in the cytoplasm, and its mRNA and protein expressions are decreased in pancreatic cancer (Li et al. 2022a, 2022b).

KLF10 has been shown to be regulated by epigenetic modulation in the pancreatic cancer process (Lin et al. 2021). The promoter region of the *KLF10* gene is methylated by DNMT1, which silencing itself triggers TGF-β1 signaling (Chang et al. 2012). The mentioned changes contribute to different stages in carcinogenesis, including improvement of proliferation, cellular migration, and invasion of cancer cells (Tsai et al. 2023).

SATB1 (specific AT-rich sequence binding protein 1) affects gene expression by causing epigenetic modifications that cause the tumor to show more aggressive and poor progression (Zelenka and Spilianakis 2020). In addition, regulating related genes also provides the classification of tumors in operations to prevent the risk of distant metastasis (Wei et al. 2018).

8.3.2 Histone Modification

Histone modification is defined as an essential part of the posttranslational epigenetic process that modulates genetic events like alterations of chromatin structure, remodeling, and enhancement of transcriptional component amounts (Zaib et al.

2022). It has been suggested that abnormal changes in histone modification types may cause pancreatic cancer progression (Biterge and Schneider 2014). The best-described modification types are methylation, demethylation, phosphorylation, acetylation, and ubiquitination (Menon and Howard 2022). Histone methylation has an essential role in the arranging of gene transcription (Yoshizawa-Sugata and Masai 2023).

During the posttranslational modification, the methyl groups are transferred to the lysine (K) and arginine (R) amino acids selectively to the cytoplasmic tails of histone 3 (H3) and histone 4 (H4) proteins (Black et al. 2012). The residues of these amino acids can be single, double, or triple at lysine residues only to activate or suppress gene expression due to the region and number at which methylation will occur (Michalak et al. 2019). Studies on the methylation of lysine residues of histone proteins have shown that the H3 protein is 4, 27 (Laugesen et al. 2019). It shows that gene expression is regulated in the direction of activation of gene expression at lysine residues at positions 36 and 36 of the H4 protein and at position 12 of the H4 protein but suppressed gene expression at positions 7, 9, and 56 of the H3 protein and at lysine residues at positions 5 and 20 of the H4 protein (Hyun et al. 2017). It has been determined that the eighth and 17th positions of the H3 protein and the arginine residues in the third position of the H4 protein are activated downstream of the signal transduction pathways (Collins et al. 2019).

Similarly, in DNA methylation, a methyl group is attached by methyltransferase enzymes, subtracted by demethylase enzymes, and read by methyl-binding motif proteins (Greenberg and Bourc'his 2019). Methyl-binding motifs help define histone methylation sites and assist in correctly inserting or removing methyl groups (Du et al. 2015a, 2015b). Site-specific enzymes work effectively in different histone methylation regions (Nishiyama et al. 2016). The stabilization between histone methylation and demethylation provided by specific enzymes has been shown to affect many physiological functions, especially embryonic development (Cavalli and Heard 2019). Anomalies or functional failures in the enzyme structure start the development of life-threatening diseases (Silverman and Shi 2016). Recent studies support that histone methylation is effective in the progression of pancreatic cancer; therefore, related methyl transferase and demethylase enzyme suppressors can be used to treat pancreatic cancer (Husmann and Gozani 2019). The methylation of histone proteins at lysine locations is catalyzed by the enzyme family (KMT), defined as lysine (K) methyltransferase (MT). The KMT3E (lysine methyltransferase 3E) enzyme, which is defined to be influential in the development of pancreatic cancer, functions through SYMD3, and the KMT6 (lysine methyltransferase 6) enzyme functions through the EZH2 protein complex (Roth et al. 2018).

8.3.2.1 Lysine Residue-Specific Methylation

The SMYD enzymes containing two important domains as SET and MYND domain structures are presented by five family members (SMYD 1–5) (Rueda-Robles et al. 2021). While the SET domain in this structure shows lysine residue-specific methyltransferase activity, the MYND domain involves a zinc-finger motif that can associate with proline-rich regions, enable protein-protein interactions, and bind proteins to DNA (Rubio-Tomás 2021). The SMYD enzyme family acts on histone and nonhistone targets for regulating various biological processes, especially cancer (Padilla et al. 2023).

It has been reported that the *SMYD 3* gene is highly expressed in Ras signal activation-stimulated cancer types, and the condition indicates a poor prognosis, especially for pancreatic cancer (Mazur et al. 2014). A significant increase in caspase-3 and matrix metalloproteinase-2 gene expressions was also detected in parallel with these cases (Zhu and Huang 2020). In cases where *SMYD3* gene expression decreased, there was a decrease in cancer cell proliferation, and regression in metastasis was detected (Wang et al. 2019a, 2019b). In this case, there is a significant decrease in the *MMP-2* gene expression and protein synthesis expression, but no significant change in the caspase-3 level has been detected (Huang and Xu 2017). Due to the stated results, it is suggested that SMYD dominant molecules can be used in antitumor therapy (Sun et al. 2021).

The SMYD3 unique small inhibitor molecule BCI-121, whose preclinical testing phases are ongoing, has been reported to reduce pancreatic cancer cell proliferation in cell lines (Peserico et al. 2015).

EZH2 (Enhancer of zeste homolog 2) is originally the 27th of the Histone 3 protein. It is known as the subunit of the PRC2 (polycomb repressive complex 2) structure with a catalytic feature, which mediates the suppression of target genes by adding three methyl groups to the lysine residue (H3K27me3, trimethylation) (Hanaki and Shimada 2021). An increase in EZH2 gene expression is associated with pancreatic cancer. FBW7, a ubiquitin ligase enzyme, suppresses EZH2 enzyme activity in pancreatic cancer cells through ubiquitination and degradation reactions (Zhang et al. 2022). In addition, CDK5 kinase activation carries out the phosphorylation reaction necessary for FBW7-mediated enzyme degradation. Therefore, the decrease in FBW7 expression has accelerated pancreatic cancer development stimulation by causing high levels of EZH2 accumulation (Jin et al. 2017). Noncoding RNA molecules use the EZH2 enzyme to modify the 27 trimethylations of histone H3 lysine (H3K27me3) of the target genes involved in the flow below in signal transmission. BLACAT1 accelerates the migration of pancreatic cancer cells by suppressing the long-coding RNA (lncRNA) molecule, EZH2-stimulated trimethylation CDKN1C expression (Zhou et al. 2020). A decrease in miR-139-5p, which has negatively correlated with high levels of EZH2 in pancreatic cancer tissues, has been stated to be related to poor prognosis (Ma et al. 2018).

In case EZH2 decreases and miR-139-5p increases, it is stated that epithelial-mesenchymal transformation is prevented in pancreatic cancer cells, and there is a decrease in metastasis (Ma et al. 2018). Because it suppresses the expression of miR-139-5p by increasing the trimethylation (H3K27me3) of EZH2, the DZNeP inhibitor molecule (3-deazaneplanocin A) is shown to reduce the expression EZH2 and trimethylation (Özel et al. 2021). With the combined use of a DZNeP inhibitor and gemcitabine, a significant increase in the apoptosis rates of pancreatic cancer cells has been detected, promising the development of new molecules for cancer treatment (Hung et al. 2013).

8.3.2.2 Lysine-Specific Demethylation

The reaction of removing methyl groups in lysine residues from recycled histone proteins is catalyzed by the histone lysine (K) demethylase (DM) enzyme family (KDM) (Jambhekar et al. 2017). Depending on the mechanism that KDM enzymes will show in the relevant field of action, it is divided into two different groups: FAD (flavin adenine dinucleotide)-dependent and Fe (II) and 2-oxoglutarate (2-OG) dependent (Zheng et al. 2022).

The enzyme KDM1 is the unique FAD-related enzyme associated with pancreatic cancer, which has two subtypes of the enzyme KDM1, KDM1A, and KDM1B (Hou et al. 2021). Both enzymes increase in pancreatic cancer samples, but the role of KDM1A in developing pancreatic cancer is still not clarified. However, due to the suppression of KDM1B, the homologs of this enzyme, cell proliferation in pancreatic cancer cell lines decreased, and a significant increase in the direction of cells' apoptosis was detected (Wang et al. 2019a, 2019b).

The JMJD Domain-Included Protein Family includes the area called Fe (II) and a-ketoglutarate-dependent dioxygenase Jumonji C (JMJD), the other class of the KDM family (Manni et al. 2022). The change in the activities of proteins in this class may be directly related to the development of the tumor. In a study with pancreatic cancer, amplification or overexpression of enzymes (KDM2B, KDM3A, or KDM4) that provide dimethyl groups in the 9th or 36th lysine of the Histone 3 protein had a positive effect on the development of pancreatic cancer. The enzyme H3K27, responsible for removing the methyl group in the 27th lysis of the Histone 3 protein, can also suppress tumors. The KDM2B enzyme, on the other hand, can direct the tumorigenic feature of its cells by ensuring the differentiation of pancreatic cancer cells through two different mechanisms. In the initial regions of transcription, the density rate of PcG proteins suppresses genes involved in the cell cycle and senescence events involving development (Tzatsos et al. 2013). KDM2B also activates the transcription of its genes that provide metabolic balance along with the MYC, a crucial oncogene, and the enzyme KDM5A, a histone demethylase. KDM 3A enables the expression of DCLK1, which correlates with the morphology of pancreatic cancer, to be increased by an epigenetic mechanism (Dandawate et al. 2019).

DCLK1 reveals the structural and functional differences of pancreatic cancer cells by being in stemlike but morphologically different cells in pancreatic cancer (Yang et al. 2022). The KDM4 family consists of four demethylase enzymes A, B, C, and D. All KDM4 subfamily members are associated with pancreatic cancer, except for subtype C (Lee et al. 2020). The regulatory factor X-related protein (RFXAP), a component for the transcription of MHC II molecules, stimulates transcription by binding to the promoter region of KDM4A, thus removing the 36th lysis-linked (H3K36) methyl region of histone three protein. The fisetin molecule causes DNA damage in pancreatic cancer cells through RFXAP/KDM4A-induced demethylation to suppress cellular proliferation (Ding et al. 2020).

KDM4B demethylates histone three-ninth protein lysis to activate ZEB1 transcription during the epithelial-mesenchymal transformation. ZEB1 is an E transcription factor, binding E-box, reported to reduce the expression of E-cadherin epigenetically (Zhang et al. 2015).

Elevated KDM4D expression in surgically removed pancreatic cancer samples is markedly related to bleak disease-independent survival, which could be a free indication of the relapsing risk in pancreatic cancer cases (Isohookana et al. 2018). However, its natural inclusion in pancreatic cancer is not elucidated. It is pointed out that KRAS mutations, which are oncogenic, can be screened in approximately whole pancreatic tumors. KDM6B, which plays a role in downstream KRAS signaling, is regulated down in pancreas cells with a reduced expression level on a poorly differentiated pancreas cancer (Yamamoto et al. 2014). KDM6B degradation can suppress the expression of the *CEBPA* gene to increase the tumor advancement of pancreatic cancer cells (Jiang et al. 2013). Further studies are needed for the work of the KDM5 family in pancreatic cancer. KDM5A is also known as a demethylase enzyme working for histonH3K4. KDM5A epigenetically inhibits the MPC-1 gene expression (mitochondrial pyruvate carrier-1) to develop cell proliferation via pyruvate mitochondria metabolism in pancreatic cancer (Li et al. 2021a, 2021b).

8.3.2.3 Arginine-Specific Methylation

PRMT family (the protein arginine N-methyltransferases), responsible for adding methyl groups to arginine residues, is categorized into three enzymes due to their catalytic activity. About 90% of the arginine methylation process is regulated by the PRMT1 enzyme catalyzing the reaction of adding methyl groups to the third arginine position in histone 4 protein to activate transcription (Dai et al. 2022). PRMT1 has been extremely expressed in different cancer cells, including pancreatic cancer, associated with poor prognosis in pancreatic cancer patients. Recent studies revealed that PRMT1 improves pancreatic cancer cell proliferation and stimulates upward regulation of b-catenin. Moreover, the wnt-b-catenin pathway, which is critically crucial for cellular biological functions, was included in developing pancreatic tumorigenesis (Thiebaut et al. 2021).

The PRMT5 enzyme includes the symmetric dimethylation reaction of the eighth arginine position in the histone H3 protein (H3R8) and the third arginine position in the histone H4 protein (H4R3). PRMT5 plays a crucial role in promoting cellular growth, migration, and epithelial-to-mesenchymal transformation while activating the signal of EGFR/AKT/b-catenin in pancreatic cells (Baldwin et al. 2014).

PRMT5 has been proven to inhibit the promoter activity of FBW7 (Qin et al., 2019). This E3 ubiquitin ligase checks c-Myc protein degradation by monitoring ubiquitous occurrence and degradation. Furthermore, PRMT5 modulates c-Myc by increasing its levels, advancing proliferation and aerobic glycolysis in the pancreatic cancer cell, and stability after translation (Kim and Ronai 2020) EZP015556, a

PRMT5 inhibitor, was influential in MTAP (a gene that usually disappears in PC) negative tumors in preclinical experiments, while several ongoing clinical studies are currently working on this inhibitor (NCT03573310, NCT02783300, and NCT03614728) (Stopa et al. 2015).

Histone demethylation regarding methylation can happen in arginine and lysine residues, but further studies need more research on arginine bundles. Until today, there is no explanatory report of specific arginine demethylases. However, it is known that well-balanced arginine methylation is usually essential for cellular growth and functional specialization. As a result, while certain enzymes, such as PRTMs, catalyze arginine methylation modifications, the others included in arginine methylation have not yet been elucidated.

8.4 Conclusion

Pancreatic cancer has been thought to be a type of cancer that has been thought to be caused by genetic mutations for many years and is rapidly progressing due to these mutations. However, research in recent years shows that epigenetic changes also have an essential effect on the progression process of pancreatic cancer. It has been determined that many epigenetic regulators have mutated in the process of pancreatic carcinogenesis, and different biomarkers are effective in terms of epigenetic changes between metastatic tissues and primary tumor tissues. Studies show that uncontrolled regulation of histone modification in pancreatic cancer epigenetic change processes affects the remodeling of chromatin. When looking holistically at the topic, even though epigenetic mechanism changes are essential in developing metastases by advancing pancreatic cancer, the mechanisms underlying these events have still not been clarified. It is thought that with further studies on these mechanisms, epigenetic regulators can be used as biomarkers in terms of diagnosis, and targeted therapies can be developed, especially for the treatment of advanced pancreatic cancer patients.

Compliance with Ethical Standards The authors have no conflicts of interest to declare. All coauthors have seen and agree with the contents of the manuscript, and there is no financial interest to report. We certify that the submission is original work and is not under review at any other publication.

References

Abe K, Kitago M, Kitagawa Y, Hirasawa A (2021) Hereditary pancreatic cancer. Int J Clin Oncol 26(10):1784–1792. https://doi.org/10.1007/s10147-021-02015-6

Ali S, Rehman MU, Yatoo AM, Arafah A, Khan A, Rashid S, Majid S, Ali A, Ali MN (2023) TGF-β signaling pathway: therapeutic targeting and potential for anti-cancer immunity. Eur J Pharmacol 947:175678. https://doi.org/10.1016/j.ejphar.2023.175678

Angeloni A, Bogdanovic O (2019) Enhancer DNA methylation: implications for gene regulation. Essays Biochem 63(6):707–715. https://doi.org/10.1042/EBC20190030

Babar Q, Saeed A, Tabish TA, Pricl S, Townley H, Thorat N (2022) Novel epigenetic therapeutic strategies and targets in cancer. Biochim et biophys Acta Molecul Basis Dis 1868(12):166552. https://doi.org/10.1016/j.bbadis.2022.166552

Bailey P, Chang DK, Nones K et al (2016) Genomic analyses identify molecular subtypes of pancreatic cancer. Nature 531(7592):47–52. https://doi.org/10.1038/nature16965

Baldwin RM, Morettin A, Côté J (2014) Role of PRMTs in cancer: could minor isoforms be leaving a mark? World J Biol Chem 5(2):115–129. https://doi.org/10.4331/wjbc.v5.i2.115

Bannoura SF, Uddin MH, Nagasaka M, Fazili F, Al-Hallak MN, Philip PA, El-Rayes B, Azmi AS (2021) Targeting KRAS in pancreatic cancer: new drugs on the horizon. Cancer Metastasis Rev 40(3):819–835. https://doi.org/10.1007/s10555-021-09990-2

Biterge B, Schneider R (2014) Histone variants: key players of chromatin. Cell Tissue Res 356(3): 457–466. https://doi.org/10.1007/s00441-014-1862-4

Black JC, Van Rechem C, Whetstine JR (2012) Histone lysine methylation dynamics: establishment, regulation, and biological impact. Mol Cell 48(4):491–507. https://doi.org/10.1016/j.molcel.2012.11.006

Botrus G, Kosirorek H, Sonbol MB, Kusne Y, Uson Junior PLS, Borad MJ, Ahn DH, Kasi PM, Drusbosky LM, Dada H, Surapaneni PK, Starr J, Ritter A, McMillan J, Wylie N, Mody K, Bekaii-Saab TS (2021) Circulating tumor DNA-based testing and actionable findings in patients with advanced and metastatic pancreatic adenocarcinoma. Oncologist 26(7):569–578. https://doi.org/10.1002/onco.13717

Cavalli G, Heard E (2019) Advances in epigenetics link genetics to the environment and disease. Nature 571(7766):489–499. https://doi.org/10.1038/s41586-019-1411-0

Chakma K, Gu Z, Abudurexiti Y, Hata T, Motoi F, Unno M, Horii A, Fukushige S (2020) Epigenetic inactivation of IRX4 is responsible for acceleration of cell growth in human pancreatic cancer. Cancer Sci 111(12):4594–4604. https://doi.org/10.1111/cas.14644

Chang VH, Chu PY, Peng SL, Mao TL, Shan YS, Hsu CF, Lin CY, Tsai KK, Yu WC, Ch'ang HJ (2012) Krüppel-like factor 10 expression as a prognostic indicator for pancreatic adenocarcinoma. Am J Pathol 181(2):423–430. https://doi.org/10.1016/j.ajpath.2012.04.025

Cheng Y, He C, Wang M, Ma X, Mo F, Yang S, Han J, Wei X (2019) Targeting epigenetic regulators for cancer therapy: mechanisms and advances in clinical trials. Signal Transduct Target Ther 4:62. https://doi.org/10.1038/s41392-019-0095-0

Ciernikova S, Earl J, García Bermejo ML, Stevurkova V, Carrato A, Smolkova B (2020) Epigenetic landscape in pancreatic ductal adenocarcinoma: on the way to overcoming drug resistance? Int J Mol Sci 21(11):4091. https://doi.org/10.3390/ijms21114091

Collins BE, Greer CB, Coleman BC, Sweatt JD (2019) Histone H3 lysine K4 methylation and its role in learning and memory. Epigenetics Chromatin 12(1):7. https://doi.org/10.1186/s13072-018-0251-8

Crowley F, Park W, O'Reilly EM (2021) Targeting DNA damage repair pathways in pancreas cancer. Cancer Metastasis Rev 40(3):891–908. https://doi.org/10.1007/s10555-021-09983-1

Dai W, Zhang J, Li S, He F, Liu Q, Gong J, Yang Z, Gong Y, Tang F, Wang Z, Xie C (2022) Protein arginine methylation: an emerging modification in cancer immunity and immunotherapy. Front Immunol 13:865964. https://doi.org/10.3389/fimmu.2022.865964

Dandawate P, Ghosh C, Palaniyandi K, Paul S, Rawal S, Pradhan R, Sayed AAA, Choudhury S, Standing D, Subramaniam D, Padhye SB, Gunewardena S, Thomas SM, Neil MO, Tawfik O, Welch DR, Jensen RA, Maliski S, Weir S, Iwakuma T et al (2019) The histone demethylase KDM3A, increased in human pancreatic tumors, regulates expression of DCLK1 and promotes tumorigenesis in mice. Gastroenterology 157(6):1646–1659.e11. https://doi.org/10.1053/j.gastro.2019.08.018

Dardare J, Witz A, Merlin JL, Gilson P, Harlé A (2020) SMAD4 and the TGFβ pathway in patients with pancreatic ductal adenocarcinoma. Int J Mol Sci 21(10):3534. https://doi.org/10.3390/ijms21103534

Dawson MA, Kouzarides T (2012) Cancer epigenetics: from mechanism to therapy. Cell 150(1): 12–27. https://doi.org/10.1016/j.cell.2012.06.013

Ding G, Xu X, Li D, Chen Y, Wang W, Ping D, Jia S, Cao L (2020) Fisetin inhibits proliferation of pancreatic adenocarcinoma by inducing DNA damage via RFXAP/KDM4A-dependent histone H3K36 demethylation. Cell Death Dis 11(10):893. https://doi.org/10.1038/s41419-020-03019-2

Du J, Johnson LM, Jacobsen SE, Patel DJ (2015a) DNA methylation pathways and their crosstalk with histone methylation. Nat Rev Mol Cell Biol 16(9):519–532. https://doi.org/10.1038/nrm4043

Du Q, Luu PL, Stirzaker C, Clark SJ (2015b) Methyl-CpG-binding domain proteins: readers of the epigenome. Epigenomics 7(6):1051–1073. https://doi.org/10.2217/epi.15.39

Dutruel C, Bergmann F, Rooman I et al (2014) Early epigenetic downregulation of WNK2 kinase during pancreatic ductal adenocarcinoma development. Oncogene 33(26):3401–3410. https://doi.org/10.1038/onc.2013.312

Edwards JR, Yarychkivska O, Boulard M, Bestor TH (2017) DNA methylation and DNA methyltransferases. Epigenetics Chromatin 10:23. https://doi.org/10.1186/s13072-017-0130-8

Feng F, Zhao Z, Cai X, Heng X, Ma X (2023) Cyclin-dependent kinase subunit2 (CKS2) promotes malignant phenotypes and epithelial-mesenchymal transition-like process in glioma by activating TGFβ/SMAD signaling. Cancer Med 12(5):5889–5907. https://doi.org/10.1002/cam4.5381

Feng L, Lou J (2019, 1894) DNA methylation analysis. Methods in Molecul Biol (Clifton, N. J.):181–227. https://doi.org/10.1007/978-1-4939-8916-4_12

Ferlay J, Colombet M, Soerjomataram I, Parkin DM, Pineros M, Znaor A, Bray F (2021) Cancer statistics for the year 2020: an overview. Int J Cancer 149:778. https://doi.org/10.1002/ijc.33588

Fitz-James MH, Cavalli G (2022) Molecular mechanisms of transgenerational epigenetic inheritance. Nat Rev Genet 23(6):325–341. https://doi.org/10.1038/s41576-021-00438-5

Galbraith K, Snuderl M (2022) DNA methylation as a diagnostic tool. Acta Neuropathol Commun 10(1):71. https://doi.org/10.1186/s40478-022-01371-2

Ganji C, Farran B (2022) Current clinical trials for epigenetic targets and therapeutic inhibitors for pancreatic cancer therapy. Drug Discov Today 27(5):1404–1410. https://doi.org/10.1016/j.drudis.2021.12.013

Greenberg MVC, Bourc'his D (2019) The diverse roles of DNA methylation in mammalian development and disease. Nat Rev Mol Cell Biol 20(10):590–607. https://doi.org/10.1038/s41580-019-0159-6

Gurreri E, Genovese G, Perelli L, Agostini A, Piro G, Carbone C, Tortora G (2023) KRAS-dependency in pancreatic ductal adenocarcinoma: mechanisms of escaping in resistance to KRAS inhibitors and perspectives of therapy. Int J Mol Sci 24(11):9313. https://doi.org/10.3390/ijms24119313

Hanahan D (2022) Hallmarks of cancer: new dimensions. Cancer Discov 12(1):31–46. https://doi.org/10.1158/2159-8290.CD-21-1059

Hanaki S, Shimada M (2021) Targeting EZH2 as cancer therapy. J Biochem 170(1):1–4. https://doi.org/10.1093/jb/mvab007

Hayashi A, Hong J, Iacobuzio-Donahue CA (2021) The pancreatic cancer genome revisited. Nat Rev Gastroenterol Hepatol 18(7):469–481. https://doi.org/10.1038/s41575-021-00463-z

Hill PW, Amouroux R, Hajkova P (2014) DNA demethylation, Tet proteins and 5-hydroxymethylcytosine in epigenetic reprogramming: an emerging complex story. Genomics 104(5):324–333. https://doi.org/10.1016/j.ygeno.2014.08.012

Hou X, Li Q, Yang L, Yang Z, He J, Li Q, Li D (2021) KDM1A and KDM3A promote tumor growth by upregulating cell cycle-associated genes in pancreatic cancer. Exp Biol Med (Maywood) 246(17):1869–1883. https://doi.org/10.1177/15353702211023473

Huang L, Xu AM (2017) SET and MYND domain containing protein 3 in cancer. Am J Transl Res 9(1):1–14

Hung SW, Mody H, Marrache S, Bhutia YD, Davis F, Cho JH, Zastre J, Dhar S, Chu CK, Govindarajan R (2013) Pharmacological reversal of histone methylation presensitizes pancreatic cancer cells to nucleoside drugs: in vitro optimization and novel nanoparticle delivery studies. PLoS One 8(8):e71196. https://doi.org/10.1371/journal.pone.0071196

Husmann D, Gozani O (2019) Histone lysine methyltransferases in biology and disease. Nat Struct Mol Biol 26(10):880–889. https://doi.org/10.1038/s41594-019-0298-7

Hyun K, Jeon J, Park K, Kim J (2017) Writing, erasing and reading histone lysine methylations. Exp Mol Med 49(4):e324. https://doi.org/10.1038/emm.2017.11

Ilango S, Paital B, Jayachandran P, Padma PR, Nirmaladevi R (2020) Epigenetic alterations in cancer. Front Biosci (Landmark edition) 25(6):1058–1109. https://doi.org/10.2741/4847

Isohookana J, Haapasaari KM, Soini Y, Karihtala P (2018) KDM4D predicts recurrence in exocrine pancreatic cells of resection margins from patients with pancreatic adenocarcinoma. Anticancer Res 38(4):2295–2302. https://doi.org/10.21873/anticanres.12474

Jambhekar A, Anastas JN, Shi Y (2017) Histone lysine demethylase inhibitors. Cold Spring Harb Perspect Med 7(1):a026484. https://doi.org/10.1101/cshperspect.a026484

Javle M, Li Y, Tan D, Dong X, Chang P, Kar S, Li D (2014) Biomarkers of TGF-β signaling pathway and prognosis of pancreatic cancer. PLoS One 9(1):e85942. https://doi.org/10.1371/journal.pone.0085942

Jiang T, Wei F, Xie K (2022) Clinical significance of pancreatic ductal metaplasia. J Pathol 257(2):125–139. https://doi.org/10.1002/path.5883

Jiang W, Wang J, Zhang Y (2013) Histone H3K27me3 demethylases KDM6A and KDM6B modulate definitive endoderm differentiation from human ESCs by regulating WNT signaling pathway. Cell Res 23(1):122–130. https://doi.org/10.1038/cr.2012.119

Jin X, Yang C, Fan P, Xiao J, Zhang W, Zhan S, Liu T, Wang D, Wu H (2017) CDK5/FBW7-dependent ubiquitination and degradation of EZH2 inhibits pancreatic cancer cell migration and invasion. J Biol Chem 292(15):6269–6280. https://doi.org/10.1074/jbc.M116.764407

Kastenhuber ER, Lowe SW (2017) Putting p53 in context. Cell 170(6):1062–1078. https://doi.org/10.1016/j.cell.2017.08.028

Khan AA, Liu X, Yan X, Tahir M, Ali S, Huang H (2021) An overview of genetic mutations and epigenetic signatures in the course of pancreatic cancer progression. Cancer Metastasis Rev 40(1):245–272. https://doi.org/10.1007/s10555-020-09952-0

Kim H, Ronai ZA (2020) PRMT5 function and targeting in cancer. Cell Stress 4(8):199–215. https://doi.org/10.15698/cst2020.08.228

Kimura H, Klein AP, Hruban RH, Roberts NJ (2021) The role of inherited pathogenic CDKN2A variants in susceptibility to pancreatic cancer. Pancreas 50(8):1123–1130. https://doi.org/10.1097/MPA.0000000000001888

Kolbeinsson HM, Chandana S, Wright GP, Chung M (2023) Pancreatic cancer: a review of current treatment and novel therapies. J Investig Surg 36(1):2129884. https://doi.org/10.1080/08941939.2022.2129884

Kung HC, Yu J (2023) Targeted therapy for pancreatic ductal adenocarcinoma: mechanisms and clinical study. MedComm 4(2):e216. https://doi.org/10.1002/mco2.216

Laugesen A, Højfeldt JW, Helin K (2019) Molecular mechanisms directing PRC2 recruitment and H3K27 methylation. Mol Cell 74(1):8–18. https://doi.org/10.1016/j.molcel.2019.03.011

Law PP, Holland ML (2019) DNA methylation at the crossroads of gene and environment interactions. Essays Biochem 63(6):717–726. https://doi.org/10.1042/EBC20190031

Lee DH, Kim GW, Jeon YH, Yoo J, Lee SW, Kwon SH (2020) Advances in histone demethylase KDM4 as cancer therapeutic targets. FASEB J 34(3):3461–3484. https://doi.org/10.1096/fj.201902584R

Li F, Liang Z, Jia Y, Zhang P, Ling K, Wang Y, Liang Z (2022b) microRNA-324-3p suppresses the aggressive ovarian cancer by targeting *WNK2*/RAS pathway. Bioengineered 13(5):12030–12044. https://doi.org/10.1080/21655979.2022.2056314

Li H, Peng C, Zhu C, Nie S, Qian X, Shi Z, Shi M, Liang Y, Ding X, Zhang S, Zhang B, Li X, Xu G, Lv Y, Wang L, Friess H, Kong B, Zou X, Shen S (2021b) Hypoxia promotes the metastasis of pancreatic cancer through regulating NOX4/KDM5A-mediated histone methylation modification changes in a HIF1A-independent manner. Clin Epigenetics 13(1):18. https://doi.org/10.1186/s13148-021-01016-6

Li S, Xie K (2022) Ductal metaplasia in pancreas. Biochim Biophys Acta Rev Cancer 1877(2): 188698. https://doi.org/10.1016/j.bbcan.2022.188698

Li X, He J, Xie K (2022a) Molecular signaling in pancreatic ductal metaplasia: emerging biomarkers for detection and intervention of early pancreatic cancer. Cell Oncol (Dordr) 45(2): 201–225. https://doi.org/10.1007/s13402-022-00664-x

Li Y, Chen X, Lu C (2021a) The interplay between DNA and histone methylation: molecular mechanisms and disease implications. EMBO Rep 22(5):e51803. 10.15252/embr.202051803

Lin J, Zhai S, Zou S, Xu Z, Zhang J, Jiang L, Deng X, Chen H, Peng C, Zhang J, Shen B (2021) Positive feedback between lncRNA FLVCR1-AS1 and KLF10 may inhibit pancreatic cancer progression via the PTEN/AKT pathway. J Exp Clin Cancer Res: CR 40(1):316. https://doi.org/10.1186/s13046-021-02097-0

Liu B, Pilarsky C (2018) Analysis of DNA Hypermethylation in pancreatic cancer using methylation-specific PCR and bisulfite sequencing. Methods Mol Biol (Clifton, N.J.) *1856*:269–282. https://doi.org/10.1007/978-1-4939-8751-1_16

Liu X, Wang W, Liu X, Zhang Z, Yu L, Li R, Guo D, Cai W, Quan X, Wu H, Dai M, Liang Z (2022) Multi-omics analysis of intra-tumoural and inter-tumoural heterogeneity in pancreatic ductal adenocarcinoma. Clin Transl Med 12(1):e670. https://doi.org/10.1002/ctm2.670

Lo EKW, Mears BM, Maurer HC, Idrizi A, Hansen KD, Thompson ED, Hruban RH, Olive KP, Feinberg AP (2023) Comprehensive DNA methylation analysis indicates that pancreatic intraepithelial neoplasia lesions are acinar-derived and epigenetically primed for carcinogenesis. Cancer Res 83(11):1905–1916. https://doi.org/10.1158/0008-5472.CAN-22-4052

Lomberk G, Dusetti N, Iovanna J, Urrutia R (2019) Emerging epigenomic landscapes of pancreatic cancer in the era of precision medicine. Nat Commun 10(1):3875. https://doi.org/10.1038/s41467-019-11812-7

Lyko F (2018) The DNA methyltransferase family: a versatile toolkit for epigenetic regulation. Nat Rev Genet 19(2):81–92. https://doi.org/10.1038/nrg.2017.80

Ma J, Zhang J, Weng YC, Wang JC (2018) EZH2-mediated microRNA-139-5p regulates epithelial-mesenchymal transition and lymph node metastasis of pancreatic cancer. Mol Cells 41(9): 868–880. https://doi.org/10.14348/molcells.2018.0109

Mangelinck A, Mann C (2021) DNA methylation and histone variants in aging and cancer. Int Rev Cell Mol Biol 364:1–110. https://doi.org/10.1016/bs.ircmb.2021.06.002

Manni W, Jianxin X, Weiqi H, Siyuan C, Huashan S (2022) JMJD family proteins in cancer and inflammation. Signal Transduct Target Ther 7(1):304. https://doi.org/10.1038/s41392-022-01145-1

Mattei AL, Bailly N, Meissner A (2022) DNA methylation: a historical perspective. Trends in Genet TIG 38(7):676–707. https://doi.org/10.1016/j.tig.2022.03.010

Mazur PK, Reynoird N, Khatri P, Jansen PW, Wilkinson AW, Liu S, Barbash O, Van Aller GS, Huddleston M, Dhanak D, Tummino PJ, Kruger RG, Garcia BA, Butte AJ, Vermeulen M, Sage J, Gozani O (2014) SMYD3 links lysine methylation of MAP3K2 to Ras-driven cancer. Nature 510(7504):283–287. https://doi.org/10.1038/nature13320

Meng H, Cao Y, Qin J, Song X, Zhang Q, Shi Y, Cao L (2015) DNA methylation, its mediators and genome integrity. Int J Biol Sci 11(5):604–617. https://doi.org/10.7150/ijbs.11218

Menon G, Howard M (2022) Investigating histone modification dynamics by mechanistic computational modeling. Methods Mol Biol (Clifton, N.J.) 2529:441–473. https://doi.org/10.1007/978-1-0716-2481-4_19

Michalak EM, Burr ML, Bannister AJ, Dawson MA (2019) The roles of DNA, RNA and histone methylation in ageing and cancer. Nat Rev Mol Cell Biol 20(10):573–589. https://doi.org/10.1038/s41580-019-0143-1

Miranda Furtado CL, Dos Santos Luciano MC, Silva Santos RD, Furtado GP, Moraes MO, Pessoa C (2019) Epidrugs: targeting epigenetic marks in cancer treatment. Epigenetics 14(12): 1164–1176. https://doi.org/10.1080/15592294.2019.1640546

Mishra NK, Guda C (2017) Genome-wide DNA methylation analysis reveals molecular subtypes of pancreatic cancer. Oncotarget 8(17):28990–29012. https://doi.org/10.18632/oncotarget.15993

Montalvo-Javé EE, Nuño-Lámbarri N, López-Sánchez GN, Ayala-Moreno EA, Gutierrez-Reyes G, Beane J, Pawlik TM (2023) Pancreatic cancer: genetic conditions and epigenetic alterations. J Gastrointestinal Surg 27(5):1001–1010. https://doi.org/10.1007/s11605-022-05553-0

Natale F, Vivo M, Falco G, Angrisano T (2019) Deciphering DNA methylation signatures of pancreatic cancer and pancreatitis. Clin Epigenetics 11(1):132. https://doi.org/10.1186/s13148-019-0728-8

Nishiyama A, Nakanishi M (2021) Navigating the DNA methylation landscape of cancer. Trends Genetics TIG 37(11):1012–1027. https://doi.org/10.1016/j.tig.2021.05.002

Nishiyama A, Yamaguchi L, Nakanishi M (2016) Regulation of maintenance DNA methylation via histone ubiquitylation. J Biochem 159(1):9–15. https://doi.org/10.1093/jb/mvv113

Özel M, Kilic E, Baskol M, Akalın H, Baskol G (2021) The effect of *EZH2*Inhibition through DZNep on epithelial-mesenchymal transition mechanism. Cell Reprogram 23(2):139–148. https://doi.org/10.1089/cell.2020.0073

Padilla A, Manganaro JF, Huesgen L, Roess DA, Brown MA, Crans DC (2023) Targeting epigenetic changes mediated by members of the SMYD family of lysine methyltransferases. Molecules (Basel, Switzerland) 28(4):2000. https://doi.org/10.3390/molecules28042000

Pandey S, Gupta VK, Lavania SP (2023) Role of epigenetics in pancreatic ductal adenocarcinoma. Epigenomics 15(2):89–110. https://doi.org/10.2217/epi-2022-0177

Peixoto P, Cartron PF, Serandour AA, Hervouet E (2020) From 1957 to nowadays: a brief history of epigenetics. Int J Mol Sci 21(20):7571. https://doi.org/10.3390/ijms21207571

Peserico A, Germani A, Sanese P, Barbosa AJ, Di Virgilio V, Fittipaldi R, Fabini E, Bertucci C, Varchi G, Moyer MP, Caretti G, Del Rio A, Simone C (2015) A SMYD3 small-molecule inhibitor impairing cancer cell growth. J Cell Physiol 230(10):2447–2460. https://doi.org/10.1002/jcp.24975

Pidugu LS, Bright H, Lin WJ, Majumdar C, Van Ostrand RP, David SS, Pozharski E, Drohat AC (2021) Structural insights into the mechanism of base excision by MBD4. J Mol Biol 433(15): 167097. https://doi.org/10.1016/j.jmb.2021.167097

Qureshi MZ, Sabitaliyevich UY, Rabandiyarov M, Arystanbekuly AT (2022) Role of DNA methyltransferases (DNMTs) in metastasis. Cell Molecul Biol (Noisy-le-grand, France) 68(1): 226–236. https://doi.org/10.14715/cmb/2022.68.1.27

Rah B, Banday MA, Bhat GR, Shah OJ, Jeelani H, Kawoosa F, Yousuf T, Afroze D (2021) Evaluation of biomarkers, genetic mutations, and epigenetic modifications in early diagnosis of pancreatic cancer. World J Gastroenterol 27(36):6093–6109. https://doi.org/10.3748/wjg.v27.i36.6093

Rao XF, Wan LH, Jie ZG, Zhu XL, Yin JX, Cao H (2019) Upregulated miR-27a-3p indicates a poor prognosis in pancreatic carcinoma patients and promotes the angiogenesis and migration by epigenetic silencing of GATA6 and activating VEGFA/VEGFR2 signaling pathway. Onco Targets Ther 12:11241–11254. https://doi.org/10.2147/Ott.S220621

Ross SE, Bogdanovic O (2019) TET enzymes, DNA demethylation and pluripotency. Biochem Soc Trans 47(3):875–885. https://doi.org/10.1042/BST20180606

Roth GS, Casanova AG, Lemonnier N, Reynoird N (2018) Lysine methylation signaling in pancreatic cancer. Curr Opin Oncol 30(1):30–37. https://doi.org/10.1097/CCO.0000000000000421

Rubio-Tomás T (2021) The SMYD family proteins in immunology: an update of their obvious and non-obvious relations with the immune system. Heliyon 7(6):e07387. https://doi.org/10.1016/j.heliyon.2021.e07387

Rueda-Robles A, Audano M, Álvarez-Mercado AI, Rubio-Tomás T (2021) Functions of SMYD proteins in biological processes: what do we know? An updated review. Arch Biochem Biophys 712:109040. https://doi.org/10.1016/j.abb.2021.109040

Silverman BR, Shi J (2016) Alterations of epigenetic regulators in pancreatic cancer and their clinical implications. Int J Mol Sci 17(12):2138. https://doi.org/10.3390/ijms17122138

Simpson DJ, Olova NN, Chandra T (2021) Cellular reprogramming and epigenetic rejuvenation. Clin Epigenetics 13(1):170. https://doi.org/10.1186/s13148-021-01158-7

Smith HL, Southgate H, Tweddle DA, Curtin NJ (2020) DNA damage checkpoint kinases in cancer. Expert Rev Mol Med 22:e2. https://doi.org/10.1017/erm.2020.3

Stopa N, Krebs JE, Shechter D (2015) The PRMT5 arginine methyltransferase: many roles in development, cancer and beyond. Cell Mol Life Sci CMLS 72(11):2041–2059. https://doi.org/10.1007/s00018-015-1847-9

Sun J, Li Z, Yang N (2021) Mechanism of the conformational change of the protein methyltransferase SMYD3: a molecular dynamics simulation study. Int J Mol Sci 22(13):7185. https://doi.org/10.3390/ijms22137185

Sun L, Zhang H, Gao P (2022) Metabolic reprogramming and epigenetic modifications on the path to cancer. Protein Cell 13(12):877–919. https://doi.org/10.1007/s13238-021-00846-7

Syren P, Andersson R, Bauden M, Ansari D (2017) Epigenetic alterations as biomarkers in pancreatic ductal adenocarcinoma. Scand J Gastroenterol 52(6–7):668–673. https://doi.org/10.1080/00365521.2017.1301989

Takeda H, Rust AG, Ward JM, Yew CC, Jenkins NA, Copeland NG (2016) Sleeping beauty transposon mutagenesis identifies genes that cooperate with mutant Smad4 in gastric cancer development. Proc Natl Acad Sci U S A 113(14):E2057–E2065. https://doi.org/10.1073/pnas.1603223113

Thiebaut C, Eve L, Poulard C, Le Romancer M (2021) Structure, activity, and function of PRMT1. Life (Basel, Switzerland) 11(11):1147. https://doi.org/10.3390/life11111147

Timar J, Kashofer K (2020) Molecular epidemiology and diagnostics of KRAS mutations in human cancer. Cancer Metastasis Rev 39(4):1029–1038. https://doi.org/10.1007/s10555-020-09915-5

Tsai YC, Cheng KH, Jiang SS, Hawse JR, Chuang SE, Chen SL, Huang TS, Ch'ang HJ (2023) Krüppel-like factor 10 modulates stem cell phenotypes of pancreatic adenocarcinoma by transcriptionally regulating notch receptors. J Biomed Sci 30(1):39. https://doi.org/10.1186/s12929-023-00937-z

Tzatsos A, Paskaleva P, Ferrari F, Deshpande V, Stoykova S, Contino G, Wong KK, Lan F, Trojer P, Park PJ, Bardeesy N (2013) KDM2B promotes pancreatic cancer via Polycomb-dependent and -independent transcriptional programs. J Clin Invest 123(2):727–739. https://doi.org/10.1172/JCI64535

Ushijima T, Clark SJ, Tan P (2021) Mapping genomic and epigenomic evolution in cancer ecosystems. Science 373(6562):1474–1479. https://doi.org/10.1126/science.abh1645

Wang D, Wu W, Callen E et al (2022) Active DNA demethylation promotes cell fate specification and the DNA damage response. *Science* 378(6623):983–989. https://doi.org/10.1126/science.add9838

Wang L, Tang J (2023) SWI/SNF complexes and cancers. Gene 870:147420. https://doi.org/10.1016/j.gene.2023.147420

Wang SS, Xu J, Ji KY, Hwang CI (2021) Epigenetic alterations in pancreatic cancer metastasis. Biomol Ther 11(8). https://doi.org/10.3390/biom11081082

Wang Y, Sun L, Luo Y, He S (2019a) Knockdown of KDM1B inhibits cell proliferation and induces apoptosis of pancreatic cancer cells. Pathol Res Pract 215(5):1054–1060. https://doi.org/10.1016/j.prp.2019.02.014

Wang Y, Xie BH, Lin WH, Huang YH, Ni JY, Hu J, Cui W, Zhou J, Shen L, Xu LF, Lian F, Li HP (2019b) Amplification of SMYD3 promotes tumorigenicity and intrahepatic metastasis of hepatocellular carcinoma via upregulation of CDK2 and MMP2. Oncogene 38(25):4948–4961. https://doi.org/10.1038/s41388-019-0766-x

Wei L, Ye H, Li G, Lu Y, Zhou Q, Zheng S, Lin Q, Liu Y, Li Z, Chen R (2018) Cancer-associated fibroblasts promote progression and gemcitabine resistance via the SDF-1/SATB-1 pathway in pancreatic cancer. Cell Death Dis 9(11):1065. https://doi.org/10.1038/s41419-018-1104-x

Wong KK (2020) DNMT1 as a therapeutic target in pancreatic cancer: mechanisms and clinical implications. Cell Oncol (Dordr) 43(5):779–792. https://doi.org/10.1007/s13402-020-00526-4

Wu X, Zhang Y (2017) TET-mediated active DNA demethylation: mechanism, function and beyond. Nat Rev Genet 18(9):517–534. https://doi.org/10.1038/nrg.2017.33

Xiao M, Liang X, Yan Z, Chen J, Zhu Y, Xie Y, Li Y, Li X, Gao Q, Feng F, Fu G, Gao Y (2022) A DNA-methylation-driven genes based prognostic signature reveals immune microenvironment in pancreatic cancer. Front Immunol 13:803962. https://doi.org/10.3389/fimmu.2022.803962

Yamamoto K, Tateishi K, Kudo Y, Sato T, Yamamoto S, Miyabayashi K, Matsusaka K, Asaoka Y, Ijichi H, Hirata Y, Otsuka M, Nakai Y, Isayama H, Ikenoue T, Kurokawa M, Fukayama M, Kokudo N, Omata M, Koike K (2014) Loss of histone demethylase KDM6B enhances aggressiveness of pancreatic cancer through downregulation of C/EBPα. Carcinogenesis 35(11): 2404–2414. https://doi.org/10.1093/carcin/bgu136

Yang L, Zhang Q, Yang Q (2022) KDM3A promotes oral squamous cell carcinoma cell proliferation and invasion via H3K9me2 demethylation-activated DCLK1. Genes Genomics 44(11): 1333–1342. https://doi.org/10.1007/s13258-022-01287-0

Ying L, Sharma A, Chhoda A et al (2021) Methylation-based cell-free DNA signature for early detection of pancreatic cancer. Pancreas 50(9):1267–1273. https://doi.org/10.1097/MPA.0000000000001919

Yoshizawa-Sugata N, Masai H (2023) Histone modification analysis of low-Mappability regions. Methods Mol Biol (Clifton, N.J.) 2519:163–185. https://doi.org/10.1007/978-1-0716-2433-3_18

Zhang L, Lu Q, Chang C (2020) Epigenetics in health and disease. Adv Exp Med Biol 1253:3–55. https://doi.org/10.1007/978-981-15-3449-2_1

Zhao R, Choi BY, Lee MH, Bode AM, Dong Z (2016) Implications of genetic and epigenetic alterations of CDKN2A (p16(INK4a)) in cancer. EBioMedicine 8:30–39. https://doi.org/10.1016/j.ebiom.2016.04.017

Zhao LY, Song J, Liu Y, Song CX, Yi C (2020) Mapping the epigenetic modifications of DNA and RNA. Protein Cell 11(11):792–808. https://doi.org/10.1007/s13238-020-00733-7

Zhang Y, Sun Z, Jia J, Du T, Zhang N, Tang Y, Fang Y, Fang D (2021) Overview of histone modification. Adv Exp Med Biol 1283:1–16. https://doi.org/10.1007/978-981-15-8104-5_1

Zhang ZM, Lu R, Wang P, Yu Y, Chen D, Gao L, Liu S, Ji D, Rothbart SB, Wang Y, Wang GG, Song J (2018) Structural basis for DNMT3A-mediated de novo DNA methylation. Nature 554(7692):387–391. https://doi.org/10.1038/nature25477

Zemojtel T, Kielbasa SM, Arndt PF, Chung HR, Vingron M (2009) Methylation and deamination of CpGs generate p53-binding sites on a genomic scale. Trends in genetics: TIG 25(2):63–66. https://doi.org/10.1016/j.tig.2008.11.005

Zelenka T, Spilianakis C (2020) SATB1-mediated chromatin landscape in T cells. Nucleus (Austin, Tex) 11(1):117–131. https://doi.org/10.1080/19491034.2020.1775037

Zaib S, Rana N, Khan I (2022) Histone modifications and their role in epigenetics of cancer. Curr Med Chem 29(14):2399–2411. https://doi.org/10.2174/0929867328666211108105214

Zhu CL, Huang Q (2020) Overexpression of the SMYD3 promotes proliferation, migration, and invasion of pancreatic cancer. Dig Dis Sci 65(2):489–499. https://doi.org/10.1007/s10620-019-05797-y

Zhang J, Kong DH, Huang X, Yu R, Yang Y (2022) Physiological functions of FBW7 in metabolism. Hormone and metabolic research = Hormon- und Stoffwechselforschung = Hormones et metabolisme 54(5):280–287. https://doi.org/10.1055/a-1816-8903

Zhou X, Gao W, Hua H, Ji Z (2020) LncRNA-BLACAT1 facilitates proliferation, migration and aerobic glycolysis of pancreatic cancer cells by repressing CDKN1C via EZH2-induced H3K27me3. Front Oncol 10:539805. https://doi.org/10.3389/fonc.2020.539805

Zheng YC, Liu YJ, Gao Y, Wang B, Liu HM (2022) An update of lysine specific demethylase 1 inhibitor: a patent review (2016-2020). Recent Pat Anticancer Drug Discov 17(1):9–25. https://doi.org/10.2174/1574892816666210728125224

Zhang P, Sun Y, Ma L (2015) ZEB1: at the crossroads of epithelial-mesenchymal transition, metastasis and therapy resistance. Cell Cycle (Georgetown, Tex) 14(4):481–487. https://doi.org/10.1080/15384101.2015.1006048

Chapter 9
Current Preclinical Applications of Pharmaco-Epigenetics in Cardiovascular Diseases

Chiara Papulino, Ugo Chianese, Lucia Scisciola, Ahmad Ali, Michelangela Barbieri, Giuseppe Paolisso, Lucia Altucci, and Rosaria Benedetti

Abstract Epigenetics is closely related to heart diseases. Genome-wide studies have highlighted the complexity of cardiovascular disease and involvement of epigenetic processes, including DNA methylation, histone modification, and non-coding RNA, on the onset and progression of cardiovascular diseases. In this chapter, we provide an overview of the key regulatory mechanisms in cardiovascular disease discussing on epigenetic machinery dysregulations in clinical conditions and the role of epigenetic modulators in cardiovascular activity. Furthermore, we point

---

Chiara Papulino and Ugo Chianese contributed equally with all other contributors.

C. Papulino · U. Chianese · A. Ali
Department of Precision Medicine, University of Campania 'Luigi Vanvitelli', Naples, Italy

L. Scisciola · M. Barbieri
Department of Advanced Medical and Surgical Sciences, University of Campania "Luigi Vanvitelli", Naples, Italy

G. Paolisso
Department of Advanced Medical and Surgical Sciences, University of Campania "Luigi Vanvitelli", Naples, Italy

Mediterranea Cardiocentro, Naples, Italy

UniCamillus, International Medical University, Rome, Italy

L. Altucci
Department of Precision Medicine, University of Campania 'Luigi Vanvitelli', Naples, Italy

Biogem Institute of Molecular and Genetic Biology, Ariano Irpino, Italy

Institute of Endocrinology and Oncology 'Gaetano Salvatore' (IEOS), Naples, Italy

R. Benedetti (✉)
Department of Precision Medicine, University of Campania 'Luigi Vanvitelli', Naples, Italy

Azienda Ospedaliera Universitaria "Luigi Vanvitelli, Medical Epigenetics Program, Naples, Italy
e-mail: rosaria.benedetti@unicampania.it

out on possible epigenetic biomarkers that could be useful in the new era of precision medicine to improve diagnosis and prognosis of cardiovascular disease patients.

Keywords Cardiovascular diseases · Epigenetics · Methylation · DNMT · HDACs · miRNAs · Acute myocardial infarction · Heart failure

Abbreviations

AMI	Acute myocardial infarction
ATP	Adenosine triphosphate
CREB	cAMP-response-element-binding protein
CVD	Cardiovascular diseases
EGR-1	Early growth response-1
HDACs	Histone deacetylases
HIF-1α	Hypoxia-inducible factor-1α
MI	Myocardial infarction
MPTP	Mitochondrial permeability transition pores
ROS	Reactive oxygen species
SIRT1	Sirtuin 1
TNF-alpha	Tumor necrosis factor-alpha
TSA	Trichostatin A

9.1 Cardiovascular Diseases: From Ischemic Injuries to Heart Repair

Acute myocardial infarction (AMI) is considered a dangerous heart condition compromising the human health worldwide (Mechanic et al. 2023). Restoring coronary blood flow as soon as possible by revascularization is the most effective therapy currently known. However, delayed revascularization may result in reperfusion damage, diminishing the therapeutic advantages of revascularization (Naito et al. 2020). There is currently no effective therapy for myocardial ischemia/reperfusion damage (Fernandez Rico et al. 2022).

Emerging data suggest that epigenetics is connected to the pathophysiology of cardiac ischemia/reperfusion injury, suggesting its potential role to treat or prevent ischemia/reperfusion injury (Wang et al. 2021a). Starting 2 h after the initial ischemic insult, AMI promotes cardiomyocyte loss due to ischemic necrosis, which peaks after 24 h and decreases 72 h later (Pulido et al. 2023). During this time, gap junctions release substances causing necrosis. Ischemic cells shift from an aerobic to anaerobic metabolism to survive in such hypoxic conditions (Kaya et al. 2020). In cardiomyocytes, anaerobic glycolysis determines H^+ accumulation that is removed by sarcolemma ion pumps in exchange for Na++, leading to an intracellular Na++ overload.

Cytoplasmic Ca^{2+} levels increase when the Na^{2+}/Ca^{2+} exchanger is operating in reverse mode. Uniporter transports Ca^{2+} into mitochondria, promoting Ca^{2+}-dependent dehydrogenase activity; this reduce NADH and ATP levels altering the electron transport chain leading to increase reactive oxygen species (ROS) (Pittas et al. 2018). The rise in Ca^{2+} in the mitochondrial matrix finally attains a plateau under hypoxia. Reperfusion rescues oxygen and ATP generation as well as mitochondrial membrane potential, which regenerates Ca^{2+} ion gradient into mitochondria (Crola Da Silva et al. 2023). Together, these effects cause the enlargement of mitochondria, thus leading to the opening of the mitochondrial permeability transition pores (MPTP), which eventually cause cellular necrosis (Pulido et al. 2023). Interestingly, MPTP opening links cardiomyocyte necrosis and I/R injury (Crola Da Silva et al. 2023). As a matter of fact, blocking late Na + channels can prevent necrosis. Additionally, cellular components released from mitochondria promote cell necrosis and apoptosis. During ischemia/reperfusion, cell apoptosis and necrosis are reduced as MPTP is inhibited (Crola Da Silva et al. 2023). Inducing autophagy at the cardiac cell level (Matsui et al. 2007; Yan et al. 2005; French et al. 2010) as a pro-survival mechanism, it eliminates unwanted or damaged cellular components while preserving metabolic homeostasis in cells. During this process, misfolded proteins and some damaged organelles are delivered into lysosomes, and then these components are broken down into nutrients like amino acids that can be recycled and reused by the cells. BNIP38 and AMPK play major roles in ischemia- or hypoxia-induced autophagy (Russell 3rd et al. 2004).

An extended period of ischemia leads to a dysfunctional autophagic response. Despite restoring oxygen and nutrients, autophagy is further upregulated in reperfusion (Matsui et al. 2007; Russell 3rd et al. 2004). In contrast to ischemia, sustained autophagy activation occurs during reperfusion through different mechanisms. The maintenance of autophagy during reperfusion may be significantly influenced by oxidative stress, Ca^{2+} overload, ER stress, and mitochondrial damage/BNIP3 (Popov et al. 2023; Liu et al. 2023). During I/R, the autophagy increase can have a dual role, beneficial or detrimental, while in mild to moderate ischemia, it has a protective role (Matsui et al. 2007). In the peri-infarct zone, where cardiomyocytes suffer a sublethal injury autophagy could have a significant impact. Autophagy, however, may help with postinfarction remodeling. As a matter of fact, some studies assess that substances regulating autophagy play a protective role (Buss et al. 2009; McCormick et al. 2012). In myocardial infarction, apoptosis displays a relevant role in the death of cardiomyocytes in the peri-infarct region (Chen et al. 2004; Daugas et al. 2000) given that is may determine infarct size, the extent of cardiac remodeling, and the development of heart failure (HF) after AMI (Parra et al. 2011, 2008). Extrinsic and the intrinsic apoptosis pathways converge on the mitochondria. The interaction between the Fas ligand and tumor necrosis factor-alpha (TNF-alpha) activates the extrinsic pathway. TNF levels and its receptors have the ability to predict the size of infarcts, left ventricular dysfunction, and prognosis (Ponpuak et al. 2015), (Zhang et al. 2018). Fas (apoptosis-stimulating fragment) is the primary factor stimulating the extrinsic apoptotic pathway during MI (Tsuchiya et al. 2018). The intrinsic way has been reported to have a significant role in regulating the death of cardiomyocytes during ischemia (Tsuchiya et al. 2018). The apoptotic cascade is amplified and causes intracellular caspase activation by proapoptotic proteins BAX and BH3, thus speeding the death of cardiomyocytes (Pott et al. 2018).

The ischemic-/hypoxia-driven events activate pro-inflammatory signaling in cardiomyocytes during AMI via hypoxia-inducible factor-1α (HIF-1α) (Zhao et al. 2023; Delgobo et al. 2023; Zhuang et al. 2022). Additionally, early growth response-1 (EGR-1) is induced by hypoxic activation of the PKC and AGEs/RAGE/PKCII/c-Jun pathways and is involved in controlling the TNF-α gene in endothelial cells. TNF-α signaling can trigger cell death and apoptosis in addition to controlling pro-inflammatory stimuli. Nitric oxide (NO) levels in cardiomyocytes are also elevated by TNF due to the activation of iNOS (inducible nitric oxide synthase). TNF-α, IL-6, and IL-10 via nuclear translocation of NF-kB are increased as a result of accumulation of NO and ROS by the activation of hypoxia-induced PKC-dependent signaling (Dong et al. 2022).

Cell injury is also caused by interleukins, cytokines, and ROS. Interleukin and TNF-α are primarily accumulated in the peri-infarct zone. Mast cell activation and TNF, histamine, tryptase, and chymase degranulation may exacerbate cardiomyocyte damage. Cardiomyocyte oxidative damage and infarct size can be reduced by preventing mast cell degranulation. There have been a number of experimental efforts to develop a cell transplantation-based strategy for repairing damaged hearts, including in vitro generation of new cardiomyocytes using different approaches such as peptides, recombinant proteins, plasmids, viruses, miRNA, and RNAi (Magadum 2022; Yap et al. 2023). Particularly, three families of extracellular signaling molecules Wnt, FGF, and TGF-α are the main regulators of cardiac progenitor cells. The long noncoding RNA Braveheart (Bvht), via Wnt signaling, mediates the cardiac lineage specification epigenetic changes (He et al. 2022). Bvht induces transcription factor Mesp1 expression (He et al. 2022) leading in turn to upregulating key cardiac-specific transcription factors, such as Gata4, Isl1, Mef2c, and Nkx2.5, factors that play crucial role in cardiogenesis. Finally, it should also be pointed out that chromatin remodeling also occurs along with the initiation of the cardiomyocyte differentiation.

9.2 The Impact of Epigenetics on Cardiovascular Activity

As a constant functioning organ, the heart requires continuous energy supply to work properly (Trifunovic-Zamaklar et al. 2022). To maintain the heartbeat and the contractile function, it consumes larger amounts of energy in comparison with other organs. The heart can use different substrates such as fatty acids (FA), glucose, lactic acid, ketones, and amino acids to produce enough adenosine triphosphate (ATP) (Lopaschuk et al. 2021) to produce a sufficient level of adenosine triphosphate (ATP), indicating that the heart is able to convert energy efficiently. Additionally, these metabolites affect cardiac structure and function altering physiological signaling (Gibb and Hill 2018). Heart for contractility and blood pumping across the organism need the right energy balance, as it uses FA that are substrates with high energy content (Volpe et al. 2023). Before the birth, in fetus, the heart produces energy via glycolysis, while after birth, its metabolism changed from the nonoxidative to the oxidative stage (Piquereau and Ventura-Clapier 2018). The heart can use various energetic substrates in response to external cues. For example, because lipid metabolism requires more oxygen than glycolysis does for ATP production, the heart will switch to glycolysis (Heather et al. 2013; Cole et al. 2016; Ng et al. 2023). This pathway, called "glucose-fatty acid cycle," allows the heart to switch its energy sources (Malandraki-Miller et al. 2018). Although several pathways have been identified, the mechanism is not completely understood. Upon low oxygen availability, the heart acquires fuel via acetyl CoA pathway in the liver that produces water soluble ketones (Puchalska and Crawford 2017). Several transcriptional factors control the transition from glycolysis to oxidative phosphorylation during the development of the heart (Piquereau and Ventura-Clapier 2018; Kreipke and Birren 2015). During cardiomyocyte development due to an increase in the number of mitochondria, there is a greater demand for energy (Persad and Lopaschuk 2022). Histone deacetylases (HDACs) are epigenetic enzymes able to remove acetyl groups from lysine residues of histone and nonhistone proteins (Milazzo et al. 2020). In addition, thanks to their activity, these enzymes influence posttranslational modifications, such as ubiquitination, methylation, and gene transcription (Seto and Yoshida 2014; Caron et al. 2005; Fischle et al. 2003). Therefore, HDACs being involved in several signaling pathways play an important role in cardiovascular activity (Bagchi and Weeks 2019; Li et al. 2020). Gene silencing studies have shown that HDAC1 and HDAC2 are critical for the heart's regular morphogenesis and growth. In fact, during perinatal period, mice not expressing HDAC2 (HDAC2-knocked out) can survive even though reporting several cardiac defects, such as hyperplasia and bradycardia (Montgomery et al. 2007). Simultaneous depletion of both HDAC1 and HDAC2 leads to neonatal death with arrhythmia, dilated cardiomyopathy, and other cardiac defects, due to upregulation of genes encoding for contractile proteins and Ca^{2+} channels (Montgomery et al. 2007). As a matter of fact, the lack of HDAC1 and HDAC2 induces an upregulation of Ca^{2+} channels and increase the expression of skeletal muscle-specific isoforms of TnI (Parmacek and Solaro 2004). Cardiovascular defects are also present in knockout HDAC7 in mice (Wei et al. 2018). The normal function of the heart is sustained by a correct cardiomyocyte gene expression programming; therefore, its alterations induce the loss of cardiac homeostasis and heart dysfunction. Deregulation in histone methylation induces developmental defects and diseases (Papait et al. 2020). Heart homeostasis and hypertrophy are regulated by this epigenetic mechanism. Indeed, the transcriptional program in healthy cardiomyocytes depends on PAX-interacting protein 1, a cofactor for histone methylation (Stein et al. 2011). Moreover, a genome-wide study demonstrated that methylation at different lysine residues can modulate gene expression reprogramming in pressure-overload hypertrophy (Papait et al. 2013). Furthermore, JMJD2A, which catalyzes the demethylation of H3K9me3, is crucial for cardiac hypertrophy (Zhang et al. 2011). This enzyme works together to SRF and myocardin to regulate FHL1, involved in mediating the hypertrophic response (Zhang et al. 2011; Antignano et al. 2014). A muscle-specific histone methyltransferase, Smyd1, has a role in skeletal and cardiac muscle development (Liang et al. 2020). Additionally, mitochondrial oxidative phosphorylation proteins induced by SMYD1 that promotes the expression of peroxisome proliferator-activated receptor-gamma coactivator (PGC-1α) is via H3K4me3. SMYD1 knockdown decreases the expression of cardiac energetics master regulators, such as PGC-1α, PPARα, and RXRα, thus leading to ATP production impairment due to a metabolic shift from oxidative phosphorylation to anaerobic glycolysis (Warren et al. 2018). DNA methylation relevance in cardiac gene expression is also widely recognized (Pepin et al. 2019a, 2019b; Nothjunge et al. 2017), acting a key role in the regulation of genes involved in tricarboxylic acid cycle, oxidative phosphorylation, and fatty acid oxidation (Nothjunge et al. 2017). Furthermore, among DNMTs, DNMT3A is responsible for DNA methylation levels in ACSL1, ACSL2, HADHA, and NDUFA5 genes, demonstrating that DNMT3A downregulates oxidative metabolic gene expression in cardiomyocytes (Chen et al. 2005).

9.3 Epigenetic Machinery in Cardiovascular Diseases

Epigenetic mechanisms act in the regulation of several cell processes: differentiation, development, homeostasis, aging, and disease. Unlike the "static" genome, epigenetic landscape is dynamic throughout replicative and chronological aging. The epigenetic marks are defined during differentiation and development to determine cell fate and to define differentiated cells. Importantly, the epigenome is sensitive to cellular cues such as redox and metabolic or neurohumoral signaling. Genome remodeling by DNA methylation, histone modification, noncoding RNA, and RNA modification profiles are involved in the cardiovascular disease (CVD) (Berulava et al. 2020; Hermans-Beijnsberger et al. 2018). Epigenetics has mostly been studied in cancer, diabetes, neurological and imprinting disorders, immunological illnesses, and aging and has been considered indispensable for the scientific research and advancement (Schiano et al. 2015). Epigenetics and CVD are linked as epigenetic related enzymes are involved in cardiovascular system processes (Shi et al. 2022; Schiano et al. 2015).

9.3.1 DNA Methylation Role in CVD

A pivotal epigenetic mechanism, playing a crucial role in heart diseases, is DNA methylation. This mechanism is regulated by a variety of enzymes and proteins such as DNA methyltransferases (DNMTs) involving in this process from development to the end of postnatal growth. DNMT3A and DNMT3B are involved in de novo DNA methylation, e.g., during gametogenesis and embryogenesis (Schubeler 2015). This process occurs predominantly in the context of symmetrical CpG dinucleotides, enriched in CpG islands where methyl groups are moved to the fifth cytosine (Greenberg and Bourc'his 2019). Depending on the site of the methylation, DNA methylation is very important for inhibiting transcription. Gene silencing is frequently related with increased methylation in CpG-enriched areas in the gene promoter region (Zhao et al. 2022a, b); this epigenetic process is crucial for DNA and protein binding to regulate transcription. Hypermethylation leads to transcriptional repression, whereas hypomethylation means activation of transcription. Abnormal methylation status is also involved in CVD development and is considered useful to evaluate the progression of the pathology. Under physiological conditions, non-promoting CpG regions are methylated, whereas promoter CpG islands are typically hypomethylated. The phenomenon of hypomethylation of non-promoter regions of DNA can cause instability and structural changes in chromosomes altering normally silent regions and provoking transcription in wrong sites. This could cause a potential harm or overexpression of normally silenced genes. Initially, the changed DNA methylation landscape was recognized as an epidemiological marker key underlying several human disorders, particularly in the process of carcinogenesis (Nishiyama and Nakanishi 2021). According to strong evidence, the pathophysiology of myocardial remodeling, linked to various etiologies, such as DCM, ischemic cardiomyopathy (ICM), and pressure disorders, is strongly supported by genome-wide studies that have been conducted over the past 10 years (Pepin et al. 2019a, b; Movassagh et al. 2010; Zhao et al. 2022a, b). In 2010 for the first time, it was shown that a significant portion of CpG and promoters is hypomethylated in the last stages of HF (Pepin et al. 2019b). According to a study published in 2019, DCM hearts have more total CpG methylation than those ones under control (Cheedipudi et al. 2019). In the hearts of DCM patients, nuclear DNA methylation is altered (enhanced) in a cardiomyocyte-specific manner (PMID: 31971668). In 2013, Haas et al. identified various DNA methylation patterns in left ventricular tissues from DCM patients and replicated the epigenetic regulation patterns of several genes, including lymphocyte antigen 75 (LY75), tyrosine kinase-type cell surface receptor HER3 (ERBB3), homeobox B13 (HOXB13), and adenosine receptor A2A (ADORA2A). Those functions were previously unknown in DCM (Cheedipudi et al. 2019). The functional association of these discovered genes was then further established in zebrafish, promising for a diagnostic role primarily in DCM but also in HF. DNA methylation study in conjunction with transcript mRNA expression discovered a reduction in oxidative cellular respiration, while anaerobic glycolysis was increased. Additionally, dysfunction in myocardial tissue was observed in ischemic hearts of patients with end-stage HF compared to nonischemic control hearts, mediated by KLF15 and polycomb methyltransferase enhancer of zeste 2 polycomb. Another epigenetic process in addition to DNA methylation is the hydroxymethylation of 5-methylated cytosines mediated by TET family; isoforms 1, 2, and 3 can oxidize 5mC to 5hmC and catalyze the conversion of 5mC to 5-formylcytosine (5fC) and 5-carboxylcytosine, which can be replaced by unmethylated cytosine (Branco et al. 2011). Considering the oxidation of 5mC and the reversal of DNA methylation-induced gene repression, they may be a significant target for future therapeutics since TET-mediated DNA hydroxymethylation has been discovered to control the hypertrophic mice in genome-wide mapping studies in both adult neonatal and cardiomyocytes (Prasher et al. 2020).

9.3.2 Histone Modifications in CVD

Histone modifications appear to have a more subdued function in epigenetic controls than DNA methylation does. Histone modification modulates target gene expression in a specific way depending on the cell type and epigenetic mark (Li et al. 2019a, b). Acetylation, methylation, phosphorylation, ribosylation, and other histone posttranslational modifications occur at various amino acid residues and possibly alter chromatin architecture and/or affect the expression of genes by attracting various regulatory molecules, such as transcription factors, chromatin regulators, and other histone modifiers (Zhao et al. 2022a, b). The modification of histones (e.g., methylation or acetylation) affects the progression of various CVD (Qadir and Anwer 2019). It is reported that among 1,109 differentially regulated genes in adult mouse cardiomyocytes under hypertrophic remodeling, 596 have at least 1 histone modifier at the promoter region, suggesting a role for the epigenetic landscape in reprogramming the transcriptome of hypertrophic cardiomyocytes (Yang et al. 2021). Histone acetylation and deacetylation through the activity of certain HATs and HDACs are implicated in CVD. In normal condition, histone acetylation by HATs may "relax" chromatin structures and activate transcription by interfering with connections between and within nucleosomes. Histone deacetylases (HDACs), instead, deacetylate histones, increasing the association between histone DNA, which results in chromatin concentration and gene repression (Gray and Teh 2001). P300/EP300, essential for cardiac homeostasis and both healthy and pathological hypertrophic development, is the most extensively researched HAT in the heart and circulatory system. P300 collaborates with the transcriptional coactivator phosphorylated CBP, also known as CREB (cAMP-response-element-binding protein). Lysine residues in histones and nonhistone proteins, such as the transcription factor GATA4, serum response factor, and myocyte enhancer factor 2C, are acetylated by P300/CBP (Backs and Olson 2006). CBP and p300 upregulation in cardiomyocytes provokes hypertrophy, whereas overexpression of their mutant form lacking HAT activity does not (Abi 2014).

9.3.3 Histone Deacetylases in CVD

HDAC is an enzymatic family removing acetyl groups from lysine residues on histone tails by using either zinc-(Zn-) or NAD+ cofactors (Seto and Yoshida

2014). Mammalian HDAC family members number at least 18, and they can be broadly categorized into 4 groups:

- Class I: 1, 2, 3, and 8
- Class II: 4, 5, 6, 7, 9, and 10
- Class III: known as Sirtuins
- Class IV: 11

Class II HDACs are found in nuclear and cytoplasmic environment, in contrast to Class I HDACs, which are mostly nuclear (Milazzo et al. 2020). Generally, it is reported that HDAC1 and HDAC2 cause pathological stress-induced cardiac hypertrophy since they have been detected in the heart in response to pressure overload. In contrast, it is reported that HDAC3 has a protective role in hypertrophy. In the absence of additional clinical stress, cardiomyocyte-specific deletion of HDAC3 resulted in increased heart weight, reduced contractility, overexpression of fetal genes associated with hypertrophy, disorder of myofibril and mitochondria, and an increase in adult mortality of 12 weeks (Wu et al. 2023). HDAC2 and HDAC3 are important in atherosclerosis initiation and progression (Chen et al. 2020a, b). The most extensively researched HDACs are Class II HDACs that play key functional role in the heart and vasculature. There are six members of the Class II family, which are further split into Class IIa (HDAC4, HDAC5, HDAC7, and HDAC9) and Class IIb (HDAC6 and HDAC10) (Milazzo et al. 2020; Morris and Monteggia 2013). Class II HDACs play a crucial role in controlling heart hypertrophy and the body's reaction to pathogenic stressors. In fact, Class II HDACs are extremely sensitive to the heart's adrenergic stimulus (Ooi et al. 2015). Mice with genetically inactivated Class II HDACs exhibit hypersensitivity to the onset of cardiac hypertrophy as well as an impaired response to pharmacological and biomechanical pro-hypertrophic stimuli such as pressure overload and calcineurin activation (Oka et al. 2007). MEF2C, a transcriptional factor that encourages the expression of pro-hypertrophy genes, is bound and inhibited by HDAC5 and HDAC9. CaMK and protein kinase D (PKD), inducted to stress, phosphorylate HDAC5 and HDAC9 in an HDAC4-dependent manner, in response to a stimulus that causes hypertrophy. The chaperone protein 14-3-3 subsequently binds to these phosphorylated HDACs, facilitating their transfer from the nucleus to the cytoplasm (He et al. 2020; Abend et al. 2017). For Class III deacetylases, Sirtuins, several mutations have been identified and linked to myocardial infarction (MI) susceptibility, particularly in the SIRTs that have a wide cardioprotective activity, such as SIRT1, SIRT2, SIRT3, and SIRT6 (Wu et al. 2022), (Yamac et al. 2019). As deacetylases, SIRTs move acetyl group to NAD+ generating nicotinamide giving a significant contribution in metabolism and oxidative stress (Houtkooper et al. 2012). As a result, they not only depend on NAD+ but also lower the NAD+/NADH ratio. The widely expressed SIRT1 is mainly cytoplasmic in adult cardiomyocytes, although it is nuclear in other tissues. SIRT1 changes location between the nucleus and cytoplasm in response to stress condition (Wang et al. 2021b) due to dual location, and SIRT1 may target both histone and nonhistone proteins. In rodents and big animals, SIRT1 expression is increased in pressure overload-induced hypertrophy and HF (Matsushima and Sadoshima 2015).

9.3.4 Histone Methylation in CVD

Histone methyltransferase transfers the methyl group from SAM to lysine or arginine residues on histone tails. They are separated into two major classes based on whether they have the SET domain or not (Sawan and Herceg 2010). Depending on the change, histone methylation is typically related to repressive chromatin activity (Greer and Shi 2012). Particularly marked by methylation in H3K27, H3K9, and H4K20 is heterochromatin, which is composed of closely packed and suppressed DNA site (Miller and Grant 2013). Equally, active promoters are connected to methylated H3K4, H3K36, and H3K79 (Zhang and Liu 2015). Many HMTs are necessary for survival and growth. Importantly, a variety of disorders with heart problems are brought on by genetic flaws or a deficiency in HMTs (Shi et al. 2022; Yang et al. 2021), for example, for the following syndromes:

- Wolf-Hirschhorn: with a developmental delay and congenital cardiac abnormalities, correlated with impaired immune system by haploinsufficiency of the H3K36 HMT and nuclear SET domain 2 (Gavril et al. 2021)
- Kleefstra: microdeletions in 9q34.3 causes a multifarious congenital disorder which leads to cardiac problems, due to alteration in a histone methyltransferase EHMT1, and other genetic conditions (Campbell et al. 2014)
- Kabuki: caused by mutations in a domain for a histone methyltransferase KMT2D/MLL2 that methylating H3K4me2 results in atrial and ventricular septal abnormalities and aorta defect in 70% of afflicted persons (Schwenty-Lara et al. 2020)

9.3.5 Noncoding RNAS in CVD

Different studies were carried out in the latest years to better understand the pathophysiological processes behind the onset and development of CVD due to their rising incidence in the world (Roth et al. 2020). Aberrant proliferation, migration, autophagy, apoptosis, and necrosis are related to altered metabolic and endocrine profile with hypoxia and oxidative stress damage due to heart and vascular exposure (Guo et al. 2022). For instance, ongoing or cyclic hypoxic circumstances can cause the cascade-like activation or inhibition of several genes, which is known as a gene regulatory network mediated by hypoxia (Chen et al. 2020a). The irreversible processes that emerge from the cellular functioning changes might have physiological and even pathological effects (Checa and Aran 2020).

In line with this, different RNA transcripts that are not translated to proteins have a huge impact on CVD (Dorn et al. 2019). These RNA molecules, known as noncoding RNAs (ncRNAs), have lately been identified as significant epigenetic regulators. ncRNAs are split into two categories: tiny ncRNAs comprise transcripts like microRNAs and short-interfering or silencing RNAs (siRNAs) and long ncRNAs (Prasher et al. 2020; Devaux et al. 2017). The ncRNA expression might vary depending on the illness stage; the distinction between myocardial ischemia/ reperfusion injury and acute AMI in terms of ncRNA profile depends on that. Additionally, maximal reperfusion is attained in a mild ischemic damage, even though the ischemic region grows in proportion to the length and severity of blood flow reduction (Marinescu et al. 2022). MicroRNAs are created in the cytoplasm after being generated in the nucleus as precursors, via maturation, and perform their biological role by luring certain proteins that serve as part of the RNA-induced silencing complex (RISC). Many miRNAs are differentially expressed in vascular cells and cardiac tissue, where they have a significant regulatory role in biological processes such cell differentiation, growth, apoptosis, proliferation, and contractility. MI and end-stage cardiomyopathy have both been linked to abnormal miRNA expression (Chistiakov et al. 2016). Several miRNAs, including miRNA-1, miRNA133a, miRNA-20a/b, and miRNA-499, are also thought to be unique signaling molecules that are highly expressed in the myocardium (Marinescu et al. 2022). In a murine and porcine model, miR-15 expression was greater in the damaged tissue, and notably PDK4 and SGK1, key mediators of the miR-15 family actions, have a role in mitochondrial activity and cardiomyocyte apoptosis. Additionally, two miRNAs, miRNA-21-5p and miRNA-126-3p, contribute to the onset and progression of CVD (Ultimo et al. 2018). Also, many lncRNAs, particularly those involved in AMI, have been demonstrated to be crucial in CVD (C. Chen et al. 2019). During an infarction, myocardial hypoxia causes a significant loss of viable cardiomyocytes through both necrosis and apoptosis (Lodrini and Goumans 2021). New studies have shown that lncRNA has a regulatory role in cardiac infarction-related apoptosis (Xie et al. 2021). Changes in lncRNA expression levels alter and modify paracrine communication in addition to intracellular signaling (Pardini and Calin 2019). Apoptosis, cell proliferation, and fibrotic remodeling in AMI were all substantially correlated with alteration of the myocardial infarction-associated transcript lncRNA (MIAT), which is mostly upregulated in the heart (Marinescu et al. 2022). In addition to targeting miR-24 to reduce the postinfarction myocardium during cardiac fibrosis, hundreds of lncRNAs play crucial roles in MIAT (Qu et al. 2017). Thus, MIAT in experimental investigations shows to increase cardiac fibrosis by targeting certain antifibrotic miRNAs such as miR-24, miR29, miR-30, and miR-133 (Zhao et al. 2022b). In an experimental mouse model of myocardial infarction, the conserved super enhancer-associated lncRNA Wisp2 (Wisper) RNA is described as a potent regulator of cardiac fibrosis as well as an alluring therapeutic target that lessens the pathological evolution of fibrosis in response to AMI, avoiding harmful remodeling in damaged heart tissue (Micheletti et al. 2017). For instance, the MIAT lncRNA was engaged in the control of the immediate inflammatory response after MI (Liao et al. 2016).

9.4 Epigenetic Dysregulation and Cardiovascular Disease Susceptibility.

Many factors can affect the development of CVDs, such as food, genetics, and the environment, and among these, aging is crucial. The prevalence of CVD increases in parallel with human life expectancy increases, most likely because of risk factors and aging mechanisms (Foreman et al. 2018), as described in one study showing the odds as increased with advanced age (Savji et al. 2013). Hypertrophy, altered left ventricular dysfunction, HF, and arterial and endothelial dysfunction are some of the pathological effects of normal cardiovascular aging that can change the heart and arterial system (Lakatta and Levy 2003a; Lakatta and Levy 2003b) besides increases in the prevalence of metabolic illnesses like diabetes markedly with aging and dramatic increases in CVD morbidity and mortality (Fadini et al. 2011). Metabolic disorders are associated also with aging so impacting cardiovascular system independently from natural aging. Age-related cellular dysfunction is due to senescence in tissues (McHugh and Gil 2018) that is common in cardiac aging and related diseases (Ock et al. 2016) showing contractile and mitochondrial dysfunction, genomic instability, and hypertrophic growth (Tang et al. 2020). Other mouse model in vivo evidences have identified cellular senescence markers, such as p16 (CDKN2A) and p53 and event like overgrowth and ROS production (Torella et al. 2004; Spallarossa et al. 2009). Epigenetic alterations are directly correlated to aging and age-related disease, including CVD. Epigenetic modules track, representing by enrichment for promoter-associated marker, H3K4me3, DNA methylation profiles, and components of another epigenetic age marker (Horvath 2013), such as Polycomb-group member SUZ12, have been observed in binding regions and related to developmental processes and hematopoietic stem cell (Dozmorov 2015), and to diverse biological sources of CVD risk, reporting the involvment in development- to immune-related processes (Westerman et al. 2019). Differentially methylated regions (associated with SLC9A1, SLC1A5, and TNRC6C genes) have been linked to monocyte activation in response to biological stimuli, and CpG (CG22304262) in SLC1A5 had a cause-effect with incident coronary heart disease (iCHD). CpGs are to be proximal to gene TSS, and their methylation level regulates transcript expression. Interstingly, mRNA levels have reported as suppressed in human HF samples and human failing myocardium compared with healthy controls (Kennel et al. 2019). SLC1A5 is a glutamine transporter pivotal for homeostasis, and all closely AA linked to glutamine metabolism, such as proline metabolism, also deregulated in heart failure, indicating an altered proline storage and us. CVDs are associated with a pro-inflammatory state and circulating cytokines, among them TNF-α. In vitro experiments confirmed the TNF-α decreased SLC1A5 levels and decreased cellular glutamine uptake. Metabolic imbalance in CVDs is indeed characterized by decreased oxidative metabolism and increased glycolysis with cause-effect in young people that impact in adult as well (Farlik et al. 2016; Laiosa and Tate 2015). Similarly, other studies in myocardial tissues have shown that epigenetics plays a crucial role also in the early stage of AMI. In a mouse model of AMI (Luo et al. 2022), a time points analysis of DNA methylation and mRNA expression showed altered DNA methylation profiles between pre- and post-AMI and that the most critical stage was 6 h. Specifically, Ptpn6, Csf1r, Col6a1, Cyba, and Map3k14 expression was correlated to gene methylation. For example, Ptpn6 transcription level increased significantly at 24 h after AMI correlated to low methylation status at the promoter site. Map3k14 is involved in a NF-κB pathway remarking NF-κB activation and inflammatory response as main pathophysiological process in the early stage of AMI. Cyba gene encodes for p22phox, a regulatory subunit of NADPH oxidase, involved in ROS homeostasis. Evidence underlines the role of the mitochondrion and the metabolic balance and related management of ROS and NAD production. Sirtuin family, a class of proteins that mediate posttranslational modification by regulating lysine residue acetylation specifically coupling lysine deacetylation to NAD+ hydrolysis (Tanner et al. 2000), has been linked to a variety of physiological functions and diseases and the development/progression of HF. Among sirtuin family, SIRT1 and SIRT3 attracted attention since in various animal models of HF, their resveratrol-mediated activation showed to preserve cardiac function and improve survival. Evidence linked sirtuin activity as subordinated to NAD+ levels and located its involvement in cardiomyocyte energy production, detoxification of oxidative stress, and intracellular Ca2+ handling.

9.5 Epigenetic Biomarkers in Cardiovascular Diseases

Molecules that can be quantitatively measured in natural samples are called biomarkers. These molecules can give information on the presence of condition known as individual biomarkers or for the evaluation of conditions, the prognostic biomarkers (Califf 2018). Traditionally, many cardiovascular biomarkers have been searched for in the bloodstream such as troponins, individual biomarkers of AMI are proteins found in cardiomyocytes, and these proteins are responsible for cardiac contractility (Wang et al. 2020). As a matter of fact, during heart attack or MI the rupture of blood force, due to damage of cardiac tissue, proteins are released in the blood (Marjot et al. 2017). Biomarker research has discovered that proteins and peptides act as markers, but also alterations impact epigenetic pathways in natural fluids or tissue necropsies (Garcia-Gimenez et al. 2017a). Epigenetic changes, particularly DNA methylation, are relatively stable and may be employed in both fluid and tissue samples, which are routinely used in clinical practice, without the requirement for sophisticated sample processing methods (Garcia-Gimenez et al. 2017b). However, several preanalytical issues may prevent the identification of the epigenetic biomarker: sample heterogeneity, blood cell composition, or validation in independent cohorts (Michels et al. 2013). These discrepancies may impact the epigenetic status in male and female, with altered transcriptomic profile during the CVD onset (Glinge et al. 2017). Studies evaluating epigenetic biomarkers defined that the influence of sex can be statistically adjusted, while without proper stratification, gender epigenetic effects may go unnoticed. DNA methylation can be analyzed with microarrays by Illumina, bisulfite sequencing, bisulfite whole genome sequencing, MeDIP CHIP, and MeDIP Seq (Leti et al. 2018). Illumina microarrays are accessible with the most standardized and well-known technique and for which multiple pipelines have been established. The most recent technologies make this technique more acceptable for clinical usage since it outperforms other sequencing bisulfite methods. Furthermore, numerous bioinformatics tools have been tested to address the issue of white blood cell heterogeneity (Jaffe and Irizarry 2014). Although research on DNA methylation gives a statistical correction for sex impact, a lack of adequate stratification may result in the omission of sex-specific epigenetic effects. Globally, several studies have been conducted focusing on methylation status in individuals with CVD. These analyses measure levels of 5mC by ELISA assay or evaluating DNA methylation at the whole genome at repetitive sequences such as LINE-1 and ALU genome (Bakshi et al. 2019; Povedano et al. 2018). New implicit biomarkers of CVD are noncoding RNAs, which can be detected mostly in the bloodstream (Shi et al. 2016). miRNAs, and their role in helping to personalize healthcare, are a great diagnostic target, becoming the most studied. The use of epigenetic mechanisms in designing new medicines to treat cases, as much as their use as biomarkers, is still premature, and additional work is requested to completely address their potential to be used in perfection drug (Rasool et al. 2015). In recent years, cardiovascular risk factors have been linked to epigenetic alterations in patients. Changes in the epigenetic landscape affect cardiovascular homeostasis and contribute to development of CVD (Gharipour et al. 2021). Changes in the epigenetic landscape alter cardiovascular homeostasis and contribute to the development of CVD (Wu et al. 2021). Contribution of epigenetic markers in CVD is still unclear even though the gene expression regulation by epigenetic mechanisms is well known (Soler-Botija et al. 2019). Investigating on epigenetic biomarkers could therefore elucidate the molecular processes and pathways involved in CVD. The following paragraph focuses on epigenetic biomarkers associated to important CVD such as MI, HF, and arrhythmia.

9.5.1 Epigenetic Biomarkers in MI

DNA methylation among the epigenetic processes can be considered as an indicator of MI. Several studies have reported that DNA methylation at different loci can be associated with MI. In rats with myocardial injury after MI, hypermethylation of ALDH2 start site led to its downregulation, impairing its cardioprotective role (Wang et al. 2015). Microarray analysis of DNA methylation from patients of the EPICOR research and the EPIC-NL cohort showed hypomethylation in the zinc finger and BTB domain-containing protein 12 (ZBTB12) and LINE-1, suggesting a possible signature in white blood cells detectable early than the MI development (Fiorito et al. 2014; Guarrera et al. 2015). In another study on MI patients, 211 differentially methylated CpG sites are associated to genes involved in cardiac function, CVD, cardiogenesis, and recovery from ischemia damage. For this reason, these genes may have a role in MI etiology or recovery (Rask-Andersen et al. 2016). In an epigenome-wide association study, 34 differentially methylated CpGs related to acute MI were identified. In the molecular pathways relevant to MI, these loci were notably correlated with several features such as smoking, lipid metabolism, and inflammation (Fernandez-Sanles et al. 2021). Zinc finger homeobox 3 (ZFHX3) and SWI/SNF-related two chromatin regulator subfamily a, member 4 (SMARCA4) from MI patients were reported as methylated in a genome-wide DNA methylation (Nakatochi et al. 2017). In a mouse model of AMI, DNA and mRNA analysis was performed across time. The most important stage of AMI was shown to be 6 h. During this stage, a high number of methylation modification sites were altered. PTPN6, CSF1R, COL6A1, CYBA, and MAP3K14 genes participate in AMI process via DNA methylation (Luo et al. 2022). One of the pathological hallmarks of MI is atherosclerotic plaque disruption. Cells and artery wall components become increasingly vulnerable to DNA damage as atherosclerosis advances, increasing programmed cell death and necrosis. For this reason, the damage to cardiac tissue can be assessed by measuring circulating histones and nucleosomes in blood samples (Soler-Botija et al. 2019). Histone modifications have a role in the pathological phase of MI as well. For example, HAT activity of p300 is critical for MI development (Soler-Botija et al. 2019). SIRT1 is a Class III deacetylase known to have cardioprotective properties and is downregulated following tissue damage. On the contrary, during renal ischemia/reperfusion, SIRT2 is activated which deacetylates FOXO3a, promoting its nuclear accumulation. This leads to FasL increased expression inducing FasL-mediated cell apoptosis, activating caspase8 and caspase3 (Wang et al. 2017). Furthermore, the SIRT3 increased expression leads to cyclophilin D deacetylation after myocardial ischemia/reperfusion, preventing lethal reperfusion injury (Bochaton et al. 2015). In addition, in HDAC4-transgenic mouse, it was discovered that HDAC4 overexpression led to myocardial fibrosis and enlargement and cardiac dysfunction (Zhang et al. 2018). Also, HDAC6 regulates the antioxidant protein peroxiredoxin 1 associated with CVD such as MI (Leng et al. 2018). LncRNAs can be also considered potential therapeutic targets for MI since they control processes such as autophagy and apoptosis. lncRNAs APF, CAIF, and Mirf that regulate cardiac autophagy are associated with MI lesion. In addition, the CPR ncRNA, MALAT1, and AK139128 lncRNA by regulating cell proliferation participate in cardiac repair and development of cardiac function (Shi et al. 2022). Many studies published in recent years show a link between circulating miRNA and diagnosis and prognosis of AMI. Following myocardial damage, miR-208 is raised in rat blood levels, so different studies have been conducted in AMI patients in which this miRNA was detected (Wang et al. 2010; Ji et al. 2009). When they were compared to healthy controls, the circulating miR-499-5p was more than tenfold higher (Olivieri et al. 2013). In another study, Devaux et al. (2017)discovered that patients with acute MI had greater levels of miR-208b and miR-499. The blood miR-1 level rose following AMI, peaking at 6 h (>200 times) and returning to baseline 3 days later (Cheng et al. 2010). Figure 9.1 reports the main epigenetic biomarkers in MI.

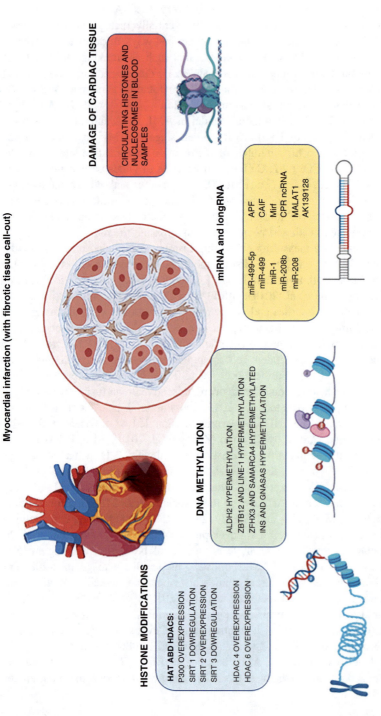

Fig. 9.1 Epigenetic biomarkers in myocardial infarction

9.5.2 Epigenetic Biomarkers in HF

Different causes can lead to HF such as hypertension, cardiomyopathy, MI, and arrhythmias (Khatibzadeh et al. 2013). There are numerous studies linking HF and epigenetic modifications (Ameer et al. 2020; Abi 2014; Kim et al. 2016). Modern technology made it possible to find regions with high density DNA methylation mapping of the whole epigenome, like epigenetic susceptibility traits and novel biomarkers, related to HF and cardiac dysfunction. Many CpG regions were identified as novel biomarkers of HF (Meder et al. 2017; Rau and Vondriska 2017). In blood leukocytes from HF patients, differentially methylated DNA regions have been identified (Li et al. 2017). Differential methylation was shown in one study of three angiogenic genes comparing left ventricular tissue from eight end-stage HF patients and controls. Specifically, hypermethylation of the platelet/endothelial cell adhesion molecule promoter and within the gene body of Rho GTPase activation protein 24 and of angiomotin was observed. DCM is a major cause of HF. Individuals with idiopathic cardiomyopathy have altered DNA methylation, which was linked to significant changes in the LY75 and ADORA2A mRNA expression (Haas et al. 2013; Soler-Botija et al. 2019). Using the same criteria, genome-wide studies have been found to enrich H3K36me3 in promoter CpG islands, genes, and intragenic CpG islands. In individuals with dilated cardiomyopathy, an altered methylation pattern is observed in regulatory areas of heart developmental genes such as T-box protein 5 (TBX5), heart and expressed neural crest derivative 1 (HAND1), and NK2 homeobox 5 (NKX2.5) (Jo et al. 2016). A computational study discovered a few gene promoters that were differently methylated (AURKB, BTNL9, CLDN5, and TK1). This study advances our understanding of DNA methylation and altered expression in dilated cardiomyopathy, which will help with treatment (Koczor et al. 2013). Epigenetic changes have been postulated to play a significant role in the evolution of HF in the pressure-overload mouse model. A decrease in sarcoplasmic reticulum Ca2+ATPase (Atp2a2) levels as well as a large increase in -myosin heavy chain mRNA (Myh7) levels was discovered. After 8 weeks of transverse aortic constriction, H3K4me2, H3K9me2, and H3K36me2 were increased, and lysine-specific demethylase KDM2A was decreased (Angrisano et al. 2014). Atp2a2 is an important element in the heart function, and its decreased activity is a distinguishing hallmark of HF. In another study by Gorski et al., acetylation at lysine 492 in Atp2a2 controlled by SIRT1 and HATp300 in HF patients dramatically decreased (Gorski et al. 2019). HDAC4 is required for histone methylation in HF, so it may be a therapeutic target (Hohl et al. 2013). Sequencing of DNA methylation showed changes in coding and noncoding RNA, and in cardiac tissue from HF patients, the hypermethylation of HEY2, MSR1, MYOM3, COX17, and miR-24-1, as well as the hypomethylation of CTGF, MMP2, and miR-155, has been discovered. As a result, a distinct set of loci has been identified as valuable diagnostic and therapeutic targets in HF (Glezeva et al. 2019). Several data from more than a decade ago showed that miRNAs are differently regulated in "weak hearts." Since then, a substantial body of evidence has been released (Soler-Botija et al. 2019).

9.5.3 Epi-Biomarkers in Arrhythmia

Atrial fibrillation (AF) is mostly studied and frequent arrhythmia. It is difficult to identify the underlying causes of AF in an individual patient, and current therapies are still not very effective (Brundel et al. 2022). In in vivo experiments, using rats demonstrates that DNMT3A is significantly overexpressed in fibrotic myocardium. In AF, this leads to alter gene expression such as Ras-association domain family 1 isoform A (Tao et al. 2014) and SUR2 overexpression (than SUR1), Pitx2, and SERCA2a (Fatima et al. 2012). Chromatin modifications, such as histone acetylation, methylation, and phosphorylation, confer information to progeny cells during AF. HDACs play a key role in calcium homeostasis and in AF genesis. For instance, murine model with an acute increase in cardiac preload exhibited nuclear export of HDAC4, H3K9 demethylation, HP1 phosphorylation dissociation from the promoter region, and activation of the ANP gene (atrial natriuretic peptide gene) (Tao et al. 2016). FA induces remodeling and altered functionality, at least HDAC6, through alteration of α-tubulin that disrupts microtubule structure of cardiomyocytes that induces remodeling and loss of contractile function in AF. HDAC6 inhibition in vivo protects against AF-related atrial remodeling (Tao et al. 2016). On the other side, sirtuins mainly play a role in mediating cell survival. In fact, Sirt1 overexpression protects myocytes from apoptosis and from the causes of modest hypertrophy (Alcendor et al. 2004). Many miRNAs have been detected to play a role in AF such as miRNA-1, miRNA-26, miRNA-133, miRNA-328, miRNA-499, and miRNA-106b-25 that all had been involved with atrial electricity remodeling. miRNA 1 appears to be lower in hospitalized patients with chronic AF compared to breast rhythm. miRNA1 and miRNA133 regulate gene function in pacemaker cell activity (Gao et al. 2013). miRNA changes have been detected, such as in miRNA-21, miRNA-133, miRNA-590, miRNA-30, miRNA-146b-5p, and miRNA-206, which are involved in structural remodeling of the atrial wall, another mechanism involved in the genesis of AF (Gao et al. 2013).

9.6 Preclinical Evidence of Pharmaco-Epigenetic Role in Cardiovascular Diseases

Drug development that targets epigenetic mechanisms, such as histone modifications, DNA methylation, and noncoding RNA, has achieved encouraging results in basic experimental research for the treatment of cardiovascular diseases, although today epi-drugs are still little used in clinical treatments. In the future, these compounds may find greater application in clinical trials, for the improvement of symptoms and prognosis in patients with cardiovascular disease (Gorica et al. 2022; Shi et al. 2022)

9.6.1 Histone Modifications

Several epi-drugs targeting HDAC activity, resulting in HDAC inhibition, were investigated as potential treatment for cardiac ischemic diseases. Ischemia/reperfusion-associated increases in HDAC activity raise the possibility that HDAC inhibition could be considered a valuable treatment for MI and I/R injury because it could attenuate cardiac fibrosis by stimulating cardiac regeneration and maintain cardiac function.

Recent studies show that Entinostat (MS 275), a Class I-specific HDAC inhibitor, reduces, through increased expression of SOD2 and catalase, via the transcription factor FoxO3a in myocardial mitochondria (Aune et al. 2014; Herr et al. 2018) MI area , improve left ventricular function and protect cardiac systolic function after ischemia–reperfusion in isolated hearts from Male Sprague Dawley rats. Results from in vivo study on mouse MI and in vitro on cultured embryonic stem cell model shows that trichostatin A (TSA), Class I and II HDAC inhibitor, reduces MI area preventing ventricular remodeling with cardioprotective effect in patients (Zhang et al. 2012a). The cardioprotective effects of TSA could be achieved by the stimulation of the AKT-1 phosphorylation, the acetylation and phosphorylation of MKK3, and the reduction of TNF-α levels in myocardium and serum. Further investigation demonstrates that TSA activated the FOXO3α signaling pathway resulting in reduction potential of mitochondrial membrane dissipation and therefore the inhibition of programmed cell death (Yu et al. 2012).

It has demonstrated that SAHA/vorinostat, a potent HDAC inhibitor, ameliorated cardiac remodeling after infarction and delayed ischemia/reperfusion injury. Indeed, SAHA increased cardiomyocyte autophagic activity in the infarct zone, exerting the cardioprotective effects during I/R (Kimbrough et al. 2018; Zhang et al. 2012a).

Therefore, it is expected that vorinostat will go under investigation in patients to treat MI (Gillette and Hill 2015).

Tubastatin A (TubA), an HDAC6 inhibitor, reduces the MI area and heart function and ROS generation through the increment in the acetylation of Prdx1 in hearts isolated from mice (Leng et al. 2018).

In addition to the inhibition, the activation of HDACs may work in the control of myocardial infarction. As a matter of fact, compounds targeting Class III HDAC improving their activity were also investigated. Resveratrol, the most potent sirtuin 1 (SIRT1), attenuated FOXO1-related proapoptotic signaling pathway, increased PCC-1α and mitochondrial biogenesis, and improved myocardial function and Ang II-induced cardiac remodeling, resulting in the protection of myocardial cells from I/R injury (Becatti et al. 2012).

SRT1720, a SIRT1 activator, alleviated mice vascular endothelial dysfunction and reduced MI in SIRT1(+/−) hearts by the activation of COX-2 signaling and the reduction of inflammatory status (Liu et al. 2017).

Nevertheless, no HDAC inhibitors have entered the clinical research phase in cardiovascular diseases, although HDAC inhibitors have been demonstrated by several in vivo and in vitro experiments to ameliorate myocardial infarction.

Histone modifications, considering their functions, can represent an effective therapeutic target to develop prevention and treatment strategies.

TSA and apicin derivative, a selective inhibitor of Class I HDACs, have been shown to improve cardiac function by preventing cardiomyocyte hypertrophy and myocardial fibrosis in mice with thoracic aortic constriction (Antos et al. 2003; Gallo et al. 2008; Kee et al. 2006).

In addition, these results have been also confirmed in mice models of cardiac hypertrophy and fibrosis where TSA and emodin, an HDAC inhibitor, have been able to improve cardiac hypertrophy (Chen et al. 2015). Mocetinostat, another Class I HDAC inhibitor, via attenuation of IL-6/pSTAT3 signaling pathway in MI rats, has been able to reverse myocardial fibrosis modulating low myocardial fibroblast activity and promoting cell cycle arrest/apoptosis (Nural-Guvener et al. 2014; Nural-Guvener et al. 2015).

Resveratrol protected H9c2 cells from Ang II-induced hypertrophy, increasing SIRT1 activity and therefore reducing IL-6 activation (Akhondzadeh et al. 2020).

Curcumin, a P300 HAT inhibitor, can prevent ventricular hypertrophy and maintain systolic function by disrupting the P300/GATA4 complex in HF of rat models (Pan et al. 2013). Moreover, it has been demonstrated that in both healthy individuals and in patients with atherosclerosis, curcumin treatment was associated with lower LDL levels and raises HDL levels (Ramirez Bosca et al. 2000).

Additional studies have shown that BRD4 inhibitor JQ1 reduces markers and extracellular matrix proteins in cardiac fibroblasts, as well as inhibiting the contractile function and β-SMA expression of myocardial fibroblasts. Interestingly, JQ1 has been shown to improve cardiac function in mice with preestablished HF or myocardial infarction, indicating its potential benefit as a drug for the clinical treatment of these conditions (Duan et al. 2017).

9.6.2 DNA Methylation

DNA methylation could be also considered a potential therapeutic target for cardiovascular diseases. Indeed, experimental evidence shows that inhibitors of DNMT could be used in HF treatments. RG108 showed to arrest myocardial hypertrophy and fibrosis (Stenzig et al. 2018). According to several studies, 5-azacytidine inhibits the DNMTs activity blocks the expression of genes involved in hypertrophy, and prevents the detrimental effects of TNF-α on SECRA2a expression (Kao et al. 2010). Interestingly, in rats, it has been demonstrated that 5-AZa-2-deoxycytidine reduces myocardial hypertrophy, improves myocardial contractility, and eliminates the susceptibility to ischemic injury (Xiao et al. 2014).

9.6.3 Noncoding RNA

Noncoding RNA and related drugs are attractive targets for potential clinical interventions for myocardial infarction. In fact, in a multicenter randomized trial, involving 18,924 patients with acute coronary syndrome and treated with alirocumab (PCSK9 inhibitor of proprotein convertase subtilisin/kexin type 9 – PCSK9), there has been a lower risk of recurrent ischemic cardiovascular events compared to the ones who received a placebo (Schwartz et al. 2018). The reduction in cardiovascular events was twice as large in patients with diabetes and LDL-C concentration of 0.65–1.30 mmol/L who were treated with alirocumab compared to those without diabetes (Ray et al. 2019).

MIAT plays a key role in the processing of Wnt7b in patients with myocardial infarction. As a matter of fact, lncRNA MIAT, targeting the miRNA-150-5p and VEGF signaling pathways, and being differentially expressed in peripheral blood, could be considered as a potential drug for MI (Liao et al. 2016).

Circulating-RNA MFACR is upregulated in MI to promote hypoxia-induced apoptosis of cardiomyocytes by downregulating miRNA-125b in plasma samples from both MI patient and healthy controls. Therefore, targeted inhibition of cirRNA MFACR could be an important therapeutic approach for treating MI and protecting myocardial cells (Wang et al. 2021c). Some drugs targeting noncoding RNA could also represent a potential treatment for HF. Currently in phase I clinical trials, MRG-110 and CDR132L indicated a significant role in targeting miRNA-92a and miRNA-132 (Huang et al. 2020).

Additionally, circRNA (HRCR) has been identified as potential target in CVD. HRCR can block the development of cardiac hypertrophy and HF and may become a controlling gene for the treatment of these conditions in the future (Wang et al. 2016). Furthermore, a study on human cell lines shows that ectopic expression of circ-FOXO3 can inhibit cell cycle progression by binding to CDK2 and p21, as well as reducing the expression of these proteins in the nucleus and promoting a cell aging phenotype. This discovery provides an alternative approach for cardioprotection (Du et al. 2016). Table 9.1 shows epigenetic drugs in preclinical and clinical trials for CVD.

9.7 Epigenetic-Based Therapies in Cardiovascular Diseases and Future Perspectives

Cardiovascular disease still represents the primary cause of death globally, affecting morbidity, quality of life, and societal costs (Roth et al. 2020). CVD preventative treatments enhance vascular outcomes in fewer than half of the patients; nevertheless, precision medicine offers an "appealing" method to fine-tuning cardiovascular disease targeting therapies to responsive people and allowing resources to be allocated more wisely and efficiently (Costantino et al. 2018). Genetic advances

Table 9.1 Epigenetic drugs in preclinical and clinical trials for CVD (Aune et al. 2014) (a); (Herr et al. 2018) (b); (Zhang et al. 2012b) (c); (Kimbrough et al. 2018) (d); (Zhang et al. 2012a) (e); (Antos et al. 2003) (f); (Kee et al. 2006) (g); (Gallo et al. 2008) (h); (Chen et al. 2015) (i) ; (Nural-Guvener et al. 2014) (j); (H. Nural-Guvener et al. 2015) (k); (Akhondzadeh et al. 2020) (l); (Pan et al. 2013) (m); (Duan et al. 2017) (n); (Stenzig et al. 2018) (o); (Xiao et al. 2014) (p); (Liao et al. 2016) (q); (Wang et al. 2021c) (r); (Huang et al. 2020) (s)

Cardiovascular disease/condition	Epigenetic target	Epigenetic drug
Myocardial infarction	Class I HDAC inhibitor	Entinostat (MS 275) (a, b)
Myocardial infarction	Class I and II HDAC inhibitor	Trichostatin A (TSA) (c)
Myocardial infarction	Pan HDAC inhibitor	Vorinostat (SAHA) (d, e)
Cardiac hypertrophy Myocardial fibrosis	Class I HDAC inhibitor	TSA, apicin derivative (f, g, h)
Myocardial infarction	HDAC6 inhibitor	Tubastatin A (TUBA) (i)
Cardiac hypertrophy	HDAC inhibitor	Emodin (j)
Myocardial fibrosis	Class I HDAC inhibitor	Mocetinostat (k)
Cardiac hypertrophy	Sirt1 activator	Resveratrol (l)
Ventricular hypertrophy	P300/HAT inhibitor	Curcumin (m)
Heart failure Myocardial infarction	BRD4 inhibitor	JQ1 (n)
Myocardial hypertrophy Myocardial fibrosis	DNMT inhibitor	RG108 (o)
Myocardial hypertrophy	DNMT inhibitor	5-Azacytidine (p)
Myocardial infarction	miRNA-150-5p	MIAT (q)
Myocardial infarction	miRNA-125b	Circulating-RNA MFACR (r)
Heart failure	miRNA-92a and miRNA 132	MRG-110 and CDR132L (s)

have uncovered novel pathways and targets that act in a variety of illnesses, paving the possibility for "precision medicine" (Schiano et al. 2015). However, the inherited genome accounts for just a portion of an individual's risk profile. Indeed, traditional genomic techniques neglect the domain of epigenetic gene control and expression. The prospect of eliminating deleterious epigenetic modifications pharmacologically to prevent illness is intriguing and gaining traction (Gorica et al. 2022; Ganesan et al. 2019). Different epigenetic elements, such as histone modifiers, are being targeted in novel techniques for the development of new treatment strategies. The use of epigenetics can help promote customized risk assessment as well as the creation and implementation of personalized CVD treatment interventions. Targeting epigenetic signals might be a promising strategy (Napoli et al. 2016). Globally in the CVD scenario, HF now affects 26 million individuals, with 15 million of them affected solely in Europe (Wenzel et al. 2022). Most critically, by 2030 its prevalence is anticipated to increase by 46% (Hu et al. 2022). Several investigations have found that non-cardiomyocyte cell populations play an important role in heart remodeling in HF. Recent research suggests that the stimulation of local fibroblasts is the root cause of fibrosis (Kwon et al. 2022). New epi-drugs have begun to acquire attention and now are under investigation in preclinical testing or clinical trials (Brookes and

Shi 2014). The application of epi-drugs in clinical practice might allow the individualized therapy, which could enhance HF care and patient prognosis. Furthermore, some known cardiovascular medicines have lately been revealed to have potential epigenetic effects (Heerboth et al. 2014).

9.8 Potential Effects of Existing Drugs on Epigenetic Modulators in Cardiovascular Disease

Repurposing "old" drugs already in use for clinical trials to treat CVD is becoming a growing trend since new drug discovery and development are an expensive, laborious, and time-consuming process (Sarno et al. 2021). Several commonly used and well-known drugs have been found to have epigenetic effects in recent years. For instance, metformin is widely used for standard therapy of type 2 diabetes (T2D) for more than 60 years but only recently has been discovered its cross talk with epi-editing machinery, and current investigations report to reduce the incidence of CVD mortality and cancer (Gu et al. 2020; Gandini et al. 2014). Metformin suppresses gluconeogenic genes through promoting AMPK phosphorylation and activation. AMPK is involved in a wide range of pathways, including epigenetic processes. By activating AMPK, this drug induces histone alterations since it increases indirectly HAT1 and SIRT1 activity and inhibits Class II HDACs (Bridgeman et al. 2018). Also, statins that are HMG-CoA reductase inhibitors lowering low-density lipoprotein cholesterol demonstrated a regression of atherosclerosis through epigenetic effect by H3 and H4 acetylation. These molecules prevent endothelial senescence via enhancing SIRT-1 expression and downregulate miR-146a/b in coronary heart disease patients (Allen and Mamotte 2017). Another class of innovative epi-drugs worth highlighting are sodium-glucose cotransporter-2 inhibitors (SGLT2i). It has been observed that they play a role in decreasing the risk of HF hospitalization in T2D patients and cause mortality (Li et al. 2019a). Dapagliflozin, an SGLT2i, has recently been proven to have cardiac and renal protective properties. This drug controls key miRNAs implicated in the pathogenesis of HF, such as miR199a-3p and miR30e-5p, which regulate PPAR levels and mitochondrial fatty acid oxidation (Gorica et al. 2022). The SGLT2i empagliflozin has also been found to improve cardiac hemodynamic in experimental HF models by increasing renal protection and preventing cardiac fibrosis (Li et al. 2019a). EMPA, mostly inhibiting SGLT2, in vitro reduced DNA methylation alterations caused by high glucose levels and revealed a novel mechanism through which SGLT2i can have cardioprotective effects (Scisciola et al. 2023).

Compliance with Ethical Standards Ethics approval and consent to participate: not applicable
Consent for publication: not applicable
Competing interests: The authors declare no conflict of interest.
The authors confirm that the figure is original.

Authors' contributions: conceptualization, C.P. and U.C.; funding acquisition, L.A. and R.B.; writing – original draft preparation, C.P. and U.C.; images, L.S. and A.A.; and writing and editing, C.P., U.C., M.B., and G.P.; G.P., L.A., and R.B. are the last and corresponding authors. All authors have read and agreed to the published version of the manuscript.

References

Abend A, Shkedi O, Fertouk M, Caspi LH, Kehat I (2017) Salt-inducible kinase induces cytoplasmic histone deacetylase 4 to promote vascular calcification. EMBO Rep. 18(7):1166–1185

Abi KC (2014) The emerging role of epigenetics in cardiovascular disease. Ther Adv Chronic Dis. 5(4):178–187

Akhondzadeh F, Astani A, Najjari R, Samadi M, Rezvani ME, Zare F et al (2020) Resveratrol suppresses interleukin-6 expression through activation of sirtuin 1 in hypertrophied H9c2 cardiomyoblasts. J Cell Physiol. 235(10):6969–6977

Alcendor RR, Kirshenbaum LA, Imai S, Vatner SF, Sadoshima J (2004) Silent information regulator 2alpha, a longevity factor and class III histone deacetylase, is an essential endogenous apoptosis inhibitor in cardiac myocytes. Circ Res. 95(10):971–980

Allen SC, Mamotte CDS (2017) Pleiotropic and adverse effects of statins-do epigenetics play a role? J Pharmacol Exp Ther. 362(2):319–326

Ameer SS, Hossain MB, Knoll R (2020) Epigenetics and heart failure. Int J Mol Sci. 21(23)

Angrisano T, Schiattarella GG, Keller S, Pironti G, Florio E, Magliulo F et al (2014) Epigenetic switch at atp2a2 and myh7 gene promoters in pressure overload-induced heart failure. PLoS One. 9(9):e106024

Antignano F, Burrows K, Hughes MR, Han JM, Kron KJ, Penrod NM et al (2014) Methyltransferase G9A regulates T cell differentiation during murine intestinal inflammation. J Clin Invest. 124(5):1945–1955

Antos CL, McKinsey TA, Dreitz M, Hollingsworth LM, Zhang CL, Schreiber K et al (2003) Dose-dependent blockade to cardiomyocyte hypertrophy by histone deacetylase inhibitors. J Biol Chem. 278(31):28930–28937

Aune SE, Herr DJ, Mani SK, Menick DR (2014) Selective inhibition of class I but not class IIb histone deacetylases exerts cardiac protection from ischemia reperfusion. J Mol Cell Cardiol. 72:138–145

Backs J, Olson EN (2006) Control of cardiac growth by histone acetylation/deacetylation. Circ Res. 98(1):15–24

Bagchi RA, Weeks KL (2019) Histone deacetylases in cardiovascular and metabolic diseases. J Mol Cell Cardiol. 130:151–159

Bakshi A, Ekram MB, Kim J (2019) High-throughput targeted repeat element bisulfite sequencing (HT-TREBS). Methods Mol Biol. 1908:219–228

Becatti M, Taddei N, Cecchi C, Nassi N, Nassi PA, Fiorillo C (2012) SIRT1 modulates MAPK pathways in ischemic-reperfused cardiomyocytes. Cell Mol Life Sci. 69(13):2245–2260

Berulava T, Buchholz E, Elerdashvili V, Pena T, Islam MR, Lbik D et al (2020) Changes in m6A RNA methylation contribute to heart failure progression by modulating translation. Eur J Heart Fail. 22(1):54–66

Bochaton T, Crola-Da-Silva C, Pillot B, Villedieu C, Ferreras L, Alam MR et al (2015) Inhibition of myocardial reperfusion injury by ischemic postconditioning requires sirtuin 3-mediated deacetylation of cyclophilin D. J Mol Cell Cardiol. 84:61–69

Branco MR, Ficz G, Reik W (2011) Uncovering the role of 5-hydroxymethylcytosine in the epigenome. Nat Rev Genet. 13(1):7–13

Bridgeman SC, Ellison GC, Melton PE, Newsholme P, Mamotte CDS (2018) Epigenetic effects of metformin: From molecular mechanisms to clinical implications. Diabetes Obes Metab. 20(7): 1553–1562

Brookes E, Shi Y (2014) Diverse epigenetic mechanisms of human disease. Annu Rev Genet. 48: 237–268

Brundel B, Ai X, Hills MT, Kuipers MF, Lip GYH, de Groot NMS (2022) Atrial fibrillation. Nat Rev Dis Primers. 8(1):21

Buss SJ, Muenz S, Riffel JH, Malekar P, Hagenmueller M, Weiss CS et al (2009) Beneficial effects of Mammalian target of rapamycin inhibition on left ventricular remodeling after myocardial infarction. J Am Coll Cardiol. 54(25):2435–2446

Califf RM (2018) Biomarker definitions and their applications. Exp Biol Med (Maywood). 243(3): 213–221

Campbell CL, Collins RT 2nd, Zarate YA (2014) Severe neonatal presentation of Kleefstra syndrome in a patient with hypoplastic left heart syndrome and 9q34.3 microdeletion. Birth Defects Res A Clin Mol Teratol. 100(12):985–990

Caron C, Boyault C, Khochbin S (2005) Regulatory cross-talk between lysine acetylation and ubiquitination: role in the control of protein stability. Bioessays. 27(4):408–415

Checa J, Aran JM (2020) Reactive oxygen species: drivers of physiological and pathological processes. J Inflamm Res. 13:1057–1073

Cheedipudi SM, Matkovich SJ, Coarfa C, Hu X, Robertson MJ, Sweet M et al (2019) Genomic reorganization of LAMIN-associated domains in cardiac myocytes is associated with differential gene expression and DNA methylation in human dilated cardiomyopathy. Circ Res. 124(8): 1198–1213

Chen C, Tang Y, Sun H, Lin X, Jiang B (2019) The roles of long noncoding RNAs in myocardial pathophysiology. Biosci Rep. 39(11)

Chen M, Zsengeller Z, Xiao CY, Szabo C (2004) Mitochondrial-to-nuclear translocation of apoptosis-inducing factor in cardiac myocytes during oxidant stress: potential role of poly (ADP-ribose) polymerase-1. Cardiovasc Res. 63(4):682–688

Chen PS, Chiu WT, Hsu PL, Lin SC, Peng IC, Wang CY et al (2020a) Pathophysiological implications of hypoxia in human diseases. J Biomed Sci. 27(1):63

Chen X, He Y, Fu W, Sahebkar A, Tan Y, Xu S et al (2020b) Histone deacetylases (HDACS) and atherosclerosis: a mechanistic and pharmacological review. Front Cell Dev Biol. 8:581015

Chen Y, Du J, Zhao YT, Zhang L, Lv G, Zhuang S et al (2015) Histone deacetylase (HDAC) inhibition improves myocardial function and prevents cardiac remodeling in diabetic mice. Cardiovasc Diabetol. 14:99

Chen ZX, Mann JR, Hsieh CL, Riggs AD, Chedin F (2005) Physical and functional interactions between the human DNMT3L protein and members of the de novo methyltransferase family. J Cell Biochem. 95(5):902–917

Cheng Y, Tan N, Yang J, Liu X, Cao X, He P et al (2010) A translational study of circulating cell-free microRNA-1 in acute myocardial infarction. Clin Sci (Lond). 119(2):87–95

Chistiakov DA, Orekhov AN, Bobryshev YV (2016) Cardiac-specific miRNA in cardiogenesis, heart function, and cardiac pathology (with focus on myocardial infarction). J Mol Cell Cardiol. 94:107–121

Cole MA, Abd Jamil AH, Heather LC, Murray AJ, Sutton ER, Slingo M et al (2016) On the pivotal role of PPARalpha in adaptation of the heart to hypoxia and why fat in the diet increases hypoxic injury. FASEB J. 30(8):2684–2697

Costantino S, Libby P, Kishore R, Tardif JC, El-Osta A, Paneni F (2018) Epigenetics and precision medicine in cardiovascular patients: from basic concepts to the clinical arena. Eur Heart J. 39 (47):4150–4158

Crola Da Silva C, Baetz D, Vedere M, Lo-Grasso M, Wehbi M, Chouabe C et al (2023) Isolated Mitochondria State after Myocardial Ischemia-Reperfusion Injury and Cardioprotection: Analysis by Flow Cytometry. Life (Basel). 13(3)

Daugas E, Susin SA, Zamzami N, Ferri KF, Irinopoulou T, Larochette N et al (2000) Mitochondrio-nuclear translocation of AIF in apoptosis and necrosis. FASEB J. 14(5):729–739

Delgobo M, Weiss E, Ashour D, Richter L, Popiolkowski L, Arampatzi P et al (2023) Myocardial milieu favors local differentiation of regulatory T cells. Circ Res. 132(5):565–582

Devaux Y, Creemers EE, Boon RA, Werfel S, Thum T, Engelhardt S et al (2017) Circular RNAs in heart failure. Eur J Heart Fail. 19(6):701–709

Dong P, Liu K, Han H (2022) The role of NF-kappaB in myocardial ischemia/reperfusion injury. Curr Protein Pept Sci. 23(8):535–547

Dorn LE, Tual-Chalot S, Stellos K, Accornero F (2019) RNA epigenetics and cardiovascular diseases. J Mol Cell Cardiol. 129:272–280

Dozmorov MG (2015) Polycomb repressive complex 2 epigenomic signature defines age-associated hypermethylation and gene expression changes. Epigenetics. 10(6):484–495

Du WW, Yang W, Liu E, Yang Z, Dhaliwal P, Yang BB (2016) Foxo3 circular RNA retards cell cycle progression via forming ternary complexes with p21 and CDK2. Nucleic Acids Res. 44(6):2846–2858

Duan Q, McMahon S, Anand P, Shah H, Thomas S, Salunga HT et al (2017) BET bromodomain inhibition suppresses innate inflammatory and profibrotic transcriptional networks in heart failure. Sci Transl Med. 9(390)

Fadini GP, Ceolotto G, Pagnin E, de Kreutzenberg S, Avogaro A (2011) At the crossroads of longevity and metabolism: the metabolic syndrome and lifespan determinant pathways. Aging Cell. 10(1):10–17

Farlik M, Halbritter F, Muller F, Choudry FA, Ebert P, Klughammer J et al (2016) DNA methylation dynamics of human hematopoietic stem cell differentiation. Cell Stem Cell. 19(6):808–822

Fatima N, Schooley JF Jr, Claycomb WC, Flagg TP (2012) Promoter DNA methylation regulates murine SUR1 (Abcc8) and SUR2 (Abcc9) expression in HL-1 cardiomyocytes. PLoS One. 7(7):e41533

Fernandez Rico C, Konate K, Josse E, Nargeot J, Barrere-Lemaire S, Boisguerin P (2022) Therapeutic Peptides to Treat Myocardial Ischemia-Reperfusion Injury. Front Cardiovasc Med. 9:792885

Fernandez-Sanles A, Sayols-Baixeras S, Subirana I, Senti M, Perez-Fernandez S, de Castro MM et al (2021) DNA methylation biomarkers of myocardial infarction and cardiovascular disease. Clin Epigenetics. 13(1):86

Fiorito G, Guarrera S, Valle C, Ricceri F, Russo A, Grioni S et al (2014) B-vitamins intake, DNA-methylation of one carbon metabolism and homocysteine pathway genes and myocardial infarction risk: the EPICOR study. Nutr Metab Cardiovasc Dis. 24(5):483–488

Fischle W, Wang Y, Allis CD (2003) Histone and chromatin cross-talk. Curr Opin Cell Biol. 15(2):172–183

Foreman KJ, Marquez N, Dolgert A, Fukutaki K, Fullman N, McGaughey M et al (2018) Forecasting life expectancy, years of life lost, and all-cause and cause-specific mortality for 250 causes of death: reference and alternative scenarios for 2016-40 for 195 countries and territories. Lancet. 392(10159):2052–2090

French CJ, Taatjes DJ, Sobel BE (2010) Autophagy in myocardium of murine hearts subjected to ischemia followed by reperfusion. Histochem Cell Biol. 134(5):519–526

Gallo P, Latronico MV, Gallo P, Grimaldi S, Borgia F, Todaro M et al (2008) Inhibition of class I histone deacetylase with an apicidin derivative prevents cardiac hypertrophy and failure. Cardiovasc Res. 80(3):416–424

Gandini S, Puntoni M, Heckman-Stoddard BM, Dunn BK, Ford L, DeCensi A et al (2014) Metformin and cancer risk and mortality: a systematic review and meta-analysis taking into account biases and confounders. Cancer Prev Res (Phila). 7(9):867–885

Ganesan A, Arimondo PB, Rots MG, Jeronimo C, Berdasco M (2019) The timeline of epigenetic drug discovery: from reality to dreams. Clin Epigenetics. 11(1):174

Gao M, Wang J, Wang Z, Zhang Y, Sun H, Xie X et al (2013) An altered expression of genes involved in the regulation of ion channels in atrial myocytes is correlated with the risk of atrial fibrillation in patients with heart failure. Exp Ther Med. 5(4):1239–1243

Garcia-Gimenez JL, Mena-Molla S, Beltran-Garcia J, Sanchis-Gomar F (2017b) Challenges in the analysis of epigenetic biomarkers in clinical samples. Clin Chem Lab Med. 55(10):1474–1477

Garcia-Gimenez JL, Seco-Cervera M, Tollefsbol TO, Roma-Mateo C, Peiro-Chova L, Lapunzina P et al (2017a) Epigenetic biomarkers: Current strategies and future challenges for their use in the clinical laboratory. Crit Rev Clin Lab Sci. 54(7-8):529–550

Gavril EC, Luca AC, Curpan AS, Popescu R, Resmerita I, Panzaru MC et al (2021) Wolf-hirschhorn syndrome: clinical and genetic study of 7 new cases, and mini review. Children (Basel). 8(9)

Gharipour M, Mani A, Amini Baghbahadorani M, de Souza Cardoso CK, Jahanfar S, Sarrafzadegan N et al (2021) How are epigenetic modifications related to cardiovascular disease in older adults? Int J Mol Sci. 22(18)

Gibb AA, Hill BG (2018) Metabolic coordination of physiological and pathological cardiac remodeling. Circ Res. 123(1):107–128

Gillette TG, Hill JA (2015) Readers, writers, and erasers: chromatin as the whiteboard of heart disease. Circ Res. 116(7):1245–1253

Glezeva N, Moran B, Collier P, Moravec CS, Phelan D, Donnellan E et al (2019) Targeted DNA methylation profiling of human cardiac tissue reveals novel epigenetic traits and gene deregulation across different heart failure patient subtypes. Circ Heart Fail. 12(3):e005765

Glinge C, Clauss S, Boddum K, Jabbari R, Jabbari J, Risgaard B et al (2017) Stability of Circulating Blood-Based MicroRNAs - Pre-Analytic Methodological Considerations. PLoS One. 12(2): e0167969

Gorica E, Mohammed SA, Ambrosini S, Calderone V, Costantino S, Paneni F (2022) Epi-Drugs in Heart Failure. Front Cardiovasc Med. 9:923014

Gorski PA, Jang SP, Jeong D, Lee A, Lee P, Oh JG et al (2019) Role of SIRT1 in modulating acetylation of the sarco-endoplasmic reticulum Ca(2+)-ATPase in heart failure. Circ Res. 124(9):e63–e80

Gray SG, Teh BT (2001) Histone acetylation/deacetylation and cancer: an "open" and "shut" case? Curr Mol Med. 1(4):401–429

Greenberg MVC, Bourc'his D (2019) The diverse roles of DNA methylation in mammalian development and disease. Nat Rev Mol Cell Biol. 20(10):590–607

Greer EL, Shi Y (2012) Histone methylation: a dynamic mark in health, disease and inheritance. Nat Rev Genet. 13(5):343–357

Gu J, Yin ZF, Zhang JF, Wang CQ (2020) Association between long-term prescription of metformin and the progression of heart failure with preserved ejection fraction in patients with type 2 diabetes mellitus and hypertension. Int J Cardiol. 306:140–145

Guarrera S, Fiorito G, Onland-Moret NC, Russo A, Agnoli C, Allione A et al (2015) Gene-specific DNA methylation profiles and LINE-1 hypomethylation are associated with myocardial infarction risk. Clin Epigenetics. 7:133

Guo J, Huang X, Dou L, Yan M, Shen T, Tang W et al (2022) Aging and aging-related diseases: from molecular mechanisms to interventions and treatments. Signal Transduct Target Ther. 7(1):391

Haas J, Frese KS, Park YJ, Keller A, Vogel B, Lindroth AM et al (2013) Alterations in cardiac DNA methylation in human dilated cardiomyopathy. EMBO Mol Med. 5(3):413–429

He T, Huang J, Chen L, Han G, Stanmore D, Krebs-Haupenthal J et al (2020) Cyclic AMP represses pathological MEF2 activation by myocyte-specific hypo-phosphorylation of HDAC5. J Mol Cell Cardiol. 145:88–98

He X, Liang J, Paul C, Huang W, Dutta S, Wang Y (2022) Advances in cellular reprogramming-based approaches for heart regenerative repair. Cells. 11(23)

Heather LC, Pates KM, Atherton HJ, Cole MA, Ball DR, Evans RD et al (2013) Differential translocation of the fatty acid transporter, FAT/CD36, and the glucose transporter, GLUT4, coordinates changes in cardiac substrate metabolism during ischemia and reperfusion. Circ Heart Fail. 6(5):1058–1066

Heerboth S, Lapinska K, Snyder N, Leary M, Rollinson S, Sarkar S (2014) Use of epigenetic drugs in disease: an overview. Genet Epigenet. 6:9–19

Hermans-Beijnsberger S, van Bilsen M, Schroen B (2018) Long non-coding RNAs in the failing heart and vasculature. Noncoding RNA Res. 3(3):118–130

Herr DJ, Baarine M, Aune SE, Li X, Ball LE, Lemasters JJ et al (2018) HDAC1 localizes to the mitochondria of cardiac myocytes and contributes to early cardiac reperfusion injury. J Mol Cell Cardiol. 114:309–319

Hohl M, Wagner M, Reil JC, Muller SA, Tauchnitz M, Zimmer AM et al (2013) HDAC4 controls histone methylation in response to elevated cardiac load. J Clin Invest. 123(3):1359–1370

Horvath S (2013) DNA methylation age of human tissues and cell types. Genome Biol. 14(10): R115

Houtkooper RH, Pirinen E, Auwerx J (2012) Sirtuins as regulators of metabolism and healthspan. Nat Rev Mol Cell Biol. 13(4):225–238

Hu Y, Tong Z, Huang X, Qin JJ, Lin L, Lei F et al (2022) The projections of global and regional rheumatic heart disease burden from 2020 to 2030. Front Cardiovasc Med. 9:941917

Huang CK, Kafert-Kasting S, Thum T (2020) Preclinical and clinical development of noncoding RNA therapeutics for cardiovascular disease. Circ Res. 126(5):663–678

Jaffe AE, Irizarry RA (2014) Accounting for cellular heterogeneity is critical in epigenome-wide association studies. Genome Biol. 15(2):R31

Ji X, Takahashi R, Hiura Y, Hirokawa G, Fukushima Y, Iwai N (2009) Plasma miR-208 as a biomarker of myocardial injury. Clin Chem. 55(11):1944–1949

Jo BS, Koh IU, Bae JB, Yu HY, Jeon ES, Lee HY et al (2016) Methylome analysis reveals alterations in DNA methylation in the regulatory regions of left ventricle development genes in human dilated cardiomyopathy. Genomics. 108(2):84–92

Kao YH, Chen YC, Cheng CC, Lee TI, Chen YJ, Chen SA (2010) Tumor necrosis factor-alpha decreases sarcoplasmic reticulum Ca2+-ATPase expressions via the promoter methylation in cardiomyocytes. Crit Care Med. 38(1):217–222

Kaya I, Samfors S, Levin M, Boren J, Fletcher JS (2020) Multimodal MALDI Imaging Mass Spectrometry Reveals Spatially Correlated Lipid and Protein Changes in Mouse Heart with Acute Myocardial Infarction. J Am Soc Mass Spectrom. 31(10):2133–2142

Kee HJ, Sohn IS, Nam KI, Park JE, Qian YR, Yin Z et al (2006) Inhibition of histone deacetylation blocks cardiac hypertrophy induced by angiotensin II infusion and aortic banding. Circulation. 113(1):51–59

Kennel PJ, Liao X, Saha A, Ji R, Zhang X, Castillero E et al (2019) Impairment of myocardial glutamine homeostasis induced by suppression of the amino acid carrier SLC1A5 in failing myocardium. Circ Heart Fail. 12(12):e006336

Khatibzadeh S, Farzadfar F, Oliver J, Ezzati M, Moran A (2013) Worldwide risk factors for heart failure: a systematic review and pooled analysis. Int J Cardiol. 168(2):1186–1194

Kim SY, Morales CR, Gillette TG, Hill JA (2016) Epigenetic regulation in heart failure. Curr Opin Cardiol. 31(3):255–265

Kimbrough D, Wang SH, Wright LH, Mani SK, Kasiganesan H, LaRue AC et al (2018) HDAC inhibition helps post-MI healing by modulating macrophage polarization. J Mol Cell Cardiol. 119:51–63

Koczor CA, Lee EK, Torres RA, Boyd A, Vega JD, Uppal K et al (2013) Detection of differentially methylated gene promoters in failing and nonfailing human left ventricle myocardium using computation analysis. Physiol Genomics. 45(14):597–605

Kreipke RE, Birren SJ (2015) Innervating sympathetic neurons regulate heart size and the timing of cardiomyocyte cell cycle withdrawal. J Physiol. 593(23):5057–5073

Kwon OS, Hong M, Kim TH, Hwang I, Shim J, Choi EK et al (2022) Genome-wide association study-based prediction of atrial fibrillation using artificial intelligence. Open. Heart. 9(1)

Laiosa MD, Tate ER (2015) Fetal hematopoietic stem cells are the canaries in the coal mine that portend later life immune deficiency. Endocrinology. 156(10):3458–3465

Lakatta EG, Levy D (2003a) Arterial and cardiac aging: major shareholders in cardiovascular disease enterprises: Part I: aging arteries: a "set up" for vascular disease. Circulation. 107(1): 139–146

Lakatta EG, Levy D (2003b) Arterial and cardiac aging: major shareholders in cardiovascular disease enterprises: Part II: the aging heart in health: links to heart disease. Circulation. 107(2): 346–354

Leng Y, Wu Y, Lei S, Zhou B, Qiu Z, Wang K et al (2018) Inhibition of HDAC6 activity alleviates myocardial ischemia/reperfusion injury in diabetic rats: potential role of peroxiredoxin 1 acetylation and redox regulation. Oxid Med Cell Longev. 2018:9494052

Leti F, Llaci L, Malenica I, DiStefano JK (2018) Methods for CpG methylation array profiling via bisulfite conversion. Methods Mol Biol. 1706:233–254

Li B, Feng ZH, Sun H, Zhao ZH, Yang SB, Yang P (2017) The blood genome-wide DNA methylation analysis reveals novel epigenetic changes in human heart failure. Eur Rev Med Pharmacol Sci. 21(8):1828–1836

Li C, Zhang J, Xue M, Li X, Han F, Liu X et al (2019a) SGLT2 inhibition with empagliflozin attenuates myocardial oxidative stress and fibrosis in diabetic mice heart. Cardiovasc Diabetol. 18(1):15

Li H, Wen Y, Wu S, Chen D, Luo X, Xu R et al (2019b) Epigenetic modification of enhancer of Zeste homolog 2 modulates the activation of dendritic cells in allergen immunotherapy. Int Arch Allergy Immunol. 180(2):120–127

Li P, Ge J, Li H (2020) Lysine acetyltransferases and lysine deacetylases as targets for cardiovascular disease. Nat Rev Cardiol. 17(2):96–115

Liang Q, Cai M, Zhang J, Song W, Zhu W, Xi L et al (2020) Role of muscle-specific histone methyltransferase (Smyd1) in exercise-induced cardioprotection against pathological remodeling after myocardial infarction. Int J Mol Sci. 21(19)

Liao J, He Q, Li M, Chen Y, Liu Y, Wang J (2016) LncRNA MIAT: myocardial infarction associated and more. Gene. 578(2):158–161

Liu C, Liu Y, Chen H, Yang X, Lu C, Wang L et al (2023) Myocardial injury: where inflammation and autophagy meet. Burns. Trauma. 11:tkac062

Liu X, Hu D, Zeng Z, Zhu W, Zhang N, Yu H et al (2017) SRT1720 promotes survival of aged human mesenchymal stem cells via FAIM: a pharmacological strategy to improve stem cell-based therapy for rat myocardial infarction. Cell Death Dis. 8(4):e2731

Lodrini AM, Goumans MJ (2021) Cardiomyocytes cellular phenotypes after myocardial infarction. Front Cardiovasc Med. 8:750510

Lopaschuk GD, Karwi QG, Tian R, Wende AR, Abel ED (2021) Cardiac energy metabolism in heart failure. Circ Res. 128(10):1487–1513

Luo X, Hu Y, Shen J, Liu X, Wang T, Li L et al (2022) Integrative analysis of DNA methylation and gene expression reveals key molecular signatures in acute myocardial infarction. Clin Epigenetics. 14(1):46

Magadum A (2022) Modified mRNA therapeutics for heart diseases. Int J Mol Sci. 23(24)

Malandraki-Miller S, Lopez CA, Al-Siddiqi H, Carr CA (2018) Changing metabolism in differentiating cardiac progenitor cells-can stem cells become metabolically flexible cardiomyocytes? Front Cardiovasc Med. 5:119

Marinescu MC, Lazar AL, Marta MM, Cozma A, Catana CS (2022) Non-coding RNAs: prevention, diagnosis, and treatment in myocardial ischemia-reperfusion injury. Int J Mol Sci. 23(5)

Marjot J, Kaier TE, Martin ED, Reji SS, Copeland O, Iqbal M et al (2017) Quantifying the release of biomarkers of myocardial necrosis from cardiac myocytes and intact myocardium. Clin Chem. 63(5):990–996

Matsui Y, Takagi H, Qu X, Abdellatif M, Sakoda H, Asano T et al (2007) Distinct roles of autophagy in the heart during ischemia and reperfusion: roles of AMP-activated protein kinase and Beclin 1 in mediating autophagy. Circ Res. 100(6):914–922

Matsushima S, Sadoshima J (2015) The role of sirtuins in cardiac disease. Am J Physiol Heart Circ Physiol. 309(9):H1375–H1389

McCormick J, Suleman N, Scarabelli TM, Knight RA, Latchman DS, Stephanou A (2012) STAT1 deficiency in the heart protects against myocardial infarction by enhancing autophagy. J Cell Mol Med. 16(2):386–393

McHugh D, Gil J (2018) Senescence and aging: Causes, consequences, and therapeutic avenues. J Cell Biol. 217(1):65–77

Mechanic OJ, Gavin M, Grossman SA (2023) Acute myocardial infarction. StatPearls, Treasure Island(FL)

Meder B, Haas J, Sedaghat-Hamedani F, Kayvanpour E, Frese K, Lai A et al (2017) Epigenome-wide association study identifies cardiac gene patterning and a novel class of biomarkers for heart failure. Circulation. 136(16):1528–1544

Micheletti R, Plaisance I, Abraham BJ, Sarre A, Ting CC, Alexanian M et al (2017) The long noncoding RNA Wisper controls cardiac fibrosis and remodeling. Sci Transl Med. 9(395)

Michels KB, Binder AM, Dedeurwaerder S, Epstein CB, Greally JM, Gut I et al (2013) Recommendations for the design and analysis of epigenome-wide association studies. Nat Methods. 10(10):949–955

Milazzo G, Mercatelli D, Di Muzio G, Triboli L, De Rosa P, Perini G et al (2020) Histone deacetylases (HDACs): evolution, specificity, role in transcriptional complexes, and pharmacological actionability. Genes (Basel). 11(5)

Miller JL, Grant PA (2013) The role of DNA methylation and histone modifications in transcriptional regulation in humans. Subcell Biochem. 61:289–317

Montgomery RL, Davis CA, Potthoff MJ, Haberland M, Fielitz J, Qi X et al (2007) Histone deacetylases 1 and 2 redundantly regulate cardiac morphogenesis, growth, and contractility. Genes Dev. 21(14):1790–1802

Morris MJ, Monteggia LM (2013) Unique functional roles for class I and class II histone deacetylases in central nervous system development and function. Int J Dev Neurosci. 31(6):370–381

Movassagh M, Choy MK, Goddard M, Bennett MR, Down TA, Foo RS (2010) Differential DNA methylation correlates with differential expression of angiogenic factors in human heart failure. PLoS One. 5(1):e8564

Naito H, Nojima T, Fujisaki N, Tsukahara K, Yamamoto H, Yamada T et al (2020) Therapeutic strategies for ischemia reperfusion injury in emergency medicine. Acute Med Surg. 7(1):e501

Nakatochi M, Ichihara S, Yamamoto K, Naruse K, Yokota S, Asano H et al (2017) Epigenome-wide association of myocardial infarction with DNA methylation sites at loci related to cardiovascular disease. Clin Epigenetics. 9:54

Napoli C, Grimaldi V, De Pascale MR, Sommese L, Infante T, Soricelli A (2016) Novel epigenetic-based therapies useful in cardiovascular medicine. World J Cardiol. 8(2):211–219

Ng SM, Neubauer S, Rider OJ (2023) Myocardial metabolism in heart failure. Curr Heart Fail Rep. 20(1):63–75

Nishiyama A, Nakanishi M (2021) Navigating the DNA methylation landscape of cancer. Trends Genet. 37(11):1012–1027

Nothjunge S, Nuhrenberg TG, Gruning BA, Doppler SA, Preissl S, Schwaderer M et al (2017) DNA methylation signatures follow preformed chromatin compartments in cardiac myocytes. Nat Commun. 8(1):1667

Nural-Guvener H, Zakharova L, Feehery L, Sljukic S, Gaballa M (2015) Anti-Fibrotic Effects of Class I HDAC Inhibitor, Mocetinostat Is Associated with IL-6/Stat3 Signaling in Ischemic Heart Failure. Int J Mol Sci. 16(5):11482–11499

Nural-Guvener HF, Zakharova L, Nimlos J, Popovic S, Mastroeni D, Gaballa MA (2014) HDAC class I inhibitor, Mocetinostat, reverses cardiac fibrosis in heart failure and diminishes CD90+ cardiac myofibroblast activation. Fibrogenesis Tissue Repair. 7:10

Ock S, Lee WS, Ahn J, Kim HM, Kang H, Kim HS et al (2016) Deletion of IGF-1 receptors in cardiomyocytes attenuates cardiac aging in male mice. Endocrinology. 157(1):336–345

Oka T, Xu J, Molkentin JD (2007) Re-employment of developmental transcription factors in adult heart disease. Semin Cell Dev Biol. 18(1):117–131

Olivieri F, Antonicelli R, Lorenzi M, D'Alessandra Y, Lazzarini R, Santini G et al (2013) Diagnostic potential of circulating miR-499-5p in elderly patients with acute non ST-elevation myocardial infarction. Int J Cardiol. 167(2):531–536

Ooi JY, Tuano NK, Rafehi H, Gao XM, Ziemann M, Du XJ et al (2015) HDAC inhibition attenuates cardiac hypertrophy by acetylation and deacetylation of target genes. Epigenetics. 10(5):418–430

Pan MH, Lai CS, Wu JC, Ho CT (2013) Epigenetic and disease targets by polyphenols. Curr Pharm Des. 19(34):6156–6185

Papait R, Cattaneo P, Kunderfranco P, Greco C, Carullo P, Guffanti A et al (2013) Genome-wide analysis of histone marks identifying an epigenetic signature of promoters and enhancers underlying cardiac hypertrophy. Proc Natl Acad Sci U S A. 110(50):20164–20169

Papait R, Serio S, Condorelli G (2020) Role of the epigenome in heart failure. Physiol Rev. 100(4): 1753–1777

Pardini B, Calin GA (2019) MicroRNAs and long non-coding RNAs and their hormone-like activities in cancer. Cancers (Basel). 11(3)

Parmacek MS, Solaro RJ (2004) Biology of the troponin complex in cardiac myocytes. Prog Cardiovasc Dis. 47(3):159–176

Parra V, Eisner V, Chiong M, Criollo A, Moraga F, Garcia A et al (2008) Changes in mitochondrial dynamics during ceramide-induced cardiomyocyte early apoptosis. Cardiovasc Res. 77(2): 387–397

Parra V, Verdejo H, del Campo A, Pennanen C, Kuzmicic J, Iglewski M et al (2011) The complex interplay between mitochondrial dynamics and cardiac metabolism. J Bioenerg Biomembr. 43(1):47–51

Pepin ME, Drakos S, Ha CM, Tristani-Firouzi M, Selzman CH, Fang JC et al (2019a) DNA methylation reprograms cardiac metabolic gene expression in end-stage human heart failure. Am J Physiol Heart Circ Physiol. 317(4):H674–HH84

Pepin ME, Ha CM, Crossman DK, Litovsky SH, Varambally S, Barchue JP et al (2019b) Genome-wide DNA methylation encodes cardiac transcriptional reprogramming in human ischemic heart failure. Lab Invest. 99(3):371–386

Persad KL, Lopaschuk GD (2022) Energy metabolism on mitochondrial maturation and its effects on cardiomyocyte cell fate. Front Cell Dev Biol. 10:886393

Piquereau J, Ventura-Clapier R (2018) Maturation of cardiac energy metabolism during perinatal development. Front Physiol. 9:959

Pittas K, Vrachatis DA, Angelidis C, Tsoucala S, Giannopoulos G, Deftereos S (2018) The Role of Calcium Handling Mechanisms in Reperfusion Injury. Curr Pharm Des. 24(34):4077–4089

Ponpuak M, Mandell MA, Kimura T, Chauhan S, Cleyrat C, Deretic V (2015) Secretory autophagy. Curr Opin Cell Biol. 35:106–116

Popov SV, Mukhomedzyanov AV, Voronkov NS, Derkachev IA, Boshchenko AA, Fu F et al (2023) Regulation of autophagy of the heart in ischemia and reperfusion. Apoptosis. 28(1-2): 55–80

Pott J, Kabat AM, Maloy KJ (2018) Intestinal Epithelial Cell Autophagy Is Required to Protect against TNF-Induced Apoptosis during Chronic Colitis in Mice. Cell Host Microbe. 23(2): 191–202 e4

Povedano E, Vargas E, Montiel VR, Torrente-Rodriguez RM, Pedrero M, Barderas R et al (2018) Electrochemical affinity biosensors for fast detection of gene-specific methylations with no need for bisulfite and amplification treatments. Sci Rep. 8(1):6418

Prasher D, Greenway SC, Singh RB (2020) The impact of epigenetics on cardiovascular disease. Biochem Cell Biol. 98(1):12–22

Puchalska P, Crawford PA (2017) Multi-dimensional roles of ketone bodies in fuel metabolism, signaling, and therapeutics. Cell Metab. 25(2):262–284

Pulido M, de Pedro MA, Alvarez V, Marchena AM, Blanco-Blazquez V, Baez-Diaz C et al (2023) Transcriptome Profile Reveals Differences between Remote and Ischemic Myocardium after Acute Myocardial Infarction in a Swine Model. Biology (Basel). 12(3)

Qadir MI, Anwer F (2019) Epigenetic modification related to acetylation of histone and methylation of DNA as a key player in immunological disorders. Crit Rev Eukaryot Gene Expr. 29(1):1–15

Qu X, Du Y, Shu Y, Gao M, Sun F, Luo S et al (2017) MIAT Is a pro-fibrotic long non-coding RNA governing cardiac fibrosis in post-infarct myocardium. Sci Rep. 7:42657

Ramirez Bosca A, Soler A, Carrion-Gutierrez MA, Pamies Mira D, Pardo Zapata J, Diaz-Alperi J et al (2000) An hydroalcoholic extract of Curcuma longa lowers the abnormally high values of human-plasma fibrinogen. Mech Ageing Dev. 114(3):207–210

Rask-Andersen M, Martinsson D, Ahsan M, Enroth S, Ek WE, Gyllensten U et al (2016) Epigenome-wide association study reveals differential DNA methylation in individuals with a history of myocardial infarction. Hum Mol Genet. 25(21):4739–4748

Rasool M, Malik A, Naseer MI, Manan A, Ansari S, Begum I et al (2015) The role of epigenetics in personalized medicine: challenges and opportunities. BMC Med Genomics. 8 Suppl 1(Suppl 1):S5

Rau CD, Vondriska TM (2017) DNA methylation and human heart failure: mechanisms or prognostics. Circulation. 136(16):1545–1547

Ray KK, Colhoun HM, Szarek M, Baccara-Dinet M, Bhatt DL, Bittner VA et al (2019) Effects of alirocumab on cardiovascular and metabolic outcomes after acute coronary syndrome in patients with or without diabetes: a prespecified analysis of the ODYSSEY OUTCOMES randomised controlled trial. Lancet Diabetes Endocrinol. 7(8):618–628

Roth GA, Mensah GA, Johnson CO, Addolorato G, Ammirati E, Baddour LM et al (2020) Global burden of cardiovascular diseases and risk factors, 1990-2019: update from the GBD 2019 study. J Am Coll Cardiol. 76(25):2982–3021

Russell RR 3rd, Li J, Coven DL, Pypaert M, Zechner C, Palmeri M et al (2004) AMP-activated protein kinase mediates ischemic glucose uptake and prevents postischemic cardiac dysfunction, apoptosis, and injury. J Clin Invest. 114(4):495–503

Sarno F, Benincasa G, List M, Barabasi AL, Baumbach J, Ciardiello F et al (2021) Clinical epigenetics settings for cancer and cardiovascular diseases: real-life applications of network medicine at the bedside. Clin Epigenetics. 13(1):66

Savji N, Rockman CB, Skolnick AH, Guo Y, Adelman MA, Riles T et al (2013) Association between advanced age and vascular disease in different arterial territories: a population database of over 3.6 million subjects. J Am Coll Cardiol. 61(16):1736–1743

Sawan C, Herceg Z (2010) Histone modifications and cancer. Adv Genet. 70:57–85

Schiano C, Vietri MT, Grimaldi V, Picascia A, De Pascale MR, Napoli C (2015) Epigenetic-related therapeutic challenges in cardiovascular disease. Trends Pharmacol Sci. 36(4):226–235

Schubeler D (2015) Function and information content of DNA methylation. Nature. 517(7534):321–326

Schwartz GG, Steg PG, Szarek M, Bhatt DL, Bittner VA, Diaz R et al (2018) Alirocumab and cardiovascular outcomes after acute coronary syndrome. N Engl J Med. 379(22):2097–2107

Schwenty-Lara J, Nehl D, Borchers A (2020) The histone methyltransferase KMT2D, mutated in Kabuki syndrome patients, is required for neural crest cell formation and migration. Hum Mol Genet. 29(2):305–319

Scisciola L, Taktaz F, Fontanella RA, Pesapane A, Surina CV et al (2023) Targeting high glucose-induced epigenetic modifications at cardiac level: the role of SGLT2 and SGLT2 inhibitors. Cardiovasc Diabetol. 22(1):24

Seto E, Yoshida M (2014) Erasers of histone acetylation: the histone deacetylase enzymes. Cold Spring Harb Perspect Biol. 6(4):a018713

Shi T, Gao G, Cao Y (2016) Long noncoding RNAs as novel biomarkers have a promising future in cancer diagnostics. Dis Markers. 2016:9085195

Shi Y, Zhang H, Huang S, Yin L, Wang F, Luo P et al (2022) Epigenetic regulation in cardiovascular disease: mechanisms and advances in clinical trials. Signal Transduct Target Ther. 7(1): 200

Soler-Botija C, Galvez-Monton C, Bayes-Genis A (2019) Epigenetic biomarkers in cardiovascular diseases. Front Genet. 10:950

Spallarossa P, Altieri P, Aloi C, Garibaldi S, Barisione C, Ghigliotti G et al (2009) Doxorubicin induces senescence or apoptosis in rat neonatal cardiomyocytes by regulating the expression levels of the telomere binding factors 1 and 2. Am J Physiol Heart Circ Physiol. 297(6):H2169–H2181

Stein AB, Jones TA, Herron TJ, Patel SR, Day SM, Noujaim SF et al (2011) Loss of H3K4 methylation destabilizes gene expression patterns and physiological functions in adult murine cardiomyocytes. J Clin Invest. 121(7):2641–2650

Stenzig J, Schneeberger Y, Loser A, Peters BS, Schaefer A, Zhao RR et al (2018) Pharmacological inhibition of DNA methylation attenuates pressure overload-induced cardiac hypertrophy in rats. J Mol Cell Cardiol. 120:53–63

Tang X, Li PH, Chen HZ (2020) Cardiomyocyte senescence and cellular communications within myocardial microenvironments. Front Endocrinol (Lausanne). 11:280

Tanner KG, Landry J, Sternglanz R, Denu JM (2000) Silent information regulator 2 family of NAD-dependent histone/protein deacetylases generates a unique product, 1-O-acetyl-ADP-ribose. Proc Natl Acad Sci U S A. 97(26):14178–14182

Tao H, Shi KH, Yang JJ, Li J (2016) Epigenetic mechanisms in atrial fibrillation: New insights and future directions. Trends Cardiovasc Med. 26(4):306–318

Tao H, Yang JJ, Chen ZW, Xu SS, Zhou X, Zhan HY et al (2014) DNMT3A silencing RASSF1A promotes cardiac fibrosis through upregulation of ERK1/2. Toxicology. 323:42–50

Torella D, Rota M, Nurzynska D, Musso E, Monsen A, Shiraishi I et al (2004) Cardiac stem cell and myocyte aging, heart failure, and insulin-like growth factor-1 overexpression. Circ Res. 94(4): 514–524

Trifunovic-Zamaklar D, Jovanovic I, Vratonjic J, Petrovic O, Paunovic I, Tesic M et al (2022) The basic heart anatomy and physiology from the cardiologist's perspective: toward a better understanding of left ventricular mechanics, systolic, and diastolic function. J Clin Ultrasound. 50(8): 1026–1040

Tsuchiya M, Ogawa H, Koujin T, Mori C, Osakada H, Kobayashi S et al (2018) p62/SQSTM1 promotes rapid ubiquitin conjugation to target proteins after endosome rupture during xenophagy. FEBS Open Bio. 8(3):470–480

Ultimo S, Zauli G, Martelli AM, Vitale M, McCubrey JA, Capitani S et al (2018) Cardiovascular disease-related miRNAs expression: potential role as biomarkers and effects of training exercise. Oncotarget. 9(24):17238–17254

Volpe M, Gallo G, Rubattu S (2023) Endocrine functions of the heart: from bench to bedside. Eur Heart J. 44(8):643–655

Wang GK, Zhu JQ, Zhang JT, Li Q, Li Y, He J et al (2010) Circulating microRNA: a novel potential biomarker for early diagnosis of acute myocardial infarction in humans. Eur Heart J. 31(6): 659–666

Wang K, Li Y, Qiang T, Chen J, Wang X (2021a) Role of epigenetic regulation in myocardial ischemia/reperfusion injury. Pharmacol Res. 170:105743

Wang K, Long B, Liu F, Wang JX, Liu CY, Zhao B et al (2016) A circular RNA protects the heart from pathological hypertrophy and heart failure by targeting miR-223. Eur Heart J. 37(33): 2602–2611

Wang P, Shen C, Diao L, Yang Z, Fan F, Wang C et al (2015) Aberrant hypermethylation of aldehyde dehydrogenase 2 promoter upstream sequence in rats with experimental myocardial infarction. Biomed Res Int. 2015:503692

Wang S, Li LL, Deng W, Jiang M (2021c) CircRNA MFACR Is upregulated in myocardial infarction and downregulates miR-125b to promote cardiomyocyte apoptosis induced by hypoxia. J Cardiovasc Pharmacol. 78(6):802–808

Wang XY, Zhang F, Zhang C, Zheng LR, Yang J (2020) The biomarkers for acute myocardial infarction and heart failure. Biomed Res Int. 2020:2018035

Wang Y, Mu Y, Zhou X, Ji H, Gao X, Cai WW et al (2017) SIRT2-mediated FOXO3a deacetylation drives its nuclear translocation triggering FasL-induced cell apoptosis during renal ischemia reperfusion. Apoptosis. 22(4):519–530

Wang YJ, Paneni F, Stein S, Matter CM (2021b) Modulating sirtuin biology and nicotinamide adenine diphosphate metabolism in cardiovascular disease-from bench to bedside. Front Physiol. 12:755060

Warren JS, Tracy CM, Miller MR, Makaju A, Szulik MW, Oka SI et al (2018) Histone methyltransferase Smyd1 regulates mitochondrial energetics in the heart. Proc Natl Acad Sci U S A. 115(33):E7871–E7E80

Wei Y, Zhou F, Zhou H, Huang J, Yu D, Wu G (2018) Endothelial progenitor cells contribute to neovascularization of non-small cell lung cancer via histone deacetylase 7-mediated cytoskeleton regulation and angiogenic genes transcription. Int J Cancer. 143(3):657–667

Wenzel JP, Nikorowitsch J, Bei der Kellen R, Magnussen C, Bonin-Schnabel R, Westermann D et al (2022) Heart failure in the general population and impact of the 2021 European Society of Cardiology Heart Failure Guidelines. ESC. Heart Fail. 9(4):2157–2169

Westerman K, Sebastiani P, Jacques P, Liu S, DeMeo D, Ordovas JM (2019) DNA methylation modules associate with incident cardiovascular disease and cumulative risk factor exposure. Clin Epigenetics. 11(1):142

Wu G, Zhang X, Gao F (2021) The epigenetic landscape of exercise in cardiac health and disease. J Sport Health Sci. 10(6):648–659

Wu QJ, Zhang TN, Chen HH, Yu XF, Lv JL, Liu YY et al (2022) The sirtuin family in health and disease. Signal Transduct Target Ther. 7(1):402

Wu YL, Lin ZJ, Li CC, Lin X, Shan SK, Guo B et al (2023) Epigenetic regulation in metabolic diseases: mechanisms and advances in clinical study. Signal Transduct Target Ther. 8(1):98

Xiao D, Dasgupta C, Chen M, Zhang K, Buchholz J, Xu Z et al (2014) Inhibition of DNA methylation reverses norepinephrine-induced cardiac hypertrophy in rats. Cardiovasc Res. 101(3):373–382

Xie L, Zhang Q, Mao J, Zhang J, Li L (2021) The roles of lncRNA in myocardial infarction: molecular mechanisms, diagnosis biomarkers, and therapeutic perspectives. Front Cell Dev Biol. 9:680713

Yamac AH, Uysal O, Ismailoglu Z, Erturk M, Celikten M, Bacaksiz A et al (2019) Premature myocardial infarction: genetic variations in sirt1 affect disease susceptibility. Cardiol Res Pract. 2019:8921806

Yan L, Vatner DE, Kim SJ, Ge H, Masurekar M, Massover WH et al (2005) Autophagy in chronically ischemic myocardium. Proc Natl Acad Sci U S A. 102(39):13807–13812

Yang Y, Luan Y, Yuan RX, Luan Y (2021) Histone methylation related therapeutic challenge in cardiovascular diseases. Front Cardiovasc Med. 8:710053

Yap J, Irei J, Lozano-Gerona J, Vanapruks S, Bishop T, Boisvert WA (2023) Macrophages in cardiac remodelling after myocardial infarction. Nat Rev Cardiol.

Yu L, Lu M, Wang P, Chen X (2012) Trichostatin A ameliorates myocardial ischemia/reperfusion injury through inhibition of endoplasmic reticulum stress-induced apoptosis. Arch Med Res. 43(3):190–196

Zhang L, Chen B, Zhao Y, Dubielecka PM, Wei L, Qin GJ et al (2012a) Inhibition of histone deacetylase-induced myocardial repair is mediated by c-kit in infarcted hearts. J Biol Chem. 287(47):39338–39348

Zhang L, Qin X, Zhao Y, Fast L, Zhuang S, Liu P et al (2012b) Inhibition of histone deacetylases preserves myocardial performance and prevents cardiac remodeling through stimulation of endogenous angiomyogenesis. J Pharmacol Exp Ther. 341(1):285–293

Zhang QJ, Chen HZ, Wang L, Liu DP, Hill JA, Liu ZP (2011) The histone trimethyllysine demethylase JMJD2A promotes cardiac hypertrophy in response to hypertrophic stimuli in mice. J Clin Invest. 121(6):2447–2456

Zhang QJ, Liu ZP (2015) Histone methylations in heart development, congenital and adult heart diseases. Epigenomics. 7(2):321–330

Zhang X, Evans TD, Jeong SJ, Razani B (2018) Classical and alternative roles for autophagy in lipid metabolism. Curr Opin Lipidol. 29(3):203–211

Zhao D, Liu K, Wang J, Shao H (2023) Syringin exerts anti-inflammatory and antioxidant effects by regulating SIRT1 signaling in rat and cell models of acute myocardial infarction. Immun Inflamm Dis. 11(2):e775

Zhao K, Mao Y, Li Y, Yang C, Wang K, Zhang J (2022a) The roles and mechanisms of epigenetic regulation in pathological myocardial remodeling. Front Cardiovasc Med. 9:952949

Zhao Y, Du D, Chen S, Chen Z, Zhao J (2022b) New insights into the functions of microRNAs in cardiac fibrosis: from mechanisms to therapeutic strategies. Genes (Basel). 13(8)

Zhuang L, Wang Y, Chen Z, Li Z, Wang Z, Jia K et al (2022) Global characteristics and dynamics of single immune cells after myocardial infarction. J Am Heart Assoc. 11(24):e027228

Chapter 10
Epigenetic Alterations in Colorectal Cancer

Brian Ko, Marina Hanna, Ming Yu, and William M. Grady

Abstract Colorectal cancer is a leading cause of cancer related deaths worldwide. One of the hallmarks of cancer and a fundamental trait of virtually all gastrointestinal cancers is altered genomic and epigenomic DNA. The genetic and epigenetic alterations drive the initiation and progression of the cancers by altering the molecular and cell biological process of the colon epithelial cells. These alterations, as well as other host and microenvironment factors, ultimately mediate the initiation and progression of cancers, including colorectal cancer. Epigenetic alterations, which include changes affecting DNA methylation, histone modifications, chromatin structure, and noncoding RNA expression, have been revealed to be a major class of molecular alteration in colon polyps and colorectal cancer over last 30 year. The classes of epigenetic alterations, their status in colorectal polyps and cancer, their effects on neoplasm biology, and their application to clinical care will be discussed.

Keywords DNA methylation · Chromatin · microRNA · Colorectal cancer · Biomarkers

---

B. Ko · M. Hanna
Department of Medicine, University of Washington, School of Medicine, Seattle, WA, USA

M. Yu
Translational Science and Therapeutics Division, Fred Hutchinson Cancer Center, Seattle, WA, USA

Public Health Sciences Division, Fred Hutchinson Cancer Center, Seattle, WA, USA

W. M. Grady (✉)
Department of Medicine, University of Washington, School of Medicine, Seattle, WA, USA

Translational Science and Therapeutics Division, Fred Hutchinson Cancer Center, Seattle, WA, USA

Public Health Sciences Division, Fred Hutchinson Cancer Center, Seattle, WA, USA
e-mail: wgrady@fredhutch.org

© The Author(s), under exclusive license to Springer Nature Switzerland AG 2023
R. Kalkan (ed.), *Cancer Epigenetics*, Epigenetics and Human Health 11, https://doi.org/10.1007/978-3-031-42365-9_10

Abbreviations

CRC	Colorectal cancer
SSL	Serrated sessile lesions
miRNAs	microRNA
MSI	Microsatellite instability
DNMTs	DNA methyltransferases
MBD	Methyl-CpG-binding domain
CIMP	CpG island methylator phenotype
TET	Ten-eleven translocation
IDH1	Isocitrate dehydrogenase
SSP	Sessile serrated polyps
FIT	Fecal immunochemical test
SWI/SNF	Switching defective/sucrosenon-fermenting complex
PR C2	Polycomb repressive complex 2
Ago-2	Argonaute-2
DMA	Direct miRNA analysis

10.1 Introduction

Colorectal cancer (CRC) is the third most common cancer in the world and a major cause of cancer related deaths, resulting in approximately 1,000,000 deaths worldwide per year (Sung et al. 2021). Colorectal cancer arises from the colon epithelium when normal epithelial cells evolve into benign, premalignant neoplasms, which include tubular adenomas and serrated polyps (aka serrated adenomas, serrated sessile lesions (SSL), that can then transform into CRC (Bettington et al. 2018; Komor et al. 2018; Bettington et al. 2013). The premalignant neoplasms, which are called polyps, evolve into CRC through a histologic process that involves the progressive acquisition of malignant traits, which is called the polyp to cancer progression sequence and typically occurs over 10–15 years (Kuipers et al. 2015).

While genetic alterations were the first and most clearly demonstrated mechanism of cancer formation, epigenetic alterations were later recognized to also play a prominent role in the pathogenesis of cancer. Epigenetics was first described by the developmental biologist Conrad H. Waddington in 1942 as the "study of heritable changes in gene expression mediated by mechanisms other than alterations in primary nucleotide sequence of a gene" (Bird 2002; You and Jones 2012). These epigenetic events have since been described and include biochemical modifications to the nucleotides in DNA, posttranslational histone modifications, nucleosome positioning, chromatin structure regulation, and microRNA expression and function. Epigenetic alterations frequently found in cancer include aberrant DNA methylation, abnormal histone modifications, and altered expression levels of various noncoding

RNAs, including microRNAs (miRNAs) (Baylin and Herman 2000; Liu et al. 2018; Tan and Davey 2011; Wolin and Maquat 2019; Calin and Croce 2006).

As our understanding of cancer epigenetics deepens, it is thought that epigenetic alterations in colorectal cancer (CRC) occur early and are more common than genetic aberrations. In addition, advances in genomic and epigenetic analysis technologies have led to the identification of specific epigenetic alterations that show potential as clinical biomarkers. This chapter will review epigenetic alterations in CRC and detail clinical applications of various epigenetic alterations as biomarkers for early detection, diagnosis, prognosis, and management of CRC patients.

10.2 Aberrant DNA Methylation

The pathogenesis of CRC at the molecular level involves discrete pathways, including chromosomal instability, microsatellite instability (MSI), epigenetic instability altered tumor microenvironments and altered metabolic states. Of these, DNA methylation is the most widely studied and best understood epigenetic state that is altered in colorectal polyps and CRC (Komor et al. 2018; Grady and Carethers 2008; Carethers and Jung 2015).

DNA methylation affects a multitude of processes in cells, such as maintenance of genome integrity, transcriptional regulation, and developmental processes (Kanwal and Gupta 2010; Holm et al. 2005). In eukaryotic cells, DNA methylation occurs at the 5' position of the cytosine ring within CpG dinucleotides and regulates transcription via methylation of promoters and noncoding DNA elements, such as enhancers. The methylation is catalyzed by a family of DNA methyltransferases (DNMTs). DNMT1 is a DNA methylation maintenance enzyme that is constitutively active in all adult replicating tissues and methylates cytosine at the 5' position in CG dinucleotides. DNMT3A and DNMT3B are de novo enzymes that initiate methylation in unmethylated CpGs and are highly active during embryogenesis and only minimally expressed later in mature tissues (Kanwal and Gupta 2010).

CpG dinucleotides, where DNA methylation occurs, are irregularly scattered across the human genome and are disproportionately concentrated in areas called CpG islands (Toyota et al. 1999; Weisenberger et al. 2006). It is estimated that the human genome contains approximately 29,000 CpG island sequences, and approximately 50%–60% of gene promoters lie within CpG islands (Kim et al. 2008). CpG islands, particularly those associated with promoters, are commonly unmethylated in normal eukaryotic cells. Approximately 6% of CpG islands become methylated during differentiation into various tissues and early development (Portela and Esteller 2010). DNA methylation can downregulate gene expression directly by inhibiting the binding of specific transcription factors and indirectly by recruiting methyl-CpG-binding domain (MBD) proteins (Leighton and Williams 2020). The role of MBD family members in recruiting histone-modifying and chromatin-remodeling complexes to methylated sites will be explained in detail in a later section.

10.2.1 DNA Methylation Alterations in Colorectal Cancer (CRC)

10.2.1.1 Hypermethylation

DNA hypermethylation is a common feature of CRC DNA and often involves CpG islands. There are typically 1000s of CpGs that become hypermethylated in the vast majority of CRCs. Hypermethylation of CpGs in regulatory regions of the DNA (e.g., promoters, enhancers) can alter the expression of the gene(s) under control of these regions. It is a mechanism for transcriptional repression of tumor suppressor genes in CRC.

Although essentially all colon polyps and CRCs have hypermethylated DNA, there is a class of CRCs characterized by a very high frequency of DNA hypermethylation that is referred to as having a "CpG island methylator phenotype (or CIMP)." CIMP was first introduced in 1999 and was characterized by having an exceptionally high frequency of hypermethylated CpG dinucleotides (Toyota et al. 1999). Weisenberger and colleagues later introduced the prevailing method used to identify CIMP in CRC based on the methylation of five genes: *CACNA1G*, *IGF2*, *NEUROG1*, *RUNX3*, and *SOCS1* (Weisenberger et al. 2006). CIMP-positive tumors exhibit unique clinicopathological and molecular features, including a predilection for proximal location of the colon, female sex, poor and mucinous histology, the presence of frequent *KRAS* and *BRAF* mutations, and frequent microsatellite instability (MSI) due to biallelic *MLH1* methylation (Leighton and Williams 2020, Toyota et al. 2000, Rijnsoever van et al. 2002, Hawkins et al. 2002, Samowitz et al. 2005).

Microsatellite instability (MSI) is a form of genomic DNA instability that occurs in a majority of CRCs because of DNA hypermethylation. MSI generally arises from inactivation of the DNA mismatch repair system, leading to an accrual of DNA replication errors in repetitive microsatellite sequences. Some of these errors are located in the exons of potential tumor suppressor genes. The inactivation of the mismatch repair system can originate from both genetic and epigenetic changes. Mutations in the genes, *MLH1*, *MSH2*, *MSH6*, and *PMS2*, account for 20% of MSI CRCs, while DNA hypermethylation constitutes approximately 80% of MSI CRCs. While MSI CRCs acocount for around 12–15% of all CRCs, they are involved in >90% of familial Lynch syndrome CRCs (Popat et al. 2005, Ward et al. 2001). Around 80% of MSI CRCs are secondary to epigenetic changes from biallelic hypermethylated *MLH1*, and approximately 10–12% are secondary to sporadic mutations (Hampel et al. 2005; Hampel et al. 2008).

10.2.1.2 Hypomethylation

Similar to DNA hypermethylation, DNA hypomethylation is another ubiquitous epigenetic alteration seen in CRC. DNA hypomethylation is hypothesized to promote carcinogenesis via chromosomal instability and loss of genomic imprinting (Holm et al. 2005; Chen et al. 1998; Gaudet et al. 2003; Suzuki et al. 2006). Hypomethylation is commonly observed in repetitive transposable elements, such as the LINE-1 or short interspersed nucleotide element (short interspersed transposable element or Alu) sequences (Yamamoto et al. 2008, Ogino et al. 2008a, b, Goel et al. 2010). Hypomethylation of LINE-1 is inversely proportional to MSI and/or CIMP. Furthermore, studies have shown the association of DNA hypomethylation with poor patient outcomes (Estécio et al. 2007, Ogino et al. 2008a, b, Antelo et al. 2012, Ahn et al. 2011, Rhee et al. 2012). A proposed rationale of this correlation is that LINE-1 hypomethylation increases the activity of proto-oncogenes, suggesting a functional role in CRC progression (Hur et al. 2014).

Although previously it was thought that hypomethylation resulted from passive DNA replication errors, recent studies have demonstrated an active process may also play a role. Two biochemically similar enzymes called ten-eleven translocation (TET) and isocitrate dehydrogenase (IDH1) regulate demethylation of DNA and are found to be mutated in certain cancers like leukemia and gliomas (Cairns and Mak 2013, Kriaucionis and Heintz 2009, Tahiliani et al. 2009). The mechanism by which TET can regulate demethylation first arises from the formation of 5-hydroxymethylcytosine (5-hmC) from 5-methylcytosine that is subsequently replaced with unmethylated cytosine by base excision repair proteins. As it pertains to CRC, studies have shown under-expression of *TET1* in the early phases of CRC, and loss of *TET1* expression was correlated with inhibition of the WNT signaling pathway and suppression of tumor proliferation (Li and Liu 2011, Neri et al. 2015). Despite the grounds of these biochemical associations, the clinical significance of 5-hmC and TET proteins in CRC is unclear, specifically with regard to hypomethylation in CRC. Additionally, mutations in *TET1*, *TET2*, or *IDH1* are not a prevailing cause of epigenetic alterations in CRC (Stachler et al. 2015).

10.2.2 DNA Methylation and the Pathogenesis of Colorectal Cancer: "Traditional" and "Serrated" Polyp Pathways

There are multiple pathways for the formation of colon polyps and CRC. Fearon and Vogelstein originally described a stepwise normal-to adenoma to cancer progression that considers adenomatous polyps as the principal preneoplastic lesions leading to CRC (Fearon and Vogelstein 1990, Vogelstein et al. 1988). The transition from normal mucosa to adenomatous polyp is characterized by both genetic and epigenetic alterations. This traditional adenoma to CRC pathway arises from alterations in the WNT signaling pathway, most commonly *APC* mutations, with subsequent mutations in genes involved in the MAPK and TP53 pathways (e.g., *KRAS* and *TP53* mutations, respectively). In addition to these genetic changes, epigenetic modifications include methylation of various genes, such as *SLC5A8*, *ITGA4*,

SFRP2, PTCH1, CDKN2A, HLTF, and *MGMT* (Li et al. 2003; Qi et al. 2006; Moinova et al. 2002; Peng et al. 2013; Esteller et al. 2000; Chan et al. 2002).

The "serrated polyp pathway" is another route of CRC development that was originally labeled as an alternative pathway to the traditional stepwise adenoma to CRC progression pathway because of the unique histological and morphological characteristics of the sessile serrated polyps that distinguish them from tubular adenomas. The serrated polyp to CRC pathway is primarily characterized by activation of the MAPK pathway by oncogenic mutations in *BRAF* and *KRAS*. Whereas the adenomatous polyp is the only precursor lesion in the traditional pathway, the serrated polyp pathway includes at least two or three precursor lesions: hyperplastic polyps, serrated polyps, and serrated adenomas (Crockett and Nagtegaal 2019). The malignant potential of these precursor lesions varies significantly 2009. Serrated adenomas classically activate the MAPK pathway via *KRAS* mutations, carry *RSPO* fusion transcripts, and typically have a low CIMP status. Serrated polyps that carry *BRAF* mutations are also commonly CIMP and can progress to microsatellite stable or unstable CRC, depending on whether epigenetic inactivation of the mismatch repair protein MLH1 occurs. In contrast to the traditional pathway, the serrated pathway is not typically characterized by genetic alterations in *APC* or *CTNNB1* and appears to activate WNT signaling late in the polyp to CRC sequence (Laiho et al. 2007; Bettington et al. 2013). Sessile serrated polyps (SSP) also more commonly evolve through epigenetic alterations of various genes belonging to the beta-catenin/WNT pathway (*SFRP* family, *CDX2*, *MCC*), p53 signaling pathway, and the DNA mismatch repair (*MLH1*) family (Dhir et al. 2011; Kohonen-Corish et al. 2007; Suzuki et al. 2010; Kriegl et al. 2011). In both traditional and serrated polyp pathways, epigenetic alterations appear to arise in the earlier phases during polyp formation unlike genetic alterations, which occur predominantly after polyp initiation.

10.2.3 Clinical Applications of DNA methylation in Colorectal Cancer

The advancement in our understanding of the molecular features of malignancies has allowed successful utilization of epigenetic modifications in the prevention and management of a variety of cancers, including GI malignancies. In particular, DNA methylation has been the most ubiquitous epigenetic alteration that these epigenetic biomarkers for early detection are based upon because epigenetic changes occur much more frequently than genetic mutations in polyps (Myint et al. 2018). The earliest application of this class of biomarkers more than two decades ago was the use of methylated *MLH1* to ascertain the likelihood of MSI CRC being sporadic or hereditary in origin (Pritchard and Grady 2011). Also, of note, the current clinically approved screening assays for CRC, the ColoGuard assay (Exact Sciences) and EpiProcolon (Epigenomics), are based on DNA methylation-based biomarkers.

10 Epigenetic Alterations in Colorectal Cancer

Table 10.1 Validated DNA methylation biomarkers for colorectal cancer

Clinical use	Biomarkers	Commercial assays	Evidence
Stool-based CRC screening	mVIM	ColoSure™	Case (N=42) control (N=241) study
	mBMP3 and mNDRG4	Cologuard® (detects mutant KRAS and includes a FIT test)	Prospective cohort based clinical trial in screening population (N=9989)
Blood-based diagnostic marker	mSEPT9	EpiproColon® 1.0; ColoVantage® ; RealTime mS9	Multiple trials: (1) prospective cohort based clinical trial in screening population (N=7941) (Church), (2) case-control study (N=269) (deVos), (3) case-control study (N=312) (Lofton-Day, 2008)
	mBCAT1 mIKZF1	Colvera	Cross-sectional study (N=220)
	Cell-free DNA	Shield™	Prospective cohort study (n=12,750)
Tissue-based prognostic markers	CIMP panel	NA	Multiple trials: (1) Case-control study from two phase I/II clinical trials (N=31) (Ogino, 2007), (2) Case-control study from phase III clinical trial (N=615) (Shiovitz et al. 2014), (3) observational cohort study (N=2050)

Other methylation-based biomarkers are now available for clinical use and listed in Table 10.1.

10.2.3.1 Stool-Based Biomarkers

The initial discovery of detecting mutant *KRAS* in fecal specimens by Sidransky and colleagues has inspired numerous studies to investigate the application of methylated DNA in stool samples for screening (Sidransky et al. 1992). A large number of hypermethylated genes, including *APC, ATM, BMP3, CDKN2A, SFRP2, GATA4, GSTP1, HLTF, MLH1, MGMT, NDRG4, RASSF2A, SFRP2, TFPI2, VIM*, and *WIF1*, have been analyzed for early detection of CRC (Leung et al. 2007; Chen et al. 2005; Wang and Tang 2008; Glöckner et al. 2009; Petko et al. 2005; Lidgard et al. 2013; Lee et al. 2009). Of this large list, *VIM, BMP3*, and *NDRG4* have shown the most robust results so far and are the only hypermethylated genes approved for clinical use. Methylated *VIM* was the first stool-based biomarker under the name ColoSure (Lab Corp, Burlington, NC) used for early detection of CRC, which is no longer available (Itzkowitz et al. 2007; Itzkowitz et al. 2008).

The current clinical stool-based biomarker assay is a multitarget (MT) stool assay that detects methylated *BMP3*, methylated *NDRG4*, mutant *KRAS*, and occult hemoglobin (Cologuard). In a large clinical trial of average-risk individuals (the Deep C trial), this MT stool assay was compared with the fecal immunochemical test

(FIT) assay and with colonoscopy (n=9989). The Deep C trial demonstrated an overall sensitivity of 92% (95% confidence interval [CI], 83%–97.5%) for CRC and 93% (95% CI, 83.8%– 98.2%) for stage I–III CRC compared with sensitivity of FIT at 74% (95% CI, 61.5%–84%) and 73% (95% CI, 60.3%– 83.9%), respectively (p = .002). For advanced adenomas and sessile serrated polyps, the sensitivity of the test increased proportionately with lesion size and grade and on average was (Imperiale et al. 2014).

10.2.3.2 Blood-Based Biomarkers

Blood-based biomarkers are invariably the most convenient and offer high patient compliance. For cancer in general, the majority of plasma biomarkers have been proteins or glycoproteins (e.g., PSA, CEA, and CA-125). For CRC, circulating tumor DNA from somatic tumor-derived mutations has displayed potential in early detection of recurrent cancer, monitoring treatment response, and prognosis (Bi et al. 2020; El Messaoudi et al. 2016; Siravegna et al. 2015; Reinert et al. 2016). However, circulating tumor DNA has not reliably been detected in the earlier stages of cancer, such as advanced polyps or early-stage CRC, in large part due to the low frequency of these mutations in the plasma (6.6 ng/mL blood, <0.1%–0.01% circulating tumor DNA) (Myint et al. 2018; Reinert et al. 2019; Tie et al. 2019; Wang et al. 2019). In contrast, because epigenetic alterations are found more frequently in advanced polyps and early-stage CRC, methylated DNA and chromatin fragmentation patterns have been detected in higher proportions of patients with early-stage CRC than DNA mutations (Nikolaou et al. 2018; Barták et al. 2017).

Several potential plasma-based diagnostic methylation biomarkers have been identified: *ALX4*, *APC*, *CDKN2A*, *HLTF*, *HPP1*, *MLH1*, *MGMT*, *NEUROG1*, *NGFR*, *RASSF2A*, *SFRP2*, *VIM*, *WIF1*, *4GAT1*, *BCAT1*, *IKZF1*, *SFRP1*, *SDC2*, and *PRIMA1* (Petko et al. 2005; Lee et al. 2009; Barták et al. 2017; Ebert et al. 2006; Leung et al. 2005; Wallner et al. 2006; Herbst et al. 2011; Picardo et al. 2019). Of these, the best studied are *mSEPT9*, *mSDC2*, and a combination of *mBCAT1* and *mIKZF1* which is marketed under the name Colvera (Clinical Genomics). Methylated Septin 9 (*SEPT9*) belongs to the gene family that encodes a group of GTP-binding and filament-forming proteins involved in cytoskeletal formation (Finger 2002; Sheffield et al. 2003). Lofton-Day and colleagues first identified methylated *SEPT9* (mSEPT9) as a noninvasive diagnostic biomarker for CRC, reporting 69% sensitivity and 86% specificity. However, a subsequent prospective CRC screening trial (PRESEPT) found lower sensitivity of 48.2% with similar 91.5% specificity (Lofton-Day et al. 2008). While subsequent studies have validated its potential as a diagnostic biomarker, the FDA only recommends it in patients who refuse to undergo other CRC screening tests because of its relatively low sensitivity for CRC (52%–72%). A larger issue is its low sensitivity for the detection of advanced adenomas (11%), and while a recent study demonstrated that the methylated SEPT9 biomarker was superior to FIT, both assays are suboptimal in diagnosing patients with advanced adenomas (Jin et al. 2015).

Combination biomarkers including methylated DNA and chromatin fragmentation patterns are being assessed in ongoing CRC screening trials at this time (Circulating Cell-free Genome Atlas study, ClinicalTrials.gov number NCT02889978; GRAIL; Session VCTPLO2, CT021-Prediction of cancer and tissue of origin in individuals with suspicion of cancer using a cell-free DNA multi-cancer early detection test, and the ECLIPSE trial of the Lunar-2 assay) (Kim et al. 2019). These assays are anticipated to improve the sensitivity of the blood-based CRC screening tests, but it is not clear if they will be accurate enough to be used in clinical care.

10.2.3.3 Epigenetic Prognostic Biomarkers

While prognostication based on pathologic staging and histologic features of the tumor is the current gold standard, the heterogeneity in survival times within same stage CRC illustrates the need to discover a more accurate prognostic system. Multiple large clinical studies have investigated specific methylated DNA biomarkers to portend prognosis in CRC (Okugawa et al. 2015).

Among all epigenetic biomarker candidates, the presence of CIMP reflects the best prospects in clinical application for CRC patients. Cancers with CIMP status are associated with an overall poor prognosis (Rhee et al. 2012; Shen et al. 2007; Ogino et al. 2007; Zlobec et al. 2010). Van Rijnsoever and colleagues found in a cohort of 206 stage III CRC patients that CIMP-positive cancer was associated with poor survival (Van Rijnsoever et al. 2003). Another study independently analyzed more than 600 CRC patients and replicated these findings that CIMP was associated an unfavorable prognosis in microsatellite stable CRC patients (Lee et al. 2008). Some studies suggested that poor prognosis in CIMP-positive CRCs is from coexisting V600E BRAF mutations (Lochhead et al. 2013; Pai et al. 2019). However, MSI status likely also accounts for differences in prognosis of CIMP-positive cancers and is an important confounding factor, and these data underline the significance of MSI status on the prognosis of CIMP-positive CRCs (Juo et al. 2014).

Aberrant methylation – both hyper- and hypomethylation – of various genes have also shown potential as prognostic biomarkers. In prospective cohort studies of CRC patients, Ogino and colleagues found a correlation between *LINE-1* hypomethylation and poor prognosis (Ogino et al. 2008a, b). Other studies have not only independently validated this finding between *LINE-1* hypomethylation and CRC prognosis but also discovered other genes that are associated with an unfavorable prognosis (Antelo et al. 2012, Ahn et al. 2011, Rhee et al. 2012, Baba et al. 2010, Nilsson et al. 2013, Nagasaka et al. 2003, Draht et al. 2014, Zhang et al. 2014, Park et al. 2015). While these studies continue to show the potential use of aberrantly methylated DNA as biomarkers, further investigation is required to develop clinically reliable, standardized assays to be used in clinical care.

10.2.3.4 Epigenetic Predictive Biomarkers for Response to Treatment

Another area of active investigation within the last decade includes predictive biomarkers for CRC patients undergoing various chemotherapeutic regimens. Of a number of methylation-based DNA markers, CIMP status has been studied intensively, but the best studies have yielded conflicting results on its use for predicting response to 5-fluorouracil (5-FU) (Min et al. 2011; Jover et al. 2011). Thus, CIMP is not used clinically for directing 5-FU-based therapy at this time. More recently, prospective studies assessing CIMP as a predictive marker for adjuvant irinotecan and oxaliplatin have shown modest prognostic effects for overall survival in stage III, microsatellite stable, and CIMP-positive CRCs, with the addition of irinotecan to adjuvant 5-FU and leucovorin, and in stage III CIMP-positive CRCs treated with 5-FU, leucovorin, and oxaliplatin (FOLFOX-4) (Gallois et al. 2018; Shiovitz et al. 2014). These studies suggest the potential of CIMP as a prognostic and possibly predictive marker to treatment but emphasize the need for further studies of the association between CIMP status and therapeutic response to various chemotherapeutic regimens.

The feasibility of using individual hypermethylated genes as predictive biomarkers has also been studied. The transcription factor AP-2 Epsilon (*TFAP2E*) has been assessed as a predictive biomarker for response to 5-FU-based chemotherapy in CRC patients (Ebert et al. 2012). Furthermore, DNA methylation microarray profiling of oxaliplatin-sensitive vs. oxaliplatin-resistant CRC cell lines revealed that oxaliplatin-resistant cells exhibited hypermethylation of the *BRCA1* interacting *SRBC* gene, which was subsequently shown to be associated with poor progression-free survival in CRC cohorts treated with oxaliplatin, although the results of the initial study have not been replicated (Moutinho et al. 2014). Studies using methylated genes as predictive markers and response predictors for CRC therapy are expected to continue.

10.2.3.5 Colorectal Cancer and "Field Cancerization"

The concept of "field cancerization" (or field effect) was first proposed in 1953 by Slaughter et al. Field cancerization is characterized by the occurrence of genetic and epigenetic alterations in histologically normal-appearing tissues (Moutinho et al. 2014). It is believed to lead to an increased risk for synchronous and metachronous primary tumors. Genetic mutations are common in cancer cells but are believed to be rare in normal cells. In contrast, somatic epigenetic dysregulation occurs not only in cancer tissues but also in preneoplastic and noncancerous tissues. Thus, epigenetic changes are potentially more promising somatic CRC risk factors (i.e., field cancerization markers) than gene mutations as they contribute to the earliest events of malignant transformation.

Methylation changes in tumor suppressor genes occur more frequently in the normal colonic mucosa of CRC patients than healthy controls, suggesting they may be one of the earliest events of malignant transformation in CRC (Luo et al. 2014; Issa et al. 1994). Loss of insulin-like growth factor-II (IGF2) gene imprinting occurs at a higher frequency in normal mucosa adjacent to cancer tissue compared with normal mucosa in patients without CRC, underscoring the potential loss of IGF2 imprinting as a biomarker to identify patients at greater risk for CRC development (Cui et al. 2002). Other studies have suggested that both hypermethylation of tumor suppressor genes, such as *SFRP*, *ESR1*, *MYOD*, *EVL*, and *MGMT*, and *LINE-1* hypomethylation in normal colonic mucosa correlate with an increased risk of CRC compared to patients without these traits (Suzuki et al. 2004; Kawakami et al. 2006; Shen et al. 2005; Kamiyama et al. 2012; Grady et al. 2008).

10.2.4 Chromatin Alterations and Histone Modifications

10.2.4.1 Overview of Histone Biology

Histones are a class of small basic proteins that include H2A, H2B, H3, and H4, and an octamer of these histone proteins forms the core of a nucleosome. A nucleosome encompasses approximately 146 bp of DNA wrapped around this octamer and consists of two subunits of each of the core histone proteins with a flexible charged NH2 terminus of the histone protein called the histone tail (Gilbert 2019). Chromatin is a highly ordered B-form structure consisting of repeats of nucleosomes connected by linker DNA. Chromatin exists in two distinct conformation states. Heterochromatin is densely compacted and transcriptionally silent, whereas euchromatin is decondensed and transcriptionally active. Thus, regulation of gene expression occurs through posttranslational histone modifications that influence chromatin conformation and determine the transcriptional status of genes within a particular region of DNA (Bersaglieri and Santoro 2019). Posttranslational modifications of the histone tails occur via covalent modifications, such as acetylation, methylation, phosphorylation, ubiquitination, sumoylation, proline isomerization, and ADP ribosylation (Cohen et al. 2011, Kouzarides 2007, Bartke and Kouzarides 2011, Tessarz and Kouzarides 2014). Euchromatin is characterized by high levels of acetylation and trimethylated H3K4, H3K36, and H3K79. In contrast, heterochromatin is characterized by low levels of acetylation and high levels of H3K9, H3K27, and H4K20 methylation (Hadley et al. 2019; Bártová et al. 2008). Chromatin alterations and histone modifications are reversible and mediated by a group of enzymes that add and remove such modifications, including histone acetyltransferases and deacetylases, methyltransferases and demethylases, kinases and phosphatases, ubiquitin ligases and deubiquitinases, and SUMO ligases and proteases, respectively (Kouzarides 2007, Tessarz and Kouzarides 2014). These enzymes generally act in complexes, such as the repressive Polycomb and activating Trithorax group complexes, which counterbalance each other in the regulation of genes important for development and which have been implicated in tumorigenic transformation (Mills 2010). Furthermore, there are histone variants that provide an additional layer of regulation, including H2A.Z, MacroH2A, H2A-Bbd, H2AvD, H2A.X, H3.3, CenH3, and H3.4 (Kamakaka and Biggins 2005). Alterations in histone modification states, activity states of the histone-modifying enzymes, and levels of the histone variants are commonly observed in the majority of cancers (Okugawa et al. 2015).

10.2.4.2 Histone Modification Alterations in Colorectal Cancer and Potential for Clinical Applications

Altered histone modifications and associated chromatin alterations are commonly found in CRCs, and some studies have demonstrated the potential for histone modifications to be utilized clinically as CRC biomarkers (Fraga et al. 2005). To date, global alterations of specific histones in primary tissues have been the focus for biomarker development in CRC. Investigation of histone modifications in circulating nucleosomes has identified reduced levels of H3K9me3 and H4K20me3 as potential diagnostic biomarkers for CRC (Leszinski et al. 2012; Gezer et al. 2013). Other studies in CRC suggest that histone modifications, such as acetylation of H3 lysine 56 and di- or trimethylation of H3 lysine 9 and 27, have potential to be prognostic markers in CRC (Fraga et al. 2005, Tamagawa et al. 2012, Baylin and Jones 2011, Benard et al. 2014a, b, Benard et al. 2015, Tamagawa et al. 2013). Additionally, studies of H3K4me2, H3K9ac, and H3K9me2 alterations detected by immunohistochemical staining of CRC liver metastases suggest that low H3K4me2 expression levels correlate with poor prognosis (Tamagawa et al. 2012). However, all of these studies are proof-of-concept phase I biomarker studies. Due to technical limitations in assessing the histone modification state in primary cancer tissues, it has been challenging to develop tests based on these alterations that are sufficiently robust to be used in clinical care, and further research is needed to determine whether any of these modifications will be clinically useful.

The expression and activity of histone-associated proteins is also altered in CRC. *ARID1A*, a critical portion of the SWItch/sucrose non-fermentation (SWI/SNF) chromatin remodeling complexes, is commonly mutated or methylated in CRC (Zhao et al. 2022). Mutations in other chromatin remodeling enzymes, such as *ARID1B*, *BCL11A*, *HDAC2*, *SMARCA2*, and *SMARCB1*, are also frequently observed in CRC (Chen et al. 2016). The polycomb group proteins have been studied the most for their biomarker potential. High expression of histone lysine N-methyltransferase EZH2 and polycomb protein SUZ12, which are both members of the polycomb repressive complex 2 (PRC2), and polycomb complex protein BMI-1 in association with expression of H3K27me3 marks has been correlated with improved prognosis in patients with CRC (Benard et al. 2014a, b). Despite these observations, to date, targeted therapies for these genes have not been shown to be effective, nor have any of these been shown to be sufficiently robust biomarkers for use in the clinic.

10.2.5 microRNA and Colorectal Cancer

10.2.5.1 Overview of miRNA Biology

miRNAs are small noncoding RNAs measuring 18–25 nucleotides in length that were first discovered in 1993 (Lee et al. 1993; Wightman et al. 1993). Premature miRNAs are transcribed from DNA sequences and transported from the nucleus to the cytoplasm and subsequently processed to form mature miRNA. Mature miRNA then binds to 3'-untranslated region of target mRNAs, which is regulated by the RNA-induced silencing complex. Depending on the degree of miRNA-mRNA sequence complementarity, this interaction can either induce mRNA degradation or suppress translation (He and Hannon 2004, Mendell 2005, Vasudevan et al. 2007).

In 2002, Croce and colleagues performed the first studies of the potential role of miRNAs in cancer by illustrating downregulated expression of miR-15 and miR-16 in chronic lymphocytic leukemia (Calin et al. 2002). This discovery then led to the discovery of several other deregulated miRNAs and their association with malignant transformation by regulating the expression of oncogenes and tumor suppressor genes (Slaby et al. 2009).

10.2.5.2 Dysregulation of miRNA Expression in "Traditional" and "Serrated" Pathways

Deregulated miRNAs appear to affect both the "traditional" (normal-adenomatous polyp-cancer) and "serrated" (normal-serrated polyp-cancer) pathways. In the traditional pathway, the miR-17-92a cluster, miR-135b, miR-143, and miR-145 regulate WNT/b-catenin signaling pathway, which regulates CRC tumorigenesis (Motoyama et al. 2009; Nagel et al. 2008; Arndt et al. 2009; Oberg et al. 2011). Specific miRNAs (e.g., miR-143, let-7, miR-21, miR-31, miR-1, miR-21, and miR-143) regulate gene expression of RAS-MAPK and PI3K/AKT cascades, which are involved in the pathways that promote progression from early adenoma to CRC (Pagliuca et al. 2013, Johnson et al. 2005, Meng et al. 2007, Xu et al. 2014, Xiong et al. 2013, Josse et al. 2014, Sun et al. 2013). Furthermore, miR-34a/b/c, miR-133a, miR-143, and miR-145 regulate p53, a commonly known tumor suppressor protein frequently involved in tumorigenesis (Pagliuca et al. 2013; Vogt et al. 2011; Tazawa et al. 2007; King et al. 2011). It has also been shown that there are genes, such as *LIN28*, promote CRC development through downregulation of let-7 miRNA formation (King et al. 2011; Tu et al. 2015). In addition, miR-21, miR-155, and miR-200 family members regulate the TGF-ß pathway (Papagiannakopoulos et al. 2008; Liu et al. 2015; Gregory et al. 2011).

In addition to the traditional pathway, miRNAs have also shown to play a prominent role in the serrated pathway. This was first introduced with the discovery that expression of miR-21 and miR-181 was increased in both hyperplastic polyps and sessile serrated adenomas (Schmitz et al. 2009). Gene expression profiling of miRNA in CRCs with or without *BRAF* mutations demonstrated an upregulation of miR-31-5p expression in cancers with *BRAF* mutations that develop from serrated polyps (Nosho et al. 2014). A significant correlation between increased miR-31 expression and specific types of serrated polyps, sessile serrated adenomas (SSAs) or traditional serrated adenomas (TSAs), underlines the possible role of miRNA in the serrated pathway (Rex et al. 2012; Ito et al. 2014). A follow-up study that analyzed 381 serrated and 222 non-serrated adenomas replicated these findings and identified an association between increased miR-31 expression and *BRAF* mutations independent of CIMP status (Ito et al. 2014). Interestingly, expression of miR-31 was directly proportional with histological features of SSAs but not of TSAs. In addition, in SSA lesions, there was a progressive rise in miR-31 expression, *BRAF* mutation, CIMP-positive status, and *MLH1* methylation from the rectum to cecum (Yamauchi et al. 2012a, b).

10.2.6 miRNAs as Diagnostic Biomarkers in Colorectal Cancer

There has been rapidly expanding interest in miRNA biomarker research because of their distinctive features in structure and their stability. The small size and hairpin-loop structure protects them from RNase-mediated degradation, which renders them remarkably convenient in permitting extraction from a wide range of clinical specimens (Creemers et al. 2012). Furthermore, cell-free miRNAs are often associated with high density lipoprotein particles, apoptotic bodies, microvesicles, and exosomes, and through their binding to argonaute-2 (Ago-2), which further enahnces their stability (Arroyo et al. 2011; Turchinovich et al. 2011; Vickers et al. 2011). In addition to their remarkable stability under a myriad of conditions, miRNAs appear to be emitted by malignant cells into blood, stool, and urine (Ogata-Kawata et al. 2014, Toiyama et al. 2013).

The success of CRC screening programs using fecal immunochemical tests (FIT) and colonoscopy, which detect premalignant colon polyps and early-stage CRC, has led to intense interest in developing screening tests that address deficiencies of the currently used tests. FIT's are not esthetically acceptable to some people and have low sensitivity for colon polyps, and colonoscopy is invasive and costly. Screening tests based on cell-free miRNA biomarkers have the potential to overcome these limitations and to be noninvasive and accurate for polyp detection. There are two classes of miRNA biomarkers that have shown some promise in their potential as diagnostic biomarkers in CRC including circulating cell-free miRNAs and fecal miRNAs.

10.2.6.1 Blood-Based Biomarkers

After the finding of miRNA in plasma, a large number of studies attempted to investigate significant miRNA expression profiles in patients with CRC. Of these, Ng and colleagues conducted the first landmark miRNA expression profiling study. In this study, they evaluated miRNA expression patterns in tissue and plasma samples from patients with and without CRC. The study demonstrated significantly elevated expression of miR-92a and miR-17-3p in patients with CRC compared to healthy controls. However, miR-92a levels were also increased in other gastrointestinal pathologies, including gastric cancer and inflammatory bowel disease, which will decrease the specificity of miR-92a based tests when used for the purpose of CRC screening. Notably, miR-92a levels dropped substantially with surgical resection of the primary tumors. Other studies have corroborated these findings (Huang et al. 2010).

Another expression profiling study suggested another miRNA, miR-21, couls be a possible biomarker. Follow-up studies have reproduced the finding that there is increased expression of miR-21 in CRC and that high miR-21 levels could be used to distinguish CRC from healthy control subjects with high sensitivity (90%) and specificity (90%). Further investigations have demonstrated unique features of miR-21 to be used as a biomarker. First, differential expression of miR-21 arises early in the carcinogenesis pathway (Schetter et al. 2008). In addition to its high expression in CRC, miR-21 is highly secreted by cancer cells which can be conveniently measured either in exosomes or in the plasma (Ogata-Kawata et al. 2014; Toiyama et al. 2013; Schee et al. 2013; Kanaan et al. 2012). Interestingly, miR-21 expression also allowed differentiation of advanced adenomas, (Toiyama et al. 2013). Similar to miR-92a, levels of miR-21 dropped after curative resection of the primary tumor (Ogata-Kawata et al. 2014, Liu et al. 2013a, b).

While these results are promising, it is unlikely that single miRNA-based tests can reliably and accurately be used for CRC screening given the molecular heterogeneity of colorectal polyps and cancers. Accordingly, several studies propose combining miRNAs into a biomarker panel to improve detection accuracy (Luo et al. 2013, Kanaan et al. 2013, Wang et al. 2014a, b, Wang et al. 2015).

There are still several challenges that remain regarding the use of miRNAs as potential diagnostic CRC biomarkers. One key issue is that there are major inconsistencies in the biomarker panels these independent studies use. Once these variabilities are resolved, validation studies utilizing standardized assays in large population studies are needed to identify optimal miRNA and biomarker panels that can be reliably used in clinical care.

10.2.6.2 Stool-Based Biomarkers

Like biomarkers in the plasma and serum, stool-based biomarkers have been intensively explored in its potential as a diagnostic tool. In 2010, using a one-step miRNA extraction and amplification method defined as "direct miRNA analysis" (DMA), a stool-based study demonstrated different expressions of miR-21 and miR-106a in CRC patients (Link et al. 2010). Another study found that fecal miR-92a was expressed differently in CRC or adenoma from healthy subjects (Wu et al. 2012). A subsequent study determined that miR-106a expression in residual fecal matter in FOBT kits improved the sensitivity of CRC screening with FOBT (Koga et al. 2013).

One of the most advanced stool miRNA-based CRC screening tests is the miRFec test. The miRFec test is a gradient boosting machine-generated algorithm that includes two fecal miRNAs (miR-421 and miR-27a-3p) and fecal hemoglobin concentration, along with age and gender. It has been shown to be more accurate than FIT for the identification of patients with advanced colorectal neoplasm (i.e., colorectal cancer, advanced adenomas, or advanced serrated lesions) among individuals participating in colorectal cancer screening programs (Duran-Sanchon et al. 2020; Duran-Sanchon et al. 2021). A current clinical trial is ongoing and aims to compare effectiveness and cost-effectiveness of the miRFec test with respect to fecal immunochemical test (FIT) for the detection of advanced colorectal neoplasm among individuals participating in colorectal cancer (CRC) screening.

10.2.7 miRNAs as Prognostic Biomarkers in Colorectal Cancer

The first study that explored the use of miRNA as prognostic biomarkers was conducted by Schetter and colleagues in 2008. This study utilized a microarray-based method to measure the expression levels of various miRNAs in CRC and matched normal tissues (Schetter et al. 2008). This study classified 37 miRNAs, including miR-20a, miR-21, miR-106a, miR-181b, and miR-203, that had significantly different expressions in CRC and healthy tissues. Additionally, levels of miR-21 in CRC patients were inversely proportional to survival. In other words, elevated expressions of miR-21 conferred poor survival (Toiyama et al. 2013, Shibuya et al. 2010, Kjaer-Frifeldt et al. 2012, Kulda et al. 2010, Bovell et al. 2013, Zhang et al. 2013a, b). While several other miRNAs have shown significant variability in CRC tissue versus normal tissue, their applicability appears limited at this time, and miR-21 is widely accepted to have the most promise clinically as a prognostic biomarker in CRC (Nishida et al. 2012, Valladares-Ayerbes et al. 2011, Yu et al. 2012, Fang et al. 2014, Tang et al. 2014, Yang et al. 2013, Igarashi et al. 2015, Liu et al. 2013a, b, Wang et al. 2014a, b, Jinushi et al. 2014).

The current guidelines recommend surgical resection of stage II CRC tumors without adjuvant chemotherapy. Unfortunately, there is a substantial percentage (approximately 15%) of patients with stage II CRC who experience recurrence or adverse outcomes despite following these recommendations (Benson et al. 2004; Figueredo et al. 2008). Thus, developing biomarkers that can help risk stratify among stage 2 CRC patients to decide who should pursue more aggressive therapy to prevent poor outcomes would be clinically useful. Schepeler and colleagues discovered that miR-320 and miR-498 miRNA may help in distinguishing these patients, as expression of these miRNAs was associated with recurrence-free survival (Schepeler et al. 2008). Similarly, miR-21 expression was found to serve the same purpose in stratifying risk among these patients (Oue et al. 2014; Hansen et al. 2014; Nielsen et al. 2011). A miRNA-based assay comprising of miR-20a-5p, miR-21-5p, miR-103a-3p, miR-106a-5p, miR-143-5p, and miR-215 was developed to discriminate risk in these stage 2 CRC patients (Zhang et al. 2013a, b).

While there has been significant progress in investigating the potential of miRNA prognostic biomarkers, there are no miRNA biomarker assays that have been shown to be reliable enough for clinical use to date.

10.2.8 Predictive Biomarkers for Response to Treatment in Colorectal Cancer

Due to development of novel targeted therapies, the treatment options for advanced CRC have improved remarkably (Bardelli and Siena 2010). Despite this advancement with monoclonal antibodies targeting several different targets including vascular endothelial growth factor (VEGFA) and epidermal growth factor receptor (EGFR) (Hurwitz et al. 2004; Saltz et al. 2004; Cunningham et al. 2004), prognosis remains poor, with the median overall survival of only 18 to 21 months (Poston et al. 2008). This dilemma led to establishing a new class of biomarkers that can predict the response to these chemotherapy regimens.

To date, only in vitro studies have demonstrated that miRNAs associate with and may mediate resistance to various chemotherapy regimens, including 5-fluorouracil (miR-10b, miR-19b, miR-20a, miR-21, miR-23a, miR-31, miR-34, miR-129, miR-140, miR-145, miR-192/-215, miR-200 family, and miR-497) (Kurokawa et al. 2012, Valeri et al. 2010, Faltejskova et al. 2012, Deng et al. 2014, Shang et al. 2014, Wang et al. 2010, Akao et al. 2011, Siemens et al. 2013, Karaayvaz et al. 2013, Song et al. 2009, Zhang et al. 2011, Boni et al. 2010, Toden et al. 2015, Guo et al. 2013), irinotecan resistance (miR-21 and miR-451), and oxaliplatin resistance (miR-20a, miR-21, miR-133a, miR-143, miR-153, miR-203, and miR-1915) (Faltejskova et al. 2012, Bitarte et al. 2011, Qian et al. 2013, Zhang et al. 2013a, b, Zhou et al. 2014, Xu et al. 2013). However, clinical studies of these miRNAs is limited and not yet able to corroborate these findings from in vitro studies. Of the limited clinical data, studies have shown that let-7g and miR-181b expression associate with predicted response to S-1 based chemotherapy, and increased miR-21 expression corresponds with poor response to 5-FU-based chemotherapy (Nakajima et al. 2006; Liu et al. 2011). Both expression and methylation levels of miR-148a are correlated with lack of response to 5-FU and oxaliplatin chemotherapies in advanced CRC patients (Takahashi et al. 2012). Lastly, miR-31-

3p was demonstrated to be a negative predictor of progression-free survival in patients with *KRAS* wild-type metastatic CRC patients treated with anti-EGFR therapy (Manceau et al. 2014).

A major limitation in our understanding of miRNA biomarkers is that the data that demonstrate these associations are solely from retrospective studies using archived tissue samples. As a result, these studies are confounded by unrecognized factors and have the other inherent biases of retrospective study designs. In addition, many of the studies have not been validated with independent studies in different cohorts. If miRNAs prove to be robust and reliable tissue-bed biomarkers, circulating blood-based miRNA biomarkers have potential to be used in biomarker assays for predicting or monitoring response to chemotherapy.

10.3 Conclusion and Future Directions

In summary, genetic and epigenetic modifications, as well as other host and environmental factors, mediate the initiation and progression of colorectal cancer. Epigenetic alterations, including DNA methylation, histone modifications, chromatin structure, and ncRNA expression changes, are common in colorectal polyps and cancer, and some of here alterations promote oncogenic behavior in the nascent cancer cells and cancer cells.

Epigenetic alterations have emerged as one of the most robust classes of biomarkers for early detection of cancer and precancerous polyps, prognosis of cancer, and prediction of response to treatment. Advancements in our understanding of these epigenetic changes have become the basis for a growing number of clinical tests for cancer screening and surveillance.

Compliance with Ethical Standards We have followed conventional ethical standards for publication and are compliant with them. There are no human subject or animal use issues.

References

Ahn JB, Chung WB, Maeda O et al (2011) DNA methylation predicts recurrence from resected stage III proximal colon cancer. Cancer. 117(9):1847–1854. https://doi.org/10.1002/cncr.25737

Akao Y, Noguchi S, Iio A, Kojima K, Takagi T, Naoe T (2011) Dysregulation of microRNA-34a expression causes drug-resistance to 5-FU in human colon cancer DLD-1 cells. Cancer Lett. 300(2):197–204. https://doi.org/10.1016/j.canlet.2010.10.006

Antelo M, Balaguer F, Shia J et al (2012) A High Degree of LINE-1 Hypomethylation Is a Unique Feature of Early-Onset Colorectal Cancer. PLOS ONE. 7(9):e45357. https://doi.org/10.1371/journal.pone.0045357

Arndt GM, Dossey L, Cullen LM et al (2009) Characterization of global microRNA expression reveals oncogenic potential of miR-145 in metastatic colorectal cancer. BMC Cancer. 9(1):374. https://doi.org/10.1186/1471-2407-9-374

Arroyo JD, Chevillet JR, Kroh EM et al (2011) Argonaute2 complexes carry a population of circulating microRNAs independent of vesicles in human plasma. Proc Natl Acad Sci. 108(12): 5003–5008. https://doi.org/10.1073/pnas.1019055108

Baba Y, Nosho K, Shima K et al (2010) HIF1A Overexpression Is Associated with Poor Prognosis in a Cohort of 731 Colorectal Cancers. Am J Pathol. 176(5):2292–2301. https://doi.org/10.2353/ajpath.2010.090972

Bardelli A, Siena S (2010) Molecular Mechanisms of Resistance to Cetuximab and Panitumumab in Colorectal Cancer. J Clin Oncol. 28(7):1254–1261. https://doi.org/10.1200/JCO.2009.24.6116

Barták BK, Kalmár A, Péterfia B et al (2017) Colorectal adenoma and cancer detection based on altered methylation pattern of SFRP1, SFRP2, SDC2, and PRIMA1 in plasma samples. Epigenetics. 12(9):751–763. https://doi.org/10.1080/15592294.2017.1356957

Bartke T, Kouzarides T (2011) Decoding the chromatin modification landscape. Cell Cycle. 10(2): 182–182. https://doi.org/10.4161/cc.10.2.14477

Bártová E, Krejčí J, Harničarová A, Galiová G, Kozubek S (2008) Histone Modifications and Nuclear Architecture: A Review. J Histochem Cytochem. 56(8):711–721. https://doi.org/10.1369/jhc.2008.951251

Baylin SB, Herman JG (2000) DNA hypermethylation in tumorigenesis: epigenetics joins genetics. Trends Genet. 16(4):168–174. https://doi.org/10.1016/S0168-9525(99)01971-X

Baylin SB, Jones PA (2011) A decade of exploring the cancer epigenome — biological and translational implications. Nat Rev Cancer. 11(10):726–734. https://doi.org/10.1038/nrc3130

Benard A, Goossens-Beumer IJ, van Hoesel AQ et al (2014a) Histone trimethylation at H3K4, H3K9 and H4K20 correlates with patient survival and tumor recurrence in early-stage colon cancer. BMC Cancer. 14(1):531. https://doi.org/10.1186/1471-2407-14-531

Benard A, Goossens-Beumer IJ, van Hoesel AQ et al (2015) Nuclear expression of histone deacetylases and their histone modifications predicts clinical outcome in colorectal cancer. Histopathology. 66(2):270–282. https://doi.org/10.1111/his.12534

Benard A, Goossens-Beumer IJ, van Hoesel AQ et al (2014b) Prognostic value of polycomb proteins EZH2, BMI1 and SUZ12 and histone modification H3K27me3 in colorectal cancer. Plos One. 9(9):e108265. https://doi.org/10.1371/journal.pone.0108265

Benson AB, Schrag D, Somerfield MR et al (2004) American Society of Clinical Oncology Recommendations on Adjuvant Chemotherapy for Stage II Colon Cancer. J Clin Oncol. 22(16):3408–3419. https://doi.org/10.1200/JCO.2004.05.063

Bersaglieri C, Santoro R (2019) Genome Organization in and around the Nucleolus. Cells. 8(6): 579. https://doi.org/10.3390/cells8060579

Bettington M, Rosty C, Whitehall V et al (2018) A morphological and molecular study of proposed early forms of traditional serrated adenoma. Histopathology. 73(6):1023–1029. https://doi.org/10.1111/his.13714

Bettington M, Walker N, Clouston A, Brown I, Leggett B, Whitehall V (2013) The serrated pathway to colorectal carcinoma: current concepts and challenges. Histopathology. 62(3): 367–386. https://doi.org/10.1111/his.12055

Bi F, Wang Q, Dong Q, Wang Y, Zhang L, Zhang J (2020) Circulating tumor DNA in colorectal cancer: opportunities and challenges. Am J Transl Res. 12(3):1044–1055

Bird A (2002) DNA methylation patterns and epigenetic memory. Genes Dev. 16(1):6–21. https://doi.org/10.1101/gad.947102

Bitarte N, Bandres E, Boni V et al (2011) MicroRNA-451 Is Involved in the Self-renewal, Tumorigenicity, and Chemoresistance of Colorectal Cancer Stem Cells. Stem Cells. 29(11): 1661–1671. https://doi.org/10.1002/stem.741

Boni V, Bitarte N, Cristobal I et al (2010) miR-192/miR-215 Influence 5-Fluorouracil Resistance through Cell Cycle-Mediated Mechanisms Complementary to Its Post-transcriptional Thymidylate Synthase Regulation. Mol Cancer Ther. 9(8):2265–2275. https://doi.org/10.1158/1535-7163.MCT-10-0061

Bovell LC, Shanmugam C, Putcha BDK et al (2013) The Prognostic Value of MicroRNAs Varies with Patient Race/Ethnicity and Stage of Colorectal Cancer. Clin Cancer Res. 19(14): 3955–3965. https://doi.org/10.1158/1078-0432.CCR-12-3302

Cairns RA, Mak TW (2013) Oncogenic Isocitrate Dehydrogenase Mutations: Mechanisms, Models, and Clinical Opportunities. Cancer Discov. 3(7):730–741. https://doi.org/10.1158/2159-8290. CD-13-0083

Calin GA, Croce CM (2006) MicroRNA signatures in human cancers. Nat Rev Cancer. 6(11): 857–866. https://doi.org/10.1038/nrc1997

Calin GA, Dumitru CD, Shimizu M et al (2002) Frequent deletions and down-regulation of micro-RNA genes miR15 and miR16 at 13q14 in chronic lymphocytic leukemia. Proc Natl Acad Sci. 99(24):15524–15529. https://doi.org/10.1073/pnas.242606799

Carethers JM, Jung BH (2015) Genetics and Genetic Biomarkers in Sporadic Colorectal Cancer. Gastroenterology. 149(5):1177–1190.e3. https://doi.org/10.1053/j.gastro.2015.06.047

Chan AOO, Broaddus RR, Houlihan PS, Issa JPJ, Hamilton SR, Rashid A (2002) CpG Island Methylation in Aberrant Crypt Foci of the Colorectum. Am J Pathol. 160(5):1823–1830. https://doi.org/10.1016/S0002-9440(10)61128-5

Chen JH, Herlong FR, Stroehlein J, Mishra L (2016) Mutations of Chromatin Structure Regulating Genes in Human Malignancies. Curr Protein Pept Sci. 17(5):411–437

Chen RZ, Pettersson U, Beard C, Jackson-Grusby L, Jaenisch R (1998) DNA hypomethylation leads to elevated mutation rates. Nature. 395(6697):89–93. https://doi.org/10.1038/25779

Chen WD, Han ZJ, Skoletsky J et al (2005) Detection in fecal DNA of colon cancer–specific methylation of the nonexpressed vimentin gene. JNCI J Natl Cancer Inst. 97(15):1124–1132. https://doi.org/10.1093/jnci/dji204

Cohen I, Poręba E, Kamieniarz K, Schneider R (2011) Histone modifiers in cancer: friends or foes? Genes Cancer. 2(6):631–647. https://doi.org/10.1177/1947601911417176

Creemers EE, Tijsen AJ, Pinto YM, van Rooij E (2012) Circulating MicroRNAs. Circ Res. 110(3): 483–495. https://doi.org/10.1161/CIRCRESAHA.111.247452

Crockett SD, Nagtegaal ID (2019) Terminology, molecular features, epidemiology, and management of serrated colorectal neoplasia. Gastroenterology. 157(4):949–966.e4. https://doi.org/10.1053/j.gastro.2019.06.041

Cui H, Onyango P, Brandenburg S, Wu Y, Hsieh CL, Feinberg AP (2002) Loss of imprinting in colorectal cancer linked to hypomethylation of H19 and IGF2. Cancer Res. 62(22):6442–6446

Cunningham D, Humblet Y, Siena S et al (2004) Cetuximab monotherapy and cetuximab plus irinotecan in irinotecan-refractory metastatic colorectal cancer. N Engl J Med. 351(4):337–345. https://doi.org/10.1056/NEJMoa033025

Deng J, Lei W, Fu JC, Zhang L, Li JH, Xiong JP (2014) Targeting miR-21 enhances the sensitivity of human colon cancer HT-29 cells to chemoradiotherapy in vitro. Biochem Biophys Res Commun. 443(3):789–795. https://doi.org/10.1016/j.bbrc.2013.11.064

Dhir M, Yachida S, Neste LV et al (2011) Sessile serrated adenomas and classical adenomas: An epigenetic perspective on premalignant neoplastic lesions of the gastrointestinal tract. Int J Cancer. 129(8):1889–1898. https://doi.org/10.1002/ijc.25847

Draht MXG, Smits KM, Tournier B et al (2014) Promoter CpG island methylation of RET predicts poor prognosis in stage II colorectal cancer patients. Mol Oncol. 8(3):679–688. https://doi.org/10.1016/j.molonc.2014.01.011

Duran-Sanchon S, Moreno L, Augé JM et al (2020) Identification and validation of microRNA profiles in fecal samples for detection of colorectal cancer. Gastroenterology. 158(4):947–957. e4. https://doi.org/10.1053/j.gastro.2019.10.005

Duran-Sanchon S, Moreno L, Gómez-Matas J et al (2021) Fecal MicroRNA-based algorithm increases effectiveness of fecal immunochemical test–based screening for colorectal cancer. Clin Gastroenterol Hepatol. 19(2):323–330.e1. https://doi.org/10.1016/j.cgh.2020.02.043

Ebert MPA, Model F, Mooney S et al (2006) Aristaless-like homeobox-4 gene methylation is a potential marker for colorectal adenocarcinomas. Gastroenterology. 131(5):1418–1430. https://doi.org/10.1053/j.gastro.2006.08.034

Ebert MPA, Tänzer M, Balluff B et al (2012) TFAP2E–DKK4 and chemoresistance in colorectal cancer. N Engl J Med. 366(1):44–53. https://doi.org/10.1056/NEJMoa1009473

El Messaoudi S, Mouliere F, Du Manoir S et al (2016) Circulating DNA as a strong multimarker prognostic tool for metastatic colorectal cancer patient management care. Clin Cancer Res. 22(12):3067–3077. https://doi.org/10.1158/1078-0432.CCR-15-0297

Estécio MRH, Gharibyan V, Shen L et al (2007) LINE-1 Hypomethylation in cancer is highly variable and inversely correlated with microsatellite instability. PLOS ONE. 2(5):e399. https://doi.org/10.1371/journal.pone.0000399

Esteller M, Toyota M, Sanchez-Cespedes M et al (2000) Inactivation of the DNA repair gene O6-methylguanine-DNA methyltransferase by promoter hypermethylation is associated with G to A mutations in K-ras in colorectal tumorigenesis. Cancer Res. 60(9):2368–2371

Faltejskova P, Besse A, Sevcikova S et al (2012) Clinical correlations of miR-21 expression in colorectal cancer patients and effects of its inhibition on DLD1 colon cancer cells. Int J Colorectal Dis. 27(11):1401–1408. https://doi.org/10.1007/s00384-012-1461-3

Fang L, Li H, Wang L et al (2014) MicroRNA-17-5p promotes chemotherapeutic drug resistance and tumour metastasis of colorectal cancer by repressing PTEN expression. Oncotarget. 5(10): 2974–2987. https://doi.org/10.18632/oncotarget.1614

Fearon ER, Vogelstein B (1990) A genetic model for colorectal tumorigenesis. Cell. 61(5): 759–767. https://doi.org/10.1016/0092-8674(90)90186-I

Figueredo A, Coombes ME, Mukherjee S (2008) Adjuvant therapy for completely resected stage II colon cancer. Cochrane Database Syst Rev. 3. https://doi.org/10.1002/14651858.CD005390.pub2

Finger FP (2002) One ring to bind them: septins and actin assembly. Dev Cell. 3(6):761–763. https://doi.org/10.1016/S1534-5807(02)00371-4

Fraga MF, Ballestar E, Villar-Garea A et al (2005) Loss of acetylation at Lys16 and trimethylation at Lys20 of histone H4 is a common hallmark of human cancer. Nat Genet. 37(4):391–400. https://doi.org/10.1038/ng1531

Gallois C, Taieb J, Le Corre D et al (2018) Prognostic value of methylator phenotype in stage III colon cancer treated with oxaliplatin-based adjuvant chemotherapy. Clin Cancer Res. 24(19): 4745–4753. https://doi.org/10.1158/1078-0432.CCR-18-0866

Gaudet F, Hodgson JG, Eden A et al (2003) Induction of tumors in mice by genomic hypomethylation. Science. 300(5618):489–492. https://doi.org/10.1126/science.1083558

Gezer U, Üstek D, Yörüker EE et al (2013) Characterization of H3K9me3- and H4K20me3-associated circulating nucleosomal DNA by high-throughput sequencing in colorectal cancer. Tumor Biol. 34(1):329–336. https://doi.org/10.1007/s13277-012-0554-5

Gilbert N (2019) Biophysical regulation of local chromatin structure. Curr Opin Genet Dev. 55:66–75. https://doi.org/10.1016/j.gde.2019.06.001

Glöckner SC, Dhir M, Yi JM et al (2009) Methylation of TFPI2 in Stool DNA: a potential novel biomarker for the detection of colorectal cancer. Cancer Res. 69(11):4691–4699. https://doi.org/10.1158/0008-5472.CAN-08-0142

Goel A, Xicola RM, Nguyen T et al (2010) Aberrant DNA methylation in hereditary nonpolyposis colorectal cancer without mismatch repair deficiency. Gastroenterology. 138(5):1854–1862.e1. https://doi.org/10.1053/j.gastro.2010.01.035

Grady WM, Carethers JM (2008) Genomic and epigenetic instability in colorectal cancer pathogenesis. Gastroenterology. 135(4):1079–1099. https://doi.org/10.1053/j.gastro.2008.07.076

Grady WM, Parkin RK, Mitchell PS et al (2008) Epigenetic silencing of the intronic microRNA hsa-miR-342 and its host gene EVL in colorectal cancer. Oncogene. 27(27):3880–3888. https://doi.org/10.1038/onc.2008.10

Gregory PA, Bracken CP, Smith E et al (2011) An autocrine TGF-β/ZEB/miR-200 signaling network regulates establishment and maintenance of epithelial-mesenchymal transition. Mol Biol Cell. 22(10):1686–1698. https://doi.org/10.1091/mbc.e11-02-0103

Guo ST, Jiang CC, Wang GP et al (2013) MicroRNA-497 targets insulin-like growth factor 1 receptor and has a tumour suppressive role in human colorectal cancer. Oncogene. 32(15): 1910–1920. https://doi.org/10.1038/onc.2012.214

Hadley M, Noonepalle S, Banik D, Villagra A (2019) Functional analysis of HDACs in tumorigenesis. In: Brosh RM (ed) Protein acetylation: methods and protocols, Methods in molecular biology. Springer, pp 279–307. https://doi.org/10.1007/978-1-4939-9434-2_17

Hampel H, Frankel WL, Martin E et al (2005) Screening for the lynch syndrome (hereditary nonpolyposis colorectal cancer). N Engl J Med. 352(18):1851–1860. https://doi.org/10.1056/NEJMoa043146

Hampel H, Frankel WL, Martin E et al (2008) Feasibility of screening for lynch syndrome among patients with colorectal cancer. J Clin Oncol. 26(35):5783–5788. https://doi.org/10.1200/JCO.2008.17.5950

Hansen TF, Kjær-Frifeldt S, Christensen RD et al (2014) Redefining high-risk patients with stage II colon cancer by risk index and microRNA-21: results from a population-based cohort. Br J Cancer. 111(7):1285–1292. https://doi.org/10.1038/bjc.2014.409

Hawkins N, Norrie M, Cheong K et al (2002) CpG island methylation in sporadic colorectal cancers and its relationship to microsatellite instability. Gastroenterology. 122(5):1376–1387. https://doi.org/10.1053/gast.2002.32997

He L, Hannon GJ (2004) MicroRNAs: small RNAs with a big role in gene regulation. Nat Rev Genet. 5(7):522–531. https://doi.org/10.1038/nrg1379

Herbst A, Rahmig K, Stieber P et al (2011) Methylation of NEUROG1 in serum is a sensitive marker for the detection of early colorectal cancer. Off J Am Coll Gastroenterol ACG. 106(6): 1110. https://doi.org/10.1038/ajg.2011.6

Holm TM, Jackson-Grusby L, Brambrink T, Yamada Y, Rideout WM, Jaenisch R (2005) Global loss of imprinting leads to widespread tumorigenesis in adult mice. Cancer Cell. 8(4):275–285. https://doi.org/10.1016/j.ccr.2005.09.007

Huang Z, Huang D, Ni S, Peng Z, Sheng W, Du X (2010) Plasma microRNAs are promising novel biomarkers for early detection of colorectal cancer. Int J Cancer. 127(1):118–126. https://doi.org/10.1002/ijc.25007

Hur K, Cejas P, Feliu J et al (2014) Hypomethylation of long interspersed nuclear element-1 (LINE-1) leads to activation of proto-oncogenes in human colorectal cancer metastasis. Gut. 63(4): 635–646. https://doi.org/10.1136/gutjnl-2012-304219

Hurwitz H, Fehrenbacher L, Novotny W et al (2004) Bevacizumab plus irinotecan, fluorouracil, and leucovorin for metastatic colorectal cancer. N Engl J Med. 350(23):2335–2342. https://doi.org/10.1056/NEJMoa032691

Igarashi H, Kurihara H, Mitsuhashi K et al (2015) Association of microRNA-31-5p with clinical efficacy of anti-EGFR therapy in patients with metastatic colorectal cancer. Ann Surg Oncol. 22(8):2640–2648. https://doi.org/10.1245/s10434-014-4264-7

Imperiale TF, Ransohoff DF, Itzkowitz SH et al (2014) Multitarget stool DNA testing for colorectal-cancer screening. N Engl J Med. 370(14):1287–1297. https://doi.org/10.1056/NEJMoa1311194

Issa JPJ, Ottaviano YL, Celano P, Hamilton SR, Davidson NE, Baylin SB (1994) Methylation of the oestrogen receptor CpG island links ageing and neoplasia in human colon. Nat Genet. 7(4): 536–540. https://doi.org/10.1038/ng0894-536

Ito M, Mitsuhashi K, Igarashi H et al (2014) MicroRNA-31 expression in relation to BRAF mutation, CpG island methylation and colorectal continuum in serrated lesions. Int J Cancer. 135(11):2507–2515. https://doi.org/10.1002/ijc.28920

Itzkowitz S, Brand R, Jandorf L et al (2008) A simplified, noninvasive stool dna test for colorectal cancer detection. Off J Am Coll Gastroenterol ACG. 103(11):2862

Itzkowitz SH, Jandorf L, Brand R et al (2007) Improved fecal DNA test for colorectal cancer screening. Clin Gastroenterol Hepatol. 5(1):111–117. https://doi.org/10.1016/j.cgh.2006.10.006

Jin P, Kang Q, Wang X et al (2015) Performance of a second-generation methylated SEPT9 test in detecting colorectal neoplasm. J Gastroenterol Hepatol. 30(5):830–833. https://doi.org/10.1111/jgh.12855

Jinushi T, Shibayama Y, Kinoshita I et al (2014) Low expression levels of microRNA-124-5p correlated with poor prognosis in colorectal cancer via targeting of SMC4. Cancer Med. 3(6): 1544–1552. https://doi.org/10.1002/cam4.309

Johnson SM, Grosshans H, Shingara J et al (2005) RAS Is Regulated by the let-7 MicroRNA Family. Cell. 120(5):635–647. https://doi.org/10.1016/j.cell.2005.01.014

Josse C, Bouznad N, Geurts P et al (2014) Identification of a microRNA landscape targeting the PI3K/Akt signaling pathway in inflammation-induced colorectal carcinogenesis. Am J Physiol-Gastrointest Liver Physiol. 306(3):G229–G243. https://doi.org/10.1152/ajpgi.00484.2012

Jover R, Nguyen T, Pérez-Carbonell L et al (2011) 5-fluorouracil adjuvant chemotherapy does not increase survival in patients with CpG island methylator phenotype colorectal cancer. Gastroenterology. 140(4):1174–1181. https://doi.org/10.1053/j.gastro.2010.12.035

Juo YY, Johnston FM, Zhang DY et al (2014) Prognostic value of CpG island methylator phenotype among colorectal cancer patients: a systematic review and meta-analysis. Ann Oncol. 25(12):2314–2327. https://doi.org/10.1093/annonc/mdu149

Kamakaka RT, Biggins S (2005) Histone variants: deviants? Genes Dev. 19(3):295–316. https://doi.org/10.1101/gad.1272805

Kamiyama H, Suzuki K, Maeda T et al (2012) DNA demethylation in normal colon tissue predicts predisposition to multiple cancers. Oncogene. 31(48):5029–5037. https://doi.org/10.1038/onc.2011.652

Kanaan Z, Rai SN, Eichenberger MR et al (2012) Plasma MiR-21: A Potential Diagnostic Marker of Colorectal Cancer. Ann Surg. 256(3):544. https://doi.org/10.1097/SLA.0b013e318265bd6f

Kanaan Z, Roberts H, Eichenberger MR et al (2013) A plasma microRNA panel for detection of colorectal adenomas: a step toward more precise screening for colorectal cancer. Ann Surg. 258(3):400. https://doi.org/10.1097/SLA.0b013e3182a15bcc

Kanwal R, Gupta S (2010) Epigenetics and cancer. J Appl Physiol. 109(2):598–605. https://doi.org/10.1152/japplphysiol.00066.2010

Karaayvaz M, Zhai H, Ju J (2013) miR-129 promotes apoptosis and enhances chemosensitivity to 5-fluorouracil in colorectal cancer. Cell Death Dis. 4(6):e659–e659. https://doi.org/10.1038/cddis.2013.193

Kawakami K, Ruszkiewicz A, Bennett G et al (2006) DNA hypermethylation in the normal colonic mucosa of patients with colorectal cancer. Br J Cancer. 94(4):593–598. https://doi.org/10.1038/sj.bjc.6602940

Kim JK, Samaranayake M, Pradhan S (2008) Epigenetic mechanisms in mammals. Cell Mol Life Sci. 66(4):596. https://doi.org/10.1007/s00018-008-8432-4

Kim ST, Raymond VM, Park JO et al (2019) Abstract 916: Combined genomic and epigenomic assessment of cell-free circulating tumor DNA (ctDNA) improves assay sensitivity in early-stage colorectal cancer (CRC). Cancer Res. 79(13_Supplement):916. https://doi.org/10.1158/1538-7445.AM2019-916

King CE, Cuatrecasas M, Castells A, Sepulveda AR, Lee JS, Rustgi AK (2011) LIN28B promotes colon cancer progression and metastasis. Cancer Res. 71(12):4260–4268. https://doi.org/10.1158/0008-5472.CAN-10-4637

Kjaer-Frifeldt S, Hansen TF, Nielsen BS et al (2012) The prognostic importance of miR-21 in stage II colon cancer: a population-based study. Br J Cancer. 107(7):1169–1174. https://doi.org/10.1038/bjc.2012.365

Koga Y, Yamazaki N, Yamamoto Y et al (2013) Fecal miR-106a Is a useful marker for colorectal cancer patients with false-negative results in immunochemical fecal occult blood test. Cancer Epidemiol Biomarkers Prev. 22(10):1844–1852. https://doi.org/10.1158/1055-9965.EPI-13-0512

Kohonen-Corish MRJ, Sigglekow ND, Susanto J et al (2007) Promoter methylation of the mutated in colorectal cancer gene is a frequent early event in colorectal cancer. Oncogene. 26(30): 4435–4441. https://doi.org/10.1038/sj.onc.1210210

Komor MA, Bosch LJ, Bounova G et al (2018) Consensus molecular subtype classification of colorectal adenomas. J Pathol. 246(3):266–276. https://doi.org/10.1002/path.5129

Kouzarides T (2007) Chromatin Modifications and Their Function. Cell. 128(4):693–705. https://doi.org/10.1016/j.cell.2007.02.005

Kriaucionis S, Heintz N (2009) The nuclear DNA Base 5-hydroxymethylcytosine is present in purkinje neurons and the brain. Science. 324(5929):929–930. https://doi.org/10.1126/science.1169786

Kriegl L, Neumann J, Vieth M et al (2011) Up and downregulation of p16Ink4a expression in BRAF-mutated polyps/adenomas indicates a senescence barrier in the serrated route to colon cancer. Mod Pathol. 24(7):1015–1022. https://doi.org/10.1038/modpathol.2011.43

Kuipers EJ, Grady WM, Lieberman D et al (2015) Colorectal cancer. Nat Rev Dis Primer. 1(1): 1–25. https://doi.org/10.1038/nrdp.2015.65

Kulda V, Pesta M, Topolcan O et al (2010) Relevance of miR-21 and miR-143 expression in tissue samples of colorectal carcinoma and its liver metastases. Cancer Genet Cytogenet. 200(2): 154–160. https://doi.org/10.1016/j.cancergencyto.2010.04.015

Kurokawa K, Tanahashi T, Iima T et al (2012) Role of miR-19b and its target mRNAs in 5-fluorouracil resistance in colon cancer cells. J Gastroenterol. 47(8):883–895. https://doi.org/10.1007/s00535-012-0547-6

Laiho P, Kokko A, Vanharanta S et al (2007) Serrated carcinomas form a subclass of colorectal cancer with distinct molecular basis. Oncogene. 26(2):312–320. https://doi.org/10.1038/sj.onc.1209778

Lee BB, Lee EJ, Jung EH et al (2009) Aberrant methylation of APC, MGMT, RASSF2A, and Wif-1 genes in plasma as a biomarker for early detection of colorectal cancer. Clin Cancer Res. 15(19): 6185–6191. https://doi.org/10.1158/1078-0432.CCR-09-0111

Lee RC, Feinbaum RL, Ambros V (1993) The C. elegans heterochronic gene lin-4 encodes small RNAs with antisense complementarity to lin-14. Cell. 75(5):843–854. https://doi.org/10.1016/0092-8674(93)90529-y

Lee S, Cho NY, Yoo EJ, Kim JH, Kang GH (2008) CpG Island methylator phenotype in colorectal cancers: comparison of the new and classic CpG Island methylator phenotype marker panels. Arch Pathol Lab Med. 132(10):1657–1665. https://doi.org/10.5858/2008-132-1657-CIMPIC

Leighton G, Williams DC (2020) The Methyl-CpG–binding domain 2 and 3 proteins and formation of the nucleosome remodeling and deacetylase complex. J Mol Biol. 432(6):1624–1639. https://doi.org/10.1016/j.jmb.2019.10.007

Leszinski G, Gezer U, Siegele B, Stoetzer O, Holdenrieder S (2012) Relevance of histone marks H3K9me3 and H4K20me3 in cancer. Anticancer Res. 32(5):2199–2205

Leung WK, To KF, Man EPS et al (2005) Quantitative detection of promoter hypermethylation in multiple genes in the serum of patients with colorectal cancer. Off J Am Coll Gastroenterol ACG. 100(10):2274

Leung WK, To KF, Man EPS et al (2007) Detection of hypermethylated DNA or cyclooxygenase-2 messenger RNA in fecal samples of patients with colorectal cancer or polyps. Off J Am Coll Gastroenterol ACG. 102(5):1070

Li H, Myeroff L, Smiraglia D et al (2003) SLC5A8, a sodium transporter, is a tumor suppressor gene silenced by methylation in human colon aberrant crypt foci and cancers. Proc Natl Acad Sci. 100(14):8412–8417. https://doi.org/10.1073/pnas.1430846100

Li W, Liu M (2011) Distribution of 5-hydroxymethylcytosine in different human tissues. J Nucleic Acids. 2011:e870726. https://doi.org/10.4061/2011/870726

Lidgard GP, Domanico MJ, Bruinsma JJ et al (2013) Clinical performance of an automated stool DNA assay for detection of colorectal neoplasia. Clin Gastroenterol Hepatol. 11(10): 1313–1318. https://doi.org/10.1016/j.cgh.2013.04.023

Link A, Balaguer F, Shen Y et al (2010) Fecal MicroRNAs as novel biomarkers for colon cancer screening. Cancer Epidemiol Biomarkers Prev. 19(7):1766–1774. https://doi.org/10.1158/1055-9965.EPI-10-0027

Liu F, Kong X, Lv L, Gao J (2015) TGF-β1 acts through miR-155 to down-regulate TP53INP1 in promoting epithelial-mesenchymal transition and cancer stem cell phenotypes. Cancer Lett. 359(2):288–298. https://doi.org/10.1016/j.canlet.2015.01.030

Liu GH, Zhou ZG, Chen R et al (2013a) Serum miR-21 and miR-92a as biomarkers in the diagnosis and prognosis of colorectal cancer. Tumor Biol. 34(4):2175–2181. https://doi.org/10.1007/s13277-013-0753-8

Liu H, Du L, Wen Z et al (2013b) Up-regulation of miR-182 expression in colorectal cancer tissues and its prognostic value. Int J Colorectal Dis. 28(5):697–703. https://doi.org/10.1007/s00384-013-1674-0

Liu K, Li G, Fan C, Zhou X, Wu B, Li J (2011) Increased expression of MicroRNA-21 and its association with chemotherapeutic response in human colorectal cancer. J Int Med Res. 39(6):2288–2295. https://doi.org/10.1177/147323001103900626

Liu Y, Sethi NS, Hinoue T et al (2018) Comparative molecular analysis of gastrointestinal adenocarcinomas. Cancer Cell. 33(4):721–735.e8. https://doi.org/10.1016/j.ccell.2018.03.010

Lochhead P, Kuchiba A, Imamura Y et al (2013) Microsatellite instability and BRAF mutation testing in colorectal cancer prognostication. JNCI J Natl Cancer Inst. 105(15):1151–1156. https://doi.org/10.1093/jnci/djt173

Lofton-Day C, Model F, DeVos T et al (2008) DNA methylation biomarkers for blood-based colorectal cancer screening. Clin Chem. 54(2):414–423. https://doi.org/10.1373/clinchem.2007.095992

Luo X, Stock C, Burwinkel B, Brenner H (2013) Identification and evaluation of plasma microRNAs for early detection of colorectal cancer. PLOS ONE. 8(5):e62880. https://doi.org/10.1371/journal.pone.0062880

Luo Y, Yu M, Grady WM (2014) Field cancerization in the colon: a role for aberrant DNA methylation? Gastroenterol Rep. 2(1):16–20. https://doi.org/10.1093/gastro/got039

Manceau G, Imbeaud S, Thiébaut R et al (2014) Hsa-miR-31-3p Expression Is Linked to Progression-free Survival in Patients with KRAS Wild-type Metastatic Colorectal Cancer Treated with Anti-EGFR Therapy. Clin Cancer Res. 20(12):3338–3347. https://doi.org/10.1158/1078-0432.CCR-13-2750

Mendell JT (2005) MicroRNAs: critical regulators of development, cellular physiology and malignancy. Cell Cycle. 4(9):1179–1184. https://doi.org/10.4161/cc.4.9.2032

Meng F, Henson R, Wehbe-Janek H, Ghoshal K, Jacob ST, Patel T (2007) MicroRNA-21 Regulates Expression of the PTEN Tumor Suppressor Gene in Human Hepatocellular Cancer. Gastroenterology. 133(2):647–658. https://doi.org/10.1053/j.gastro.2007.05.022

Mills AA (2010) Throwing the cancer switch: reciprocal roles of polycomb and trithorax proteins. Nat Rev Cancer. 10(10):669–682. https://doi.org/10.1038/nrc2931

Min BH, Bae JM, Lee EJ et al (2011) The CpG island methylator phenotype may confer a survival benefit in patients with stage II or III colorectal carcinomas receiving fluoropyrimidine-based adjuvant chemotherapy. BMC Cancer. 11(1):1–10. https://doi.org/10.1186/1471-2407-11-344

Moinova HR, Chen WD, Shen L et al (2002) HLTF gene silencing in human colon cancer. Proc Natl Acad Sci. 99(7):4562–4567. https://doi.org/10.1073/pnas.062459899

Motoyama K, Inoue H, Takatsuno Y et al (2009) Over- and under-expressed microRNAs in human colorectal cancer. Int J Oncol. 34(4):1069–1075. https://doi.org/10.3892/ijo_00000233

Moutinho C, Martinez-Cardús A, Santos C et al (2014) Epigenetic Inactivation of the BRCA1 interactor srbc and resistance to oxaliplatin in colorectal cancer. JNCI: J Natl Cancer Inst. 106(1):djt322. https://doi.org/10.1093/jnci/djt322

Myint NNM, Verma AM, Fernandez-Garcia D et al (2018) Circulating tumor DNA in patients with colorectal adenomas: assessment of detectability and genetic heterogeneity. Cell Death Dis. 9(9):1–16. https://doi.org/10.1038/s41419-018-0934-x

Nagasaka T, Sharp GB, Notohara K et al (2003) Hypermethylation of O6-methylguanine-DNA methyltransferase promoter may predict nonrecurrence after chemotherapy in colorectal cancer cases. Clin Cancer Res Off J Am Assoc Cancer Res. 9(14):5306–5312

Nagel R, le Sage C, Diosdado B et al (2008) Regulation of the adenomatous polyposis Coli gene by the miR-135 family in colorectal cancer. Cancer Res. 68(14):5795–5802. https://doi.org/10.1158/0008-5472.CAN-08-0951

Nakajima G, Hayashi K, Xi Y et al (2006) Non-coding MicroRNAs hsa-let-7g and hsa-miR-181b are associated with chemoresponse to S-1 in colon cancer. Cancer Genomics Proteomics. 3(5):317–324

Neri F, Dettori D, Incarnato D et al (2015) TET1 is a tumour suppressor that inhibits colon cancer growth by derepressing inhibitors of the WNT pathway. Oncogene. 34(32):4168–4176. https://doi.org/10.1038/onc.2014.356

Nielsen BS, Jørgensen S, Fog JU et al (2011) High levels of microRNA-21 in the stroma of colorectal cancers predict short disease-free survival in stage II colon cancer patients. Clin Exp Metastasis. 28(1):27–38. https://doi.org/10.1007/s10585-010-9355-7

Nikolaou S, Qiu S, Fiorentino F, Rasheed S, Tekkis P, Kontovounisios C (2018) Systematic review of blood diagnostic markers in colorectal cancer. Tech Coloproctology. 22(7):481–498. https://doi.org/10.1007/s10151-018-1820-3

Nilsson TK, Löf-Öhlin ZM, Sun XF (2013) DNA methylation of the p14ARF, RASSF1A and APC1A genes as an independent prognostic factor in colorectal cancer patients. Int J Oncol. 42(1):127–133. https://doi.org/10.3892/ijo.2012.1682

Nishida N, Yamashita S, Mimori K et al (2012) MicroRNA-10b is a prognostic indicator in colorectal cancer and confers resistance to the chemotherapeutic agent 5-fluorouracil in colorectal cancer cells. Ann Surg Oncol. 19(9):3065–3071. https://doi.org/10.1245/s10434-012-2246-1

Nosho K, Igarashi H, Nojima M et al (2014) Association of microRNA-31 with BRAF mutation, colorectal cancer survival and serrated pathway. Carcinogenesis. 35(4):776–783. https://doi.org/10.1093/carcin/bgt374

Oberg AL, French AJ, Sarver AL et al (2011) miRNA expression in colon polyps provides evidence for a multihit model of colon cancer. PLOS ONE. 6(6):e20465. https://doi.org/10.1371/journal.pone.0020465

Ogata-Kawata H, Izumiya M, Kurioka D et al (2014) Circulating exosomal microRNAs as biomarkers of colon cancer. PLOS ONE. 9(4):e92921. https://doi.org/10.1371/journal.pone.0092921

Ogino S, Kawasaki T, Nosho K et al (2008a) LINE-1 hypomethylation is inversely associated with microsatellite instability and CpG island methylator phenotype in colorectal cancer. Int J Cancer. 122(12):2767–2773. https://doi.org/10.1002/ijc.23470

Ogino S, Meyerhardt JA, Kawasaki T et al (2007) CpG island methylation, response to combination chemotherapy, and patient survival in advanced microsatellite stable colorectal carcinoma. Virchows Arch. 450(5):529–537. https://doi.org/10.1007/s00428-007-0398-3

Ogino S, Nosho K, Kirkner GJ et al (2008b) A cohort study of tumoral LINE-1 hypomethylation and prognosis in colon cancer. JNCI J Natl Cancer Inst. 100(23):1734–1738. https://doi.org/10.1093/jnci/djn359

Okugawa Y, Grady WM, Goel A (2015) Epigenetic alterations in colorectal cancer: emerging biomarkers. Gastroenterology. 149(5):1204–1225.e12. https://doi.org/10.1053/j.gastro.2015.07.011

Oue N, Anami K, Schetter AJ et al (2014) High miR-21 expression from FFPE tissues is associated with poor survival and response to adjuvant chemotherapy in colon cancer. Int J Cancer. 134(8):1926–1934. https://doi.org/10.1002/ijc.28522

Pagliuca A, Valvo C, Fabrizi E et al (2013) Analysis of the combined action of miR-143 and miR-145 on oncogenic pathways in colorectal cancer cells reveals a coordinate program of gene repression. Oncogene. 32(40):4806–4813. https://doi.org/10.1038/onc.2012.495

Pai RK, Bettington M, Srivastava A, Rosty C (2019) An update on the morphology and molecular pathology of serrated colorectal polyps and associated carcinomas. Mod Pathol. 32(10): 1390–1415. https://doi.org/10.1038/s41379-019-0280-2

Papagiannakopoulos T, Shapiro A, Kosik KS (2008) MicroRNA-21 Targets a Network of Key Tumor-Suppressive Pathways in Glioblastoma Cells. Cancer Res. 68(19):8164–8172. https://doi.org/10.1158/0008-5472.CAN-08-1305

Park SJ, Kim S, Hong YS et al (2015) TFAP2E Methylation Status and Prognosis of Patients with Radically Resected Colorectal Cancer. Oncology. 88(2):122–132. https://doi.org/10.1159/000362820

Peng L, Hu J, Li S et al (2013) Aberrant methylation of the PTCH1 gene promoter region in aberrant crypt foci. Int J Cancer. 132(2):E18–E25. https://doi.org/10.1002/ijc.27812

Petko Z, Ghiassi M, Shuber A et al (2005) Aberrantly methylated CDKN2A, MGMT, and MLH1 in colon polyps and in Fecal DNA from patients with colorectal polyps. Clin Cancer Res. 11(3): 1203–1209. https://doi.org/10.1158/1078-0432.1203.11.3

Picardo F, Romanelli A, Muinelo-Romay L et al (2019) Diagnostic and prognostic value of B4GALT1 hypermethylation and its clinical significance as a novel circulating cell-free DNA biomarker in colorectal cancer. Cancers. 11(10):1598. https://doi.org/10.3390/cancers11101598

Popat S, Hubner R, Houlston RS (2005) Systematic review of microsatellite instability and colorectal cancer prognosis. J Clin Oncol Off J Am Soc Clin Oncol. 23(3):609–618. https://doi.org/10.1200/JCO.2005.01.086

Portela A, Esteller M (2010) Epigenetic modifications and human disease. Nat Biotechnol. 28(10): 1057–1068. https://doi.org/10.1038/nbt.1685

Poston GJ, Figueras J, Giuliante F et al (2008) Urgent need for a new staging system in advanced colorectal cancer. J Clin Oncol. 26(29):4828–4833. https://doi.org/10.1200/JCO.2008.17.6453

Pritchard CC, Grady WM (2011) Colorectal cancer molecular biology moves into clinical practice. Gut. 60(1):116–129. https://doi.org/10.1136/gut.2009.206250

Qi J, Zhu YQ, Luo J, Tao WH (2006) Hypermethylation and expression regulation of secreted frizzled-related protein genes in colorectal tumor. World J Gastroenterol. 12(44):7113–7117. https://doi.org/10.3748/wjg.v12.i44.7113

Qian X, Yu J, Yin Y et al (2013) MicroRNA-143 inhibits tumor growth and angiogenesis and sensitizes chemosensitivity to oxaliplatin in colorectal cancers. Cell Cycle. 12(9):1385–1394. https://doi.org/10.4161/cc.24477

Reinert T, Henriksen TV, Christensen E et al (2019) Analysis of plasma cell-free DNA by ultradeep sequencing in patients with stages I to III colorectal cancer. JAMA Oncol. 5(8):1124–1131. https://doi.org/10.1001/jamaoncol.2019.0528

Reinert T, Schøler LV, Thomsen R et al (2016) Analysis of circulating tumour DNA to monitor disease burden following colorectal cancer surgery. Gut. 65(4):625–634. https://doi.org/10.1136/gutjnl-2014-308859

Rex DK, Ahnen DJ, Baron JA et al (2012) Serrated lesions of the colorectum: review and recommendations from an expert panel. Off J Am Coll Gastroenterol ACG. 107(9):1315. https://doi.org/10.1038/ajg.2012.161

Rhee YY, Kim MJ, Bae JM et al (2012) Clinical outcomes of patients with microsatellite-unstable colorectal carcinomas depend on L1 methylation level. Ann Surg Oncol. 19(11):3441–3448. https://doi.org/10.1245/s10434-012-2410-7

van Rijnsoever M, Grieu F, Elsaleh H, Joseph D, Iacopetta B (2002) Characterisation of colorectal cancers showing hypermethylation at multiple CpG islands. Gut. 51(6):797–802. https://doi.org/10.1136/gut.51.6.797

Saltz LB, Meropol NJ, Loehrer PJ, Needle MN, Kopit J, Mayer RJ (2004) Phase II trial of cetuximab in patients with refractory colorectal cancer that expresses the epidermal growth factor receptor. J Clin Oncol. 22(7):1201–1208. https://doi.org/10.1200/JCO.2004.10.182

Samowitz WS, Albertsen H, Herrick J et al (2005) Evaluation of a large, population-based sample supports a CpG Island methylator phenotype in colon cancer. Gastroenterology. 129(3): 837–845. https://doi.org/10.1053/j.gastro.2005.06.020

Schee K, Lorenz S, Worren MM et al (2013) Deep sequencing the MicroRNA transcriptome in colorectal cancer. PLOS ONE. 8(6):e66165. https://doi.org/10.1371/journal.pone.0066165

Schepeler T, Reinert JT, Ostenfeld MS et al (2008) Diagnostic and prognostic MicroRNAs in stage II colon cancer. Cancer Res. 68(15):6416–6424. https://doi.org/10.1158/0008-5472.CAN-07-6110

Schetter AJ, Leung SY, Sohn JJ et al (2008) MicroRNA expression profiles associated with prognosis and therapeutic outcome in colon adenocarcinoma. JAMA. 299(4):425–436. https://doi.org/10.1001/jama.299.4.425

Schmitz KJ, Hey S, Schinwald A et al (2009) Differential expression of microRNA 181b and microRNA 21 in hyperplastic polyps and sessile serrated adenomas of the colon. Virchows Arch. 455(1):49–54. https://doi.org/10.1007/s00428-009-0804-0

Shang J, Yang F, Wang Y et al (2014) MicroRNA-23a antisense enhances 5-fluorouracil chemosensitivity through APAF-1/Caspase-9 apoptotic pathway in colorectal cancer cells. J Cell Biochem. 115(4):772–784. https://doi.org/10.1002/jcb.24721

Sheffield PJ, Oliver CJ, Kremer BE, Sheng S, Shao Z, Macara IG (2003) Borg/septin interactions and the assembly of mammalian septin heterodimers, trimers, and filaments*. J Biol Chem. 278(5):3483–3488. https://doi.org/10.1074/jbc.M209701200

Shen L, Catalano PJ, Benson AB III, O'Dwyer P, Hamilton SR, Issa JPJ (2007) Association between DNA methylation and shortened survival in patients with advanced colorectal cancer treated with 5-fluorouracil–based chemotherapy. Clin Cancer Res. 13(20):6093–6098. https://doi.org/10.1158/1078-0432.CCR-07-1011

Shen L, Kondo Y, Rosner GL et al (2005) MGMT promoter methylation and field defect in sporadic colorectal cancer. JNCI J Natl Cancer Inst. 97(18):1330–1338. https://doi.org/10.1093/jnci/dji275

Shibuya H, Iinuma H, Shimada R, Horiuchi A, Watanabe T (2010) Clinicopathological and prognostic value of MicroRNA-21 and MicroRNA-155 in colorectal cancer. Oncology. 79(3-4):313–320. https://doi.org/10.1159/000323283

Shiovitz S, Bertagnolli MM, Renfro LA et al (2014) CpG Island methylator phenotype is associated with response to adjuvant irinotecan-based therapy for stage III colon cancer. Gastroenterology. 147(3):637–645. https://doi.org/10.1053/j.gastro.2014.05.009

Sidransky D, Tokino T, Hamilton SR et al (1992) Identification of ras oncogene mutations in the stool of patients with curable colorectal tumors. Science. 256(5053):102–105. https://doi.org/10.1126/science.1566048

Siemens H, Jackstadt R, Kaller M, Hermeking H (2013) Repression of c-Kit by p53 is mediated by miR-34 and is associated with reduced chemoresistance, migration and stemness. Oncotarget. 4(9):1399–1415. https://doi.org/10.18632/oncotarget.1202

Siravegna G, Mussolin B, Buscarino M et al (2015) Erratum: clonal evolution and resistance to EGFR blockade in the blood of colorectal cancer patients. Nat Med. 21(7):827–827. https://doi.org/10.1038/nm0715-827b

Slaby O, Svoboda M, Michalek J, Vyzula R (2009) MicroRNAs in colorectal cancer: translation of molecular biology into clinical application. Mol Cancer. 8(1):102. https://doi.org/10.1186/1476-4598-8-102

Song B, Wang Y, Xi Y et al (2009) Mechanism of chemoresistance mediated by miR-140 in human osteosarcoma and colon cancer cells. Oncogene. 28(46):4065–4074. https://doi.org/10.1038/onc.2009.274

Stachler MD, Rinehart E, Lindeman N, Odze R, Srivastava A (2015) Novel molecular insights from routine genotyping of colorectal carcinomas. Hum Pathol. 46(4):507–513. https://doi.org/10.1016/j.humpath.2015.01.005

Sun D, Yu F, Ma Y et al (2013) MicroRNA-31 Activates the RAS Pathway and Functions as an Oncogenic MicroRNA in Human Colorectal Cancer by Repressing RAS p21 GTPase Activating Protein 1 (RASA1) *. J Biol Chem. 288(13):9508–9518. https://doi.org/10.1074/jbc.M112.367763

Sung H, Ferlay J, Siegel RL et al (2021) Global cancer statistics 2020: GLOBOCAN estimates of incidence and mortality worldwide for 36 cancers in 185 countries. CA Cancer J Clin. 71(3): 209–249. https://doi.org/10.3322/caac.21660

Suzuki H, Igarashi S, Nojima M et al (2010) IGFBP7 is a p53-responsive gene specifically silenced in colorectal cancer with CpG island methylator phenotype. Carcinogenesis. 31(3):342–349. https://doi.org/10.1093/carcin/bgp179

Suzuki H, Watkins DN, Jair KW et al (2004) Epigenetic inactivation of SFRP genes allows constitutive WNT signaling in colorectal cancer. Nat Genet. 36(4):417–422. https://doi.org/10.1038/ng1330

Suzuki K, Suzuki I, Leodolter A et al (2006) Global DNA demethylation in gastrointestinal cancer is age dependent and precedes genomic damage. Cancer Cell. 9(3):199–207. https://doi.org/10.1016/j.ccr.2006.02.016

Tahiliani M, Koh KP, Shen Y et al (2009) Conversion of 5-methylcytosine to 5-hydroxymethylcytosine in mammalian DNA by MLL partner TET1. Science. 324(5929): 930–935. https://doi.org/10.1126/science.1170116

Takahashi M, Cuatrecasas M, Balaguer F et al (2012) The clinical significance of MiR-148a as a predictive biomarker in patients with advanced colorectal cancer. PLOS ONE. 7(10):e46684. https://doi.org/10.1371/journal.pone.0046684

Tamagawa H, Oshima T, Numata M et al (2013) Global histone modification of H3K27 correlates with the outcomes in patients with metachronous liver metastasis of colorectal cancer. Eur J Surg Oncol. 39(6):655–661. https://doi.org/10.1016/j.ejso.2013.02.023

Tamagawa H, Oshima T, Shiozawa M et al (2012) The global histone modification pattern correlates with overall survival in metachronous liver metastasis of colorectal cancer. Oncol Rep. 27(3):637–642. https://doi.org/10.3892/or.2011.1547

Tan S, Davey CA (2011) Nucleosome structural studies. Curr Opin Struct Biol. 21(1):128–136. https://doi.org/10.1016/j.sbi.2010.11.006

Tang W, Zhu Y, Gao J et al (2014) MicroRNA-29a promotes colorectal cancer metastasis by regulating matrix metalloproteinase 2 and E-cadherin via KLF4. Br J Cancer. 110(2):450–458. https://doi.org/10.1038/bjc.2013.724

Tazawa H, Tsuchiya N, Izumiya M, Nakagama H (2007) Tumor-suppressive miR-34a induces senescence-like growth arrest through modulation of the E2F pathway in human colon cancer cells. Proc Natl Acad Sci. 104(39):15472–15477. https://doi.org/10.1073/pnas.0707351104

Tessarz P, Kouzarides T (2014) Histone core modifications regulating nucleosome structure and dynamics. Nat Rev Mol Cell Biol. 15(11):703–708. https://doi.org/10.1038/nrm3890

Tie J, Cohen JD, Wang Y et al (2019) Circulating tumor DNA analyses as markers of recurrence risk and benefit of adjuvant therapy for stage III colon cancer. JAMA Oncol. 5(12):1710–1717. https://doi.org/10.1001/jamaoncol.2019.3616

Toden S, Okugawa Y, Jascur T et al (2015) Curcumin mediates chemosensitization to 5-fluorouracil through miRNA-induced suppression of epithelial-to-mesenchymal transition in chemoresistant colorectal cancer. Carcinogenesis. 36(3):355–367. https://doi.org/10.1093/carcin/bgv006

Toiyama Y, Takahashi M, Hur K et al (2013) Serum miR-21 as a diagnostic and prognostic biomarker in colorectal cancer. JNCI J Natl Cancer Inst. 105(12):849–859. https://doi.org/10.1093/jnci/djt101

Toyota M, Ahuja N, Ohe-Toyota M, Herman JG, Baylin SB, Issa JPJ (1999) CpG island methylator phenotype in colorectal cancer. Proc Natl Acad Sci. 96(15):8681–8686. https://doi.org/10.1073/pnas.96.15.8681

Toyota M, Ohe-Toyota M, Ahuja N, Issa JPJ (2000) Distinct genetic profiles in colorectal tumors with or without the CpG island methylator phenotype. Proc Natl Acad Sci. 97(2):710–715. https://doi.org/10.1073/pnas.97.2.710

Tu HC, Schwitalla S, Qian Z et al (2015) LIN28 cooperates with WNT signaling to drive invasive intestinal and colorectal adenocarcinoma in mice and humans. Genes Dev. 29(10):1074–1086. https://doi.org/10.1101/gad.256693.114

Turchinovich A, Weiz L, Langheinz A, Burwinkel B (2011) Characterization of extracellular circulating microRNA. Nucleic Acids Res. 39(16):7223–7233. https://doi.org/10.1093/nar/gkr254

Valeri N, Gasparini P, Braconi C et al (2010) MicroRNA-21 induces resistance to 5-fluorouracil by down-regulating human DNA MutS homolog 2 (hMSH2). Proc Natl Acad Sci. 107(49): 21098–21103. https://doi.org/10.1073/pnas.1015541107

Valladares-Ayerbes M, Blanco M, Haz M et al (2011) Prognostic impact of disseminated tumor cells and microRNA-17-92 cluster deregulation in gastrointestinal cancer. Int J Oncol. 39(5): 1253–1264. https://doi.org/10.3892/ijo.2011.1112

Van Rijnsoever M, Elsaleh H, Joseph D, McCaul K, Iacopetta B (2003) CpG island methylator phenotype is an independent predictor of survival benefit from 5-fluorouracil in stage III colorectal cancer. Clin Cancer Res Off J Am Assoc Cancer Res. 9(8):2898–2903

Vasudevan S, Tong Y, Steitz JA (2007) Switching from repression to activation: micrornas can up-regulate translation. Science. 318(5858):1931–1934. https://doi.org/10.1126/science.1149460

Vickers KC, Palmisano BT, Shoucri BM, Shamburek RD, Remaley AT (2011) MicroRNAs are transported in plasma and delivered to recipient cells by high-density lipoproteins. Nat Cell Biol. 13(4):423–433. https://doi.org/10.1038/ncb2210

Vogelstein B, Fearon ER, Hamilton SR et al (1988) Genetic alterations during colorectal-tumor development. N Engl J Med. 319(9):525–532. https://doi.org/10.1056/NEJM198809013190901

Vogt M, Munding J, Grüner M et al (2011) Frequent concomitant inactivation of miR-34a and miR-34b/c by CpG methylation in colorectal, pancreatic, mammary, ovarian, urothelial, and renal cell carcinomas and soft tissue sarcomas. Virchows Arch. 458(3):313–322. https://doi.org/10.1007/s00428-010-1030-5

Wallner M, Herbst A, Behrens A et al (2006) Methylation of serum DNA is an independent prognostic marker in colorectal cancer. Clin Cancer Res. 12(24):7347–7352. https://doi.org/10.1158/1078-0432.CCR-06-1264

Wang CJ, Stratmann J, Zhou ZG, Sun XF (2010) Suppression of microRNA-31 increases sensitivity to 5-FU at an early stage, and affects cell migration and invasion in HCT-116 colon cancer cells. BMC Cancer. 10(1):616. https://doi.org/10.1186/1471-2407-10-616

Wang DR, Tang D (2008) Hypermethylated SFRP2 gene in fecal DNA is a high potential biomarker for colorectal cancer noninvasive screening. World J Gastroenterol. 14(4):524–531. https://doi.org/10.3748/wjg.14.524

Wang J, Huang SK, Zhao M et al (2014a) Identification of a circulating microRNA signature for colorectal cancer detection. PLOS ONE. 9(4):e87451. https://doi.org/10.1371/journal.pone.0087451

Wang S, Xiang J, Li Z et al (2015) A plasma microRNA panel for early detection of colorectal cancer. Int J Cancer. 136(1):152–161. https://doi.org/10.1002/ijc.28136

Wang S, Yang MH, Wang XY, Lin J, Ding YQ (2014b) Increased expression of miRNA-182 in colorectal carcinoma: an independent and tissue-specific prognostic factor. Int J Clin Exp Pathol. 7(6):3498–3503

Wang Y, Li L, Cohen JD et al (2019) Prognostic potential of circulating tumor DNA measurement in postoperative surveillance of nonmetastatic colorectal cancer. JAMA Oncol. 5(8): 1118–1123. https://doi.org/10.1001/jamaoncol.2019.0512

Ward R, Meagher A, Tomlinson I et al (2001) Microsatellite instability and the clinicopathological features of sporadic colorectal cancer. Gut. 48(6):821–829. https://doi.org/10.1136/gut.48.6.821

Weisenberger DJ, Siegmund KD, Campan M et al (2006) CpG island methylator phenotype underlies sporadic microsatellite instability and is tightly associated with BRAF mutation in colorectal cancer. Nat Genet. 38(7):787–793. https://doi.org/10.1038/ng1834

Wightman B, Ha I, Ruvkun G (1993) Posttranscriptional regulation of the heterochronic gene lin-14 by lin-4 mediates temporal pattern formation in C. elegans. Cell. 75(5):855–862. https://doi.org/10.1016/0092-8674(93)90530-4

Wolin SL, Maquat LE (2019) Cellular RNA surveillance in health and disease. Science. 366(6467): 822–827. https://doi.org/10.1126/science.aax2957

Wu CW, Ng SSM, Dong YJ et al (2012) Detection of miR-92a and miR-21 in stool samples as potential screening biomarkers for colorectal cancer and polyps. Gut. 61(5):739–745. https://doi.org/10.1136/gut.2011.239236

Xiong B, Cheng Y, Ma L, Zhang C (2013) MiR-21 regulates biological behavior through the PTEN/PI-3 K/Akt signaling pathway in human colorectal cancer cells. Int J Oncol. 42(1): 219–228. https://doi.org/10.3892/ijo.2012.1707

Xu K, Liang X, Cui D, Wu Y, Shi W, Liu J (2013) miR-1915 inhibits Bcl-2 to modulate multidrug resistance by increasing drug-sensitivity in human colorectal carcinoma cells. Mol Carcinog. 52(1):70–78. https://doi.org/10.1002/mc.21832

Xu L, Zhang Y, Wang H, Zhang G, Ding Y, Zhao L (2014) Tumor suppressor miR-1 restrains epithelial-mesenchymal transition and metastasis of colorectal carcinoma via the MAPK and PI3K/AKT pathway. J Transl Med. 12(1):244. https://doi.org/10.1186/s12967-014-0244-8

Yamamoto E, Toyota M, Suzuki H et al (2008) LINE-1 hypomethylation is associated with increased CpG Island methylation in helicobacter pylori–related enlarged-fold gastritis. Cancer Epidemiol Biomarkers Prev. 17(10):2555–2564. https://doi.org/10.1158/1055-9965.EPI-08-0112

Yamauchi M, Lochhead P, Morikawa T et al (2012a) Colorectal cancer: a tale of two sides or a continuum? Gut. 61(6):794–797. https://doi.org/10.1136/gutjnl-2012-302014

Yamauchi M, Morikawa T, Kuchiba A et al (2012b) Assessment of colorectal cancer molecular features along bowel subsites challenges the conception of distinct dichotomy of proximal versus distal colorectum. Gut. 61(6):847–854. https://doi.org/10.1136/gutjnl-2011-300865

Yang MH, Yu J, Chen N et al (2013) Elevated MicroRNA-31 expression regulates colorectal cancer progression by repressing its target gene SATB2. PLOS ONE. 8(12):e85353. https://doi.org/10.1371/journal.pone.0085353

You JS, Jones PA (2012) Cancer genetics and epigenetics: two sides of the same coin? Cancer Cell. 22(1):9–20. https://doi.org/10.1016/j.ccr.2012.06.008

Yu G, Tang JQ, Tian ML et al (2012) Prognostic values of the miR-17-92 cluster and its paralogs in colon cancer. J Surg Oncol. 106(3):232–237. https://doi.org/10.1002/jso.22138

Zhang J, Guo H, Zhang H et al (2011) Putative tumor suppressor miR-145 inhibits colon cancer cell growth by targeting oncogene friend leukemia virus integration 1 gene. Cancer. 117(1):86–95. https://doi.org/10.1002/cncr.25522

Zhang JX, Song W, Chen ZH et al (2013a) Prognostic and predictive value of a microRNA signature in stage II colon cancer: a microRNA expression analysis. Lancet Oncol. 14(13): 1295–1306. https://doi.org/10.1016/S1470-2045(13)70491-1

Zhang L, Pickard K, Jenei V et al (2013b) miR-153 Supports Colorectal Cancer Progression via Pleiotropic Effects That Enhance Invasion and Chemotherapeutic Resistance. Cancer Res. 73(21):6435–6447. https://doi.org/10.1158/0008-5472.CAN-12-3308

Zhang R, Kang KA, Piao MJ et al (2014) Epigenetic alterations are involved in the overexpression of glutathione S-transferase π-1 in human colorectal cancers. Int J Oncol. 45(3):1275–1283. https://doi.org/10.3892/ijo.2014.2522

Zhao L, Li Y, Xu T et al (2022) Dendritic cell-mediated chronic low-grade inflammation is regulated by the RAGE-TLR4-PKCβ1 signaling pathway in diabetic atherosclerosis. Mol Med. 28(1):4. https://doi.org/10.1186/s10020-022-00431-6

Zhou Y, Wan G, Spizzo R et al (2014) miR-203 induces oxaliplatin resistance in colorectal cancer cells by negatively regulating ATM kinase. Mol Oncol. 8(1):83–92. https://doi.org/10.1016/j.molonc.2013.09.004

Zlobec I, Kovac M, Erzberger P et al (2010) Combined analysis of specific KRAS mutation, BRAF and microsatellite instability identifies prognostic subgroups of sporadic and hereditary colorectal cancer. Int J Cancer. 127(11):2569–2575. https://doi.org/10.1002/ijc.25265

Chapter 11
Epigenetic Alterations in Hematologic Malignancies

Emine Ikbal Atli

Abstract Epigenetic control is necessary for tissue homeostasis, which is preserved through the self-renewal and differentiation of somatic stem cells, as well as for development. Leukemia stem cells and self-renewing hematopoietic stem cells are both maintained by epigenetic regulators, according to mounting evidence. In hematologic malignancies, recent genome-wide comprehensive investigations have discovered mutations in genes that regulate epigenetic processes, including genes whose products change DNA and histones. Both cell-intrinsic and cell-extrinsic regulators, such as transcription factors, signal transduction pathways, and niche factors, affect hematopoietic stem cells. However, little is known about the process through which epigenetic regulators work in conjunction with these elements to maintain blood homeostasis. With an emphasis on the function of DNA-methylation modulators in hematopoietic cells and their offspring, we review current discoveries in the epigenetic control of hematopoiesis in this chapter.

Keywords DNA methylation · Decitabine · Ten-Eleven-Translocation · HDAC inhibitors

Abbreviations

AML	Acute myeloid leukemia
DNMTi	DNA-methyltransferase inhibitors
HATs	Histone acetyl transferases
KG	Ketoglutarate
Len	Lenalidomide
MDS	Myelodysplastic syndromes
MPN	Myeloproliferative neoplasms
OS	Overall survival

E. I. Atli (✉)
Faculty of Medicine, Department of Medical Genetics, Trakya University, Edirne, Turkey
e-mail: eikbalatli@trakya.edu.tr

PRC   Polycomb repressive complexes
TET   Ten-Eleven-Translocation

11.1 Introduction

The link between cancer and epigenetic control has been a promising area of study over the past 20 years (Stahl et al. 2016). In addition to genetic mutations, numerous oncogenic signals can also produce epigenetic dysregulation by changing the transcriptional patterns in transformed cells (Hayakawa et al. 2017). Patients with hematologic malignancies have altered levels of histone lysine methylation, phosphorylation, and acetylation, as well as altered DNA cytosine methylation and oxidized derivatives of methylated cytosines (Chung et al. 2012). Perhaps most significantly, epigenetic regulation has shown to be a crossroads of several crucial characteristics of cancer, including immunology, metabolism, or aging (Fig. 11.1). Although there are notable exceptions, many findings were initially presented in relation to hematologic malignancies. As a result, it has been difficult to illustrate how these discoveries apply to solid tumors. Similar to this, myelodysplastic syndromes and acute myeloid leukemias have shown to benefit from epigenetic targeting when treated with DNA hypomethylating drugs and, to a lesser extent, histone deacetylase inhibitors. Our arsenal of medications with epigenetic targets grew as our techniques for studying epigenetics advanced. The study of medications that target epigenetic regulators, including DOT1L, BET proteins, LSD1, and IDH1/2 inhibitors, is at the forefront of hematology research (Stahl et al. 2016). Numerous mechanisms such as abnormal DNA methylation, posttranslational histone tail modifications, noncoding RNAs, mutations in epigenetically regulated genes, dysregulation of enhancers, DNA-looping 3D chromosomal architecture, and RNA splicing are examples of epigenetic changes (Cruz-Rodriguez et al. 2018).

11.2 Hematological Cancers with Abnormal DNA Methylation

According to the genomic location of the methylated CpG sites, DNA methylation in humans is only found on the cytosine of CpG dinucleotides (cytosine followed by guanine). Methylation on promoter regions is typically associated with transcriptional silencing, whereas methylated gene bodies are found in transcriptionally active genes (Dimopoulos and Grønbaek 2019).

This process typically takes place in CpG islands, which are DNA areas with a high concentration of CpG sites. CpG islands are present in around 40% of the promoters of mammals and are typically unmethylated in genes with normal expression. Through the action of a family of proteins called DNA-methyltransferases

Fig. 11.1 Epigenetic dysregulation as a hallmark of cancer

(DNMTs), which consists of at least three distinct members – DNMT1, DNMT3A, and DNMT3B – promoter hypermethylation functions as a gene silencer. Cytosine residues in CG dinucleotides are methylated by DNMTs. DNMTs are crucial for hematopoietic stem cell (HSC) self-renewal, niche maintenance, and multilineage hematopoiesis differentiation in the normal hematopoiesis process. Initial research found that DNMT expression was aberrant in a variety of solid organ cancers. The discovery of DNMT3A mutations in acute myeloid leukemia (AML), however, has attracted a lot of attention from the hematology community. Recurrent DNMT3A mutations were discovered in 20% of AML patients using cutting-edge high-throughput DNA sequencing techniques. While tailored standard chemotherapy regimens with the addition of high-dose daunorubicin are associated with worse outcomes generally, overall survival (OS) is improved as a result of this connection. Uncertainty exists on how DNMT3A mutations in AML affect the sensitivity to hypomethylating drugs. Hypomethylating drugs appear to mitigate the detrimental effects of DNMT3A mutations, according to a retrospective analysis of small cohorts treated with various regimens, but more research is needed in prospective clinical trials. Somatic mutations either result in the protein being prematurely truncated or affect just one amino acid, R882, which causes a reduction in the amount of enzyme activity. It's interesting to note that heterozygous mutations are more frequent, and recent research has shown that the R882 DNMT3A mutation has a dominantly negative effect by inhibiting DNMT3A oligomerization. Patients with myeloproliferative neoplasms (MPN) and myelodysplastic syndromes (MDS) have also been found to have DNMT3A mutations, which are linked to a higher risk of developing into AML. In fact, some investigations have found that secondary AML has the same DNMT3A mutation as the antecedent hematologic illness, suggesting that these mutations may represent an early stage in the evolution of malignant clones (Fong et al. 2014).

11.2.1 Inhibitors of DNA-Methyltransferase

When it comes to treating acute leukemia, the cytidine analogs 5-azacytidine and 5-aza-20-deoxycytidine (or decitabine) first became available in the 1970s. Early clinical trials with relatively large dosages (ranging from 150 to 750 mg/m2) of the drug demonstrated modest antileukemic activity, but with considerable toxicity (Tahiliani et al. 2009; Ito et al. 2010; He et al. 2011). A revolutionary finding was made in the interim; a low dose of 5-azacytidine caused a decrease in DNA methylation in cell culture and resulted in the development of cardiac muscle cells from embryonic mouse cells, indicating that it was more than just a simple cytostatic drug because it could cause severe phenotypic changes at lower, noncytotoxic doses (Dimopoulos and Grønbaek 2019).

Later research revealed that 5-azacytidine, an epigenetic medication, had this impact via lowering DNA-methylation levels. As a result, reduced dosing regimens aiming for an epigenetic effect started to develop, and greater doses of 5-azacytidine administration used to induce a direct cytotoxic effect were abandoned. In a number of clinical trials, 5-azacytidine has since shown to be very effective in treating patients with myelodysplastic syndrome (MDS), increasing both the response rate and overall survival. Its subsequent FDA approval for the treatment of MDS patients represented the first time an epigenetic medication used in cancer therapy had received FDA approval. Due to the lack of an overall survival benefit, the FDA initially rejected the use of decitabine in clinical trials for MDS; however, the FDA has since approved the drug for the same indications as 5-azacytidine. In contrast, the European Medicines Agency (EMA) has approved both medications for different uses in Europe.

DNA-methyltransferase inhibitors (DNMTi) demonstrate their methods of action by integrating into the DNA of proliferating cells, while a smaller portion also gets metabolized to deoxycytidine derivatives that get incorporated into the DNA, where they covalently sequester DNMT1, targeting to p53. Decitabine is a deoxycytidine analog; thus, it incorporates exclusively into the DNA, whereas 5-aza cytidine includes. The initial methylation pattern gradually disappears during subsequent cell divisions because DNMT1 is primarily responsible for replicating the methylation pattern to the freshly generated DNA strand during replication. Different models have been put forth; however, it is still unclear how DNMT1 inhibition could have antitumor effects. The promoter demethylation and consequent reactivation of aberrantly silenced tumor suppressor genes have long been assumed to be the primary effects of DNMTi.

In comparison with acute myeloid leukemia (AML) or MDS, the clinical efficacy of DNMTi in lymphoid malignancies and multiple myeloma (MM) is less pronounced. Currently, decitabine is being evaluated as a single treatment for diffuse large B-cell lymphoma that has relapsed or become resistant (NCT03579082). Lack of efficacy forced the termination of a phase II, single-arm research examining the impact of 5-azacytidine in relapsed MM (NCT00412919). DNMTi's role in lymphomas and myelomas as a monotherapy is so debatable. Although further research is continuing, some studies have found that using DNMTi in addition to conventional chemotherapy can enhance clinical response in lymphomas and/or cause resensitization to earlier chemotherapy.

Another pilot study in myeloma looked at the effectiveness of using lenalidomide (Len) and 5-azacytidine as an induction therapy and autologous stem cell support in 17 individuals with newly diagnosed MM (NCT01050790).

There are other DNMTi besides 5-azacytidine and decitabine, which are FDA-approved DNMTi, that have demonstrated promise in preclinical and early clinical investigations. A new generation DNMTi called guadecitabine (SGI-1010) has more prominent immunomodulatory effects than its predecessors due to its resistance to cytidine deaminase degradation and extended half-life and in vivo exposure. Guadecitabine was well tolerated and physiologically active when administered subcutaneously to individuals with MDS and AML for 5 days, according to a phase I research. Oral decitabine (ASTX727), which combines decitabine with a cytidine deaminase inhibitor (cedazuridine or E7727) to avoid first-pass clearance and boost its bioavailability after oral intake, is the most recent development in DNMTi. The American Society of Hematology (ASH) published preliminary findings from a phase II study comparing ASTX727 with intravenous decitabine in patients with MDS in 2017, showing comparable pharmacokinetics and pharmacodynamics, safety profile, and response rates between the two treatments. These findings are still pending confirmation in larger clinical trials (Srivastava et al. 2014; Issa et al. 2015; Kantarjian et al. 2017; Pan et al. 2020).

11.2.2 The TET Enzymes and Hydroxymethylation of DNA

DNA methylation was always thought to be a rather stable DNA alteration, but new discoveries of the Ten-Eleven-Translocation (TET) enzymes and genome-wide high resolution mapping of 5mC during cellular differentiation have shown a more dynamic situation (Liang et al. 2011; Tahiliani et al. 2009; Ito et al. 2010).

The three TET enzymes (TET1–3) are Fe2+ dependent dioxygenases that catalyze the sequential oxidation of 5mC to 5hmC, 5-formylcytosine, and 5-carboxycytosine. They are -ketoglutarate (-KG) and enzymes reliant on TET (He et al. 2011; Ito et al. 2011). It is clear that the 5mC derivatives are crucial for the control of transcription, even if their precise function is still not entirely known. The binding and recruitment of chromatin regulators, such as the polycomb repressive complexes (PRC), as well as their role in the reversal of transcriptional silence have all been demonstrated to depend on them as crucial stages in both active and passive DNA demethylation. Furthermore, identification of the 5hmC gene in mouse embryonic stem cells has revealed by context-dependent promoter hypomethylation of pluripotency components or modification of PRC recruitment, its function in the formation and maintenance of pluripotency (Wu and Zhang 2011). SNP arrays that identified a minimally deleted area on chromosome 4q24.5 allowed for the initial description of TET2 mutations in myeloid malignancies in MDS and MPN. TET2 has now been discovered to be mutated in myeloid malignancies such as AML, MDS, and MPN, with a significant percentage of individuals with these diseases carrying mutations (Langemeijer et al. 2009, Solary et al. 2014, Abdel-Wahab et al. 2009). Patients with normal karyotypes are more likely to have TET2 mutations, which are linked to worse overall survival in AML and CMML but are not connected with clinical outcome in MDS and MPN. TET2 mutations may serve as a biomarker for responsiveness to hypomethylating drugs even if they do not have a good predictive link with clinical outcome in MDS (Itzykson et al. 2011).

11.3 The Relationship Between Metabolism and Epigenetics and –Ketoglutarate

Approximately 20% of AML genomes have recurrent somatic mutations of the cytosolic enzyme IDH1 or its mitochondrial counterpart IDH2, while other hematologic malignancies have these mutations less often. Specific epigenetic fingerprints are linked to these anomalies in a crucial cellular metabolic system. Isocitrate is typically converted to -KG by the actions of IDH1 and IDH2. But the most prevalent mutations in IDH1 (R132) and IDH2 (R140, R172) lead to the development of neomorphic enzyme activity, resulting in high intracellular quantities of the aberrant oncometabolite 2-hydroxyglutate (2-HG). Competitive inhibition of Fe2+ and -KG dependent demethylases, such as the TET enzymes and JmjC domain-containing lysine demethylases, is brought about by 2-HG, a structural homolog of -KG (KDMs). The effects of inhibition include altered gene expression, abnormal DNA and histone methylation, and poor lineage-specific differentiation. IDH1/2 and TET2 mutations are mutually exclusive yet linked to overlapping distinct hypermethylation signatures, supporting a shared involvement in AML etiology. The idea that "oncometabolites" like 2-HG play a significant role in carcinogenesis as a result of the identification of IDH mutations has been further supported by the combination of epigenetics and metabolomics in cancer. It has been shown that IDH1/2 mutations affect the outcomes of patients who are considered to be at low molecular risk (NPM1-mutant/FLT3-ITD negative), with IDH1-R132 and IDH2-R172 mutant alleles being linked to poor results and IDH2-R140 mutations being linked to good outcomes. Through the activation of differentiation and death in IDH mutant leukemia cell lines, a variety of new small molecule inhibitors that target the abnormal gain-of-function resulting from mutant IDH alleles have recently shown potential selective in vitro efficacy.

11.4 Histone Modification

Our genome is packaged and organized by histones, which function as spools around DNA strands to produce the nucleosome structure. Histones are highly conserved proteins. Histones are essential for controlling transcriptional activities because they compress DNA and modify chromatin structure in response to various inputs. Recently, posttranslational alterations of histones have been identified as another epigenetic modification implicated in tumor formation in addition to DNA methylation.

Acute leukemias usually contain dysregulated levels of several histone posttranslational modifications (PTMs), including acetylation, ubiquitination, phosphorylation, and methylation (Wouters and Delwel 2016, Greenblatt and Nimer 2014).

Histone acetylation is a well-studied PTM that is crucial for chromatin remodeling and is controlled by the activity of histone acetyl transferases (HATs) and histone deacetylases (HDACs). Gene expression or gene suppression is made possible by the actions of HATs and HDACs, which add and take away acetyl moiety from the lysine residues of histones, respectively (Kurdistani and Grunstein 2003).

Numerous cellular processes, including the control of gene expression, replication, and DNA repair, have been linked to histone modifications, and mutations in the genes responsible for histone lysine acetylation are seen in many forms of cancer (Bannister and Kouzarides 2011; Chalmers et al. 2017).

One of these genes, which is produced by the histone acetyltransferase CREBBP, which may acetylate different residues in a number of histones, has been linked to lymphoid leukemia on a regular basis (Mullighan et al. 2011, Pasqualucci et al. 2011).

Due to loss of HAT activity and transcriptional dysregulation, these mutations or deletions in CREBBP have been found to be very prevalent in patients with relapsed and high hyperdiploid B-cell precursor acute lymphoblastic leukemia (B-ALL). This finding raises the possibility that they play a part in chemotherapy resistance (Inthal et al. 2012, Malinowska-Ozdowy et al. 2015). It has been shown that the survival of ALL patients depends on the different expression of genes involved in histone deacetylation (Gruhn et al. 2013).

The phosphorylation and methylation of histones, like histone acetylation, are mechanisms that control gene expression by altering the chromatin structure. Less is known about the functions of these alterations in leukemia, but numerous histone methyltransferases, including MLL1, EZH2, NSD1, and SET7/9, which are often disturbed in hematologic malignancies, have been identified to be involved in the pathogenesis of B-ALL. These genes' mutations and altered expression are thought to be linked to target gene overexpression and abnormal histone methylation. Histone phosphorylation regulates a number of processes, including transcription, chromatin condensation, mitosis, apoptosis, and DNA replication, through the addition and removal of the modification by kinases and phosphatases, respectively (Bannister and Kouzarides 2011).

11.4.1 Acetylation

Lysine acetyltransferases (KATs) and histone deacetylases are two opposing groups of enzymes that dynamically regulate histone acetylation, one of the most studied histone modifications (HDACs). The acetyl group is transferred from the common cofactor acetyl CoA to the -amino group of lysine side chains in histones as a result of KATs' enzymatic activity (Shahbazian and Grunstein 2007, Kleff et al. 1995). As a result, the chromatin is in an open conformation, making it easier for chromatin-associated proteins to access it. This is functionally compatible with the finding of KATs as transcriptional coactivators. Based on their intracellular location, KATs can be classified as either type A (predominantly nuclear) or type B (predominantly cytoplasmic) subtypes. Type A KATs are the enzymes that make up the CBP/p300, MYST, and GNAT families. Many hematologic malignancies have recurrent mutations in CBP and p300, particularly lymphoid neoplasms. Similar to this, myeloid malignancies include chromosomal translocations involving KATs, such as MLL-CBP and MOZ-TIF2. In mouse models of MLL-CBP leukemia, it was specifically shown that the KAT domain and bromodomain of CBP are necessary for leukemic transition after an initial myeloproliferative phase (Lavau et al. 2000; Mullighan et al. 2011).

Similar to the previous example, the MOZ-TIF2 fusion protein is adequate for leukemic transformation due to its capacity to bind nucleosomes and attract CBP to abnormal locations, which causes the activation of a self-renewal mechanism and the acquisition of stem cell characteristics (Huntly et al. 2004).

KATs can influence protein-protein interactions and the activity of target nonhistone proteins using their acetyltransferase activity, which is not restricted to histone substrates. For instance, it has been shown that leukemogenicity and the ability to self-renew are only conferred by KAT3B (p300) acetylating the leukemic fusion protein AML1-ETO. In a mouse AML1-ETO model, pharmacological suppression of KAT3B improves survival (Wang et al. 2011). In general, KATs' limited substrate specificity and extensive participation in multi-protein complexes that characterize their molecular function have so far made therapeutic targeting of KATs difficult. It's interesting to note that C646 is a small molecule p300/CBP inhibitor that has just been discovered using a structure-based in silico technique. Through cell cycle arrest and death, C646 selectively inhibited primary human AML with the AML1-ETO translocation in vitro. This was accompanied by a dose-dependent decrease in global histone H3 acetylation as well as c-kit and bcl-2 expression. AML cells (Gao et al. 2013).

The stability of local chromatin architecture is brought about by HDACs' ability to undo lysine acetylation, which restores the positive charge and is compatible with their primary function as transcriptional repressors (Bolden et al. 2006).

On the basis of sequence homology, 18 human isoenzymes of HDACs have been discovered and are divided into four groups. HDACs, like KATs, may specifically target both histone and nonhistone proteins, with the members of component protein complexes determining the substrate (Bannister and Kouzarides 2011).

Though HDAC inhibitors have been extensively tested in a variety of malignancies, recurring mutations of HDACs are not seen in cancer genomes. This is mostly due to the improper recruitment of these cells by different oncoproteins to unsuitably start or maintain malignant gene expression programs. For example, it has been demonstrated that the leukemic fusion proteins PML-RAR and PZLF-RAR attract repressor complexes that include the HDAC enzyme, causing abnormal gene silencing. The use of HDAC inhibitors (HDACi) improves survival in mice models of APML by potentiating or reversing the retinoid-induced differentiation of retinoic acid sensitive and resistant tumors. HDACi has been proven to be effective in the treatment of cutaneous T-cell lymphoma. However, more clinical evidence supporting the use of this class of treatments in treating other hematologic malignancies is still pending. HDACi mostly work by specifically preventing the entrance of necessary cofactors to the active site (Bojang Jr and Ramos 2014).

11.4.2 Inhibitors of Histone Deacetylase

Histone acetyltransferases (HATs) and histone deacetylases are two distinct types of enzymes that regulate the acetylation of histones, a key epigenetic mark (HDACs). The acetylation of histone lysine residues by HATs results in a more "open" chromatin structure that promotes transcriptional activity by neutralizing histone lysine residues' positive charge and decreasing their interaction with negatively charged DNA strands. HDACs are an enzyme family that eliminates acetyl groups from histone lysine residues. There are 4 main subclasses among the 18 HDACs that have been found so far (Barneda-Zahonero and Parra 2012).

Class I includes the nuclear-only HDACs 1, 2, 3, and 8; class II includes the nuclear and cytoplasmically localized HDACs 4, 5, 6, 7, 9, and 10; class III includes the sirtuin protein family; and class IV includes the HDAC11, which is only found in the cytoplasm (Dimopoulos et al. 2014). HDACs interact with other proteins in addition to histones, as evidenced by the fact that they are also found in the cytoplasm. In fact, research has demonstrated that important carcinogenic proteins like p53, NF-kB, c-MYC, and STAT3 directly interact with HDACs (Ashburner et al. 2001; Gupta et al. 2012; Nebbioso et al. 2017). HDACs have pleiotropic properties in addition to being nonspecific to histones. a variety of biological processes, including cell cycle control, stress response, protein breakdown, cytokine signaling, and apoptosis. In addition to not being histone-specific, HDACs also have pleiotropic action and are involved in a wide range of physiological processes, including cell cycle control, stress response, protein breakdown, cytokine signaling, and apoptosis (New et al. 2012).

The antitumor effect of HDACi was recognized early, and preclinical evidence demonstrated a specifically improved efficacy of romidepsin and vorinostat against T-cell lymphomas (Richard et al. 2004). Preclinical data show a specifically improved efficacy of vorinostat (suberoylanilide hydroxamic acid (SAHA)) and romidepsin against T-cell lymphomas, demonstrating the antitumor activity of HDACi.

Vorinostat became the second epigenetic medication to be approved for the treatment of a hematologic malignancy as a result of being used to treat CTCL. Similar to this, romidepsin was found to be effective in two further clinical trials for CTCL, and the FDA also approved it for the treatment of relapsed/refractory CTCL (Duvic et al. 2007; Olsen et al. 2007; Mann et al. 2007; Piekarz et al. 2009; Whittaker et al. 2010). Romidepsin is now being studied in about 50 studies, either as monotherapy or in combination with other medications, primarily for the treatment of T-cell lymphomas, according to clinicaltrials.gov as of this writing.

Numerous early preclinical investigations have shown that several HDACi demonstrate strong anti-myeloma action in vitro, even at very low dosages, apart from T-cell lymphomas. Early clinical studies using HDACi as a monotherapy to treat MM, however, demonstrated poor efficacy, with only panobinostat and vorinostat exhibiting low response rates. Although more preclinical research revealed that HDACi may increase the toxicity of other medications, it was not fully abandoned that HDACi be used to treat myeloma. This strengthened the case for a combinatorial strategy, particularly when used in conjunction with proteasome inhibitors. The FDA approved panobinostat for the therapy after receiving sufficient data. Panobinostat, along with bortezomib, dexamethasone, and panobinostat, was the latest epigenetic medication to receive FDA clearance for the treatment of recurrent MM (Kuruvilla et al. 2008, Deleu et al. 2009, San-Miguel et al. 2014, Corrales-Medina et al. 2015). HDACi are effective against lymphomas and myeloma, but they appear to have little clinical effect on myeloid malignancies. Without providing any proof of effectiveness, panobinostat, vorinostat, and belinostat have all been studied as monotherapies in AML. As an alternative, the use of HDACi with a narrower target range may also enhance the antitumor activity with a better toxicity profile. HDAC6, a cytoplasmic HDAC (and not a genuine epigenetic target), is an intriguing example. It is essential for the destruction of misfolded proteins because it promotes the development of the aggresome, which is a secondary mechanism to the proteasome. Therefore, concurrent inhibition of the proteasome and HDAC6 will result in a preclinical scenario; a synergy between bortezomib and ACY-1215 (ricolinostat), a particular HDAC6 inhibitor, led to the accumulation of misfolded proteins and induced cell death (Santo et al. 2012).

Ricolinostat is now being evaluated in B-cell lymphomas (NCT0209 1063 and NCT02787369), as well as MM, in conjunction with bortezomib, lenalidomide, and pomalidomide (NCT01583283, NCT01997840).

11.5 Noncoding RNAs

Short noncoding RNAs called microRNAs (miRNAs) target particular cellular mRNA to modify gene expression patterns and cellular communication pathways. They are significant epigenetic regulators of gene expression. They work via post-transcriptional gene silencing, controlling other epigenetic regulators and modifying the expression of protein-coding genes. They are encoded inside intergenic regions, or within the introns or exons of protein-coding genes of the genome. There is growing evidence that miRNA expression varies between the healthy and disease state, suggesting disease specific methylation patterns. miRNAs are involved in a wide range of biological processes, are frequently dysregulated in human cancers, and are expressed differently in the healthy and disease states (Van den Hove et al. 2014; Piletič and Kunej 2016).

The path of lymphoid precursor maturation is influenced by miRNAs, which have been shown to express themselves differently at different phases of lymphopoiesis. For instance, changes in the expression of miRNA-150 and miRNA-155, which are involved in the differentiation of B and T cells, prevent the transition of immature hematopoietic cells into mature cells. B and T lymphoid precursors express the miRNAs miRNA-17, miRNA-18a, miRNA-19a, miRNA-20a, miRNA-19b-1, and miRNA-92-1, and their lack causes BIM, the miRNAs' target protein, to be produced at higher levels (Luan et al. 2015; Schotte et al. 2012; Ventura et al. 2008).

As a result, several miRNA modifications have been described that can be exploited for differential diagnosis, classification, prognosis, and treatment in ALL. These alterations include aberrant expression of miRNAs implicated in various tumor processes. MiRNA-708 is highly expressed in TEL-AML1, BCR-ABL, E2A-PBX1, hyperdiploid, and other B-cell malignancies in comparison with MLL-rearranged ALL and T-ALL, demonstrating the distinct miRNA signatures of each subtype, according to some authors who have discovered differential expression of miRNAs in ALL subtypes (Heo et al. 2019; Schotte et al. 2011).

miRNA signatures can be utilized for more than just the diagnosis of ALL; through controlling cell growth and death, a number of miRNAs can affect the prognosis of patients with acute leukemia. In contrast to cells from healthy donors, acute leukemias were discovered to have greater levels of miRNA-92a and miRNA-16 expression. Higher levels of miRNA-9, miR-33, miR-92a, miR-142-3p, miR-146a, miR-181a/c, miR-210, miR-215, miR-369-5p, miR-335, miR-454, miR-496, miR-518d, and miR-599 expression (Ohyashiki et al. 2010).

The phosphorylation of retinoblastoma (Rb) and enhanced cell proliferation brought on by miR-124a downregulation result in ALL patients having a worse prognosis and greater death rates. In children with ALL, higher miR-128b expression at diagnosis indicated a better prognosis and prednisolone response (Nemes et al. 2015).

Additionally, several studies show a substantial correlation between the expressions of the genes miR-10a, miR-134, miR-214, miR-221, miR-128b, miR-484, miR-572, miR-580, miR-624, and miR-627 with a positive clinical result (Wang et al. 2010).

11.6 Conclusion

More precise diagnostic and prognostic data are made possible by the growing use of high-throughput genetic technology in clinical settings. This could eventually lead to the development of customized medicine. To fully benefit from these developments, it is necessary to understand the molecular processes that lead to malignant transformation. Aberrant control of epigenetic processes has become a key unifying factor in hematologic malignancies, despite their high degree of heterogeneity. Thus, in the ongoing search for more effective treatments, the hematologic malignancies serve as useful models to examine important epigenetic pathways and nodes of regulation.

Conflict of Interest The authors declare that they have no conflicts of interest.

Compliance with Ethical Standards This article does not contain any studies involving human participants performed by any of the authors.

References

Abdel-Wahab O, Mullally A, Hedvat C et al (2009) Genetic characterization of TET1, TET2, and TET3 alterations in myeloid malignancies. Blood 114(1):144–147. https://doi.org/10.1182/blood-2009-03-210039

Ashburner BP, Westerheide SD, Baldwin AS Jr (2001) The p65 (RelA) subunit of NF-kappaB interacts with the histone deacetylase (HDAC) corepressors HDAC1 and HDAC2 to negatively regulate gene expression. Mol Cell Biol 21(20):7065–7077. https://doi.org/10.1128/MCB.21.20.7065-7077.2001

Bannister AJ, Kouzarides T (2011) Regulation of chromatin by histone modifications. Cell Res 21(3):381–395. https://doi.org/10.1038/cr.2011.22

Barneda-Zahonero B, Parra M (2012) Histone deacetylases and cancer. Mol Oncol 6(6):579–589. https://doi.org/10.1016/j.molonc.2012.07.003

Bojang P Jr, Ramos KS (2014) The promise and failures of epigenetic therapies for cancer treatment. Cancer Treat Rev 40(1):153–169. https://doi.org/10.1016/j.ctrv.2013.05.009

Bolden JE, Peart MJ, Johnstone RW (2006) Anticancer activities of histone deacetylase inhibitors. Nat Rev Drug Discov 5(9):769–784. https://doi.org/10.1038/nrd2133

Chalmers ZR, Connelly CF, Fabrizio D et al (2017) Analysis of 100,000 human cancer genomes reveals the landscape of tumor mutational burden. Genome Med 9(1):34. Published 2017 Apr 19:34. https://doi.org/10.1186/s13073-017-0424-2

Chung YR, Schatoff E, Abdel-Wahab O (2012) Epigenetic alterations in hematopoietic malignancies. Int J Hematol 96(4):413–427. https://doi.org/10.1007/s12185-012-1181-z

Corrales-Medina FF, Manton CA, Orlowski RZ, J. (2015) Chandra efficacy of panobinostat and marizomib in acute myeloid leukemia and bortezomib-resistant models. Leuk Res 39(3):371–379

Cruz-Rodriguez, N., Combita, A.L., Zabaleta, J. (2018). Epigenetics in Hematological Malignancies. In: Dumitrescu, R., Verma, M. (eds) Cancer Epigenetics for Precision Medicine . Methods in Molecular Biology, vol 1856. Humana Press, New York, NY https://doi.org/10.1007/978-1-4939-8751-1_5

Deleu S, Lemaire M, Arts J et al (2009) Bortezomib alone or in combination with the histone deacetylase inhibitor JNJ-26481585: effect on myeloma bone disease in the 5T2MM murine model of myeloma. Cancer Res 69(13):5307–5311. https://doi.org/10.1158/0008-5472.CAN-08-4472

Dimopoulos K, Gimsing P, Grønbæk K (2014) The role of epigenetics in the biology of multiple myeloma. Blood Cancer J 4(5):e207. Published 2014 May 2. https://doi.org/10.1038/bcj.2014.29

Dimopoulos K, Grønbaek K (2019) Epigenetic therapy in hematological cancers. APMIS 127(5):316–328. https://doi.org/10.1111/apm.12906

Duvic M, Talpur R, Ni X et al (2007) Phase 2 trial of oral vorinostat (suberoylanilide hydroxamic acid, SAHA) for refractory cutaneous T-cell lymphoma (CTCL) [published correction appears in Blood. 2007 Jun 15;109(12):5086]. Blood 109(1):31–39. https://doi.org/10.1182/blood-2006-06-025999

Fong CY, Morison J, Dawson MA (2014) Epigenetics in the hematologic malignancies. Haematologica 99(12):1772–1783. https://doi.org/10.3324/haematol.2013.092007

Gao XN, Lin J, Ning QY et al (2013) A histone acetyltransferase p300 inhibitor C646 induces cell cycle arrest and apoptosis selectively in AML1-ETO-positive AML cells. PLoS One 8(2):e55481. https://doi.org/10.1371/journal.pone.0055481

Greenblatt S, Nimer S (2014) Chromatin modifiers and the promise of epigenetic therapy in acute leukemia. Leukemia 28:1396–1406. https://doi.org/10.1038/leu.2014.94

Gruhn B, Naumann T, Gruner D et al (2013) The expression of histone deacetylase 4 is associated with prednisone poor-response in childhood acute lymphoblastic leukemia. Leuk Res 37(10):1200–1207. https://doi.org/10.1016/j.leukres.2013.07.016

Gupta M, Han J, Stenson M et al (2012) Regulation of STAT3 by histone deacetylase-3 in diffuse large B-cell lymphoma: implications for therapy. Leukemia 26:1356–1364. https://doi.org/10.1038/leu.2011.340

Hayakawa J, Kanda J, Akahoshi Y et al (2017) Meta-analysis of treatment with rabbit and horse antithymocyte globulin for aplastic anemia. Int J Hematol 105(5):578–586. https://doi.org/10.1007/s12185-017-2179-3

He YF, Li BZ, Li Z et al (2011) Tet-mediated formation of 5-carboxylcytosine and its excision by TDG in mammalian DNA. Science 333(6047):1303–1307. https://doi.org/10.1126/science.1210944

Heo JN, Kim DY, Lim SG et al (2019) ER stress differentially affects pro-inflammatory changes induced by mitochondrial dysfunction in the human monocytic leukemia cell line, THP-1. Cell Biol Int 43(3):313–322. https://doi.org/10.1002/cbin.11103

Huntly BJ, Shigematsu H, Deguchi K et al (2004) MOZ-TIF2, but not BCR-ABL, confers properties of leukemic stem cells to committed murine hematopoietic progenitors. Cancer Cell 6(6):587–596. https://doi.org/10.1016/j.ccr.2004.10.015

Inthal A, Zeitlhofer P, Zeginigg M et al (2012) CREBBP HAT domain mutations prevail in relapse cases of high hyperdiploid childhood acute lymphoblastic leukemia. Leukemia 26:1797–1803. https://doi.org/10.1038/leu.2012.60

Issa JJ, Roboz G, Rizzieri D, Jabbour E, Stock W, O'Connell C, Yee K, Tibes R, Griffiths EA, Walsh K, Daver N, Chung W, Naim S, Taverna P, Oganesian A, Hao Y, Lowder JN, Azab M, Kantarjian H (2015 Sep) Safety and tolerability of guadecitabine (SGI-110) in patients with myelodysplastic syndrome and acute myeloid leukaemia: a multicentre, randomised, dose-escalation phase 1 study. Lancet Oncol 16(9):1099–1110. https://doi.org/10.1016/S1470-2045 (15)00038-8. Epub 2015 Aug 19. PMID: 26296954; PMCID: PMC5557041.

Ito S, D'Alessio AC, Taranova OV et al (2010) Role of Tet proteins in 5mC to 5hmC conversion, ES-cell self-renewal and inner cell mass specification. Nature 466(7310):1129–1133. https://doi.org/10.1038/nature09303

Ito S, Shen L, Dai Q et al (2011) Tet proteins can convert 5-methylcytosine to 5-formylcytosine and 5-carboxylcytosine. Science 333(6047):1300–1303. https://doi.org/10.1126/science.1210597

Itzykson R, Kosmider O, Cluzeau T et al (2011) Impact of TET2 mutations on response rate to azacitidine in myelodysplastic syndromes and low blast count acute myeloid leukemias. Leukemia 25(7):1147–1152. https://doi.org/10.1038/leu.2011.71

Kantarjian HM, Roboz GJ, Kropf PL et al (2017) Guadecitabine (SGI-110) in treatment-naive patients with acute myeloid leukaemia: phase 2 results from a multicentre, randomised, phase 1/2 trial. Lancet Oncol 18(10):1317–1326. https://doi.org/10.1016/S1470-2045(17)30576-4

Kleff S, Andrulis ED, Anderson CW, Sternglanz R (1995) Identification of a gene encoding a yeast histone H4 acetyltransferase. J Biol Chem 270(42):24674–24677. https://doi.org/10.1074/jbc.270.42.24674

Kurdistani SK, Grunstein M (2003) Histone acetylation and deacetylation in yeast. Nat Rev Mol Cell Biol 4(4):276–284. https://doi.org/10.1038/nrm1075

Kuruvilla J, Pintilie M, Tsang R, Nagy T, Keating A, Crump M (2008) Salvage chemotherapy and autologous stem cell transplantation are inferior for relapsed or refractory primary mediastinal large B-cell lymphoma compared with diffuse large B-cell lymphoma. Leuk Lymphoma 49(7): 1329–1336. https://doi.org/10.1080/10428190802108870

Langemeijer SM, Kuiper RP, Berends M et al (2009) Acquired mutations in TET2 are common in myelodysplastic syndromes. Nat Genet 41(7):838–842. https://doi.org/10.1038/ng.391

Lavau C, Du C, Thirman M, Zeleznik-Le N (2000) Chromatin-related properties of CBP fused to MLL generate a myelodysplastic-like syndrome that evolves into myeloid leukemia. EMBO J 19(17):4655–4664. https://doi.org/10.1093/emboj/19.17.4655

Liang P, Song F, Ghosh S et al (2011) Genome-wide survey reveals dynamic widespread tissue-specific changes in DNA methylation during development. BMC Genomics 12(1):231. Published 2011 May 11. https://doi.org/10.1186/1471-2164-12-231

Luan C, Yang Z, Chen B (2015) The functional role of microRNA in acute lymphoblastic leukemia: relevance for diagnosis, differential diagnosis, prognosis, and therapy. Onco Targets Ther 8: 2903–2914. https://doi.org/10.2147/OTT.S92470. PMID: 26508875; PMCID: PMC4610789

Malinowska-Ozdowy K, Frech C, Schönegger A et al (2015) KRAS and CREBBP mutations: a relapse-linked malicious liaison in childhood high hyperdiploid acute lymphoblastic leukemia. Leukemia 29(8):1656–1667. https://doi.org/10.1038/leu.2015.107

Mann BS, Johnson JR, Cohen MH, Justice R, Pazdur R (2007) FDA approval summary: vorinostat for treatment of advanced primary cutaneous T-cell lymphoma. Oncologist 12(10):1247–1252. https://doi.org/10.1634/theoncologist.12-10-1247

Mullighan CG, Zhang J, Kasper LH et al (2011) CREBBP mutations in relapsed acute lymphoblastic leukaemia. Nature 471(7337):235–239. https://doi.org/10.1038/nature09727

Nebbioso A, Carafa V, Conte M et al (2017) C-Myc modulation and acetylation is a key HDAC inhibitor target in cancer. Clin Cancer Res 23(10):2542–2555. https://doi.org/10.1158/1078-0432.CCR-15-2388

Nemes K, Csóka M, Nagy N et al (2015) Expression of certain leukemia/lymphoma related microRNAs and its correlation with prognosis in childhood acute lymphoblastic leukemia. Pathol Oncol Res 21(3):597–604. https://doi.org/10.1007/s12253-014-9861-z

New M, Olzscha H, La Thangue NB (2012) HDAC inhibitor-based therapies: can we interpret the code? Mol Oncol 6(6):637–656. https://doi.org/10.1016/j.molonc.2012.09.003

Ohyashiki JH, Umezu T, Kobayashi C et al (2010) Impact on cell to plasma ratio of miR-92a in patients with acute leukemia: in vivo assessment of cell to plasma ratio of miR-92a. BMC Res Notes 3:347. https://doi.org/10.1186/1756-0500-3-347

Olsen EA, Kim YH, Kuzel TM et al (2007) Phase IIb multicenter trial of vorinostat in patients with persistent, progressive, or treatment refractory cutaneous T-cell lymphoma. J Clin Oncol 25(21): 3109–3115. https://doi.org/10.1200/JCO.2006.10.2434

Pan D, Rampal R, Mascarenhas J (2020) Clinical developments in epigenetic-directed therapies in acute myeloid leukemia. Blood Adv. 4(5):970–982. https://doi.org/10.1182/bloodadvances.2019001245. Erratum in: Blood Adv. 2020 Apr 14;4(7):1220. PMID: 32150613; PMCID: PMC7065485

Pasqualucci L, Dominguez-Sola D, Chiarenza A et al (2011) Inactivating mutations of acetyltransferase genes in B-cell lymphoma. Nature 471(7337):189–195. https://doi.org/10.1038/nature09730

Piekarz RL, Frye R, Turner M et al (2009) Phase II multi-institutional trial of the histone deacetylase inhibitor romidepsin as monotherapy for patients with cutaneous T-cell lymphoma. J Clin Oncol 27(32):5410–5417. https://doi.org/10.1200/JCO.2008.21.6150

Piletič K, Kunej T (2016) MicroRNA epigenetic signatures in human disease. Arch Toxicol 90(10): 2405–2419. https://doi.org/10.1007/s00204-016-1815-7

Richard LP, Robert WR, Zhan Z et al (2004) T-cell lymphoma as a model for the use of histone deacetylase inhibitors in cancer therapy: impact of depsipeptide on molecular markers, therapeutic targets, and mechanisms of resistance. Blood 103(12):4636–4643

San-Miguel JF, Hungria VT, Yoon SS et al (2014) Panobinostat plus bortezomib and dexamethasone versus placebo plus bortezomib and dexamethasone in patients with relapsed or relapsed and refractory multiple myeloma: a multicentre, randomised, double-blind phase 3 trial [published correction appears in Lancet Oncol. 2015 Jan;16(1):e6]. Lancet Oncol 15(11): 1195–1206. https://doi.org/10.1016/S1470-2045(14)70440-1

Santo L, Hideshima T, Kung AL et al (2012) Preclinical activity, pharmacodynamic, and pharmacokinetic properties of a selective HDAC6 inhibitor, ACY-1215, in combination with bortezomib in multiple myeloma. Blood 119(11):2579–2589. https://doi.org/10.1182/blood-2011-10-387365

Schotte D, De Menezes RX, Akbari Moqadam F et al (2011) MicroRNA characterize genetic diversity and drug resistance in pediatric acute lymphoblastic leukemia [published correction appears in Haematologica. 2011 Aug;96(8):1240]. Haematologica 96(5):703–711. https://doi.org/10.3324/haematol.2010.026138

Schotte D, Pieters R, Den Boer ML (2012) MicroRNAs in acute leukemia: from biological players to clinical contributors. Leukemia 26(1):1–12. https://doi.org/10.1038/leu.2011.151

Shahbazian MD, Grunstein M (2007) Functions of site-specific histone acetylation and deacetylation. Annu Rev Biochem 76(1):75–100

Solary E, Bernard OA, Tefferi A, Fuks F, Vainchenker W (2014) The ten-eleven Translocation-2 (TET2) gene in hematopoiesis and hematopoietic diseases. Leukemia 28(3):485–496. https://doi.org/10.1038/leu.2013.337

Srivastava P, Paluch BE, Matsuzaki J et al (2014) Immunomodulatory action of SGI-110, a hypomethylating agent, in acute myeloid leukemia cells and xenografts. Leuk Res 38:1332–1341

Stahl M, Kohrman N, Gore SD, Kim TK, Zeidan AM, Prebet T (2016) Epigenetics in cancer: a hematological perspective. PLoS Genet 12(10):e1006193. https://doi.org/10.1371/journal.pgen.1006193. PMID: 27723796; PMCID: PMC5065123

Tahiliani M, Koh KP, Shen Y et al (2009) Conversion of 5-methylcytosine to 5-hydroxymethylcytosine in mammalian DNA by MLL partner TET1. Science 324(5929): 930–935. https://doi.org/10.1126/science.1170116

Van den Hove DL, Kompotis K, Lardenoije R et al (2014) Epigenetically regulated microRNAs in Alzheimer's disease. Neurobiol Aging 35(4):731–745. https://doi.org/10.1016/j.neurobiolaging.2013.10.082

Ventura A, Young AG, Winslow MM et al (2008) Targeted deletion reveals essential and overlapping functions of the miR-17 through 92 family of miRNA clusters. Cell 132(5): 875–886. https://doi.org/10.1016/j.cell.2008.02.019

Wang L, Gural A, Sun XJ et al (2011) The leukemogenicity of AML1-ETO is dependent on site-specific lysine acetylation. Science 333(6043):765–769. https://doi.org/10.1126/science.1201662

Wang Y, Li Z, He C et al (2010) MicroRNAs expression signatures are associated with lineage and survival in acute leukemias. Blood Cells Mol Dis 44(3):191–197. https://doi.org/10.1016/j.bcmd.2009.12.010

Whittaker SJ, Demierre MF, Kim EJ et al (2010) Final results from a multicenter, international, pivotal study of romidepsin in refractory cutaneous T-cell lymphoma. J Clin Oncol 28(29): 4485–4491. https://doi.org/10.1200/JCO.2010.28.9066

Wouters BJ, Delwel R (2016 Jan 7) Epigenetics and approaches to targeted epigenetic therapy in acute myeloid leukemia. Blood 127(1):42–52. https://doi.org/10.1182/blood-2015-07-604512. Epub 2015 Dec 10. PMID: 26660432.

Wu H, Zhang Y (2011) Tet1 and 5-hydroxymethylation: a genome-wide view in mouse embryonic stem cells. Cell Cycle 10(15):2428–2436. https://doi.org/10.4161/cc.10.15.16930